MEDIEVAL CHRONICLES
AND THE ROTATION OF THE EARTH

MEDIEVAL CHRONICLES AND THE ROTATION OF THE EARTH

Robert R. Newton

THE JOHNS HOPKINS UNIVERSITY PRESS

BALTIMORE AND LONDON

Copyright © 1972 by The Johns Hopkins University Press
All rights reserved
Manufactured in the United States of America

The Johns Hopkins University Press, Baltimore, Maryland 21218
The Johns Hopkins University Press Ltd., London

Library of Congress Catalog Card Number 78-39780
ISBN 0-8018-1402-2

PREFACE

The main purpose of this work is to obtain
and assess a large body of observations of solar
eclipses from medieval records, and to use the observations in improving our knowledge of the motion
of the solar system. In the course of studying the
medieval sources in order to find records of solar
eclipses, I realized that they contain a large a-
mount of data that might be useful in other types
of scientific research. I therefore prepared cata-
logues of various types of data that are given in
appendices to the main body of the work. The data
include but are not limited to lunar eclipses, comets, novae, famines, earthquakes, and a few oddities that are too amusing to pass over without mention.

It is customary to consider that there are
three main types of historical work, namely annals,
chronicles, and histories. I have used sources of
all three types, but for brevity I have mentioned only
chronicles in the title. The sources used are
European, including Iceland, and with the understanding that the Eastern Roman (Byzantine) Empire
is included regardless of its geographical extent
at any particular epoch. Histories of the Holy Land
by contemporaneous European writers are also used.

The term "medieval" is used here to mean
roughly the period from 400 to 1200, with the rec-
ognition that it has been used with other meanings
in other places. The reason for choosing this particular time interval will be explained in the
first chapter. Observations lying somewhat outside
this interval are included if they occur in works
that also contain observations lying within it.

I hope that this work will be useful both
to scientists and to students of medieval history
and literature. I have therefore tried to avoid
jargon and methods of presentation that are peculiar to one field. I have especially tried to a-
void the methods of citing literature that are peculiar to one field. As a consequence, the bibli-

ographic entries are more detailed than those usually found in scholarly works.

When sources are taken from several languages, it is always necessary to make some arbitrary decisions about citations and quotations.
I give names of authors and their works in the
languages that they used, but with Greek names
usually translated into Latin. I hope that this
will help to avoid some of the confusion found in
the literature, in which the same work is often
given apparently different titles. I make exceptions for works by British authors; it seems un-
necessarily pedantic to write "Florentius Wigorniensis" rather than "Florence of Worcester", for
example, in a work written in English. I make a
few other exceptions when there are well-establish-
ed English forms.

On the other hand, I give quotations only in
the form of English translations, unless the exact
wording of the original is crucial to the argument.
The translations are mine unless there is a specific
attribution.

A work is cited by giving the name of the
author underlined (or the name of the work if it
is anonymous, as so many medieval sources are),
followed by the year of writing enclosed in square
brackets; sometimes both name and year are enclosed. This practice instantly gives the reader
the two most important items of information about
a source, without the necessity of finding a footnote or of tracing back through a long line of
op. cit.'s. When a work is cited frequently within a short space, the year is usually omitted after
the first appearance. If the title of an anonymous
work is long, I use only a short but characteristic part in the citation.

When it is necessary to specify a particular place within a reference, such as a page or
chapter, the designation of the place follows the
year within brackets.

Many annals and chronicles were written

year by year; in citing these I use the last year
appearing in the work, preceded by "ca." Some
works of this sort were kept by a known person up
to a certain point and then continued by other
people who are frequently unknown. The continuations may or may not be distinguished from the
original in citations. I have generally followed
the practice of the editor of the printed edition
that I have used. If the editor presents the continuations as a separate work, I usually give separate citations for the continuations.

        In assessing the records of solar eclipses,
it is necessary to use a large amount of background
information, such as the dates of reigns of various
monarchs or the places of battles. For such information I have relied upon what I call the "standard
sources", and I have listed the sources called "standard" in a note at the head of the list of references. It would be excessively tedious to acknowledge each item of information obtained from these
sources. They are cited specifically only where
there is a particular point of interest connected
with the information.

        The dates of the Popes are taken from
<u>Mercati</u> [1947].

        I thank Mr. Clyde Holliday and Mr. I. H.
Schroader of The Applied Physics Laboratory for
advice about visual phenomena connected with solar
eclipses. I thank Dr. R.P. Rich, also of The Applied
Physics Laboratory, and Professor Emeritus R.E. Parker
of The University of Tennessee for frequent help concerning problems of translation and for general discussions regarding the background of this study. I
particularly thank Dr. Rich for help regarding lit-
urgical and similar matters in the Roman Catholic
Church, and I thank Dr. S.M. Krimigis, of The Applied
Physics Laboratory, for similar help regarding practices in the Greek Orthodox Church. Mr. R.E. Jenkins
and Dr. W.H. Guier, of The Applied Physics Laboratory,
have provided valuable advice connected with the mathematical portions of the analysis. Professor Henry
Kratz of the University of Tennessee helped me find
much of the material relating to the Icelandic and
Norse literature.

I gratefully acknowledge the support of this research by the Department of the Navy through its contract with The Johns Hopkins University. I also thank the Director's Office of The Applied Physics Laboratory for encouraging me in the study.

I also especially thank Mrs. Mary Jane O'Neill and Mr. J. W. Howe, of The Applied Physics Laboratory, for the skill and dedication with which they prepared this work for the press.

CONTENTS

CONTENTS (continued)

CONTENTS (continued)

CONTENTS (continued)

FIGURES

TABLES

TABLES (continued)

TABLES (continued)

MEDIEVAL CHRONICLES
AND THE ROTATION OF THE EARTH

CHAPTER I

BACKGROUND AND PURPOSES OF THE WORK

1. Motivation for the Work

The road that has led me to the present work is rather unusual, and I hope the reader will forgive me for taking a few paragraphs to describe it. Even though it is not needed for understanding the work, this description will help in understanding the choice of time period of the work. The starting point was an attempt to decrease the costs of operating and using the Transit Navigation Satellite System, which was developed for the Department of the Navy under the technical direction of The Applied Physics Laboratory of The Johns Hopkins University.

If a navigator is going to use any object for the purpose of finding out where he is, he must know where the object is at the time he uses it. If the object is moving, the navigator must content himself with using a predicted position. At the time the development of the Transit system was started, it was inconceivable to anyone involved that the position of an artificial satellite could be predicted with acceptable accuracy for more than a day, and many people were not willing to concede that even that limited amount of prediction was possible. The need for communicating predictions to a navigator anywhere on earth with a delay of less than a day was responsible for a substantial part of the cost of using and operating the system.

After the success of the system on the basis of daily predictions was well established, I began to study the possibilities of yearly predictions, which held the promise of substantial reductions in costs. During a year, a satellite in the Transit system travels more than 200,000,000 kilometers. Even if the goal is to predict with

an accuracy of no more than 1 kilometer, the relative accuracy needed is high. Accurate prediction
is possible only if the forces acting on the satellites are well known.

Anyone who lives on a seacoast knows that
the oceans are distorted in shape by the tides,
with their twice-daily (at most places) rise and
fall. Perhaps fewer people know that the apparently rigid earth is also distorted by the tides, and
that the elevation of a point on earth rises and
falls by about half a meter twice a day. The tidal
distortions in the shape of the earth and oceans
affect the path of a satellite by their gravita-
tional attraction, and by an amount that is significant over a period of a year.

The rise and fall of the tides can be measured with suitable instruments, but the measurements have not been made at enough places to give
us a good picture of the world-wide tidal distor-
tion and its change with time of day. Since available tide measurements do not give enough information to let us calculate the effects on a satellite,
the only way to proceed is to find the effects on
the motions of certain satellites, and to construct
a theory by which the results can be used for any
satellite. Kozai [1967] and I [Newton, 1968] have
done this independently.

After we had done this, I wanted an independent test of the accuracy of the results, but
a test cannot be made with ordinary tidal measurements. However, the tides act as brakes upon the
motion of the moon and the rotation of the earth
and hence they change the lengths of the day and
the month. In principle, one can use astronomical
data from ancient times, deduce from them the
changes that have occurred in the lengths of the
day and month, and hence deduce properties of the
tides that will test some of the results obtained
from artificial satellites.

There have been many attempts to find
changes in the day and month from ancient astronomical data. Since solar eclipses are striking

phenomena that have been observed by many people, and not just by professional astronomers, solar eclipses have played a large role in such attempts. When I looked into these attempts, I was astonished by what I found. Many of them, including all uses of solar eclipses that I have seen, were based upon the logical fallacy of reasoning in a circle. Specifically, most reports used could not be dated on the basis of their texts or their historical contexts. The workers thereupon assigned dates by finding which ones led to accelerations that agreed most closely with assumed values. It is not surprising that the resulting "data" were self-consistent.†

I thereupon did my own investigation of the ancient data, using data from -762 to 1241,‡ and without making the circular assumptions that the effects of the tides have been small and constant. I found [Newton, 1970], in a work that I shall cite as AAO for brevity, that the tidal effects have been about twice as large as current geophysical theory can explain and that they have changed by large amounts within historic times, also in contradiction to existing theory. These results are startling, and I have hesitated to accept them myself without additional confirmation. This has brought me to the present work, and it only remains to explain the choice of time period.

The results of AAO (Chapter XIV, particularly Figure XIV.2) indicate strongly that there was a sudden change in the properties of the tides around the 7th or 8th century. Interestingly, this includes the time, about 709, when the Normandy coastline suddenly and drastically changed, convert-

_____

† See Newton [1969], or Section IV.1 below, for more details.

‡ A year written without a sign is a year of the Christian era. The year before the year 1 will be called 0, the preceding year will be called -1, and so on. -762 is the year sometimes written as 763 B.C. I shall follow the astronomical practice of writing the year, then the month, and then the day, when a complete date is to be given.

ing Mont S. Michel† from a part of the mainland
into "Mont S. Michel au péril de la mer". My
original goal for the present work was to obtain
enough data for the period around 700 to see
whether the preliminary finding of AAO would prove
valid.

The time of interest lies in what has often been called the "Dark Ages". Fortunately,
the age was not as dark as it has often been pic-
tured. I was taught, in my formal education, that
everyone before Columbus thought that the world is
flat, and I suspect that many people still have
this distorted view of scientific history. I know
now, as many writers have pointed out before me,
that this is not so. The doctrine that the earth
is spherical (and departures from a sphere are
trivial for present purposes) was well established
by the time of Aristotle, and Aristotle [ca. -325]
gives some surprisingly modern arguments for a
spherical earth, including one that would be called a "group theoretical" argument in modern scientific jargon. This knowledge, and much other
scientific knowledge, has never been lost to those
who were interested. Of course, a poor farmer in
the Middle Ages, bound by law or poverty to a
small parcel of land, probably never saw beyond
the boundaries of his field and probably thought
that the earth was flat, if such a question ever
entered his mind at all.

In fact, there always remained, even in
the "Dark Ages", a surprising amount of learning
and a surprising amount of mobility in Europe, but
not of course at the level that prevailed in the
Indian summer of the Roman Empire under the adopt-
ive emperors.

However, scientific research did almost
completely die out, and its decline preceded the
fall of the Empire. Theon made measurements (AAO,

---

† I use "S." throughout to denote "Saint" in all
its various inflectional forms, when it occurs
as part of a name.

Section V.5) of the solar eclipse of 364 Jun 16 at
Alexandria. From that time until the beginning of
Islamic scientific activity about 5 centuries later,
there was practically no new scientific work or
scientific measurement in the European or Mediterranean worlds. So far as I can find out, little or
nothing of scientific interest happened in China
during that period. It is possible that there was
scientific activity in India, but so far I have
found no details, if indeed there are any to be
found.

Because of the decline of scientific activity, we must rely for most of our scientific
data during the period of interest upon observations made by people who were not scientists. For
the sake of having a fairly homogeneous work, therefore, I have decided to restrict this work entirely
to observations made by people whose interests were
historical rather than scientific. Almost all of
the observations appearing here were recorded in
monasteries or other church establishments. Islamic data appear during the latter part of the
period of current interest, but they will not be
used here. Islamic data that were not used in
AAO will be used in a later work.

2. Plan of this Study

With one exception, the only observations
found in the medieval documents that are useful
in studying the day and month are observations of
solar eclipses. The reasons for this will be explained in Appendix I. As a consequence, observations of solar eclipses will receive most of the
attention of this study. Little will be done with
other observations beyond cataloguing them, but
the observations of solar eclipses will be analyzed in detail.

In order to use an observation of a solar
eclipse, three things must be known. First, we
must know the "magnitude" of the eclipse, that is,
the maximum fraction of the sun that was obscured

-5-

during the eclipse.† The magnitude is 1 for a total eclipse and it is between 0 and 1 for a partial
eclipse. The magnitude is the least important of
the three things needed, in the sense that any reasonable errors made in deciding upon the magnitude
tend to cancel for the whole set of eclipses. Besides the magnitude, we need the date of the eclipse
and the place where it was observed.

The analysis of an eclipse observation
thus has two distinct parts. The first part is the
deduction of the three items of information from
the record. The second is the use of the information in some astronomical calculations. The
second part is straightforward, although it may
be incomprehensible except to the trained astronomer. The first part requires textual analysis
of the records and is considerably more difficult.

Textual analysis of the records was also
needed for the eclipse records used in AAO, but the
problems to be solved there were mostly quite different from those posed by the medieval chronicles.
In a few cases, the records in AAO were so ambiguous that it was hard to decide whether an eclipse
was being described or not. In most cases of difficulty, however, the problem was to decide whether
the record was truthful or not. Many records that
had been soberly used in astronomical research
turned out to be propaganda, for example, or literary inventions which their authors probably never
intended to be taken seriously. There are only a
few cases in the medieval records in which we have
to worry about the basic truthfulness of the observation.

With some of the medieval records, determining the date is a problem, particularly with
records from the earlier part of the period. In
these cases, a date is given in some form or another, such as by a statement of a contemporaneous
event, but there was no possible eclipse. In most
cases, we can establish the nature of the dating

---

† Magnitude will be discussed in more detail in
Section IV.5.

error from internal evidence and can restore the correct date without ambiguity. A few records will turn out to be unusable because no certain date can be determined. Determining the place of the observation is a major problem with many and probably most of the eclipse observations, both those used in AAO and those to be used here. It was a rare observer who stated where an observation was made. Observers in different times and places had various reasons for noting the occurrence of an eclipse, but the idea that a 20th-century scientist might use their observations was apparently never among them. To the observers, the fact of an eclipse, not its place, was the important thing.

Sometimes we can infer with high confidence that the writer of a record saw the eclipse personally. We may have detail of a type that would probably not be recorded except by an eyewitness. Occasionally we have an explicit statement. For example, William of Malmesbury [ca. 1142] wrote of the eclipse of 1133 Aug 2: "In the eclipse I saw myself the stars around the sun . . ." We can be almost certain that this eclipse was total or nearly so at Malmesbury.

Most of the eclipse observations occur in the records or annals of a monastery or other church institution. If we could be sure that the observation was recorded contemporaneously, we could be fairly sure that the observation was also local, although we can certainly not bar the possibility that correspondence or a visitor brought the news. Because of the nature of the annals-making process, we must proceed carefully before we decide that a record was independent, contemporaneous, and hence probably local.

Most monastic institutions, for example, felt impelled to have their own chronicles, annals, or histories, presumably for the purpose of teaching their novices. A few were content to begin their records with the establishment of the institution, but most went back to the beginning of the Christian Era and some went back to the Creation,

-7-

that is, to the events of Genesis. When a new
institution was established, it typically copied
the records of a sister or neighbor institution
until the record was up to date, and maintained
independent records from that point on. Since we
usually know the date of establishment, one might
think that the decision about the independence of
of records is simple: The records are independent
and probably local beginning a safe interval after
establishment (one must allow some time for getting
started), and the records are derivative before
that time. Unfortunately, the process is not that
simple.

For one thing, an institution may not have
begun the keeping of its own records for a long
time. It may have lost its records in a fire and
had to start over. Either of these reasons would
make its records derivative for a long time after
establishment. On the other hand, a record that
is derivative in the sense that it was copied from
some other record may be independent for the purposes of this study. This would happen if the copy
has survived while the original has not. When this
happens, we can take the copy as an independent useful record if we can safely deduce the provenance
of the lost original from internal evidence.

All chronicles and other materials used in
this study have been edited by excellent scholars
who have gone extensively into the identification
of the independent portions and into the identification of original sources and their provenance for
the derivative portions. These studies have aided
the present study enormously, but they cannot be
taken as sufficient.

For one thing, I believe that the editors
have occasionally made mistakes in their analysis,
either because they overlooked material, or because
new material has come to light which forces changes
in their conclusions.

A more significant defect from the viewpoint
of this study is that most editors have not been interested in the scientific data, particularly the

astronomical observations. Some editors even appear to resent the fact that the chronicler wasted
space upon such uninteresting items. It is necessary to test all conclusions that were based upon
the overall texts specifically for each scientific
observation, and I have therefore had to do independent textual analysis in the present work.

I do not wish to give the impression that
I compare the textual analysis given here in importance with that which has been done previously,
from the standpoint of the general historian. However, the analysis of the eclipses yields some information that may interest the historian. It allows us to make corrections or refinements in the
chronology of some of the sources. It occasionally
yields new information about the relations between
historical sources.

For example, three Irish annals will be
cited and studied in Section VII.1. The editor of
one of them, the Chronicum Scotorum, thinks that
it is a copy of a source used by the compilers of
the Annals of the Four Masters, and does not mention any connection between it and the Annals of
Ulster. His statement may represent the situation
correctly so far as the general historical material
is concerned, but the eclipse records give a different impression. So far as eclipses are concerned, Scotorum is almost identical with Ulster and
has nothing in common with Four Masters.

In summary, the first task is to study the
texts for the purpose of determining which are original sources of observations of solar eclipses. This
information is needed both to eliminate duplicate
observations and to determine the places of observation; occasionally a detailed study is needed in
order to find the date of an eclipse. After this
study is complete, the independent observations
will be used in calculations whose goal is to study
changes in the lengths of the day and month. Finally, other types of scientific data will be discussed briefly in appendices.

The textual studies will be preceded by a short discussion of the development of medieval annals, chronicles, and histories, which is intended to help in understanding the nature of the sources. That discussion must be preceded by a discussion of the "Easter problem", that is, the problem of deciding when Easter should occur in any given year. The Easter problem is intimately connected with the sources of data in three different ways. First, some of the sources owe their very existence to the Easter problem; indeed they sometimes share the same physical medium with Easter documents. Second, some of the observations may have been motivated by the Easter problem. Third, the Easter problem bears upon questions in dating or otherwise understanding many of the records.

I have found only two earlier works that have appreciable overlap in purposes or results with this study. Johnson [1889] summarizes a large number of eclipses from pre-Christian times through the 18th century, but he does not use them for astronomical purposes. He mentions an earlier collection by Tycho Brahe called Historia Celestis that I have not tried to locate, and he apparently relies heavily upon Tycho's collection. Johnson's work has been of some value in this study; in particular it kept me from overlooking the record† 348 Oct 9 M,B from the Byzantine source Theophanes [ca. 813]. Johnson shares the common characteristic, common particularly among 19th-century writers on the subject, of accepting "literary" and "magical" eclipses‡ seriously. However he at least sternly rejects the idea of an eclipse at the Crucifixion of Jesus and refers to the "shallowness of argument" of those who try to identify the Gospel account with the "eclipse

---

† In Section XV.4.

‡ See the discussion by Newton [1969]. Also see Section IV.6 below.

of Phlegon".[†] He makes a few mistakes in the dates
of records of real eclipses.

Ginzel [1882 and 1884] made an extensive
compilation of records of solar eclipses in ancient
and medieval sources. He combined 23 medieval records with 6 ancient ones in making estimates of
the astronomical accelerations. Since he used only
a few of the medieval records that he compiled, he
carried out only a small amount of textual analysis. His work has been of great value to me. It
has led me to records that I should otherwise have
overlooked. Further, his extensive list of the
places where records were made, and of their geographical coordinates, has been invaluable. His
work has been the only source of geographical information available to me in many instances. However, the compilation made here is not a mere duplicate of his. I have not tried to locate all the
sources that he found, and I have found others that
he did not use. The reader should note that Ginzel
used the old astronomical day that began at noon.
Hence his dates disagree by a day with those given
here about half the time.

3. A Few Conventions to be Used in Writing

I have already mentioned in a footnote some
of the conventions that will be used in writing
dates. Dates will always be in the Julian calendar, with the first of the year always assumed to
be Jan 1 regardless of local custom at the place
and time of a record. In writing a date, I shall
always use only the first 3 letters of the name of
a month, without the use of a period following.
However, the name of a month occurring in discussion or in a quotation will usually be written in
full.

It is probable that all numerals in the
original medieval sources were either written out

[†] See the discussion of the "eclipse of Phlegon"
on pp. 110-113 of AAO.

in words or were written in Roman numerals.[†] It
was common if not universal to put a period at the
end of a numeral; thus a medieval annalist would
probably write "after iii. days"[‡] rather than "after iii days" or 'after 3 days", for example. Many
editors have followed the practice of converting
all numerals to Arabic numerals. This is unfortunate, because many numerals are in error. It would
help us in trying to correct the error if we knew
the exact form used in writing the original.

       In translations, I have changed to Arabic
numerals. When I give a direct quotation because
a passage poses a problem, I first give the numeral
in the form given by the editor of the work being
quoted. If this is not in Roman numerals, I then
speculate upon the original form when it is relevant.

       In most of the sources, the days of the
year are written in terms of the Roman calendar.
We still use the Roman months but we state dates
in a quite different fashion. In the Roman calendar, each month had 3 principal days that had
names. The "calends" was the first day of the
month. The "nones" was either the 5th or the 7th,
depending upon the month, and the "ides" was either
the 13th or the 15th. I shall not bore the reader
by giving the rules for fixing the nones and ides;
he should have no trouble if he wants to find them.

       The Romans specified a day other than a
named day by counting backward from the next succeeding named day. The counting was inclusive.
That is, the named day was counted as "1" and the
day in question also had to be counted. The speci-
fication of a day thus had 3 parts. First was the

[†] Or in Greek numerals in Byzantine sources.

[‡] Except that he did not write modern English, of
course.

number[†] obtained by counting. Second was the name of the reference named day. Third was the name of the month to which the named day belonged. In translating a date, I shall first give the 3 parts, but in English translation. I shall then state the equivalent date in our calendar.

For an example, a text might have "iiii. non. Iul." I shall render this as "4th nones July (= Jul 4)". Bastille Day is "ii. ides Iul." or "2nd ides July" or Jul 14. As a final example, consider Christmas, which is Dec 25. This is "8th calends January" in the Roman calendar.

The last example illustrates a particular hazard in using the Roman calendar, and one that has caused many errors, but not as many as one might expect. More than half the days in any month must be dated with the use of the name of the next month, the month to which the following calends belongs. There is a strong temptation to use "calends" followed by the name of the current month.

Dates in the last part of December[‡] pose not merely a hazard but an ambiguity. Consider Christmas Day in the year 1969, that is, 1969 Dec 25. Would a person using the Roman calendar write it as "8th calends January 1969" or "8th calends January 1970"? In other words, is the year to go with the calends used or with the current year? I believe that the convention most commonly used was to use the number of the current year,[*] contrary

---

[†] The number is in ordinal form, at least implicitly. I shall use the cardinal form when quoting from the Latin, because the ordinal form was usually inflected.

[‡] I write for simplicity as if the year always began on Jan 1. If the year began on some other day, the problem is merely changed in terms but not in substance.

[*] That is, to write "8th calends January 1969" for 1969 Dec 25.

to the rule followed for the name of the month. However, I am by no means sure that this convention was universal. Even if it were intended that this should be the rule, the intention does not prevent the occurrence of errors.

When use is made of a particular part of a reference, it is usually necessary to indicate the part by a chapter number, a page number, or the like. With most of the observations to be used in this study, this is not necessary. What is necessary is to give the year or other chronological designation that the writer assigned to an event; this is necessary in order to identify the event. Most of the sources are arranged chronologically and spend less than a page on the events of each year. In these cases, the year is a better indication of where to find the passage than the page number would be, and any further indication of location will be omitted.

CHAPTER II

THE EASTER PROBLEM

1. Statement of the Problem

Most people, if asked to say when Easter falls in any year, would probably answer something like the following: Easter is the first Sunday after the first full moon after the vernal equinox (or, perhaps, after the beginning of spring). This statement is approximately correct. It is exact only if most of the terms in it are given definitions that differ from both their ordinary civil meanings and their astronomical meanings.

Thus we shall have to define an "ecclesiastical equinox". We shall also have to define a hypothetical "ecclesiastical moon" whose motion differs from that of the real moon. The reader might then think that "full moon" means the time when the ecclesiastical moon would be full if it were the real moon, but he would be too hopeful. "Full moon" in this context means the 14th day of a certain Jewish lunar month. More precisely yet, it means the 14th day of what that Jewish lunar month would be if it were governed by the Christian ecclesiastical moon and equinox, and by Christian rather than Jewish rules. I shall give the necessary definitions in full in later sections.

Although the statement is correct when the terms are properly defined, it has not always been so. There has been an Easter controversy, that is, an argument about what the rules should be, on top of the question of finding what day satisfies the rules. The controversy was a lively affair until about the end of the 7th century. Apparently it then smouldered, with an occasional flare-up, until the adoption of the Gregorian calendar by the Roman church in 1582 revived it. Even today the observance of Easter is not the same in all major churches.

2. The Easter Controversy

Jones [1943] gives the most extensive discussion of the Easter controversy that I have seen. His account is excellent and readable (considering the complexity of his subject), and much of this chapter is based upon his writing.

Plummer, in the notes to his edition of Bede [734?],[†] objects to the "tediousness" of the Easter controversy (v. 2, p. 348). I feel, as Jones (p. 4) also seems to do, that this objection is somewhat misplaced. It is important to know the subject matter of major controversies; to know what man argues about is to know something about man himself. Besides, holy days and holidays are not trivial matters. As Jones (p. 81) reminds us, those of us who went through it should remember the excitement, argumentation, and genuine inconvenience caused by the "Thanksgiving controversy" in the United States in the 1930's and 1940's.

Jesus of Nazareth was a Jew, and the first Christians regarded themselves as forming a part of the Jewish religion. If Christianity had not been quickly adopted and then dominated by Gentiles, the Easter controversy might never have arisen.

I believe that all authorities now agree that Jesus was crucified on a Passover[‡] that was also the 6th feria[*], in a year that is uncertain.

[†] The question mark following "734" is explained under Bede in the references.

[‡] More accurately, I mean that the authorities agree that the chronological clues furnished in the Gospels indicate this date. Not all authorities agree that the Crucifixion was an historical event. For our purposes, it does not matter whether it was or not; it only matters that medieval Christians thought that it was and the Gospels provide the historical data.

[*] The Hebrews numbered the days of the week. Most Christians in the Middle Ages did the same, scorning the names of the days derived from pagan gods. Instead of writing out "day of the week" I shall use the shorter term "feria", which is equivalent in medieval usage if not in literal meaning. This

Passover is Nisan 14, the 14th day of the Hebrew month Nisan, which is the first month of the Hebrew year.†

The Hebrew calendar is a mixed calendar in which the months are lunar. The beginning of the month was once determined by direct observation of the moon; the first day of the month began when the crescent of the new moon could be seen at sunset. Since this is usually about a day after the astronomically defined new moon, full moon usually occurs about the 14th day of the month.

Twelve lunar months amount to about 354 days. The Hebrew calendar is kept in average adjustment with the solar year by varying the number of months in a year. In an ordinary or common year,

---

practice will preserve some of the flavor of the medieval writing. "6th feria" is our Friday.

†Almost no statement about the Hebrew calendar applies at all places and at all times within the historical period. The first month may be Nisan in the spring or Tishri in the autumn; for simplicity I shall call Nisan the first month. "Passover" is used to denote a feast lasting many days that incorporates two main elements of probably quite different origins. One is the Paschal sacrifice on Nisan 14, which is probably of pastoral origin. The other is the Feast of Unleavened Bread on Nisan 15, which is probably of agricultural origin. "Passover" may be used to denote either of these elements, or the day on which either occurs. In current usage, at least in the United States, the Paschal sacrifice plays little if any part. The day listed as the first day of Passover in the American Ephemeris and Nautical Almanac, for example, is the day corresponding to Nisan 15. The present chapter is mostly concerned with early Christian usage, in which Passover meant the sacrifice on Nisan 14. Hence I shall use "Passover" to mean Nisan 14.

there are 12 months, of which Nisan is the first.
In slightly more than one year out of three, a
13th month is inserted as the last month in the
year; a year with 13 months is called embolismic.

Thus Nisan 14 has no fixed date in the
purely solar calendar that the Romans used and
that we still use. It may occur on any day of
the week.

Certain early Christians observed Passover, that is, Nisan 14, as the day of the Crucifixion. I have seen no definite statement on the
matter, and the facts may not be known, but I imagine that these were mostly the Christians who also
regarded themselves as Jews. Other early Christians observed Easter as the 1st feria[†] following
Nisan 14. I imagine that these Christians were
mostly Gentiles. The disagreement between these
two groups was the first version of the Easter
controversy.[‡]

This version of the controversy was set-
tled officially by the Council of Nicaea, which
ruled [Eusebius, ca. 338] that all Christians must
celebrate Easter on one and the same day, and that
they must never "keep company" (societatem habere)

[†]The 1st feria, which is Sunday, is ordinarily
called Die Dominica, the Lord's Day, in medieval
usage. Except in direct quotations, I shall call
it the 1st feria in order not to be invidious.

[‡]The reader should notice that there is a fundamental inconsistency. If the Crucifixion were on
Nisan 14, the Resurrection was on Nisan 16. For
consistency, then, one would expect Easter to be
Nisan 16, or the 1st feria thereafter for those
who wished to adhere to the 1st feria. Above,
with regard to the "Nisan 14 adherents", I said
for simplicity that they observed the Crucifixion
on that day. I am not sure whether in fact they
observed Nisan 14 as the day of the Crucifixion
or the Resurrection.

with the Jews.† This was apparently taken to mean
that Easter, rather than the day that commemorates
the Crucifixion,‡ must never coincide with Passover,
and that Easter must be celebrated on the 1st
feria.

The "extirpation" of the "Nisan 14 heresy"
did not end the Easter controversy. As the Christians found out, it is easier to make a rule than
to enforce it. Arguments about the rules governing Easter continued for centuries. Jones (p. 23)
wryly observes that all churches adopted the same
rules "except the Churches of Ireland, Africa,
probably Spain, and others in outlying provinces."
He might have added "and except the Church of Rome".
The Roman church insisted for several centuries
[Jones, p. 26] that April 21 be the latest possible date of Easter, and vigorously contradicted
itself by insisting that March 22 be the earliest
date.* In doing so, it also implicitly contra-

---

† It is frequently said that the Council of Nicaea
laid down the rule given at the beginning of this
chapter. This cannot be explicitly denied, but
there is no known evidence that the Council did
so. The "sense of the Council" seems to have been
that Christians should be uniform in their practices and that they should avoid the practices of
the Jews.

‡ This continues the inconsistency noted in an earlier footnote. Easter is a day of rejoicing. The
day of the Crucifixion is a day of mourning; it
can also be described as a day of sacrifice. Nisan 14 is also a day of sacrifice. Thus there
would be an obvious motive in forbidding "Good
Friday" to coincide with Nisan 14, but this is
not what the Council did. In effect, the Council
forbade Christians to rejoice on the Jewish day
of sacrifice. It is doubtful that this was the
purpose of the Council. The reasons for the detailed actions of the Council remain obscure.

* If March 22 is the earliest date, April 25 must
be the latest date, provided that Easter is both
to be on the 1st feria and to be governed by a
lunar calendar.

dicted the Council of Nicaea.

The rule stated at the beginning of this
chapter is the one finally adopted by the Roman
church and by all of western Europe. The last
major western dissent from the rule was apparently
that of certain Celtic churches, who did not all
adopt the Roman rule until the 9th century. I am
not certain whether the eastern Orthodox churches
in medieval times ever adopted exactly the same
rules as the Roman church. Apparently the differences, if any, after the 7th or 8th centuries were
minor until the Roman church adopted the Gregorian
calendar in 1582.

The eastern Orthodox churches and many of
the western Protestant churches refused for varying
lengths of time to adopt the Gregorian calendar.
I believe that the calendar and the rules for Easter are now the same in all western Christian
churches, but there is still a difference between
western and some eastern practice.

A Congress of eastern Churches held in
Istanbul in 1923 [Explanatory Supplement, 1961,
p. 413] adopted a minor modification of the Gregorian calendar, which will not differ from the
western calendar until 2800.† The Explanatory
Supplement (p. 413) also says that the Congress
adopted the determination of Easter by means of
the "astronomical moon for the meridian of Jerusalem".

The differences between the rules just
stated and the rules used in the west would rarely
produce a difference in the date of Easter. It is
a matter of observation, however, that the Greek
Easter differs from the western Easter most of the
time, so the Greek Orthodox Church presumably did
not ratify the action of the Congress. I have not
found a full explanation of the Greek rules, but
it is plausible that the Greek church uses the

---

† The year 2800 will be a leap year in the west but
not in the east, unless something is done.

"Theophilan moon" (see Section II.4 below) combined in some way with the Gregorian calendar.

Even after a general rule governing Easter is accepted, there is still the problem of determining the date of Easter in any year. This problem is essentially astronomical. It is plausible that the relation of astronomy to the date of Easter helped keep astronomical knowledge alive during the early medieval period.

There is no doubt that a fair amount of astronomical knowledge was preserved throughout the medieval period. Bede [725], for example, discussed such matters as eclipses and the variation of the length of the day with the seasons, and even of the relation between the moon and the tides. However, it is clear that much of the learning of the Alexandrian astronomers was not available to him.

The Easter controversy and the Easter problem may account for some of the astronomical observations found in the medieval records. Jones (p. 119) mentions the chroniclers Hydatius (see Chapter XIV) and Marcellinus (see Section XV.3), and infers "that the increasing interest of Idatius[†] and Marcellinus in celestial phenomena, eclipses, comets, and even earthquakes" arose in part from the Easter problem.

Certainly there is no question that interest in celestial phenomena varied widely from one chronicler or annalist to another. For example, the Anglo-Saxon Chronicle (see Section VI.1) records the lunar eclipses of 796 Mar 28, 800 Jan 15, 802 May 21, and 806 Sep 1, and the solar eclipse of 809 Jul 16. The unknown recorder of these eclipses must have had an unusual interest in eclipses; such a density of eclipse observations is rare in any source. His interest may have been stimulated by the Easter problem, although this is clearly just a guess. After 809 Jul 16, the next eclipse is the

------

[†] An alternate spelling of Hydatius used by Jones.

lunar eclipse of 828 Dec 25; it is possible that
this one was recorded only because it coincided
with Christmas.† The next eclipse after 828 that
I noticed comes in 878, half a century later.

The Easter calculation is fundamentally an
astronomical calculation even though the entities
involved in the rule for Easter must be taken as
ecclesiastical rather than astronomical. The ecclesiastical entities arose as convenient approximations to the astronomical ones.

3.   The Ecclesiastical Equinox

Astronomically, the equinox‡ is either an
instant in time or a direction in the heavens, according to the context. It is the instant when the
sun crosses the equator going north in the spring,
and it is also the direction of the sun in the heavens at that instant. The zodiacal sign Aries begins at the direction of the equinox. Thus the sun
enters Aries at the instant of the equinox, and
the direction of the equinox is sometimes called
the first line of Aries or the first degree of
Aries.

The ecclesiastical equinox, by contrast,
means a certain day in the official calendar.

The traditional Roman dates of the equinoxes and solstices were [Pliny, ca. 77, Chapters
59, 66, 68, 79] the "viii. Kalends" of January,
April, July, and October. These dates have a sim-

---

†But the chronicler missed the lunar eclipse of
809 Dec 25, which is recorded in Welsh sources
(see Section VII.3).

‡Equinox, when used alone, will mean the vernal
equinox.

plicity[†] that is lost when they are stated in our calendar; they are December 25, March 25, June 24, and September 24. Three of these dates are important in the Church calendar.

June 24 is the festival of S. John the Baptist;it is also Midsummer Day. September 24 is marked in the Church calendar as the Conception of S. John. If June 24 is his birthday, September 24 is, so to speak, a biological necessity according to medieval ideas. I am not aware that September 24 is an important day in the calendar.

The festival of the Nativity of Jesus is now celebrated on December 25, but it has not always been. In several places [Jones, p. 7, for example], I have seen an allegorical explanation of why the Church adopted this date in the absence of historical evidence: Jesus was born to bring light into a world of great darkness. From what I understand of the human mind, I believe that many people can accept such a statement as a proof, but to me the statement smacks more of justification of an accomplished fact than of the basis for a decision. The day of the winter solstice, long before the time of Christ (see [Frazer, 1922, Chapter XXXVII], for example), was celebrated in many places as the birthday of the sun, or of a solar deity, or of a human vicar for the sun, and it was an important festival. I suspect that the Church adopted

---

[†]Pliny may have been more interested in simplicity than in consistency. The intervals between these dates are 91 (or 92 for leap years), 92, 92, and 90 days, respectively. However, in Chapter 59, Pliny gives the intervals as 90 1/8, $93\frac{1}{2}$, $92\frac{1}{2}$, and 89 1/8 days, respectively. Even when we allow for rounding off the fractions, the inconsistency is striking. I do not know the source of his astronomical intervals. The correct intervals in his time were about 90.2, 94.0, 92.4, and 88.7 days, respectively. Thus the errors in the astronomical times that led to Pliny's astronomical intervals must not have been more than a few hours.

the date in order to facilitate proselyting and
explained it later.

March 25 is the Feast of the Annunciation.
This date, like the date of September 24 for S.
John, is a necessity if December 25 is the Nativity. Some early Christians [Frazer, 1922, Chapter
XXXVII but see also Chapters XXXIII through XXXVI]
observed Easter on March 25, that is, at the equinox. I suspect that these Christians were adherents of Gnosticism, which became one of the most
execrated of early heresies.

All discussions that I have seen which mention the matter say the following: When Julius
Caesar reformed the Roman calendar, he adjusted it
so that the equinox would fall on the traditional
date of March 25. By the time of the Council of
Nicaea in 325, the date of the equinox had shifted
to March 21 because of the error in the length of
the Julian year. When the Gregorian calendar was
adopted, it was adjusted so as to restore the equinox to March 21, the date it had at the Council.
None of these statements is correct, although the
last one is close enough for ordinary purposes.

Caesar adjusted the calendar by making the
year -45 have 445 days [Explanatory Supplement,
1961, p. 410], and the year -44 was intended to be
the first year[†] for using the Julian calendar. In
-44, a leap year, I calculate that the equinox was
on March 22 at about 23 hours; the error in this
calculation probably does not exceed about 1 hour.
During the next three years, the equinox advanced
to about 17 hours on March 23, before falling back
to March 22 in the next leap year.

Similarly, in the years around 325, the
equinox fell on March 20, at times ranging from
about 3 hours to about 21 hours; it did not come
on March 21 in any nearby year. In 1582, the

[†] Because of confusion about what was intended, it
apparently took several decades before the Julian
calendar was in routine correct use.

equinox (in the Gregorian calendar) was a few minutes before midnight on March 20, according to my
calculation. In 1583, it advanced to about 6 hours
on March 21, but occurred before noon on March 20
in the following leap year.

The times have been given for the meridian
of Greenwich. About 1 hour should be added for
the meridian of Rome and about 2 hours for the meridian of Alexandria. Thus, if the Gregorian calendar
was calculated for Rome, the equinox around 1582
fell on March 20 about half the time and on March
21 about half the time, in terms of the day that
begins at midnight.

It seems plausible that the Julian calendar
was actually adjusted to make the traditional "viii.
Kalendas" as accurate as possible for all the cardinal times of the year and not just for the vernal
equinox. For -44, I calculate the following hours
for the cardinal times adjusted to the meridian of
Rome: just before midnight on March 22, just before midnight on June 24, at about 9 hours on September 25, and at about 2 hours on December 23.
This represents a reasonable compromise with the
traditional dates.

At the Council of Nicaea, however, the interest was specifically in the vernal equinox, and
the question is why it was stated to be on March
21 when it was always on March 20 at that time in
history. There is a plausible explanation.

In the fourth century, the astronomers of
Alexandria were about the only professionals left
in the Christian world. The astronomer with the
highest repute at that time was Ptolemy, who flourished in the second century. Ptolemy [ca. 152,
Book III, Chapter II][†] gave the length of the year
as 365 + $\frac{1}{4}$ - (1/300) days. He also said that he
measured the equinox of the year 140 with great
care and found that it occurred on March 22 at 13
hours. It is possible that no Alexandrian astronomer after Ptolemy tried to improve upon his values.

[†] Also see AAO, Section II.2.

The errors in all of Ptolemy's equinox and solstice data average about 30 hours; a reasonable error in measurement in his time would have been a few hours (see the discussion about Pliny in the third footnote back). In AAO (Section II.2), I showed almost beyond question that these "observations", which Ptolemy claimed that he made with great care, were not observed at all but were fudged with the help of Hipparchus' tables. If the astronomers who fixed the date of the equinox at the time of the Council of Nicaea had calculated the equinox of 324 from the values in Ptolemy's work, instead of finding it by observation, they would have got March 21 at about 22 hours.

In summary, the date March 21 has been the ecclesiastical equinox since the 4th century, whether the calendar has been Julian or Gregorian. The original choice of date is not consistent with astronomical observations of any reasonable level of accuracy. The date perhaps rested upon a faked observation of the equinox found in the writings of Ptolemy.[†]

One should not suppose that medieval writers dealt clearly with the equinox, perhaps because it meant different things to them in different contexts. A church calendar that I keep on my desk illustrates the confusion. This is a copy of the "Sherborne calendar" [Wormald, 1934], which was prepared about 1060. I acquired it because it is one of the earliest church calendars that lists S. Olaf's Day, and it furnishes part of the evidence that the famous "eclipse of Stiklestad" did not occur at the death of S. Olaf (see Section XIII. 2; also see Section IV.4 of AAO). I use it because it seems appropriate to use a medieval church calendar when working with medieval church records.

[†] There may be an "unscientific" explanation of the date. The astronomically correct date was the 13th calends April, while the adopted date was the 12th calends. 13 is an unlucky number, while 12 is almost a "sacred" number.

The Sherborne calendar lists the equinox on "xii.KL April" (= Mar 21), and it also lists the equinox, without comment on the inconsistency, on "viii.KL April" (= Mar 25). It thus preserves both the old Roman date and the later Christian ecclesiastical date; even Bede [725, Chapter XXX] felt compelled to give both equinox dates. The calendar also marks "xv. KL April" (= Mar 18) as the day when the sun enters the sign of Aries. Since entering Aries is synonymous with being at the equinox, the calendar in effect gives three dates for the equinox. In fact, it gives three dates for each equinox and each solstice.

In listing March 18 as the entrance into Aries, the calendar is again following Bede [725, Chapter VI], who in turn is following Pliny [ca. 77]. I offer no explanation of this separation of the beginning of Aries from the equinox.[†] If the date were intended to mark the entry of the sun into the constellation Aries, as contrasted with the zodiacal sign of the same name, entrance into Aries would be after the equinox. There may have been some numerological significance to the number 8 in the Roman mind; according to Pliny [ca. 77], the equinoxes and solstices occurred in the 8th degrees of the appropriate signs, as well as being on the 8th calends of the appropriate months.[‡]

[†] Jones (p. 21) says: "There is no ancient support for the belief that the equinoctial point must be in the first degree of Aries." However, this is not a belief; it is the definition of "the first degree of Aries". Further, this definition is at least as old as Ptolemy [ca. 152, Book II, Chapter VII], and it had probably been adopted well before the time of Pliny.

[‡] The recurrence of the number 8 in both uses may also be a coincidence. According to Neugebauer [1957, p. 188], the 8th degree of Aries is frequently used as the vernal point in astrology. Neugebauer traces the use of this point back to Babylonian ephemerides prepared in the -2nd century or thereabouts.

4. The Ecclesiastical Moon

Here I shall deal only with the ecclesiastical moon that was used in connection with the Julian calendar. The Gregorian calendar was introduced centuries after the last observations that will be used in this work.

This is a good point to raise the following question: Why use the ecclesiastical equinox and moon at all? Why not use direct observation of the real things in fixing the date of Easter?

Direct observation is the simplest procedure and was presumably the method that all societies once used in their calendrical problems. It is feasible and convenient for a compact group of people. It would have been awkward and inconvenient for medieval Christians who were widely spread geographically but who wanted uniformity in their ecclesiastical practices.

Because of errors in observation, the use of direct observation would have required the designation of an official Church astronomer; otherwise, there would not have been the necessary uniformity. His results would then have had to be communicated to all Christians in a timely fashion. Consider, for example, an English Christian who wanted to make a pilgrimage to Jerusalem, and who wanted to schedule his arrival in time for Easter. He would have needed many months for the trip. The announcement of the date of Easter would have had to leave Rome, say, for England some months earlier yet. In order to be satisfactory, the designation of Easter would have had to be made about a year in advance. This means that, at best, observation would have had to be supplemented by a considerable theoretical structure.

The most satisfactory solution for the Church was therefore the adoption of an official set of tables for the moon and the equinox. These tables then defined the ecclesiastical equinox and moon implicitly. The table of the ecclesiastical equinox, as we have just seen, was simple: the

ecclesiastical equinox was the fixed date March 21.
The principles of the ecclesiastical moon will be
outlined in a moment.

Church practice in this regard did not differ in principle from ordinary scientific practice
today; it differed only in accuracy. We tend to
think that the positions of celestial bodies tabulated in the major national ephemeris publications
are based upon the most recent and accurate observations possible, but this is not so. Instead, the
"official" British and American sun and equinox are
"Newcomb's sun and equinox" [Explanatory Supplement,
1961, Chapter 4], which were promulgated in 1895.
The "official" moon is "Brown's moon", which was
promulgated in 1919 and modified slightly in 1954.
The differences in accuracy between Newcomb's and
Brown's quantities and the corresponding medieval
ecclesiastical quantities result primarily from the
different times at which they were devised.

The first several centuries [Jones, 1943,
pp. 11-104] saw the production of many ecclesiastical lunar tables. Some of these were adopted by
certain churches, and the variety of early ecclesiastical moons accounts for some of the variety
of early Easter observances. The ecclesiastical
moon that was finally adopted by the Roman Church,
and that was thereafter the European standard until the adoption of the Gregorian calendar, apparently originated with Theophilus, Bishop of
Alexandria (385-412) [Jones, p. 29], who prepared
a table for the years 380-479. A table called the
Cyrillan table extended Theophilus' table through
531; Jones (pp. 39-54) discusses the lively controversy about the author of the Cyrillan table.
A person called Dionysius Exiguus [Jones, p. 69]
extended the table through 626 and someone unknown
extended it through 721 [Jones, p. 74].

I do not know when the Roman church accepted this ecclesiastical moon and the dates of
Easter that go with it. Since these tables allow
Easter to be as late as April 25, the Romans could
not have accepted them as long as they insisted
that April 21 must be the latest possible Easter.

Pope Symmachus refused [Jones, p. 67] to accept 501
Apr 22, which was observed as Easter by the Alexandrian church, and substituted March 25. Bede [725,
Chapter XXX], however, accepted the limits March 22
through April 25, with the statement that these
dates had been established by the Council of Nicaea.
Since Bede adhered firmly to the Roman church, it
is almost certain that the Roman church had accepted the enlarged limits by 725. Further, Bede's
work played a large part in the continued use by
the Church of the table being discussed. So we
can conclude that Theophilus' moon and his principles for finding Easter became Catholic sometime
between 501 and 725.

In the rest of this work, "ecclesiastical
moon" will mean the moon tabulated by Theophilus,
which remained the standard throughout the medieval
period. The ecclesiastical moon is based upon the
19-year Metonic cycle, named for the Athenian astronomer Meton who discovered it around -430. The
cycle is based upon the observation that 19 years
equal 235 months almost exactly. Thus, the relative positions of the sun and moon repeat almost
exactly after 19 years, and the lunar months occur
at almost exactly the same times in the solar calendar after 19 years.[†]

In terms of astronomical quantities, 19
years = 6939.602 days and 235 months = 6939.688
days. The discrepancy is 0.086 days or about 2
hours. Theophilus' contribution was to make 19
years = 235 months[‡] exactly in terms of the Julian
year and the (invented) ecclesiastical moon.

A common lunar year (in the Christian but

[†] Equivalently, the position of the sun repeats
almost exactly in the lunar calendar; we tend to
forget this in our preoccupation with the solar
calendar.

[‡] Theophilus probably did not originate this equal-
ity; it seems to have been used in the tables devised by one Anatolius about 270 [Jones, p. 20].

not the Jewish calendar) contains 6 months of 29
days each and 6 months of 30 days each, for a total of 354 days. An embolismic year contains an
extra month of 30 days [Bede, 725, Chapter XLV],
for a total of 384 days. The cycle of 19 years is
made to contain 12 common years and 7 embolismic
years, for a total of 6936 days. However, 19 years
contain $4\frac{3}{4}$ leap years, on the average, and hence
they contain $4\frac{3}{4}$ intercalary days,[†] for a total of
$6940\frac{3}{4}$ days. 19 Julian years contain only $6939\frac{3}{4}$
days. The discrepancy was corrected by simply
omitting the last day of the last lunar month in
the cycle. Thus the ecclesiastical moon had a
jump in its motion, which was called the saltus
lunae; in earlier tables of the ecclesiastical
moon, the saltus lunae occurred at other times.

The exact details of how the embolismic
years and the various months were distributed are
given implicitly by tables of the ecclesiastical
moon, in various forms. Bede [725, Chapter XX],
for example, gives a table of the day of the lunar

[†] We are accustomed to thinking of the extra day as
being connected with the solar calendar, not a lunar one. However, in the Roman solar calendar, the
intercalary day was a repeated day, that is, a day
with the same designation as a regular day. This
contrasts with our practice of having a separate
day called February 29. The day of the week was
not repeated. In the Roman calendar, the intercalary day was "vi. kal. Martii" [Bede, 725, Chapter XXXVIII], which is our February 24 in a non-
leap year. Suppose that the vi. kal. Martii first
arrived on the 3rd feria in some leap year, for
example. Then there would be another day, also
called vi. kal. Martii, but the 4th feria. The
date, being simply repeated, affected the lunar
year in which it occurred as well as the solar
year. Because the vi. kal. (of March) was repeated, a leap year was often called a bis-sextile
year, that is, a year with two 6's. The fact that
the intercalary day was the 6th calends March suggests that the Roman year once began on the following day.

month that comes on the first day of each calendar
month of the solar year, throughout the 19-year
cycle.[†] The Explanatory Supplement [1961, p. 422]
gives a table of the day in each calendar month on
which the ecclesiastical "luna 1" occurs; that is,
it gives the first day of each lunar month. The
tables are equivalent in principle; either can be
calculated from the other. I checked a few values,
but did not undertake a systematic comparison of
the two tables.

The average length of an ecclesiastical
month was $6939\frac{3}{4}$ days divided by 235, or 29.530 851
days. This is too large by about 0.000 262 days
or about 23 seconds. In a century, there are about
1237 months. Hence the lunar calendar generated
by the ecclesiastical moon accrues error at the
rate of 1237 × 0.000 262 = 0.32 days per century.
It does a better job than the Julian solar calendar,
which accrues error at the rate of about 0.78 days
per century.

The preceding statement applies if we are
thinking of the error in days. If we think of the
error in the position of the moon or sun on a given
day, the opposite is true because of the more rapid motion of the moon. By the time of Bede the
true moon was already running about two days (about
24° in position) ahead of the ecclesiastical moon
on the average, and the error in the lunar tables
should have become fairly obvious. If early observers noted an error in the ecclesiastical tables,
they did not often have the temerity to point it
out. Marianus Scotus [ca. 1082], in the last paragraph of his chronicle, made the first specific

---

[†]More accurately, Bede gives a set of rules in
Chapter XX for finding the day of the lunar month
for each "calends". In Jones's edition, this information is displayed in tabular form in his
notes, p. 356. It was not clear to me whether
the actual table was due to Bede or not. See
Section V.4 for further discussion.

reference to an error that I have noted† (see
Section XI.1).

However, the error in the ecclesiastical
moon did often produce an odd result that is shown
in the writing of Fulcher [ca. 1127], among many
examples. Fulcher observed the lunar eclipse of
1117 Jun 16, but he did not call it an eclipse. He
said that if the following day had been the 14th
of the moon, he would have known that the darkening of the moon was an eclipse. However, the following day was not the 14th of the moon and hence
the darkening was an omen instead of an eclipse.
Thus the error, in conjunction with some knowledge
of astronomy, helped to preserve superstition.

5.   The Ecclesiastical Full Moon

The ecclesiastical full moon means the 14th
day of a lunar month. Here we are concerned specifically with the Paschal full moon, which means the
14th day of the lunar month Nisan. Nisan is the
first month of the Jewish calendar‡ and comes in
the spring. As I said in the first section of this
chapter, Nisan in the context of the Easter problem does not necessarily mean this Hebrew month.
Instead it means what early Gentile Christians said
the Jewish month Nisan was.

In every position that has been taken with
regard to the Easter problem, Easter has been tied
in some way to the date Nisan 14. Assuming agreement on a rule for relating the two, one would
think that the logical procedure for the Christians

---

† Bernoldus [1100], in connection with the solar
eclipse record 1093 Sep 23c E,CE in Section IX.3,
implied that Herimannus [1054] noted the existence of an error in connection with the eclipse
of 1033 Jun 29. I have been unable to find this
passage in Herimannus.

‡ But see the footnote about the Jewish calendar in
Section II.2 above.

would be to find out from the Jews how the month
Nisan was determined. In a situation so charged
with emotion, it is doubtful that the logical approach was ever tried.† If it had been tried, it
is doubtful that it would have succeeded,‡ at
least not until after the Council of Nicaea had
long passed.

Jones [1943, p. 8] writes: "But there is
no definite evidence how the early Hebrews determined which new moon began the month Nisan." If
he means the Hebrews in Mosaic times, this is
certainly correct, but we are not concerned with
such early times; we are concerned with the first
few centuries of the common era. For this period,
there is definite evidence, according to Jacobs
and Adler [1906], who say that the following rule
was adopted "even before the destruction of the
Temple": If the sun (in the month following the

---

† Jones (p. 59) cites a sixth century source that I
have not consulted directly which says in part:
"The Jews are as ignorant of the Passover as they
are of God." Such an attitude does not encourage
cooperation.

‡ It was the duty of the Jewish priesthood to determine the beginning of each month and the beginning of the year (Nisan 1). In Mosaic times, the
determination was surely by observation. At some
time, calculation from tables was doubtless substituted. Use of approximate tables may have begun during the Exile, which could have brought the
Jews into close contact with Babylonian astronomy.
The calendrical rules of the Jews were priestly
secrets for a long time. After the Dispersion,
and particularly after the persecutions of the
Jews under the Christian Roman emperors, it became difficult for the priesthood to communicate
the calendar to its followers, and the priesthood
made the rules public. According to Jacobs and
Adler [1906], the rule governing the intercalation
of the 13th lunar month was first published by the
Patriarch Hillel II because of the persecution during the reign of Emperor Constantius (ruled 337-
361).

-34-

12th month) will not reach the vernal equinox before the 16th day, the month is to be called Adar
Sheni (second Adar or Veadar) and the following
month is Nisan. In other words, the equinox must
occur on or before Nisan 16. A 13th month is to
be intercalated as necessary in order to make this
happen.[†] Jacobs and Adler did not say in any place
that I noticed how the date of the equinox was
determined.

　　　The Christian rule is that Nisan 14, not
Nisan 16, is to be on or after the equinox.

6.　The Jewish Calendar

　　　An ordinary Jewish year nominally contains
6 months of 30 days each and 6 months of 29 days
each, for a total of 354 days. An embolismic year
nominally contains 7 months of 30 days and 6 months
of 29 days for a total of 384 days. The calendar
is based upon a 19-year cycle; the 3rd, 6th, 8th,
11th, 14th, 17th, and 19th years in each cycle are
embolismic. If all years had their nominal length,
the full cycle would contain 6936 days. The average length of a year would be $6936 \div 19 = 365.0526$
days and the average month would contain $6936 \div 235$
$= 29.5149$ days. Both these values are considerably
too small; the values given by modern astronomy are
365.242 199 days and 29.530 589 days, respectively.

　　　So far in the discussion the Jewish calendar and the Christian ecclesiastical calendar are
identical except for small details. Their fundamental differences arise from the way in which they
alter the nominal values in order to give more
accurate values to the average month and year. The
Christians met this problem, as we have seen, by
making the year be the Julian year of $365\frac{1}{4}$ days.

　　　In the Jewish calendar, the problem is met
by forbidding certain calendar dates to fall on

[†]
The fact that the intercalary month is added just
before Nisan suggests that Nisan was the first
month at the time the rule was adopted.

-35-

certain days of the week. A date is prevented
from falling on a forbidden day by either of two
devices. The 9th month Kislev may be shortened
from 30 days to 29 days, yielding a "defective"
year of 353 or 383 days, depending upon the number
of months in it. Alternately, the 8th month
Heshvan may be lengthened from 29 to 30 days,
yielding a "perfect" or "abundant" year of either
355 or 385 days. A year with the nominal number
of days, either 354 or 384 according to the number
of months, is called "regular". Only about 1 year
in 3 is regular.

Whether a year that is not regular is to
be defective or abundant depends upon a complex
set of rules that are stated in full by Resnikoff
[1943]. Because of the complexity of the rules,
there is no simple perpetual Jewish calendar. The
net effect of the rules is to give more abundant
than defective years. The average length of the
year is 365.246 828 days and that of the month is
29.350 594 days. The year is too long by about
0.0046 days, making the equinox fall later than
the correct value by about one day in 216 years.
The month is too long by about 0.000 005 days,
making the full moon fall too late by about 1 day
in 200 000 months, about 16 000 years.

This calendar was perhaps promulgated
about the year 350 by the Patriarch Hillel II
already mentioned. It will be a long time before
the average error in the full moon becomes significant for calendrical purposes, although there is
necessarily an occasional error of a day, because
of rounding numbers. The equinox corresponding to
the calendar is now too late by 7 or 8 days, and
comes on the equivalent of about March 27 or 28.

Because of the complex way in which the
rules governing defective and abundant years are
stated, it is hard to say to what extent the accuracy of the year and month is accidental and to
what extent it is purposeful. The accuracy of the
average year is about what we would expect of
astronomical knowledge in the 4th century. The
fantastic accuracy of the month seems to exceed

the level of knowledge in the 4th century and is
probably accidental.

7. Comparison of Passover and Easter

        Although the Council of Nicaea enjoined
Christians from "keeping company" with the Jews
in their holy days, the rules adopted for determining Easter did not necessarily satisfy the in-
junction. In effect, what the Christians did was
to make up their own set of rules for finding
Passover; let me use the term "Christian Passover"
for the day defined by the Christian rules. The
Christians then adopted rules that would not let
Easter coincide with the Christian Passover.

        There are several reasons why the Passover
and the Christian Passover may not coincide.

        a. Aside from the long-term growth of er-
ror in the medieval ecclesiastical moon, there are
detailed differences between this moon and the Jewish calendrical moon.[†] This is shown, for example,
by the fact that all lunar years in the Christian
ecclesiastical lunar year are regular although only
about one Jewish year in three is regular.

        b. The latest date for the equinox is
Nisan 16 in the Jewish calendar, according to Jacobs

[†]
Jones (p. 55) cites a fifth century source that
I have not consulted directly which refers to Hebrew tables unknown to Romans (meaning Roman
Christians, I believe). Jones then adds: "It
could not have been an ancient Hebrew, or Mosaic,
table, because there never was such a thing;..."
Here and in many other places Jones seems to take
"Mosaic" literally; that is, he seems to think
that Jewish religious practices in early Christian
times were identical with what they had been in
the time of Moses. Tradition, as well as errors
in the Jewish date of equinox, indicate an origin
around the 4th or 5th century for the present Jewish calendar, and it is plausible that there were
earlier tables that are now lost.

and Adler [1906], but it is Nisan 14 in the Christian version of the Jewish calendar. This difference by itself has no effect unless a full moon
comes close to the equinox. When a full moon is
close to the equinox, the difference tends to separate Passover from the Christian Passover by a
lunar month.

     c. Around the 4th or 5th centuries, the
Christian and Jewish equinoxes were close together
if not coincident. The Jewish equinox moves steadily later in the solar calendar and is now about
8 days late. The Christian equinox moved steadily
later and was about 10 days late when it was corrected by the introduction of the Gregorian calendar.† The present difference in the two "religious"
equinoxes gives rise to the peculiar result that
Easter now comes before Passover about 1 year in 4,
although the original intention was that Easter
should be the 1st feria after Passover.

     The result of the differences in rules is
that Passover and Christian Passover may differ by
about 1 lunar month, they may differ by 1 or 2 days,
or they may coincide. Easter and Passover may coincide, but I do not know the years in which this has
happened. In order for Easter and Passover to coincide, it is necessary for Passover to come on the
1st feria, and for the Christian Passover to be a
day or so earlier than Passover. Easter, being the
1st feria after the Christian Passover, will then
coincide with Passover. Because of the drifting
apart of the equinoxes, coincidences of Easter and
Passover must be less frequent now than in the centuries immediately after the Council of Nicaea.

8. The Great Easter Cycle

     The cycle of 19 Julian years contains $6939\frac{3}{4}$
days. It takes four such cycles, lasting 76 Julian
years (27 759 days), to contain an integral number

---
† It would now be about 13 days late if the calendar had not been reformed.

of days; if a cycle is considered to begin at midnight, say, it is 76 years before another cycle can be considered to do so. Since 27 759 is not a multiple of 7, the day of the week at the beginning of a cycle is not the same as it is 4 cycles later. After 7 instances of 4 cycles, however, there have been an integral number of days and of weeks; that is, the days of the week begin to repeat.

Thus the ecclesiastical moon and the dates of Easter, according to the medieval rules, were strictly cyclic in the Julian calendar, with a repetition period of $7 \times 4 \times 19 = 532$ years. The repetition took place with regard to solar dates, lunar dates, and days of the week.

There is more to the "great Easter cycle" than this period; the cycle was considered to have definite years of beginning and ending. By coincidence, Dionysius (Section II.4 above) began his extension of the Easter tables with the year 532. Thus the idea arose [Bede, 725, Chapter LXV, for example] that the year 532 began[†] the second "great Easter cycle" of 532 years, which ended in 1063, followed by the third cycle, and so on.

The periodicity makes it particularly easy to find the date of Easter in any year in the Julian calendar: Either form of the table of the ecclesiastical moon [Bede or Explanatory Supplement, see Section II.4 above] allows us to find the date of the Paschal full moon; this table needs to run for only 19 years. In order to find the date of Easter, we then need only the day of the week of the Paschal full moon. A table that repeats every 28 years gives the relation between dates and the days of the week. In order to find Easter in any year, we have to divide the number of the year by

---

[†] The error involved in this idea should not surprise us. I write this shortly after New Year's Day of 1970. Almost everyone thinks that this is the first year of a new decade instead of what it really is, the last year of the seventh decade (of the 20th century).

19 and by 28 and keep the remainders. Use of the remainders in the tables on pp. 422-423 of the Explanatory Supplement immediately gives the date of Easter. The calculation for the Gregorian calendar is but slightly more complex; the needed tables are found on pp. 425-428. There is no simple "great Easter cycle" in the Gregorian calendar.

9. Easter Tables and Computi

Certainly after the work of Bede, and perhaps before, it was a fairly simple matter to calculate the date of Easter in any year. Lethargy, however, is a powerful force, and tables that listed the date of Easter for each year continued to be a popular item in medieval religious libraries. Such Easter tables saved the labor of computation, but the connotations of this remark may be unjustified. Easter tables also helped insure the uniform observance of Easter, at least within a coherent group of communicants.

An Easter table was often combined with a short treatise on astronomy and a description of the methods by which the table was calculated. A work of this sort, with perhaps additional information to suit the fancy of the writer, is called a computus [Jones, 1943, p. 75]. Computi were popular during the Middle Ages, although the name is apparently a recent invention.

In its simplest form, an Easter table has two columns. The first column gives the year and the second gives the date of Easter for that year. The first column is not necessary. The dates themselves are sufficient provided that none are omitted and that they occur in order. However, some designation of the year provides insurance against omission or inversion of the order of dates.

Some Easter tables gave considerably more information of a calendrical or astronomical sort. Poole [1926], in the frontispiece of his book Chronicles and Annals, gives a reproduction of part of a 10th-century Easter table used in the

-40-

monastery of Einsiedeln in Switzerland. It has ten columns. (Poole said eight columns, but two of the eight can each be considered as two individual columns written together without a break.) The columns other than those of the year and the date of Easter are irrelevant to this chapter, and I shall not describe them. With one exception, the columns give information connected with calendar problems, such as the age of the ecclesiastical moon on New Year's Day, but most of the information is not needed in finding the date of Easter.†

Knowledge of the Easter problem and related matters is necessary in order to understand many of the medieval records of solar eclipses. This is the reason I have spent so much time discussing it. Easter tables are important in the development of medieval historical writing. This development will be discussed in the next chapter.

---

† Only two quantities are needed to determine Easter in the Julian calendar. They are the remainders mentioned in the preceding section. The remainder after division by 19, with 1 added, is called the Golden Number. The remainder after division by 28 is often replaced by the Dominical Letter, which identifies the days that come on the 1st feria (Die Dominica).

CHAPTER III

ANNALS, CHRONICLES, AND HISTORIES

1. Lack of Precise Definitions

The use of records of solar eclipses from
medieval historical writings requires some discussion of the nature and development of those
writings as part of the general background.

Historical writings can perhaps be put
into one of the three categories of annals, chronicles, or histories. I doubt that it is either possible or desirable to adopt strict definitions of
the terms; instead, the various classes of writing
merge into each other. I propose the following
loose definitions of the terms:

Annals: a collection of brief notices of
events, arranged chronologically by year. An individual entry is an annal.

Chronicle: a collection of extensive descriptions of events, arranged chronologically, and
perhaps including attempts to trace individual
themes through a series of events.

History: a study of the connections between events having some common element, such as a
relation to a geographical area, with emphasis on
causal relations. The arrangement may be chronological or topical.

Length of individual entries provides a distinction
between annals and chronicles. The extent of topical or thematic emphasis provides a distinction between chronicles and histories.

Some writers have tried to give more precise definitions, and some have tried to provide

an evolutionary pattern of the various historical forms, but it does not seem possible to maintain these definitions and patterns over a wide range of material.

For example, there is a type of historical record associated with Easter tables. Poole [1926, p. 26] wrote that these "represent the earliest type of medieval Chronicle, which it is convenient to distinguish from the more elaborate works into which they developed by the name of Annals." To me, this implies that medieval chronicles developed from annals, but Poole seems to contradict this on the next page where he wrote that the medieval chronicle developed from the Chronicle of Jerome[†], which is earlier than what Poole had just called the "earliest type of medieval Chronicle".

Some writers have distinguished between chronicles and histories on the basis of their elegance of treatment or their loftiness of theme. It seems to me that this distinction is invidious and that it cannot be sustained.

2. Annals and Easter Tables

I wrote (Section II.9) that an Easter table needs only two columns, and that one of these could conceivably be eliminated. On most writing media, a table with only two columns leaves considerable blank space. In many medieval church institutions, the blank space was used to make annalistic entries.

[†] Presumably Poole meant the translation, revision, and continuation of Book II of the Chronicon of Eusebius [ca. 325] that Jerome (Eusebius Sophronius Hieronymous) made about 375. Book I of the Chronicon was apparently not popular, and Book II may have been all of it that was known to medieval writers. It may be all that was known to Poole, who wrote (p. 10) that Eusebius called his work "Chronological Tables" rather than a "Chronicle". This is correct if it is applied to Book II only, but it is not correct for the whole work.

In other words, medieval annals were often written on Easter tables.

I would not go as far in emphasizing this point as Jones [1943], who writes on p. 114: "An aesthetic urge influences the writer to use those excessive margins for additional matter, ...", and who also writes on p. 117, with respect to the expanded tables with eight columns: "... the list, despite its now traditional eight columns, still left ample margin."

I doubt that the urge to use the margins is best described as aesthetic. The urge may simply have been the urge to "doodle", or it may have been a matter of convenience, or of saving paper. The "ample margin" is also open to question. The first annals-Easter table manuscript page that Poole [1926] reproduced in his frontispiece has eight columns, and there is definitely not ample margin. An annal is written between the lines of the table when it is short enough; when it is too long, a reference mark transfers it to the narrow margin at the left, where the annal is continued vertically like another column.

Even when the table had fewer columns, the space for an annal was often inadequate. The second page that Poole reproduced had only four columns. Here an annal continues a year line; that is, it starts in the margin to the right of the table proper. Many entries are too long for one line and have to be crowded into the lines that belong with other years.

It seems to me that Jones and Poole go to considerable lengths to establish that the Easter table was the origin of annals, and that annals could have had no other origins. For example, Jones (p. 116) writes: "These Easter-tables, or lists of moveable feasts, were an innovation in the West. Although the Romans had calendars in pagan times and also had historical lists with political records on them, history was not, in the main, forced into annalistic form by a required annual list. The introduction of such a

-45-

list . . . eventually changed the whole course of historical writing."

By the Roman "lists with political records", Jones probably meant the consular lists[†] that I shall describe in the next section. Poole (p. 8) also mentions consular lists, which he calls Fasti Consulari. He clearly means the set of consular lists that is often miscalled the Annals of Ravenna. He says that they "served in time to form the basis of rudimentary Chronicles" and that they would have developed into chronicles if they had not stopped too soon. Poole also said that a chronicle or annals that lists every year even when it does not list an event for that year is ultimately derived from an Easter table.

Jones's statements about the importance of annals seem too strong to me. History has not been, and it was not in the medieval period, "forced into annalistic form". Neither the Easter table nor any other annual list changed the "whole course of historical writing". The writing of histories (histories by anyone's definition, I think) has been essentially continuous in the western world since Herodotus, although the rate of production of histories in western Europe declined during the early medieval period.

Annals certainly have been written upon Easter tables, but that does not mean that they originated from the tables. Some people keep diaries, and some companies manufacture printed forms for convenience in keeping diaries, but that does not mean that these companies originated diaries. The most one can say in either case, it seems to me, is that the recording function used a convenient available form, not that the function followed the form.

There are historical records, which I think are properly called annals, that antedate the medieval period by a considerable margin. I shall des-

---

[†] It is also possible that Jones meant the Annales Maximi, described in the next section.

cribe three of these in the next section. The
oldest set of these annals was not known in the
medieval period. It could not have influenced
the medieval annalists directly, but it could
easily have been part of a tradition that continued unbroken into the Middle Ages. The other two
were definitely known in the medieval period, and
one of them was famous then. I cannot prove that
they directly led to medieval annals, but I think
their existence means that there would almost
surely have been medieval annals, Easter tables
or no.

In fact, I doubt that one can ascribe any
definite origin to annals, chronicles, or histories. They must all be older than writing itself.

3.   Annals Before the Medieval Period

From the standpoint of form, an Easter
table is a table with two columns. Its essential
feature is that it gives a single item of information for each year, and this information looks
arbitrary to a casual reader. One column gives
the arbitrary information. The other column identifies the year in some way; this column could
be omitted if there were assurance that the succession of years for the arbitrary information
was unbroken. There are many ways of identifying
the year. One way is by a number giving a count
of years from some epoch. Another is to give
the name of a reigning monarch or other official,
along with the number of years that he has been in
office. There are many circumstances that call
for the existence of an annual list.[†]

[†] The Jews had a problem analogous to the Easter
problem but stated in a quite different form.
Passover, as well as the other festivals, has a
fixed date in the Jewish calendar, unlike Easter
in the Roman calendar. However, it is necessary
to determine for each year whether it is to be
ordinary or embolismic and whether it is to be
defective, regular, or abundant (Section II.6). One

Under the Roman republic, the two consuls were equal heads of government with annual terms of office. Consuls continued to exist under the Empire. Although their position became mostly decorative, they apparently always retained some official functions. Their terms might be greater or less than a year, and some years there was only one consul. Acts of the Senate and other governmental actions, particularly under the Republic, were often identified by means of the current consuls. Thus there was a need for a consular list.

Each administrative or judicial district probably needed its own consular list, and a number of lists have survived. Fasti Vindobonenses [ca. 576] is typical.[†] It occurs in the collection that Jones [1943, p. 116] perhaps had in mind when he referred to Roman political lists. This is the same collection that Poole [1926, p. 8] called Fasti Consulares and to which he gave the status of a rudimentary chronicle (see the preceding section).

Poole and Jones demur from flatly using the words "annals" or "chronicles" in connection with the consular lists. To my eye, Fasti Vindobonenses have all the appearance of annals. They are based upon arbitrary annual information, namely the identification of the consuls, which they give in tabular form. In the remaining space they often

---

would suspect that this need would give rise to tables that list these properties for each year and that might furnish the medium for annals. I have not found such tables or annals, nor have I found any clues that such annals ever existed.

[†] Fasti Vindobonenses will be discussed in more detail in Section XII.1. It is one of the consular lists that the editor of the edition cited has collated under the general term Consularia Italica, although there are other Italian consular lists. The Consularia Italica are apparently the collection that was once published under the title Annals of Ravenna. As the editor points out, the term Annals of Ravenna is much too restrictive; the various annals could have been prepared in many places.

give additional brief notices of events.† These
events may be political or scientific. Several
records of solar eclipses from Fasti Vindobonenses
will be used later in this work, and there are
other types of scientific record as well. Several lists are both consular lists and Easter
tables.

In a preceding section I mentioned the
Chronological Tables of Eusebius [ca. 325], which
form the second book of Eusebius' Chronicon. Jones
(p. 118) refers to Eusebius and Jerome as "chron-
ographers" but does not describe the type of work
that the Chronicon represents, at least not in any
place that I noticed. Poole (p. 27) calls the
work the foundation of medieval chronicles, but
implicitly denies it a part in the development of
annals.

However, Eusebius' Chronological Tables
in fact have the two main characteristics of annals,
and they have no other important characteristics.
First, they give arbitrary annual information.
This information is the conversion from the year
of the "era of Abraham" to the year according to
several other chronological schemes such as the
regnal years of the emperors or the years of the
Egyptian dynasties. Second, there is an annal
parallel to most but not all of the arbitrary annual entries. For examples, Eusebius notes the
births of Socrates and Jesus, and he notes the
"eclipse of Thales" and the "eclipse of Phlegon".‡

Thus I do not see how anyone can deny
the title of "annals" to Eusebius' Chronological
Tables. One cannot deny the title by the claim

---

† If Jones (see the preceding section) is right that
"an aesthetic urge influences the writer to use
those excessive margins for additional matter,"
then consular lists must necessarily have given
rise to annals, contrary to Jones's statement that
"Easter table" annals were an innovation.

‡ See AAO, pp. 94-96 and 110-113.

that genuine annals must be contemporaneous; these tables, along with Jerome's continuation, are contemporaneous for their last century. Further, it seems to me that one must admit at least the possibility that the Chronological Tables influenced medieval annalistic writing. There is little question that they were famous in early medieval times. Bede [725, Preface, Chapter IX, and probably other places that I did not note], for example, refers to the chronological work of Eusebius. The Chronological Tables were an ever-present example that could have inspired imitation by anyone because of the high repute of their author.[†]

There is no specific instance in which Fasti Vindobonenses is known to have influenced medieval annals directly. However, it is of a form that appears to have been common in Italy. Bede [734?] certainly used annals that were contemporaneous with Fasti Vindobonenses and that probably came from Italy.[‡] Thus Fasti Vindoboneneseses is part of a continuous tradition of annals that leads directly into the Middle Ages, whether it had specific individual influence or not. Thus it is plausible to allow consular lists a role in the development of medieval annals.

There is a weak point in the argument. In their present physical form, the consular lists date from a period later than the first annals that were derived from Easter tables. Thus the annals based upon consular lists may have been compiled by combining consular lists with annals from Easter tables. The argument would be stronger if there were annals known in, say, republican Rome.

[†] Book I of Eusebius would be classed somewhere between a chronicle and a history, according to the approximate definitions that I have adopted. It was probably lost during the Middle Ages; if so, it did not influence medieval historical writing.

[‡] See the records 538 Feb 15 E,I and 540 Jun 20 E,I in Section XII.2.

Cicero [-54] says that the pontifex maximus, from the earliest of Roman times down to the pontifex P. Mucius,† posted an annual notice of the main events of each year at his home. These annals had presumably been collected and were known in Cicero's time as the Annales Maximi. The authenticity of these annals has been doubted. It is unlikely that the annals from earlier times survived the looting by Brennus about -390, if the stories about Brennus' raid are true. It is unlikely that the custom of keeping the annals would have stopped at the time stated. Forgers of history commonly attribute their product to ancient sources that they have found. It is plausible that someone about -100, say, forged an early history of Rome in the form of annals and attributed them to earlier pontifices maximi who were no longer around to correct the attribution.

From the standpoint of the development of the form, it does not matter whether the Annales Maximi were genuine or not. They prove that the idea of annals existed in late republican Rome and that documents accepted as annals existed then and there. Six centuries later we find annals still being kept in connection with consular lists. In between we have the class of historians known as annalists, who kept the form of annals alive, whether there were any annals being kept continually as a living form or not. In sum, we find an unbroken tradition of annals from republican Rome into the Middle Ages, and these annals have nothing to do with Easter tables.

There is a known set of annals that is older than Rome itself. It comes from Assyria. The Assyrians [Smith, 1958], like the Romans, had an official for whom each year was named. This official was called a limmu; there was one limmu per year. An Assyrian limmu list was a necessity, just as a Roman consular list was. Smith says that there were two classes of limmu lists: ". . the first class consists of a simple list of names, the second gives in a column opposite the name

---

† Publius Mucius Scaevola, who was consul in -132 and who became pontifex maximus in -129.

an entry concerning the events of the year." The
second class is thus a set of annals; it combines
an annual bit of necessary arbitrary information
with an annal for the year.

One set of limmu annals, often called the
"eponym canon", contains a record of a solar eclipse,
along with the month in which it occurred. This eclipse, often called the "eponym canon eclipse", can
be dated with virtual certainty as the eclipse of
-762 Jun 15 (AAO, Section IV.2). This fixed date,
along with continuous limmu lists, allows the fixing of Assyrian chronology from -889 to -647.
Broken lists which cannot be dated exactly go
back at least to -1200.

The Assyrian limmu lists were unknown in
the Middle Ages and could not have influenced medieval annalists directly. However, there were
close cultural contacts between Babylonia-Assyria
and the Mediterranean world, and there has been a
high civilization at least somewhere in that general area since long before the Assyrians. The
limmu annals are part of a series of historical
writings that has continued unbroken from ca. -1200
through the Middle Ages to today.

In sum, annals were being kept more than
1500 years before the beginning of the medieval
period, and continuous tradition connects the culture in which they were written with the Middle
Ages. There is no need to assume that annals were
invented because of the stimulus of an Easter table.

I quoted Poole [1926] as saying that a set
of annals that lists every year, even when it does
not list an event for every year, is ultimately derived from an Easter table.[†] The Chronological Tables
of Eusebius (with their continuation by Jerome) furnish a counter-example. They list the number of
every year of Abraham through 2395. They do not
list an event for every year, even during the peri-

[†] Poole used this idea as a basis for deciding
which of the versions of the Anglo-Saxon Chronicle
are oldest in their original forms.

-52-

od when they are contemporaneous with their authors.

To be sure, the Chronological Tables may give each year in order to maintain the flow of arbitrary information (the conversion between various chronological systems). One could regard this as an extension of Poole's argument. However, I doubt that the desire to have an unbroken flow of information is needed in order to have the number of every year listed.

A medieval annalist needed to leave space for revision and for addition of individual annal items. When he was preparing a set of annals for past years from various sources, for example, he might always come across an interesting item in a year that previously lacked one; thus he may have prepared for this possibility by putting down the number of the year in advance. When he was maintaining a contemporaneous set of annals, he had to allow for the possibility of the late arrival of news. It was customary to note the deaths and accessions of popes, for example. An English annalist might hear of such an event six months after it happened, when he might have already gone on to the next year.

I am not sure that even these precautionary measures are needed in order to account for the listing of a year without a corresponding event. Some people simply like to make lists.

4. The Christian Era

It is not necessary when preparing an Easter table to count the years from some fixed era, but it is helpful. Most Easter tables [Jones, 1943] have used an era. The era of the accession of the emperor Diocletian was perhaps the most common until the time of Bede. At least one early table [Jones, p. 64] used the era of the Passion, which at that time was placed in the year that we call 28.

The first known use of the Christian era occurred in the Easter table prepared by Dionysius Exiguus [Jones, p. 69; also see Section II.4 above], who was one of the early users of the ecclesiastical moon that became the medieval standard. Dionysius' table ran from 532 through 626. The earlier table, which Dionysius simply extended by using the same principles, had used the era of Diocletian. Jones (p. 69) quotes Dionysius as saying that he changed the reference year because "I did not wish to preserve the memory of the impious persecutor in my cycles"; he wanted instead to denote times from the birth of Jesus.[†]

Bede [734?] adopted the Christian era in his great history. He was not the first to use it in historical writing [Jones, p. 70], but his history influenced later medieval writers greatly, and there is little question that Bede's influence helped bring about the universal (universal in the west, that is) adoption of the Christian era for historical purposes.

Jones (p. 69) says that the use of the era for history arose only "from using the margins of

---

[†] The year in which Jesus was born is unknown, and the accuracy of Dionysius' era, which is the one that we still use, has been questioned for a long time. It is interesting that the day called 1 Dec 25 in the Julian calendar fell on the 1st feria, or die Dominica. A certain type of mind would accept this fact as "proof" that the year chosen was correct, but it is doubtful that Dionysius based his choice upon this fact. Apparently the year 1 was intended to be the year after Jesus was born. Syncellus [ca. 810] makes this intention explicit. In Syncellus' chronological tables, Jesus was born in the year 5500 after the Creation, and the year 5501 after the Creation is equated to the year 1 of Christ. Thus the intended date for Jesus' birth is the date that we call 0 Dec 25 in the Julian calendar, and it came on the 7th feria.

the Easter-tables as a place for preserving annual notices." I hesitate to accept this statement. Jones's argument, I believe, is that Bede's influence led to adopting the Christian era, and that Bede derived his dates from annals written on Easter tables.

Jones (p. 121) himself suggests another explanation that sounds more plausible to me. In an earlier work, Bede had estimated the age of the world and had used the "era of the world" in giving his dates. Bede's estimate differed considerably from earlier estimates, and as a result he had been charged with heresy. The charge was not pressed formally, but Bede always remained sensitive to it. Jones's suggestion is that he changed his era in order not to revive the heresy charge.

This suggestion does not by itself account for Bede's adoption of the Christian era. The most convenient era would be the era[†] most frequently used in his sources. According to Jones, some English sources of the 7th century used the Christian era; I have not seen evidence suggesting that the Christian era was the most popular one to be found in Bede's sources. If Bede felt compelled to change eras, he might have chosen the Christian era for the reason that led Dionysius to use it in the first place: He wanted to preserve the memory of the birth of Jesus.[‡]

Poole [1926, p. 25] writes with regard to the Christian era: "It was the discovery of this Era that made the revival of historiography possi-

---

[†] If any; his sources would not necessarily have used an era. They could have used regnal years, for example.

[‡] Dionysius was the first to use what Christians call years A. D., but Bede was apparently the first to use "B.C.". In Book I, Chapter II and Book V, Chapter XXIV, Bede mentions Caesar's invasion of Britain and puts it in the 60th year before the Incarnation.

ble, and it was beyond question an English discovery." I am unable to agree with this statement.
I do not believe that it was an English discovery
and I do not believe that it or any other era is
necessary for historiography. Even if an era were
necessary, there were suitable eras that had been
extensively used before the first known use of the
Christian era.

To take the simplest point first, Dionysius
Exiguus made the earliest known use of the Christian era. Dionysius spent his adult life in Rome,
and he was probably Scythian by birth. Victoris
Tonnennensis [ca. 567] made the first known use of
the Christian era in historical writing; Victoris
was from Africa.

Next, eras had been in common use before
the introduction of the Christian era. Most of
them would be as convenient for historiography as
the Christian era. Dating by means of the Olympiads was common, and this is one of the systems in
Eusebius' tables.[†] Hydatius (see Chapter XIV),
writing in Spain in the fifth century, still used
the Olympiad system. The choice of the Christian
era came from religious or political motives and
not from its importance in historical writing.[‡]

[†] I have long had an impression that the Romans
from the time of the later Republic dated from
the founding of the city (A.U.C.), which they assigned to the year equivalent to our -752. Poole
[1926, p. 8] says, of the consular lists at least,
that A.U.C. dates occur only where they have been
added by modern editors. I have not tried to investigate this point. Eusebius [ca. 325] does
not include the A.U.C. system in his tables. The
Roman system that he includes applies only to the
Empire; it is the system of regnal years of the
emperors.
[‡] Eusebius [ca. 325], as we saw above, used the "era
of Abraham" in his chronological work, but he apparently had few imitators in this regard. Byzantine historians (see Section XV.1) used what modern references call the "Mundane Era of Constan-

-56-

Finally, it does not seem that an era is needed for historiography. Historians need dates so that they may put events in order, so that they may correlate events in different areas, and so that they can find approximate time intervals between events. Use of an era simplifies the task of finding long time intervals accurately. The ability to do this is important to the astronomer and perhaps to other scientists, and I am glad that most medieval records used a particular era. The historian needs to find short intervals conveniently, and regnal years, among other possibilities, meet this need. I can see the importance of knowing that the United States was involved in the Second World War for slightly less than four years; this gives some idea of the balance of the opposing forces. It does not seem important for history that there were 165 years, 5 months, and 3 days between the Declaration of Independence and entry into the Second World War; it would be sufficient to know that it was a bit more than $1\frac{1}{2}$ centuries. The need of the historian for an accurate estimate of a long interval, if it exists, is rare; he would not be seriously impeded in his efforts if he had to find this interval by adding up lengths of reigns.

5. The Ontogeny of Medieval Annals

We have seen in earlier sections that institutions in many times and places have found it useful to prepare and maintain annals. Medieval religious institutions shared in this process. Easter tables were not needed in order for annals to develop, but they furnished a convenient medium for the keeping of annals. The Easter problem that lay back of Easter tables provided and to some extent enforced a uniform chronological system for almost all Christians. This system ultimately led to the adoption of the Christian era,

tinople", although they themselves did not use this term. The Mundane Era was the standard era in Byzantine historical writing for many centuries.

whose use is a convenience but not a necessity.

Although Easter tables provided a uniform chronological system, their form also encouraged errors in the use of that system, as I suggested in Section III.2; see Section V.5 below.

There is no fixed pattern to the development of individual sets of annals. Perhaps the most common pattern is the following: A newly-founded institution borrowed annals or chronicles from one or more older institutions and copied or merged them in order to form a history up to the present. Subsequently various persons in the institution maintained the annals, in a reasonably current fashion, for varying lengths of time. Some sets of annals were maintained continuously for several centuries.

The pattern of handwriting in many sets of annals shows this development, according to their editors. We often find that all entries up to a certain point are in the same hand; these represent the copied portion of the annals. Entries after that tend to change hands at reasonable intervals. This probably means that one person in the institution was assigned or assumed responsibility for keeping the annals. He did so for a number of years, after which death or transfer or some other incident caused the responsibility to pass to another. A variety of hands suggests but does not prove that the annals were being kept on a current basis and hence that they were independent.

There were many variants of this pattern. In one variant, an institution may have deferred the preparation of annals for a considerable time, during which it accumulated miscellaneous records of various sorts. At some time it may then have inaugurated the pattern just described. The first compiler would then have started the set of annals by merging the early local records with the borrowed annals. Thus we often find independent local material in the middle of largely derivative annals.

A complete discussion of the patterns of annals would be prohibitively long, and I shall mention only one more example. After a set of annals had been started, a new person may have become interested in their earlier periods; perhaps he had just arrived from another institution with new material. He might then go back into the earlier portions of the annals and add material. In this case, we would have a portion of the annals in two or more hands, but the portion would be entirely derivative in spite of this fact.

6. Development of Chronicles and Histories

It is not necessary to say much about the development of these types of historical writing. They existed long before the medieval period, and we do not need to assume that they developed out of the simpler annals.

The Iliad, for example, can fairly be called a history, I believe. It is a history that was first composed for oral transmission and that was not written down for several centuries. It was known in the medieval period, as was much else in Greek history and mythology. Christians had another and more obvious source of history in the Bible. Genesis and Exodus are histories by any definition, even by the one that requires loftiness of theme.

Chronicles provided an obvious example of the chronicle for imitation.

The Byzantine Empire was a direct continuation of the ancient Mediterranean and eastern civilizations, and the production of histories and chronicles by individuals continued essentially unbroken in the Byzantine Empire throughout the medieval period. The writing of histories and chronicles also continued in western Europe, although probably on a reduced scale. Even if we confine ourselves to works that will be used later in this study, we can cite the following histories

or chronicles written by western Europeans in the
early medieval period: Gregory of Tours [ca. 592],
Isidorus [ca. 624], Bede [734?], and Paulus Diaconus
[ca. 787]. These and other writings kept the historical tradition alive in the west.

Chronicles and histories were nearly always written using material from other historical
records. Annals, on the other hand, when they are
original, are usually written almost contemporaneously with the events; they are something like
diaries. For this reason, annals tend to be the
most useful form for the purposes of this study;
they are the form that tends to contain the original information while chronicles and histories
tend to be derivative.

Annals tend to be the most useful for another reason. They tend to record events that the
writer found interesting, as opposed to events
that the writer thought would impress posterity;
there are exceptions, of course. Thus an annalist often recorded a solar eclipse or other interesting physical phenomenon. Many historians, on
the other hand, scorned to give space to such
events on the grounds that they are not important.

There has been some discussion about
whether annals developed into chronicles and histories. Amusingly, histories and chronicles often
developed into annals. Some of the medieval histories or chronicles became quite popular in their
own time and were widely copied. Their presence
in a monastic library, for example, often inspired
a local resident to continue them by making annalistic entries at the end of the original manuscript.

7. Dating Errors

Casual inspection shows that medieval historical writing contains many errors in dates.
We can tell this by means of eclipses that can be
dated accurately from modern calculation. We can
also tell by inconsistencies in dates assigned to
the same event in different sources.

When I wrote AAO, I assumed that the errors in dates usually arose with the writer of the original record. I assumed that he might often be ignorant of the year because he had lost count of the years, and that, in an age of poor communications, he had no easy way to check the current year number and no particular reason to do so. I realize now that I did not understand the problem because I was not aware of the importance of the Easter problem.

Easter rarely if ever comes on the same date twice within a decade. Therefore a knowledge of the date of Easter is equivalent to knowledge of the exact year, unless a writer has lost track of the years to a serious extent. The date of Easter was a serious doctrinal problem, and celebrating Easter on the correct date (as chosen by a particular branch of the Church) was an important matter. The date of Easter was also the subject of fairly regular correspondence between church institutions.

Thus there were strong reasons for every church institution to pay close attention to Easter dates. The writers of most sources used in this work were associated with the Church. Therefore they always knew the current year, and they would not make a mistake in the current year except by a mere blunder such as a typographical error.

I imagine that simple blunders by the original writer were comparatively rare and that most dating errors are basically copying errors. Annalistic entries (see Section III.2) often spill over the space allotted for a year. This causes confusion about the year to which an entry belongs, and it also causes confusion in adjacent years. A chronologist or historian almost certainly wrote from notes, annals, or other records, and he had ample opportunity to make a mistake in the year that went with a particular event.

Section V.5 contains a statistical study of the dating errors in medieval records of solar

eclipses. It is plausible that the dating errors
in other types of event are approximately the same
size.

CHAPTER IV

GENERAL REMARKS ABOUT RECORDS OF
LARGE SOLAR ECLIPSES

1. Textual and Astronomical Analysis

I remarked in Chapter I that two different types of analysis are needed for the medieval records. One type is textual. It is necessary to decide from the text (which often exists in several variant forms) what the text means and what are the specific astronomical data that it contains. The other type is astronomical. It is necessary to use the data in a theoretical analysis in order to estimate important astronomical quantities.

I also remarked in Chapter I that all uses of old records of solar eclipses that I had seen before writing AAO were based upon a logical fallacy, that of reasoning in a circle. The students of the old records had allowed their textual interpretations to be governed by the results of astronomical calculations, in which they had to assume the very results they were trying to obtain. After discarding all records, or interpretations thereof, which did not agree with their assumptions, they then calculated the quantities that were originally assumed.† Not surprisingly, they always

---

† In particular, most of the eclipse records used did not give the date of the eclipse. Instead of ignoring such records, the writers proceeded to find which possible date required the smallest values of the astronomical accelerations. This date was then taken as the "proved" date of the eclipse, and the record with this "proved" date was then used to estimate the accelerations. If no date yielded accelerations that the investigator found plausible, the record was ignored. Further, most of the records used were of "literary" or "magical" eclipses and not of real ec-

found excellent confirmation of their assumptions from the "data".

The best way to avoid any possibility of this kind of circular reasoning is to separate strictly the textual analysis from the astronomical analysis. Therefore I shall do here what I did before in AAO. I shall do all the textual analysis and write it down before doing any astronomical analysis. After the initial writing of the textual analysis, I shall not make any changes in it except those needed for clarity of expression. In particular, I shall not make any changes that could affect the astronomical results.

2. Some Properties of Solar Eclipses

It is unfortunately necessary to give some technical information about solar eclipses at this point.

As the sun and moon sweep through space, the moon is always casting its shadow into some part of space. The shadow has three main portions, which are shown in Figure IV.1; the figure is not to scale. M represents the moon and S represents the sun.

The umbra is the portion of the shadow shown in black and lying between the moon and the point marked V. No light from the sun penetrates into this region.

Surrounding the umbra is the region shown by cross-hatching in Figure IV.1, called the penumbra. Light from some but not all of the sun reaches any point within the penumbra.

---

lipses, in my opinion. Whether I am right about this latter point or not, the accelerations cannot be estimated validly by means of circular calculations. See the discussion of the "identification game" in Newton [1969].

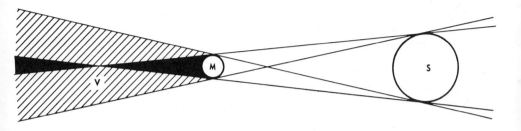

Figure IV.1.   The geometry of a solar eclipse.

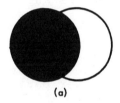

(a)

(b)

Figure IV.2.   Appearances (a) during a partial
eclipse and (b) during an annular eclipse.

Finally, there is the region shown in black but lying on the other side of V from the moon. There is no standard name for this region. I shall call it the outer umbra, and I shall use the term inner umbra for the region that I called simply the umbra above. I shall use umbra to apply to either the inner or outer umbra; most of the time it will not be necessary to say whether the inner or the outer umbra is meant.

Some light from the sun reaches any point within either the penumbra or the outer umbra. The moon and sun present different appearances within the two regions, however, as we can see with the aid of Figure IV.2. Within the penumbra, they give the appearance of part (a) of the figure; the dark moon cuts into one rim of the sun by a greater or less amount. How far the moon cuts into the sun depends upon where one is in the penumbra. Within the outer umbra, the moon and sun give the appearance shown in part (b) of the figure; the dark moon always appears surrounded by the bright rim of the sun.

Whenever any part of the earth enters any part of the moon's shadow, there is a solar eclipse somewhere on earth. Solar eclipses must be classed on the basis of two different principles.

If any part of the earth enters the penumbra, but if no part of it touches or enters either part of the umbra, I shall call the eclipse "penumbral".[†] If any part of the earth touches or enters either part of the umbra, I shall call the eclipse "umbral".

Division of eclipses into umbral and penumbral does not involve a particular point on earth.

[†] There are no standard terms for the classes of eclipses. Some writers use "partial" for what I call penumbral and "central" for what I am about to call umbral. However, partial is also used in a different sense that will be explained below. I prefer to reserve "partial" for this yet-to-be-defined sense.

The other type of eclipse classification does involve the location of the observer on the surface of the earth.

By an astronomical accident, the distance from the moon to earth is about the same as the distance from the moon to point V in Figure IV.1. During an umbral eclipse, then, the part of the earth's surface that is inside the umbra is a small circular or oval region. The size and shape of the region depend upon the exact circumstances, but the region within the umbra is typically a few hundreds of kilometers across. As this region sweeps over the surface of the earth during the progress of an eclipse, it sweeps out a narrow zone that I shall call the umbral zone; it is called the central zone by some writers.

The distance between the earth and the moon is not always the same.[†] Sometimes the surface of the earth during an umbral eclipse is closer to the moon than point V is, and sometimes it is farther away. If the surface is closer than V, an observer inside the umbra sees the moon block the sun completely from view; when this happens, the eclipse is called total. If the surface of the earth is farther away than V, the moon and sun give the appearance of part (b) of Figure IV.2 at some time during the eclipse. Because the sun shows up as a ring around the moon, such an eclipse is called annular.

Very little of the sun's disk is visible at the peak of an annular eclipse. I do not know when astronomers learned to distinguish between annular and total eclipses. The astronomical tables of Ptolemy [ca. 152] are incompatible with the occurrence of annular eclipses. This suggests that annular eclipses were unknown to him; however, I have not studied his work in enough detail to know whether he did know the difference or not. In most of

---

[†] Neither is the distance between the earth and the sun, but that point is not important in this discussion.

this work, I shall not try to maintain the distinction, and I shall let "total" represent either type of umbral eclipse except when great accuracy is needed.

If an observer is inside the penumbra, but never inside the umbra, the sun and moon give the appearance shown in part (a) of Figure IV.2 even at the peak of the eclipse. For such an observer, the eclipse is called partial. An umbral eclipse is partial to observers outside the umbral zone; since the umbral zone is narrow, this includes most observers. A penumbral eclipse is partial to every observer who sees it. An observer who sees a partial eclipse cannot tell from the appearance whether it is umbral or penumbral.

3. Assigning Dates to Eclipse Records

The amount of information given about a solar eclipse varies considerably from one record to another. Loosely speaking, there tends to be a moderate amount of detail if the record was contemporary and a small amount of detail if the record was old when compiled into the form that we now have. Still speaking loosely, this means that the more recent records tend to have more detail than the older ones.

This seems like a natural tendency. If a copier were not interested in solar eclipses, and if he were paying attention to the sense of what he was copying, he would tend to omit detail that was dull to him. Since omitted information cannot be restored, we tend to have less information in a record the more times it is copied. In turn, this means that the records of earlier eclipses tend to have less detail.[†]

[†] Beckman [1912, p. 32] says (my translation): "If different annals have one a longer, the other a shorter, record of the same event, the shorter one must usually be the original." I do not understand his conclusion. It is possible that he meant

The barest form of record merely asserts
that an eclipse took place during a certain year.
Additional detail may take several forms. It is
fairly common for the complete calendar date to
be given. It is also fairly common for the hour of
the day, or for a range of hours, to be given.
Interestingly, quite a few reports give the hour
but not the complete date.

In addition to information about the time
of the eclipse, a record may give information
about its size. Some records assert complete totality, some say that stars could be seen during
the eclipse, while others may give reactions of
people to the eclipse. For example, the Anglo-
Saxon Chronicle says that people had to light candles to eat by during the eclipse of 1140 Mar 20
(see Section VI.3). A reaction of this kind indicates a large eclipse but not necessarily a total
one. Some records of partial eclipses include an
attempt to describe how much of the sun was covered by the moon.

I have noticed only two instances[†] in which
two solar eclipses were visible at the same place
within a year of each other. With a trivial number of exceptions, then, we could say that the
year would be the only information needed in order
to specify an eclipse uniquely, if there were no
errors in the records. Unfortunately, there are
many errors, and the identification of an eclipse

that the first written record of an event (in the
Icelandic literature) was always a short notice,
while details were preserved for a long time only
in the oral literature before they were written
down, and that he did not mean for the conclusion
to have a wider application. Even in this restricted situation, I find it difficult to accept the
conclusion as "usually" being so. Certainly a long
record cannot be validly derived from a shorter
record.

[†] The eclipses of 810 Jul 5 and 810 Nov 30 were
recorded in sources from western Europe. The eclipses of 812 May 14 and 813 May 4 were recorded
in Constantinople.

date is frequently a problem. The dating problems
are varied, and no single approach solves all of
them. The approach to be described is probably
the one that I have used most commonly.

When I find a record of an eclipse, I begin by looking for other events nearby in the historical writing that can be dated independently.
If such events can be found, they often establish a
pattern of dating. They may show that the dates
are mostly right in the vicinity of the record, for
example, or they may show that the dates are mostly
two or three years late, to give another example.
This step, when it can be taken, supplies what we
may call an historical identification of the year
of the eclipse.

The other step is to obtain an astronomical
identification of the year of the eclipse. If the
historical and astronomical identifications agree
within a credible margin of error, the identification of the eclipse can be taken as secure in all
but a small number of cases.

The Canon der Finsternisse by Oppolzer
[1887] is the standard aid in identifying eclipses.
The Canon, as I shall identify it for brevity,
gives the calculated circumstances of 8000 solar
eclipses between -1207 Nov 10 and 2161 Nov 17; the
Canon includes all the eclipses for which Oppolzer's
calculations indicated visibility anywhere on earth
between these dates. The Canon also gives the circumstances of 5200 lunar eclipses between -1206 Apr
21 and 2163 Oct 12; these were all the eclipses between the dates given, according to Oppolzer's
calculations.[†]

_____

[†] Ginzel [1899] gives similar information to Oppolzer's Canon for the time period -900 to 600 and
for the areas of "classical antiquity" only.
Schroeter [1923] covers Europe only for the period 600 to 1800. Within their restricted regions,
Ginzel and Schroeter are probably more accurate
than Oppolzer. Oppolzer is accurate enough, however, for the purpose of eclipse identification,
and it is the only work of the sort that I have
used in this study.

Maps in the Canon show the umbral zones (or umbral paths, or paths of totality, as they are sometimes called) of all umbral eclipses within the given dates, for all the earth north of latitude 30° S. The zone maps also show the points where the eclipse was total at local sunrise, local noon, and local sunset. From these maps, we can judge the extent of an umbral eclipse, and the time at which the eclipse was greatest, at any point of interest. Because Oppolzer necessarily had to make approximations in order to get his Canon done at all, the information given by the maps is approximate. However, it is probably as accurate as many medieval records.

In order to find an astronomical identification of an eclipse, I see whether there is a unique eclipse in Oppolzer which satisfies the record within a credible number of years of the year stated in the record. If there is more than one possible eclipse within a reasonable time, I conclude that the eclipse cannot be identified. If there is only one eclipse, but if its date disagrees with the historically identified year by an unreasonable amount, I again conclude that the eclipse cannot be identified. I adopt an identification only if the astronomical identification is unique and if it agrees well with the date reached on historical grounds alone.

Detail in a record beyond a mere statement of a year is almost always helpful. It may allow making a unique choice between two or more eclipses that might agree with the year alone. It may not agree with any eclipse inside the allowable time period; if so, it has probably saved us from making a wrong identification. Finally, if all the detail agrees with a single choice, it increases our confidence in the identification.

Many if not most records contain a considerable amount of detail about an eclipse, such as the exact date, the day of the week, the age of the moon, or the hour of the day. When the details correspond uniquely and with considerable accuracy to a plausible eclipse, I usually omit the step of finding an historical identification.

There are almost surely some errors of identification in this study. One error that I almost made is connected with an entry under the year 951 in the Belgian source Blandinienses [ca. 1292]: "The sun was eclipsed at the 3rd hour on the 14th calends August (= Jul 19) and a glowing star appeared in the east until the month of September." There was no eclipse on 951 Jul 19 but there were possible eclipses on 939 Jul 19 and 958 Jul 19, so the record presumably referred to one of these. The eclipse of 939 Jul 19 is reported in other sources as occurring at the 3rd hour in places near Belgium, and the hour looks reasonable on the basis of Oppolzer [1887]. The penumbral eclipse of 958 Jul 19 is not otherwise reported, and the 3rd hour is possible but not likely. Thus I tentatively concluded at one point that the record in Blandinienses refers to 939 Jul 19 and that there was a dating error of 12 years.

By accident, I discovered the Danish source Esromenses [ca. 1307] as I was in the process of writing Chapter VIII dealing with reports from Belgium. Esromenses has a report of the eclipse of 418 Jul 19 in words that are virtually identical with those in Blandinienses for 951. I would not normally seek to correlate reports as far apart in time as 418 and 951. I did so, and thus correctly identified the record in Blandinienses, only by the accident of reading the two records within a few hours of each other. I discuss the origin of such a large dating error under the designation 418 Jul 19 E,BN in Section VIII.2.

An occasional error in dating will not affect the accuracy of the astronomical deductions appreciably if there is no pattern of bias. Errors of the sort just described have no bias. I have thought of only one type of bias that is likely in the dating, but I believe that it will have no serious astronomical consequences.

Unless it is certain that an eclipse was large, we must consider penumbral eclipses as well as umbral ones. Unfortunately, the information in the Canon tends to focus our attention upon umbral eclipses, and it is easy to overlook a penumbral eclipse that might satisfy all the conditions of a record. In other words, the process of identification is biased toward umbral eclipses. This in turn means that we tend to put the moon farther south during eclipses, on the average, than it really was. However, it was just as likely to be going toward the south as toward the north, and putting it too far south does not tend to change its longitude on the average. The longitude of the moon is the quantity that dominates the astronomical calculations, and it is not biased by the accidental substitution of umbral for penumbral eclipses.

Unless a record gives the exact date of an eclipse with almost sure accuracy, there is some uncertainty in identifying an eclipse as a penumbral eclipse. The reason is that Oppolzer's Canon does not give much detail about penumbral eclipses. In many cases, it is not possible to tell whether a penumbral eclipse occurred during daylight at a particular place or whether, even if it certainly occurred during daylight, it would have been visible there.

In the astronomical part of this study, I shall first calculate the magnitude and other local circumstances of all eclipses taken to be penumbral in the first part. If the circumstances do not agree reasonably well with the record, I shall discard the identification. I shall decide upon this point before making any calculations about the accelerations of the earth and moon; hence the decisions about penumbral eclipses will be unbiased.

It will appear that the dates, or other numerical information, appearing in a number of records must be in error. We must then decide whether the error is a credible copying or writing error or not. If we decide that the error could have happened easily, there is no particular reason to question the general validity of the record. If we decide that the error is unlikely to have happened accidentally, the record must be discarded or, if used, it must be used with great care. In deciding about the plausibility of an error, we must remember that numerals were Roman or Greek numerals, although I may use Arabic numerals in quotations unless the form of the numeral is needed in the investigation of an error. It must also be remembered that most dates are given by means of the Roman calendar. I shall always give the date in the same manner as the source; if a date is not in terms of the Roman calendar in a quotation, it was not in terms of that calendar in the original.

4. Assigning Places of Observation

It was pointed out in Chapter I that most records of a solar eclipse do not say explicitly where the observation was made. In the absence of a specific statement, we must go on the assumption that the observation was made at the place where the original record was written. This assumption is valid unless a person observed an eclipse in one place and then travelled to another place, where he or another person made the first record.

Thus the process of assigning a place of observation to a record of an eclipse is the same as the process of deciding where the original record was written, if we bar the case in which information was transmitted by some means other than the written word. The process involves two main parts. First, we must decide whether a record is original. Second, if we decide that a particular record is original, we must then decide where it was written.

The records pose many problems with regard to originality, and here we can only state some general and obvious rules. If two records agree word for word, it is almost sure that the newer one is a copy; if two such records seem to be of about the same age, I generally use both but with lowered reliability (see Section IV.6 below for the meaning of "reliability"). If one record contains information that is not in a second record, the first cannot be copied from the second. A record that is not contemporaneous (at least within the time span of accurate personal memory of the writer) cannot be original.

A distinction must be made between independent records and original ones. We frequently find that a record contains information about an eclipse that is not found in any other record. Such a record is obviously independent of other known records, but it does not need to be the original record; the original may have been lost. If the independent record was not written down until long after the event, it is obviously not the original.

After we decide that a record is original, we must still decide where it was written. Most records falls into one of two classes. One class is a record kept as part of the annals of an institution such as a monastery, and the other is the personal writing of a known individual. With the former class, it is natural to assume that the observation was made at the physical site of the institution. With the second class, we must find where the writer was at the time. In the absence of specific information, it is usually safe to assume that the writer was at his current home.

The process of assigning a place of observation is obviously not free of error. The biggest source of error is probably failing to find an original record without realizing it, and thereby mistaking a secondary source for an original.

In several cases, I decide that a record is independent but not original. If the record seems to be part of a body of information with a reasonable amount of detail, and if the detail refers to a known place or region, I assume that the observation originated in that place or region. The writing of Theophanes [ca. 813]† provides many examples. Theophanes gives many records that are obviously not original, but that seem to be based upon detailed information that is local to Constantinople. If a record is not original, however, there is always the

---

† See Sections XV.3 and XV.4.

possibility that the original came from elsewhere, and I never give the same confidence to an independent but not original record that I give to one that seems to be original.

If an independent record has little detail to help in locating the observation, I either ignore it or I use it with low confidence.

With one exception, I have worked entirely from printed versions of the sources rather than from manuscripts. Manuscripts provide detailed information, such as forms of letters, special symbols, and arrangement on the page, that is usually not available from the printed versions. It is conceivable that the information available from manuscripts might have caused me to reach different conclusions in a few cases.

Various printed versions of a source do not always agree. Disagreement sometimes results from errors made by the editor who was responsible for a printed version. Sometimes disagreement results from the nature of the source. Medieval writers, like modern writers, frequently revised their work. In modern publications, revisions would appear as different editions. In the medieval manuscripts, revisions are often not specifically identified, and it may be necessary to accept variant versions as equally authentic. When there are serious discrepancies between printed versions of a record, I usually ignore it.

When I conclude that there is an error in a record, this means of course that the error is in the printed version or versions that I used. For simplicity in discussion, I shall usually attribute the error to the source. I usually have no way to tell whether the error was actually present in the source or whether it arose in the process of preparing the printed version.

The astronomical conclusions to be drawn from the eclipses will not be affected seriously if the errors in assigning places of observation are random. During the early part of the medieval

period, movement of information tended to be more
from the east to the west than from west to east.
During this time, we must be especially cautious
in assigning places, because longitude errors tend
to produce bias. From the eighth century on, however, and perhaps earlier, information flowed
rather freely in both directions, and errors in
position are likely to be random. Errors in the
north-south direction do not tend to bias the
astronomical conclusions.

Most editors have made detailed studies
regarding the originality of the sources that they
have edited, and these studies have great value for
this work. However, I have regarded the studies of
the editors only as guides to help me in doing independent study, and I have always tested the relevant
conclusions of the editors against the results of
independent investigation. I disagree with the
editors in a few cases; disagreement happens just
often enough to make it clear that independent investigation is necessary.

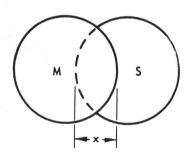

Figure IV.3. Geometry to illustrate the definition
of magnitude.

5. Magnitude of an Eclipse

The magnitude is a number that describes the extent of an eclipse. If an eclipse is total, the magnitude[†] is 1. If an eclipse is partial, the magnitude is between 0 and 1. The meaning of the magnitude can be seen with the aid of Figure IV.3.

In Figure IV.3, the moon's disk on the left is shown overlapping the sun's disk on the right by the amount x. If D is used to denote the diameter of the sun, the magnitude μ is the fraction:

$$\mu = x/D. \qquad (IV.1)$$

In analyzing an eclipse report, it is necessary to assign a value to the magnitude of the eclipse that was seen by the observer, and to estimate what size of error in the assigned magnitude would be reasonable. It is necessary to do this even if the report does not say anything about the magnitude; failure to assign a magnitude and an error is tacitly equivalent to saying that the eclipse was total, and this is an assignment of magnitude. Luckily, it is not vital that the assigned values should be accurate.

For example, consider three hypothetical reports of different eclipses. Suppose that the first merely notes the occurrence of an eclipse, with no information about the magnitude, while the second explicitly states that the eclipse was total and the third explicitly states that the eclipse was partial and gives an estimate of the magnitude. Suppose further that all these eclipses were treated as total in the analysis. The statistical expectation is that the resulting errors in

---

[†] If an eclipse is annular rather than being rigorously total, the magnitude is somewhat harder to define. In any actual annular eclipse, almost all of the sun is covered by the moon. For the purposes of this section, we can neglect annular eclipses and treat them as if they were total.

the astronomical accelerations would be zero; an
error in the magnitude tends to increase the estimated accelerations in some cases and to decrease
them in others. However, errors in the assigned
magnitudes would affect what statisticians call
the confidence limits of the accelerations found.
Thus it is desirable to exercise some care in assigning magnitudes and errors therein.

Some reports say quite clearly that an
eclipse was total. If the report seems to be reliable, it is safe in such cases to assign a magnitude of 1 and a standard deviation of the magnitude
("standard deviation" means an error that is probably exceeded no more than about 1 time in 3) of
0.

Some reports do not say that an eclipse
was total but do give some details that indicate
a large eclipse. A detail that occurs frequently
is a statement that stars became visible. In AAO,
I assigned a magnitude of 1 and a standard deviation of 0.01 to such reports. I based the estimate
of 0.01 upon observations of stars near sunset.
This estimate was not necessarily reliable, because the conditions of sky illumination are different at sunset and during an eclipse. I used
0.01 because I could find no other information.

If a report said nothing about the magnitude, I assumed that the eclipse was large enough
to be apparent to a person who was not an astronomer and who was not alerted to the occurrence of
an eclipse. I assumed in AAO that an eclipse usually had to reach a magnitude of about 0.9 in order
to intrude upon the notice of a person who was not
concerned with it. The evidence for this value
was even more flimsy than the evidence regarding
visibility of stars.

Since writing AAO, I have discovered a
statement by Beckman [1912; p. 17], who said that
people normally notice an eclipse only when the
magnitude reaches 3/4. Beckman's source for the
statement was a private communication from F.K.
Ginzel. Ginzel apparently based the statement

upon agreement of eclipse records with his calculations of eclipses, and not upon direct evidence.

The total solar eclipse of 1970 Mar 7 provided an opportunity to test the assumptions made
in AAO. The path of this eclipse, after crossing
the Caribbean and Florida, went up the eastern coast
of the United States. I watched it, with the assis-
tance of another observer, from Assateague Island,
on the very edge of the path of totality. I had the
help of a corps of observers, who were stationed in
Knoxville, Tennessee, where the magnitude reached
0.90 (but cloud cover prevented useful observation),
in Roanoke, Virginia, where the magnitude reached
0.94, near Washington, D.C., where the magnitude
reached 0.95, and in Millington, Maryland, where
the magnitude reached 0.97.[†] The observers reached
the following conclusions:

1. It is unlikely that an observer who was
not expecting an eclipse would notice one unless
the magnitude reached at least 0.9. There was some
question about whether he would be likely to notice
one at a magnitude of 0.95.[‡] My own suspicion is
that he would probably notice something unusual at
magnitude 0.95; the quality of the light at that
magnitude seemed quite striking. He might not necessarily realize that an eclipse was the cause.

2. Venus could not be seen until the magnitude exceeded 0.98. None of the observers mentioned
above who were outside the zone of totality saw
Venus or any other "star". We on Assateague could

[†] The observer in Millington was J. D. Trimmer and
the one near Washington was R. E. Jenkins. Otherwise, since the observers served without pay, I
practiced nepotism. R. C. Newton observed in
Knoxville, L. M. Newton observed in Roanoke, and
F. P. Newton helped me on Assateague. I sincerely
thank all these observers.

[‡] My reactions suggested that a person would notice
the lack of warmth before he would notice the lack
of light.

not see Venus two minutes before totality, but we could see it one minute before.

3. An observer could not tell the shape of the sun by looking at it with the unaided eye except where there was just the right amount of cloud cover. This was an unexpected result. An observer who looked at the sun long enough to form an after-image on the retina and who then conscious-ly considered the after-image might realize that he had seen a crescent sun.

Fotheringham [1920] says that there should be no difficulty in seeing Venus when the magnitude is only 7/8. Johnson [1889, p. 32] says that plan-ets, in the plural, should become visible when the magnitude is only 2/3. Conclusion 2 above is con-trary to these statements. I find it hard to be-lieve that any planet, even Venus, can be seen when the magnitude is as small as 7/8, much less 2/3, except perhaps under highly unusual viewing condi-tions. Experience of all observers on 1970 Mar 7 confirms that 0.01 is a reasonable value for the standard deviation of the magnitude when "stars" can be seen[†]; the first stars seen are probably planets. I shall continue to use 0.01 for the standard deviation when stars are mentioned.

Conclusion 3 poses a problem with regard to many medieval records. A number of medieval records describe the shape of the partially ec-lipsed sun. They may refer to a crescent or to horns. They often give information about the orientation of the horns, and the information has been correct in almost every case that I have test-ed.

[†] Several records that mention stars refer to "stars around the sun", which suggests that "stars" near the sun are easiest to see. This may be so. Venus should probably be the easiest to see unless it is quite close to conjunction with the sun. Mercury is usually hard to see, but it should become easy to see under eclipse conditions, again except in close conjunction. Venus and Mercury are both "stars near the sun".

The accuracy of the information suggests that many medieval observers had visual aids that allowed resolution of the sun's shape by direct viewing.[†] If the observers had such visual aids, one would expect to find a few incidental remarks about them, but I have not found any. I am unable to offer a plausible explanation for the ability to describe the shape of the sun; it seems unlikely that the cloud cover would have been just right so many times.[‡]

In AAO,I used 0.06 for the standard deviation of magnitude when a report gave no information about the degree of totality. This value is fairly consistent with Conclusion 1 reached from the observations of 1970 Mar 7. However, it seems to me that there was a definite increase in the number of partial eclipses reported in going from classical antiquity to medieval times. I agree with the statement quoted above, which implies that a medieval observer often noticed an eclipse at a magnitude of around 3/4 or 4/5, and in fact I increased the standard deviation in AAO to 0.1 for a few of the later eclipses.

The conclusion that an eclipse is likely to be noticed at a magnitude of, say, 0.8 seems to conflict with Conclusion 1 from 1970 Mar 7. There are several possible explanations, and I do not know which one is correct. It may be that there was an increased awareness of eclipses and when they occur in medieval times. Observers may have watched for eclipses because of interest in the

---

[†] Viewing by reflection inverts the sun and hence the orientation of the horns, for example. The observer could have remembered this fact and given the correct orientation in the record, but one would expect to find many errors if this is what had happened.

[‡] However, Mr. D.G. King-Hele of The Royal Aircraft Establishment believes that the cloud cover often provides the necessary conditions in the British Isles. (Private communication.)

Easter question (Section II.2). If they did, they would have found quickly that the tables of the ecclesiastical moon had serious errors, and one would expect much comment on this fact; such comment is almost absent, perhaps from fear of charges of heresy. Finally, there may be a fair probability that some person in a monastery, say, will notice an eclipse even though there is only a small probability that a given individual will notice it.

In this study, I shall use 0.1 for the standard deviation of the magnitude when a record says nothing about the degree of totality. This implies a reasonable probability of noticing an eclipse with a magnitude of 0.8.

Records which say that an eclipse was partial and which give an estimate of the magnitude will have to be dealt with individually. Some records say that an eclipse was partial without giving any clue to the magnitude. I shall use a magnitude of 2/3 for such eclipses, with a standard deviation of 1/6.†

In dealing with partial eclipses, it will be necessary to make an assumption about which side of the path the point of observation was on. The assumption made will be based upon the charts in Oppolzer [1887]. Because of the approximations that Oppolzer made in order to perform his mammoth task, his paths sometimes have appreciable errors, and the assumed relation of the path to the observer may consequently be wrong. If it turns out, on the basis of detailed calculations made in the astronomical part of this study, that a wrong assumption has been made for this reason, it will be legitimate to change the assumption before the final calculations are made.

---

†This implies, among other things, a low probability that the magnitude was greater than about 0.9. If an eclipse had had a greater magnitude, it would probably have been described in more definite terms than "a partial eclipse", because the eclipsing would have been considerable.

The reader should remember that errors in assigned values of magnitude, if they are random, will not tend to bias the astronomical conclusions. The best way to insure that the errors are random is to assign the magnitudes before making any astronomical calculations.

6. Reliability of the Eclipse Records

Many ancient and a few medieval descriptions of solar eclipses belong in a class that I have called "wrong but romantic" [Newton, 1969]. Wrong but romantic[†] eclipses include "literary" eclipses and "magical" eclipses (AAO, Chapter III). An example of a literary eclipse is the "eclipse of Plutarch" (AAO, Section IV.5); this is a description of an eclipse that Plutarch inserted into one of his dialogues for literary effect. Magical eclipses include the "eclipse of Stiklestad" which "occurred" at the death of S. Olaf (AAO, Section IV.4; also Section XIII.2 below), and the eclipse that Herodotus used as an omen (AAO, Section IV.5) at the beginning of Xerxes' campaign against Greece. It is unfortunately true that most studies of ancient eclipses have relied heavily if not exclusively upon "wrong but romantic" eclipses.

Luckily, the sources that will be used in this study were mostly intended to be sober history, and it is rare that we have to question the basic veracity of the records. Sometimes we find superstitions that have been recorded by credulous writers or even by cautious writers. Examples of superstitions in otherwise sound works include the discovery of the body of Gawain[‡] [William of Malmesbury, ca. 1125, p. 342] and the magical darkening of the sun for 17 days in 798 recorded by Theophanes [ca. 813; see Section XV.4 below].

[†] I originally borrowed the adjectives from Chapter XXXV of Sellar and Yeatman [1931].

[‡] I hasten to add that William apparently had strong reservations about the truth of this discovery.

Veracity does not necessarily mean reliability, however. Many records were copied at least once between the original writing and the text that we now have. Each copying introduces the chance of error.

A single typographical error is not likely to cause trouble: Because eclipses are relatively rare, the correct date can almost always be restored in the presence of a single error. A more serious problem is a mistake in meaning, or the omission of material, or the insertion of material by accident, all of which can drastically alter the sense of a passage. Even modern translations must be read sceptically; the two different translations into modern English of a Latin record of the lunar eclipse of 755 Nov 23 (the record 755 Nov 23 B,E in Section V.1 below) illustrate this point nicely.

Because of danger in using translations, I have worked from the original language except for the few passages in Celtic. I have often consulted translations of other records, but only for help in locating observations or to test the translation in a difficult case.

As I did in AAO, I shall attach a number to each record that will be called the reliability. The statistical weight to be given each record when the accelerations are estimated will be directly proportional to the reliability. If a record is almost surely the original, and if the author seems to be basically accurate, I shall give the value of 1 to the reliability. If a record has been copied, or if there is reason to suspect the writer of exaggeration, I shall lower the reliability at least to 0.5 and even to zero in extreme cases.

Astrology, which had been an influential superstition during much of classical antiquity, had relatively little influence during the early medieval period, but became important again toward

the end of the period of this study.† Astronomical
phenomena used in astrology are usually calculated
and not observed. There is a danger that an astrological calculation will be recorded by a writer
who did not note that it was a calculation rather
than an observation; indeed, the distinction may
not have seemed important to him. There are a few
references to eclipses that were probably the records of astrological calculations.

Astrological calculations are likely to
contain detail, such as times quoted to the minute
or even to the second, that are absent from genuine
observations (see Thorndike [1923] for more discussion). Thus we should be suspicious of records
that give an unusual amount of detail, particularly
if the accuracy implied by the detail was beyond
the observing ability of the times. This conclusion is opposite to that of MacCarthy [1901, p. civ],
who says that minute detail proceeds "from no other
than an eye-witness."

The main reason for lowering the reliability of a record that is a copy in the oldest known
form is not necessarily doubt about the veracity.
In most copies, so far as one can judge, the copier
(by which I mean a person who is copying old records in order to compile a chronicle, not a person
who is simply preparing a copy of a manuscript) did
an honest job. However, he may have made a simple
error or he may have misunderstood his source.
What is even more serious for the purposes of this

<hr>

† One of the earliest references to astrology occurs
in Reicherspergenses [ca. 1279; see Section IX.1].
There the annalist, in an entry that appears under
1145 but that was probably written about 1170,
discusses the effect of a conjunction of Saturn
with the sun. Then he says that this is "according to astrologers, who study the nature of things
in the motions of the stars." Marbacenses [ca.
1375] has another interesting reference that is
mentioned under the discussion of this source in
Section XI.1.

TABLE IV.1

PROVENANCES AND SUBDIVISIONS USED IN THIS WORK

Provenance	Subdivision	Symbol	Chapter or Section Where Discussed
British Isles	England	B,E	VI
	Ireland	B,I	VII.1
	Scotland and Man	B,SM	VII.2
	Wales	B,W	VII.3
Europe	Belgium and The Netherlands	E,BN	VIII
	Central Europe[†]	E,CE	IX
	France	E,F	X
	Germany	E,G	XI
	Italy	E,I	XII
	Scandinavia	E,Sc	XIII
	Spain	E,Sp	XIV
Mediterranean (eastern parts only)	Byzantine Empire	M,B	XV.4
	Holy Land	M,HL	XV.5

[†]Includes Austria, Czechoslovakia, and Switzerland.

study, he did not realize the importance to 20th-
century science of knowing where an eclipse was
observed. In copying, the medieval chronicler
rarely preserved the location of an observation.
Thus the mere fact that the record is a copy low-
ers the reliability of assigning the place, even
if the copy is exact.

7. Designation of Eclipse Records

        In AAO I designated eclipse records by a
combination of the date and the provenance. Six
provenances were used in AAO; these six, along
with their identifying symbols, are as follows:
B, British Isles; BA, Babylonia and Assyria; C,
China; E, continental Europe excluding Greece and
European Turkey; Is, Islamic countries; and M,
Mediterranean countries, including Greece, and including pre-Islamic records from areas that later
became Islamic.

        There are so many eclipse records in this
study that the division by provenance used in AAO
is no longer adequate. Accordingly, I shall divide each provenance into smaller units. Each
subdivision will be denoted by two symbols separated by a comma. The first symbol will denote
the main area, with the same divisions and symbols
as in AAO. The second symbol will denote the subdivision. The subdivisions to be used are listed,
along with their symbols, in Table IV.1. The table
also lists the chapter or section of this work in
which the eclipse records for each subdivision are
discussed.

        Even with the use of subdivisions, there
are often two or more records of the same eclipse
from the same subdivision. The scheme of designation must distinguish between such records.

        The designation of an eclipse record con-
tains three parts. The date, in terms of the Julian calendar, is the first part. If there are two
or more records of the same eclipse from the same
subdivision of the provenance, individual records

will be distinguished by a lower-case letter following the date. This is the second part of the designation; it is blank if there is only one record. The third part contains the symbols for provenance and the subdivisions.

For example, the designation 664 May 1b B,I refers to the second record of the eclipse of 664 May 1 from the Irish subdivision of the British provenance. The designation 878 Oct 29 E,Sc refers to the only record found of the eclipse of 878 Oct 29 in sources from the Scandinavian subdivision of the European provenance. The lower-case letters that distinguish different records were assigned in the order in which I discovered the records; they have no other significance.

A few records found in the literature from one subdivision indicate that the place of observation was in a different subdivision. In such cases, the subdivision used in the designation of the record is that of the place of observation.

The provenances and their subdivisions serve a useful purpose beyond helping to designate an eclipse report. The subdivisions have been chosen in such a way that there is relatively little interaction between reports from different subdivisions. In discussing the eclipse reports, I shall devote a separate chapter or section, according to the number of records involved, to each subdivision. The chapters or sections will be arranged in the order that the subdivisions are listed in Table IV.1.

8. Hours of the Day (Unequal Hours)

This topic does not particularly belong here, but there is no other convenient place to put it.

When an hour during daylight is stated in a medieval source, it is often identified as an "hour of the day", and we can be almost certain that this is meant even when the qualifying "of the day" is omitted. In this system of time keeping, the interval from sunrise to sunset was divided into 12 parts, each of which was an hour of the day. Thus the length of an hour depended upon the season. In London, for example, an hour of the day was about 80 of our minutes at the summer solstice and only about 40 of our minutes at the winter solstice. This system of measuring hours was used by the Romans, who probably derived it indirectly from Babylonia.

The fact that they used this system does not mean that medieval people did not know about the variation in the length of the day.[†] Astronomers in Hellenistic times already used an hour that is the same as ours, and the two kinds of hour were often distinguished by calling them "equal hours" and "unequal hours". I believe that the use of "unequal hours" in Europe persisted in common usage to about the time of the Renaissance or perhaps even later, but I have not tried to investigate the question. Certainly "unequal hours" or "hours of the day" were in use throughout the period covered by this work.

The night was similarly divided into 12 hours of the night. Thus the length of an hour was different between day and night except at the equinoxes.

[†] The Sherborne calendar [Wormald, 1934], which was prepared about 1060 and which I mentioned in Section II.3, gives, for each month, the average length of a day in terms of modern hours.

CHAPTER V

RECORDS OF PARTICULAR INTEREST

1. Eclipse Records of More than Routine Interest

In Chapters VI through XV, I shall discuss the sources and the records of solar eclipses from the areas listed in Table IV.1. It will develop that a large number of records pose problems, usually about the place of origin. Many of the problems were interesting to solve, but the discussions of them will prove tedious to the reader and there is no point in pretending the contrary.

There are a few records of solar eclipses that I feel should be brought to the attention of the reader for a variety of reasons. It is too much to expect that many readers would discover them by the process of reading the following chapters, therefore I shall mention such records briefly in this section. I shall precede their mention by a discussion of the one record of a lunar eclipse that will be used in the astronomical analysis. The reasons for not using the many other records of lunar eclipses are presented in Appendix I.

755 Nov 23 B,E. Reference: Simeon of Durham [ca. 1129]. Simeon writes under 755: "The moon moreover was covered over with a blood-red color on the 8th calends December (= Nov 24) at age 15,[†] that is, full moon; and then, with the darkness decreasing little by little, it returned to its original brightness. And remarkably indeed a bright star following the moon itself passed through it, and after the illumination it preceded the moon by as much space as it had followed the moon before the eclipse."

[†] That is, the 15th day of the ecclesiastical moon.

Since the heavens appear to rotate from
east to west, "to precede" in astronomic parlance
means "to be west of" and "to follow" means "to be
east of". Thus the last sentence says that a bright
star was east of the moon before the eclipse, that
it seemed to pass through the moon, that it was west
of the moon after the eclipse, and that its westward
distance after the eclipse equalled its eastward distance before the eclipse.[†]

Since the moon travels from west to east
with respect to the stars, this is a correct description of an occultation that coincided with an
eclipse. Rather surprisingly, this is the only such
coincidence that I have noticed; it should happen
fairly often.

It will be necessary to identify the star
by calculation. The uncertainty in the moon's position because of not knowing the acceleration of the
moon is less than a degree for a time in 755. Unless the moon happened to be in a particularly crowded part of the sky, it should be possible to identify
the star with no ambiguity. The rest of the discussion will proceed on the assumption that the identification can be made.[‡]

[†] I believe that the famous scholar Joseph Stevenson overlooked the astronomical meanings of precede and follow. In his translation of Simeon
(in The Church Historians of England, v. III,
part II, Seeleys, London, 1855), Stevenson translated the last sentence as follows: "For, very
remarkably, a bright star following the moon itself, and passing across it, excelled it in bril-
liancy, as much as it was inferior before the
moon's obscuration." I give the original in order that the reader may judge: Nam mirabiliter,
ipsam lunam sequente lucida stella et pertrans-
eunte, tanto spatio eam antecedebat illuminatam,
quanto sequebatur antequam esset obscurata.
Johnson [1889, p. 37] gives a translation that
essentially agrees with mine.

[‡] Johnson [1889, p. 37] said the star was Aldebaran,
but he did not give a basis for the statement
that I noticed.

The record implies that the longitude of
the moon at the center of the eclipse was the same
as the longitude of the star. Thus the longitude
of the moon at the center of the eclipse is known.
The longitude of the sun must differ by exactly
180°, so it is known also. But the apparent yearly motion of the sun furnishes the measure of ephemeris time (ET).[†] Hence the observation gives us
the position of the moon at a particular value of
ET. This is the oldest observation of the moon at
a given value of ET that I know of.[‡] This makes
the observation of great potential value.

Unfortunately, it is likely that the errors
of observation will be large. In AAO (p. 204) I
estimated that the error in an observation of position comparable with this one is about 0°.06 when
made by a professional astronomer. Since this observation was probably not made by a professional,
we might expect the error to be perhaps 0°.1. It
is helpful to confirm this estimate by an independent one.

From the tables in Oppolzer [1887] we find
that the duration of the eclipse, including all
its partial phases, was 218 minutes, and the error
in his calculation is likely to be small. During
this time the moon moved about 2°. The width of
the moon is about 0°.5. Thus the sum of the dis-

[†] Ephemeris time is to be contrasted with either
solar time or Universal Time (UT), which are
measured by the rotation of the earth upon its
axis.

[‡] So far as this point is concerned, the oldest
data that Spencer Jones [1939] used in his study
of the lunar motion come from about 1700 or
slightly earlier. It should also be noted that
the magnitude of a partial lunar eclipse (see
AAO, Chapter IX, for example) gives the latitude
of the moon at the time of the eclipse, but the
value of ET at the center of the eclipse is not
known a priori.

tances from the star to the moon before and after
the eclipse is about $1°.5$. If the longitude of
the star differed from that of the moon at the
center of the eclipse by $0°.25$, the distances
would be $0°.5$ and $1°.0$. This makes the distances
have the ratio 2:1 instead of being equal as the
record says. It is unlikely that the error is
this large, and it is unlikely that the observer
was aware of the entire duration of partialness.

The independent estimates of error in the
moon's position are thus $0°.1$ and something rather
less than $0°.25$. If we choose 1/6 of a degree, we
may expect to have a reasonable estimate.

The same error applies to the sun's position. An error of 1/6 degree in the position of
the sun is equivalent to an error of about 4 hours
in ET, and this is the most serious component of
error in the observation. An observation with this
much time error will not contribute much to astronomical problems, but a moderate number of such
observations would be quite valuable.

The "reliability" does not mean the same
thing for this record that it does for a record of
a solar eclipse. In using a solar eclipse, the
place of observation is highly important. The
reliability of a record of a solar eclipse must
reflect not only the basic veracity of the record
but the probability that we have identified the
original record and hence the place of observation.
Here the place of observation is not needed accurately. On the assumption that the moon is above
the horizon, the place enters only into the calculation of the parallax. An error of 1000 kilometers in the position would be acceptable. I shall
take Durham as the place, and it is unlikely that
this is in error enough to matter.

Simeon is a compiler of a chronicle who
worked long after the event, and his account is
certainly not original. Our concern is about the
accuracy with which his copy represents the original. The only test that we can apply is that of
the date. The date given, and the accompanying

age of the moon, are those of the day after the
eclipse. That is, they apply to the day after if
the day is considered to begin at midnight. The
original writer may well have followed a different
convention and 755 Nov 23 may well have been the
correct date in his own terms. Hence the account
seems basically reliable, but I shall lower the
reliability to 0.5 because the record is not original.

418 Jul 19 E,BN and 418 Jul 19b E,I. The
first of these is from the Belgian source Blandinienses [ca. 1292] and will be discussed more fully
in Section VIII.2. The other is from the Italian
source Farfenses [ca. 1099] and it will be discussed
in Section XII.2. These two records have the two
largest dating errors that I have found. The first
has an error of 533 years and the second an error
of 550 years. I do not believe that it is a coincidence that the largest errors both deal with the
same eclipse. First there was an error of 532 years,
the length of the great Easter cycle, combined with
an accidental error of 1 year, and the compiler of
Blandinienses used a record with this error present.
The compiler of Farfenses probably started with a
record that was already wrong by 532 years, and
made an additional error of 19 years (one cycle of
the Golden Number),[†] as well as an accidental error. The errors indicate that the compilers worked
from sources in which the year was specified by
giving "Easter cycle" data such as the Golden Number
and the Dominical Letter.

497 Apr 18 B,I(?) and 512 Jun 29 B,I. Both
these records are found in the Irish sources Ulster
[ca. 1498] and Scotorum [ca. 1650], and they will
be discussed in Section VII.1. They have caused a
considerable amount of speculation. However, they
are almost identical with records found in the Byzantine source Marcellinus [ca. 534], and there

---

[†]See the footnote at the end of Section II.9.

seems to be little doubt that Marcellinus is their
origin. The reports in Marcellinus appear as the
records 497 Apr 18 M,B(?) and 512 Jun 29 M,B in Section XV.4.

534 Apr 29 E,I, 538 Feb 15 E,I, and 540 Jun
20 E,I. The first of these is from the Italian consular list Fasti Vindobonenses [ca. 576]. The other
two are by the English writer Bede [734?], but it is
likely that he derived them from Italian sources.
All three reports are discussed in Section XII.2.
Bede's eclipses have caused much speculation. Every
discussion that I have seen concludes that he probably obtained his records of the 538 and 540 eclipses from Italy. Since the record from Fasti Vindobonenses is the only Italian record of an eclipse
near 538 or 540, there have been many attempts to
link it with Bede's eclipses. The data in Fasti Vindobonenses are not compatible with either 538 Feb 15
or 540 Jun 20. Some writers have ignored this awkward fact, while others have explained it away by
ingenious arguments. Since the data in Fasti Vindobonenses fit the eclipse of 534 Apr 29, there seems
to be no reason to search farther for an explanation.
There is still no clue to Bede's sources.

664 May 1b E,I. Here we have a reverse flow
of information, from England to Italy. This record,
to be discussed in Section XII.2, is from the Italian source Paulus Diaconus [ca. 787]. It seemed to
me that Paulus put the eclipse in the year that we
call 680. Paulus's editors, however, said that the
eclipse was in 679. I believe that there is no
doubt that Paulus derived his information from the
record of the eclipse of 664 May 1 in Bede.[†] Paulus
dated by means of the Indiction,[‡] which is at best
confusing to use. I think he then made an error by
1 cycle of the Indiction, which is 15 years. He
may or may not have made a further error of 1 year

_____

[†] This is the record 664 May 1 B,E in Section VI.3.
[‡] See Section XV.1.

because of confusion about calculating the Indiction.

807 Feb 11 E,G. This interesting record appears in several sets of German annals and is discussed in Section XI.2. It includes records of the solar eclipse of 807 Feb 11 and of the lunar eclipses of 806 Sep 1, 807 Feb 26, and 807 Aug 21. It records an occultation of Jupiter by the moon on 807 Jan 31. It records what seems to be a great auroral display on 807 Feb 26, on the same night as a lunar eclipse, and it contains what seems to be the first European observation of a sunspot, seen on 807 Mar 14 and for 8 days thereafter. The spot was slightly above the sun's equator.

934 Apr 16 E,BN(?). This record, from the Belgian source Sigebertus [ca. 1111], is discussed in Section VIII.2. I doubt that it records a solar eclipse, and I have assigned a date only in order to have a way to cite the record. It records an eruption and a darkening of the sun in 937. We may make the presumption that the darkening was a consequence of the eruption. The puzzling feature is that Sigebertus claims that the eruption was of the mountain in Quedlinburg, Germany. It seems unlikely that there was an active volcano in Germany in 937. Further, the local annals Quedlinburgenses [ca. 1025] have no record of such an event. I suspect that Sigebertus became confused about geography, just as he is often confused about chronology. However, his chronology has sometimes been wrongly maligned. Sigebertus has what seems to be a correct record of the eclipse of 1009 Mar 29, but Ginzel identified it as 1010 Mar 18 in one of his rare mistakes.

1030 Aug 31. Snorri [ca. 1230] has caused much confusion by putting a magical eclipse at the death of S. Olaf, at the battle of Stiklestad in Norway. Except for a vague reference in a passage attributed to the Skald Sigvat, Snorri's is the earliest account that contains an eclipse. In contradiction to Snorri, all earlier records put the

battle and Olaf's death on 1030 Jul 29. The idea
that Olaf died during the eclipse of 1030 Aug 31
is almost inconceivable to me. The best one can
say for the idea is that it is contrary to the pre-
ponderance of the evidence now known. Unfortunately,
some scholars have what <u>Mattingly</u> [1959, p. xviii]
calls a "cheerful disregard for accuracy". The
evidence has not kept some scholars, even within the
past decade, from asserting the idea as an unques-
tioned and unquestionable fact.

Sigvat's vague account appears as the record 1030 Aug 31 E,Sc in Section XIII.2. The only
clear and specific record of the eclipse of 1030
Aug 31 that I have found is the Irish record 1030
Aug 31 B,I in Section VII.1.

1064 Apr 19 E,Sc(?). This record, which appears in Section XIII.2, is actually from Greenland
and is unique in its origin. I inferred a year and
a place from the record, and there was indeed an
eclipse of the sun that was annular in the year and
near the place inferred. In spite of this I do not
dare to use the record because of its fanciful aspects.

1091 May 21 M,B(?). <u>Anna Comnena</u> [ca. 1120]
wrote a history of the reign of her father, the
Byzantine emperor Alexius I Comnenus. Into her
history she inserts this "literary" eclipse, which
is discussed in Section XV.4. Although no eclipse
fits the circumstances that she describes, some
historians have postulated a date for the eclipse
and have then used it in order to fix some of the
chronology of Byzantine history. I do not know
the extent to which the "date" of this literary
eclipse has worked its way into the standard sources.
It is interesting that Anna's account is an example
of the well-known device of predicting an eclipse
in order to confound an uneducated opponent. One
thing does seem to be established by her account:
The accurate prediction of eclipses was accepted
as commonplace in Constantinople in 1100; Anna
takes it for granted that her readers will look
upon the prediction as an ordinary matter. I use

the date 1091 May 21 only in order to designate
the record. I do not at all mean to imply any
suggestion that it is the correct date.

1133 Aug 2. This must have been a spectacular eclipse. Its path started at dawn near James
Bay. It sped across the Atlantic and reached the
British Isles before midday. From there it swept
across central Europe, narrowly missing Constantinople, and ended in the Indian Ocean near the tip
of India. I have found more independent reports
of this eclipse than of any other eclipse among
medieval sources. Of the 379 records that will be
used in the astronomical analysis in this study,
36 are of the eclipse of 1133 Aug 2.

In the record 1133 Aug 2d E,F, to be discussed in Section X.3, the French writer Honorius
[ca. 1137] describes this eclipse and the recovery
from it. Of the recovery, he says that the sun
"first appeared in the manner of a star, then like
a new moon, . . ." I believe that he first saw
the returning sun as a point of light before he
saw it in a crescent shape. If so, he has written
the first account I have noticed that mentions the
effect often described as the "diamond ring".

1140 Mar 20a E,BN and 1147 Oct 26a E,BN.
We do not know who wrote these records from Gembloux, Belgium, but the wording suggests that they
were written by the same person [Gemblacensis, ca.
1148]. Both records describe the sun as being
circumfusus caligine. Circumfusus means "encompass-
ed" or "surrounded". The most natural reading of
the passages is that the sun was surrounded by a
mist or vapor, although other readings are perhaps
possible.

The corona is mentioned in most modern discussions of total solar eclipses, and to most people it is probably the typical and spectacular
sight associated with a total eclipse. In view of
this, it is surprising to see how little the corona
appears in ancient or medieval accounts. Plutarch

-99-

[ca. 90, Chapter 19] mentions "a kind of light" during a solar eclipse that "keeps the shadow from being profound and absolute". This suggests that Plutarch knew of the corona,[†] but we have no specific observations of the corona in or before his time. There are various medieval accounts that may refer vaguely to the corona. These accounts by the unknown annalist of Gembloux are the earliest I have seen that strongly suggest, at least to me, a specific observation of the corona.[‡]

The Canon of Eclipses [Oppolzer, 1887] presents an obstacle to this interpretation. According to it, the eclipse of 1140 Mar 20 was total but the eclipse of 1147 Oct 26 was only annular. The corona should not be visible during an annular eclipse. Because of the approximations that he was forced to make, it is possible that Oppolzer mistook a total eclipse for an annular eclipse. The calculations that will be made later in this study should give a more accurate resolution of the type of eclipse, and I shall return to these records in Section XVII.6.

1239 Jun 3 and 1241 Oct 6. Celoria [1877a and 1877b] collected all the reports of these eclipses that he could find, and I considered his

---

[†] However, [AAO, p. 114], there is reason to believe that the distinction between total and annular eclipses was not known to Plutarch. Thus the "kind of light" may have been the rim of the sun that remains visible during an annular eclipse.

[‡] In the record 1133 Aug 2b B,E, William of Malmesbury [ca. 1142] says that the sun was covered with "gloomy rust"; William borrowed the phrase from Virgil. In AAO (p. 54) I suggested that this might be an attempt to describe the corona. Further reflection makes me doubt this. William says that the sun was covered over, not surrounded by, the rust. It seems more likely that William was merely indulging in elegant writing. The Gembloux annalist, in contrast, uses a word that normally means "surrounded" rather than "covered over".

collections in AAO (pp. 87 - 90). I concluded that one could not use his records for 1239 because the geography of the umbral path forced the places of observation to lie almost entirely on one side of the path, but I concluded that the 1241 reports were probably unbiased in their distribution.

Here I use some but not all of the reports given by Celoria. I use only those in sources that I have examined. For these sources, I use all the reliable reports that they contain. There is no bias in using all reports from a given source, even if the reports for a given eclipse are often necessarily biased in their geographical distribution.

1261 Apr 1 M,B(?). Pachymeres [ca. 1308, Book I, Chapter 13] says that the death of the emperor Theodore II Lascaris was foretold in the following manner: "For the sun was eclipsed at the third hour of the 6th day and deep darkness seized everything, so that stars appeared." According to the standard sources, Theodore came to the throne in 1254 and he died in either 1258 or 1259; there is some uncertainty about the date of his death.

The editor of the cited edition of Pachymeres identified the eclipse as that of 1255 Dec 30. The eclipse of 1255 Dec 30, as I point out in Section XV.4, was on the 5th feria, and it was maximum in Constantinople at about the 11th hour. These data clearly contradict the record. The editor forces them into agreement by making the day be the 6th day after the coronation of Theodore and the hour be the 3rd hour post meridiem. If this be correct, the usage differs from that in every other Byzantine record I have read. Further, this interpretation puts the coronation on 1255 Dec 25, in contradiction to the accepted date.

The eclipse is a "magical" eclipse. No eclipse during Theodore's reign fits the circumstances given. If the eclipse were assimilated, we can look outside Theodore's reign for identification. The eclipse of 1261 Apr 1 was nearly total at about the 5th hour of the 6th feria; thus the 3rd hour may have been meant to refer to the visible beginning of the eclipse. No other eclipse fits the record at all. For safety, I shall not use this record in the astronomical calculations.

2. Novae

There are numerous Chinese records of novae, beginning possibly as early as about the year -1400. Hsi [1958] has prepared the most recent catalogue of the Chinese observations. His catalogue includes 90 observations dating from ca. -1400 to 1690 Oct 18, although the first two are possibly different observations of the same event. Hsi and others have often remarked that it is sometimes difficult to tell whether an object is a comet or a nova. Hsi has eliminated from his catalogue several objects that have previously been listed as novae but that, on the basis of his study, are almost surely comets. He warns that several objects included in his catalogue may turn out to be comets when the reports are studied more fully.

There are relatively few medieval European records of novae. This situation has caused much comment and speculation, and it is worth giving the matter some attention here; an extensive study would be outside the scope of this work. Since the medieval Europeans did not tell us why they did not record the occurrence of novae, any explanation of why they did not is necessarily somewhat speculative.

The following explanation has found considerable favor: ". . the failure of medieval Europeans and Arabs to recognize such phenomena was due, not to any difficulty in seeing them, but to prejudice and spiritual inertia connected with the groundless belief in celestial perfection."[†]

[†] This is usually attributed to Sarton, but I do not know where it occurs in Sarton's writings. The quotation is taken from Needham [1959, p. 428], who attributes it to Sarton but does not give where it is found.

As I understand the matter,[†] the doctrine
that the heavens are perfect, unchanging, and un-
changeable is an Aristotelian doctrine. The doctrine is in a part of Aristotle's works that reached western Europe about 1200 and that became a part
of "standard" medieval philosophy about 1250. This
is about the close of the time period covered in
this study. Thus something besides the suggested
explanation is needed to account for the paucity of
novae in the works used here.

The usual question is: "Why did medieval
Europeans fail to record novae?" Perhaps a more
significant question is: "Why did anyone record
novae anywhere at anytime?" There are about 1000
stars easily visible to the naked eye in the North
Temperate Zone. Who would notice and record that,
for a short span of time, there were 1001?

One thing that leads people to record novae
is a strong superstitious belief that the heavens
control, or at least predict, human fate. There
was a strong belief in China that events in the
heavens predicted the fate of the realm. Thus the
government long maintained an organized service to
watch and to record changes in the heavens, including eclipses, comets, and other phenomena, and
to notify the court of any important omen. We may
speculate that this organized superstitious concern is the reason for the voluminous Chinese records.

Medieval Europeans were indeed often superstitious about astronomical happenings, and eclipses
and comets and other happenings are often mentioned
as omens. However, the records that have been studied for this work show little trace of an organized celestial search, and many if not most of the
records in fact show little superstition connected
with eclipses and so on; of course the superstition may have been there without being patent.
Thus we may speculate that the relative paucity of
early European records of novae is due to relative

[†] See, for example, the discussion by Dreyer [1905].

freedom from an organized superstitious concern with the heavens.

This discussion is not intended to belittle the value of the Chinese records for modern research. The records are of great value, regardless of the motives that lay behind the observations. Nor is it intended to imply that the Chinese were more superstitious than their European contemporaries. The point is the organization of effort. The superstitious concern in China was organized by and for the state while in Europe it was individual. Perhaps the main point to be made is that it is hard to infer the motives that lead to records from the mere existence of the records, and it is also hard to infer motives from the absence of records.

Since a belief in celestial perfection does not account for the European situation before 1250, it may not be needed in order to explain the same situation after 1250.

We now turn to medieval European sources that may record novae. It is frequently difficult to know whether an apparition in the records is a comet or a nova, or even a meteor in some cases. Further, the dating is as uncertain for these records as for the eclipse records that will be studied in later chapters. I am not trying to make a careful study of these records here. I am only trying to make them known so that others may study them more carefully.

Humboldt [1850] gives a list that includes some of these observations. Needham [1959, p. 172] says that only 9 novae were recorded in the west between Hipparchus and Tycho Brahe, and some of them are questionable. He refers to A.M. Clerke, The System of the Stars, Black, London, 1905 and to F.J.M. Stratton, Handbuch d. Astrophysik, 6, 251, Springer, Berlin, 1928. I have not consulted the latter two references.

334. Reference: Theophanes [ca. 813]. Theophanes records that at Antioch, in 334, in the

-104-

middle of the day, there was seen a star in the eastern part of the sky, and dense smoke poured out from the star from the 3rd to the 5th hour of the day. Theophanes' chronology is rather accurate[†] wherever I have been able to test it, and it is unlikely that the year is wrong by more than 1 or 2. This does not read much like a nova, but it does not read much like anything else either. Comets were familiar phenomena and it is unlikely that a comet would be described in such terms. Hsi [1958] records novae in 305 and 358 but none between.

388 or 389. References: Philostorgius [ca. 425] and Marcellinus [ca. 534]. The date could probably be determined more accurately by careful study. The contemporaneous historian Philostorgius uses the title "New Star" for Chapter 9 of his Book X. This was an exceedingly bright star that first appeared in the circle of the Zodiac near the position of Venus. For some time it rose and set with Venus. Then it began to move northward and reached Ursa Major in about 40 days, after which it vanished. It is startling to see this called a "new star". Since it moved, it is probably a comet rather than a nova. Since it was not described as a comet, it probably lacked a tail.

Marcellinus, writing a century later than Philostorgius, said that there was a star in 389 in the north that shone like Venus for 26 days and vanished. Because of the coincidence in date, this is probably the same as Philostorgius' object. Marcellinus' account does not seem to be derived from Philostorgius.

Hsi has a nova in 386 near the position of

[†] That is, it is accurate if Theophanes' chronological system is translated into ours by means of the rules given in Section XV.1 below. The dates assigned by the editor of Theophanes' work are uniformly wrong by 7 or 8 years.

the winter solstice and another in 393 about 15°
farther south.

ca. 522. Reference: Georgios Hamartolos
[ca. 842]. During the reign of Justin I, and
therefore between 518 and 527, Georgios records a
star that appeared over the bronze gate and that
shone for 27 nights. A person well acquainted with
the topography of imperial Constantinople could
probably assign a position to the star if he knew
the time of year. Cedrenus [ca. 1100] records what
is probably the same object in the 7th year of Justin I and hence in about 524. There is not enough
information to let us judge the nature of this object with confidence. Hsi has novae in 437 and
538 and none between.

614. Reference: Scotorum [ca. 1650].
This source says, under 614, that a star was visible at the 7th hour of the day. Such a star could
well be a nova. However, we cannot put much confidence in such a brief entry in a late compilation,
and the information is not found in any other source
that I have seen. Hsi has novae in 588 and 639 and
none between.

984. Reference: Quedlinburgenses [ca.
1025]. This set of annals says that a bright star
appeared at the axis of heaven in the middle of
the day in 984. There is probably enough auxiliary
information to let us derive an accurate date. Again
we have what may well be a nova, but there is insuf-
ficient evidence to allow a confident judgment.
This object was taken to be a sign. Hsi has novae
in 930 and 1006 and none between.

1006 Apr 8. References: Sangallenses [ca.
1056] and Quedlinburgenses [ca. 1025]. Here we
have what is almost certainly a nova and one that
is almost certainly recorded in the Chinese records.
Sangallenses has the following under 1006: "A new
star appeared with an unusual magnitude, brilliant

in appearance and striking to the eye, not without
dread. In a marvellous fashion it sometimes grew
dimmer and sometimes grew brighter and sometimes
even vanished. It was seen moreover for three
months in the farthest limits of the south, beyond
all signs that are seen in the sky."[†] Humboldt
[1850, p. 223] cited this record in his list of
novae.[‡]

Quedlinburgenses has, under 1008: "A
star was seen in the middle of the day on the 6th
ides April (= Apr 8), on the second feria in Easter week." The data are not consistent with any
likely year. Easter was on Apr 21 in 1006, on Apr
6 in 1007, and on Mar 28 in 1008. 1007 is the only
one of these years in which Apr 8 was in Easter
week and it was then the 3rd feria rather than the
2nd. 1006 was the only year in which Apr 8 was on
the 2nd feria. Of course "Apr 8" may itself be in
error. It is unlikely that this star is different
from those in Sangallenses and in the Chinese records, and the year 1006 seems well established.
Thus it is probable that 1006 Apr 8, the 2nd feria,
is correct and that the annalist accidentally wrote
"in Easter week" in place of "in Easter season" for
example.

Hsi has a nova in 1006, with no time of

[†] The set of annals called Sangallenses is often
attributed, apparently without justification, to
a person named Hepidannus; see Section IX.3 for
further discussion. The date given in the annals
is 1012 rather than 1006. A large block of annals starting about 975 is uniformly dated 6 years
too late for some reason.

[‡] Humboldt quoted the Latin of the original and then
gave a translation into German. His quotation of
the original is correct but he has inserted some
extraneous material into the translation. He has
added the statement that the star was seen in
Aries from the end of May. This was probably a
statement meant to appear somewhere else and
accidentally transferred to this point of the
text in the typesetting.

the year given. Thus the European records seem to
give some information not in the Chinese record.
Hsi gives the approximate position Right Ascension
15^h, Declination -50°, probably with respect to the
coordinate system based upon the equator of 1950.0.[†]
Hsi has deduced this position from the description
of the position in the Chinese record. This confirms
the statement in Sangallenses that the star was extremely far to the south. In fact, Hsi's position
seems too far south since the star was seen in Quedlinburg, which has a latitude of 51°.8N. A careful
calculation, including the precession of the equinox,
would be needed in order to decide if Hsi's approximate position needs to be shifted slightly to the
north. Quedlinburgenses also says that the star was
seen "in the middle of the day" there. This phrase
is often used to mean nothing more precise than
broad daylight. Calculation, including the precession, would be needed in order to decide what limits
of position would be imposed by the requirement of
being above the horizon in daylight in Quedlinburg
in April of 1006.

        In sum, it is almost certain that there
was a nova visible from about 1006 Apr 8 to, say,
about 1006 Jul 1. It is doubtful that "three months"
is meant to be accurate to the day.

        Thietmarus [ca. 1018, p. 839] records a
star in the middle of the day near Easter of a year
that could be either 1013 or 1014. It is hard to
believe that this is different from the star of 1006.
It is hard to explain the error in year, since Thietmarus was writing contemporaneously.[‡] Perhaps the
question of the correct year needs to be left open.

[†] Since the equatorial plane shifts in the heavens
because of the precession of the equinoxes, it is
necessary to associate a date with a coordinate
system.

[‡] His dates are 976 - 1018.

1023. Reference: Ademarus [ca. 1028].
The French historian Ademarus (ca. 988 - ca. 1030)
records that two stars were seen in 1023 in the
south, in the sign of Leo. They fought between
themselves the entire autumn, with the larger and
brighter to the east and with the smaller to the
west. It is possible that this is some flight of
fancy, but it is also conceivable that it records
a nova of greatly fluctuating intensity. Hsi has
novae in 1011 and 1032 with none between. Since
Ademarus records the solar eclipse of 1023 Jan 24
in the same paragraph, it is not likely that his
year is in error.

1077. Reference: Sigebertus [ca. 1111].
Sigebertus says that a star appeared at about the
6th hour on Palm Sunday in 1077. His passage has
been copied by other annalists. Since errors of
as much as 5 years are common[†] in his writing, it
is not worth the trouble to calculate the date of
Palm Sunday in 1077. Hsi records novae appearing
in December of 1070 and in January of 1080. Unless one of them lasted for several months before
being noticed in Europe, it is not likely that
either is the object recorded by Sigebertus.

1106 Feb 2. Reference: Sigebertus [ca.
1111]. Sigebertus also says that a star was seen
from the 3rd to the 9th hour of the day on 1106
Feb 2, and that it was "about a cubit" from the
sun. His statement has been copied extensively.
There was a comet recorded in many sources[‡] for
1106. The date of its first sighting, when given,
is later than Feb 2, so Sigebertus' object is probably not the comet. Hsi has novae in 1080 and 1138,
with none between.

1112 Jun 20. Reference: Beneventani
[ca. 1130]. Both related but not identical copies

[†] See the discussion of his work in Section VIII.1.
[‡] See Appendix II.

of this Italian source say that a very bright star
appeared on 1112 Jun 20. Thus this object is distinct from the one recorded by Sigebertus, and it is
also distinct from any listed in Hsi.

    1131 Feb 22 and 1131 Jul 25. Reference:
Wissegradensis [ca. 1142]. The unknown canon of
the monastery of Wissegrad in Bohemia was thoroughly
excited by two remarkable stars that appeared
in 1131. The first appeared on Feb 22 and the canon
says that "you, O Bohemians" called it Venus; apparently the canon does not agree. The second appeared on Jul 25. It gradually diminished and did
not appear on Jan 12 (presumably of 1132). It seems
possible to me that both objects could be the same
star, but I have not tried to work out the details
of position and time carefully. The annals date the
solar eclipse of 1133 Aug 2 correctly, so the year
1131 is probably correct. Hsi gives novae in 1080
and 1138, but with none between. See the following
discussion.

    1136 and 1137. Reference: Wissegradensis
[ca. 1142]. The annalist continues his state of excitement arising from the appearance of new stars.
He says that Venus appeared on 1136 Jul 16 at the
"winter rising of the sun".[†] After some time a new
star appeared near Venus. Two more new stars appeared in 1137. One rose before dawn on 1137 Sep 11
in the place where the sun rises when it is in Leo.
The other rose before dawn on 1137 Dec 28; it was
called Venus, presumably by others but not by the
annalist. The annalist also says that the star that
appeared near Venus the year before could no longer
be seen.

[†] Presumably he means that Venus rose at the place
on the local horizon where the sun rises at the
winter solstice. Since the sun on the 26th of
July rises near the point of "summer rising", the
observation seems to put Venus at an unlikely
distance from the sun. I have not tested the
point by calculation. Perhaps the annalist meant to
say "summer" rising.

Altogether the annalist at Wissegrad records five new stars in the years 1131 to 1137. It
is unlikely that they were distinct objects. It
seems more plausible that the annalist saw one
or more fluctuating novae. Hsi records a nova
appearing in Aries in June of 1138 and another in
Virgo in March of 1139. It seems superficially
plausible that the objects in Wissegradensis could
all be at one or the other of the two positions
given by Hsi, but I have not tested the matter by
calculation.

In any case, prejudice did not keep the
annalist of Wissegrad from recording what he believed to be new stars.

1179 Aug 1. Reference: Colonienses
[ca. 1238]. The annals say that a star was seen
near the sun at the 6th hour of the day on 1179 Aug
1. Hsi records a nova in August of 1175 and another
in August of 1181. Neither was at all close to the
sun.

1222. Reference: Pragensium [ca. 1283].
This source says that a star of unusual brightness
appeared "in the west of the world" in 1222, but
gives no other details. Hsi lists novae in 1203
and 1224 and none between. It is possible that the
date in Pragensium is in error and that the object
is the nova of 1224, but there is no sure way to
judge.

1245 May. Reference: Albertus [ca. 1256].
Albertus writes that a star appeared near Ascension
Day in 1245 (= 1245 May 4). It was toward the south
in Capricorn. It was large and bright but red. He
says that it could not be Jupiter, because Jupiter
was then in Virgo. He also says that many claimed
that it was Mars because of its color. After 1245
Jul 25 it was no longer bright, and it continued
to lose brightness day by day. The dating seems to
be beyond question, because Albertus has a correct
record of the solar eclipse of 1245 Jul 25 in the

middle of the passage about the star.

A nova seems to be the most plausible explanation of this object. However, before a definite judgment is reached, the positions of the planets should be calculated. If the position given for Jupiter proves correct, that would indicate ability to keep up with the planets rather well and would make it unlikely that a planet was mistaken for a new star. Further, the gradual loss of brightness is typical of a nova, and does not sound like appropriate behavior for either a comet or a planet. Tentatively, we have a well-dated and well-located nova that does not appear in Hsi.

1265. Reference: Mellicenses [ca. 1564]. These Austrian annals say that, during the entire autumn of 1265, a star of unusual brightness could be seen from shortly after midnight until dawn; thus it was not bright enough to be visible in the daytime. It is unlikely that the dating in this part of the annals is in error by more than a year if at all. The star poured out smoke like a furnace. This reminds us of Theophanes' record for 334. It is an unusual way to describe either a comet or a nova. Hsi has novae in 1240 and 1297 and none between.

1282 Mar 30. Reference: Pachymeres [ca. 1308]. Pachymeres (ca. 1242 - ca. 1310) was a high official of the established church of the Byzantine Empire.[†] He was also a historian and his history is one of our main sources for the history of the Byzantine Empire for the period 1260 to 1310 approximately. The passage that concerns us here consists of all of Chapter 16 of Book VII[‡] of his

---

[†] That is, what we in the West often call the Eastern or Greek Orthodox Church.

[‡] The editor calls this Book I of Andronicus Palaeologus rather than Book VII of Michael and Andronicus Palaeologus. Division into two works seems to be the act of the editor rather than of the author.

history, plus the last few sentences of Chapter 15. The passage can be translated thus:[†]

> While these things were happening, the main feast of the Resurrection of the Lord arrived, when by the nature of their faith it behooves all Christians and not just ministers to profess brotherhood with the kiss of love. Therefore both the bishops and the clergy are called together on the day following the main feast of Easter and all, exchanging kisses as the time and occasion demand, declare their brotherly love for each other.
>
> But what happened next is like a joke or a farce. In the middle of the day appeared a star that was called Saturn by those who know about such things; and, so that I am not telling about a vision that appeared to a certain one of the chief clergy in his sleep, when all were assembled in a great synod, they jumped up from their long bench and threw themselves on the floor in an unseemly manner, gesturing to each other for mercy.
>
> But let us return to the phenomenon of the star. In the middle of the day, with the sun shining, near the beginning of April, when the sun near its vernal point shines brighter since the clouds of winter have broken up, the star shone forth in the middle of the sky; shining, I say, although one does not expect planets to shine brightly when they are nearby. Nevertheless at that time that star was shining, . . .

Some speculations about the cause of such an apparition follow.

If this is not the account of a person who was present and saw the star, Pachymeres had a

---

[†] I am indebted to Dr. S.M. Krimigis of the Applied Physics Laboratory for invaluable help in translating this passage.

vivid literary imagination. The passage poses
difficulties in translation, and others may disagree with some aspects of the translation given
above. However, I believe the translation preserves the main facts that relate to the star.

Pachymeres puts the event on the day after Easter of the year that we call 1283. Easter
in 1283 was on April 18. It is unlikely that
Pachymeres would have said that April 19 was near
the beginning of April, or that he would have said
that the sun on that day was near its vernal point.
In 1282, in contrast, Easter was on March 29, and
1282 Mar 30 fits all the conditions excellently.
Therefore it is almost surely the correct date.

Since "those who know about such things"
called the star Saturn, it is likely that the star
was near the calculated position of the planet for
that date. The apparent position of Saturn on
1282 Mar 30, referred to the coordinate system for
1950.0, is about 19^h in right ascension and $-23°$ in
declination. This position fits the observation
excellently. An object in this position would be
near the western horizon in the middle of the morning, which is the most probable time for the cere-
mony described, and thus it would be prominent to
a casual observer. Further, the position is fairly
close to the center of our galaxy and it is thus
near the most probable position for a nova.

Hsi has novae in 1240 and 1297 and none
between.

3. Oddities.

Some records are odd because of the nature
of the events that they relate. With other records,
the oddity is that the writer chose to record an
apparently trivial event. This class of record reminds us of something that we often tend to forget.
Wars, revolutions, and struggles for political power
are the stuff of conventional history, but they are
not the stuff of ordinary life. Only rarely does
an historian relate an "unimportant" event. These

records remind us that an important fact of history is that daily life goes on with amazingly little reference to the "important" facts of history.

There is no sharp division between the two types of odd records. Other records that might be considered "odd" are given in Appendix V. I have not made an effort to find the originals of the following odd records, and I may well attribute them to a copy rather than to the original source.

One class of odd record concerns miraculous events, particularly the discovery of sacred relics. With these, we are amazed that people were ever so gullible as to accept them. In Christian tradition, this class of story perhaps begins with the alleged discovery of the True Cross by S. Helena, the mother of Constantine the Great. It perhaps peaks in the 12th and 13th centuries when the wholesale manufacture of "ancient" relics was a sizeable industry. Only a few examples will be given.

Thus Matthew Paris [ca. 1250] records that the body of Gawain was discovered in Wales in 1087. This story was already known to William of Malmesbury [ca. 1125], who encourages us by showing the critical faculty needed to doubt the truthfulness of the find. In 1162, according to Lambertus Parvus [ca. 1193], the bodies of the three kings who followed the star to find the infant Jesus were brought from Milan to Cologne; I do not know how they got to Milan in the first place. In 1181, according to Villola [ca. 1376], an angel showed a peasant near Antioch where to find the spear used to wound Christ on the Cross.

In another class of miracle, the incorruptibility of the body was sometimes cited as a proof of sanctity. Nithardus [ca. 843, p. 671] does this when he tells us that the body of his father S. Angilbert, on 842 Nov 5, was found to be still uncorrupted 29 years after his death.

There is also what we may call the "propaganda" miracle. For example, even a reputable his-

torian like Theophanes [ca. 813, p. 732]† feels
impelled to pass on the story that the sun "gave
off no rays" for 17 days in 798, during the reign
of the empress Irene. Irene was greatly disliked,
we gather justifiably so, and the story is probably a piece of propaganda directed against her,
although it may have been suggested by some physical
event.

William Rufus was another unpopular monarch. William of Malmesbury was among those who
did not seem to like him. The king was buried in
a church in Winchester. According to William of
Malmesbury [ca. 1125, p. 379], the church tower
promptly collapsed; in "the opinion of some" this
was for the sin of burying William Rufus there.

As a final example, we learn from Scotorum
[ca. 1650] that an Irishman named Lughaidh was
struck down by lightning in 507 for denying S. Pat-
rick.

Dogs are the subjects of several stories.
Georgios Hamartolos [ca. 842] found it appropriate
to devote an entire chapter (Chapter 255 of Book
IV) to the story of a faithful dog in the reign of
Leo IV, emperor from 775 to 780. A robber killed
the dog's master. The keeper of a nearby inn found
the body and gave it burial. Out of gratitude the
dog adopted the innkeeper as his new master. Many
years later the robber entered the inn, whereupon
the dog seized him. Interrogation into the matter
led to the confession and execution of the robber.

The most chilling story that I have found
in the medieval records concerns another dog. This
story is the only thing that Cedrenus [ca. 1100]
records for the 17th year of Justinian, and hence
in about 543. The passage can be translated as
follows: "There was a certain copper worker named
Andreas. He had a reddish and black dog who made
known unnatural things. With a group standing by,

---

† See also the discussion under the designation
798 Feb 20 M,B in Section XV.4.

the worker received from the bystanders, without
the dog knowing, rings of gold, silver, and iron,
and covered them over. Thereupon the dog upon
command restored to each his own. He could even
distinguish coins of different emperors mixed together. Indeed, with a group of men and women
standing by, he pointed out pregnant women, forni-
cators, adulterers, the compassionate, and the
base without error." One shudders to think how
many innocent people were the victims of Andreas
who, in all probability, gave the fatal signals to
his dog.

I shall close this section with two amus-
ing oddities. <u>Ammianus Marcellinus</u> [ca. 391, Section XVIII.7] gives us an early ecology lesson.
According to him, the entire Orient would be over-
run by lions if it were not for gnats. The gnats
torment the lions around the eyes. Thereupon the
lions either jump in the river to escape and are
consequently drowned, or they scratch their eyes
out and cease to be a menace.

<u>Marianus Scotus</u> [ca. 1082] says that
there was a famous bearded clerk in Ireland named
Aed, who ran a school there. One day in 1053 he
directed that the girls should cut their hair in
the same way as the boys. Therefore he was thrown
out of Ireland.[†]

4. <u>Bede's Use of Zero</u>

The concept of zero as a number does not
seem to be one of the simplest number concepts to
man. The simplest concept seems to be that of the
"counting" or positive integers 1, 2, 3, and so on.
Other numbers have arisen from the desire to perform various operations on the integers, and from
other mathematical problems.

[†] It has taken 900 years to vindicate Aed.

Thus the product of two integers is always an integer, but the division of one integer by another rarely yields an integer result. The rational numbers, or fractions, were thereupon invented in order that the quotient should always have a meaning in our number system. Similarly, the sum of two positive integers is always a positive integer but the difference is not necessarily one. Hence zero and the negative integers were added to the set of positive integers to form the enlarged set called simply the integers. In the enlarged set, both the sum and the difference of two integers are always integers. Other numbers, such as the irrationals[†] and the complex numbers, have risen in response to other needs.

Not much is known about the history of the concept of zero, and I shall not try to go into it here. It is likely that the concept was not created full-grown in one single act of generation, but that it evolved gradually. The discovery of zero is usually attributed to India in perhaps the 7th century. It is interesting to find that Bede, writing in England in the early 8th century, had a useful concept of zero.

Bede's usage of zero occurs in Chapter XX of De Temporum Ratione [Bede, 725]. In this chapter, Bede develops a simple and fairly accurate method of estimating the age of the ecclesiastical moon[‡] on the first day of the month. He does so by the use of two tables. The first gives a

---

[†] The irrationals were in fact invented, or discovered if the reader pleases, long before zero and the negative integers.

[‡] In using this term, I do not mean to imply that Bede was aware of the distinction between the "ecclesiastical" moon and the "real" moon. It is desirable, however, to keep in mind that Bede worked only with tables of the "ecclesiastical" moon, so far as we know, and not with tables of the "real" moon.

TABLE V.1

A FUNCTION USED BY BEDE IN FINDING THE AGE OF THE MOON

January	viiii	July	xiii
February	x	August	xiiii
March	viiii	September	xvi
April	x	October	xvi
May	xi	November	xviii
June	xii	December	xviii

For use with the Gregorian calendar in the 20th century, subtract vii.

TABLE V.2

A TABLE OF BEDE'S EPACT

Year of the 19-year cycle	Bede's Epact	Year of the 19-year cycle	Bede's Epact
i	nulla	xi	xx
ii	xi	xii	i
iii	xxii	xiii	xii
iiii	iii	xiiii	xxiii
v	xiiii	xv	iiii
vi	xxv	xvi	xv
vii	vi	xvii	xxvi
viii	xvii	xviii	vii
viiii	xxviii	xviiii	xviii
x	viiii		

number associated with each calendar month of the
year. This table is reproduced in Table V.1 with
the Roman numerals that Bede used. I have not
given the tabulated number any physical label, but,
as Bede called it and as we shall see in a moment,
it is the age of the moon on the 1st of each month
in the first year of the 19-year cycle.

The second table appears in Table V.2.
It gives what I have called "Bede's Epact" for each
year of the 19-year cycle.[†] Let C denote the
year of the cycle; C is now called the "Golden
Number", but this name apparently was not used until long after Bede's time. To find C, divide the
number of the year by 19 and add 1 to the remainder.
For example, C is 15 for 1971.

In order to find the age of the moon on
the first (calends) of any month, find C for that
year. Then find Bede's Epact from Table V.2; it is
4 for 1971, for example. Then add the epact to the
value from Table V.1 for that month, and the sum is
the quantity sought. For 1971 Mar 1, for example,
add the epact 4 to the value for March from Table
V.1, which is 9, and the age of the ecclesiastical
moon is 13. It should be emphasized again that this
applies to the ecclesiastical moon as used in the
medieval period in combination with the Julian calendar. A trivial adjustment, to which I shall turn
in a moment, allows us to use Bede's tables for the
real moon today with considerable accuracy.

A notable feature of Table V.2 is the appearance of nulla for Bede's Epact in the first
year of the cycle. This means that we add 0 to
the numbers in Table V.1 and hence it means that
the numbers in Table V.1 are the ages of the moon
on the first days of the months in the first year
of the cycle. Bede did not use the symbol 0, of
course. He wrote the number out using the Latin
word nulla, as Table V.2 shows.

[†] Note that Bede's Epact is generated by adding 11
to the preceding value, and subtracting 30 whenever the new value exceeds 30.

Bede's use indicates, I think, that he had a concept of nulla as a number, and that he did not simply use it in its ordinary meaning of nothing. There are two reasons that lead to this conclusion.

First, Bede specifically instructs the reader to add the epact to the number for the month, and he works out some examples. Thus he is contemplating the possibility of adding zero to a positive integer.

The second reason is connected with Bede's unique definition of epact. The epact for a given year is usually taken to mean the age of the moon on Jan 1 of that year; when the epact is tabulated in Easter tables, it always has this meaning so far as I know. Bede gives "epact" a different meaning, and that is why I call it "Bede's Epact". Bede (Chapter L) defines "epact" to be the age of the moon on Mar 22 rather than Jan 1. He probably chose a date in March in order to have the maximum possible accuracy of his method in relation to finding Easter. Why he chose the 22nd rather than some other day in March is another matter.

In every other usage that I have seen, the age of the moon is an integer that starts with 1 and goes to 29 or 30, according to whether the formal lunar month is short or long. The age then reverts to 1, not 0. In March of the 1st year of the cycle, the age of the moon is 1 on the 23rd day of the month. Bede could have met his problem by defining the epact to be the age on Mar 23. He could then have instructed the user to add the epact to the number from Table V.1 and subtract 1 from the result. More simply, he could have decreased the numbers in Table V.1 by 1, thus removing the need for the subtraction and simultaneously removing the simple interpretation of the numbers.

However, Bede did not do this. He went backward from Mar 23 to Mar 22, and he realized that in doing so he took the age of the moon back from the integer 1 to nulla. This gave him an elegant solution to his problem. It would be necessary only to add two numbers, without the need for a subsequent subtraction, and yet the values for the months in Table V.1 would still be real ones for the first year of the cycle.

In sum, Bede seems to have been aware of two important properties of nulla or 0. First, if we count backward[†] through the positive integers and pass 1, we reach 0. Second, if we add 0 to an integer we leave the integer unchanged. It is interesting that Bede did not find it necessary to explain these properties of 0. He apparently expected his readers to understand them.

A trifling modification makes Bede's tables useful today. We saw in Section II.4 that the medieval lunar tables accrued error at the rate of about 0.32 days per century. In Bede's time the error was about 2 days and in the ensuing 12 centuries it has grown to 6 days. Thus we can use Bede's method by adding 6 to the numbers in Table V.1 for use in the 20th century with the Julian calendar. The discrepancy between the Julian and Gregorian calendars in the 20th century is 13 days, so we subtract 7 (= 13 - 6) from the numbers in Table V.1 in order to use it with the Gregorian calendar in the 20th century. Thus, for 1971 Mar 1, we get 6 for the age of the moon rather than the value 13 that we obtained earlier.

In fact, we can easily use Bede's tables to find the age of the moon on any date. The reader should be able to convince himself that the following rule works: Take the number for the month from Table V.1, add Bede's Epact for the year from Table V.2, add the day of the month, and subtract 8 rather than 7. For example, what is the age of the moon on 1971 Mar 25? 1971 is the 15th year of the cycle, as we have said, and Bede's Epact

---

[†]The reader should also remember that Bede was able to deal with years "B.C." (Section III.4 above), and he may have been the first person to do so. This is not quite the same as having a concept of negative integers, but it is coming close.

is 4. The number for March is 9. 9 + 4 + 25 - 8
= 30, so that 1971 Mar 25 is the last day of the
moon, according to Bede's method. In fact, 1971
Mar 26 is the date of an astronomical new moon.
The error in Bede's method, as modified, should
rarely be more than 1 day.

5. A Summary of Dating Accuracy in Medieval Records of Solar Eclipses

I have restricted this study of dating
accuracy to the records of solar eclipses because
they are the easiest to study from this viewpoint.
There is no obvious reason why the accuracy of eclipse records should differ from the accuracy of
other types of record.

We cannot rigorously define what we mean
by an error in dating. Several types of problems
may arise in assigning an error, of which the following are only two examples. (1) Suppose that
the data in a record are so badly garbled that we
cannot identify the eclipse.[†] What numerical value
do we assign to the error in this record? (2) The
year of an eclipse may be given correctly but auxiliary data may be wrong; the month may be given as
March rather than May, say. If we know the writer's
chronological system, we can correct the month.
However, if we were trying to determine the writer's
chronological system from the eclipse, we would
make a wrong identification in order to find an
eclipse in March instead of May. What do we call
the error?

In studying the errors in the records,
I have put the unidentifiable records into a separate category so that the reader may handle them
as he chooses. In assigning errors in other ques-

[†] This problem is not unique to eclipses. For example, in giving the dates of accession of popes,
Sigebertus [ca. 1111] sometimes has them so thoroughly confused that we cannot tell which pope is
meant.

TABLE V.3

A SUMMARY OF DATING ERRORS IN MEDIEVAL
RECORDS OF SOLAR ECLIPSES

Error, Years	Number of Records with this Error	Error, Years	Number of Records with this Error
0	472	9	2
1	88	10	1
2	21	12	1
3	16	16	1
4	6	33	1
5	5	99	1
6	3	533	1
8	1	550	1
Unidentifiable	8		

tionable cases, I have tried to assign the error
that I believe I would have made if I had been
trying to use the record for chronological purposes.
This choice is not unique, and others might make
different choices. It is doubtful that the questionable cases affect the final conclusions about
accuracy to an appreciable extent.

We must also decide what sample of records to use. Do we consider copies in the study
of accuracy, or do we consider only originals?
Arguments can be made out on both sides, and I have
adopted an intermediate position, as follows: In
accumulating the records to be used in the astronomical calculations, I examined about 2000 records of solar eclipses. About 2/3 of these were
so obviously copies that I made no notes about
them. I filed 629 records for further study. Of
these, I shall conclude in succeeding chapters
that 379 are probably independent of other known
sources† and include enough authentic information
to be useful in the calculation of the accelerations. The assessment of accuracy will be based
upon the sample of 629. The reader can base his
own assessment upon the sample of 379 if he wishes.

I may have sometimes assigned an error
of a year because I did not understand the writer's
system of dating. This is probably compensated by
the times when I took the error to be 0 because
the editor had already corrected the source.

The errors are summarized in Table V.3.
One can still get a variety of answers for the
standard deviation of the error according to the
assumptions one wishes to make. If the unidentifiable eclipses are omitted, the standard deviation
of the error is 31.5 years. The reader may object
to including the errors of 533 and 550 years.
These arose primarily because of the "great Easter
cycle" of 532 years, and, if that cycle had not been

---

† This circumlocution is needed because it is almost certain that the original has been lost in
a number of cases.

involved in early medieval chronology, the errors would have been only 1 year and 18 years, respectively.

Most chronological systems involve a cyclical or at least a repetitive[†] element. There is always the possibility of making an error of one or more cycles. However, the cycle of 532 years is unusually long, and errors as large as 550 years might not arise in most systems.

Perhaps a reasonable course is to assign a numerical value of error to the eclipses that could not be identified and then to omit the errors directly associated with the great Easter cycle. A reasonable number to assign to the unidentifiable records is 20 years; this is approximately the average interval between eclipses that are likely to have been recorded at a given place. If we do this, we find the standard error to be 3.0 years.

In sum, the standard error in the dates of solar eclipses in the medieval records is about 30 years for all records. If the two errors directly connected with the great Easter cycle of 532 years are omitted, the standard error is about 3 years.

The preceding study has been concerned only with errors in the year of an eclipse. There are also errors in the day within the year. I have not summarized the errors in the day; they are numerous but they are probably not as numerous as errors in the year. One point about errors in the day needs comment.

Bede [734?] recorded the eclipse of 664 May 1, but he recorded it under the date of 664

[†] For instance, it is common to date by the years of a ruler's reign. An error in naming the ruler can displace a date by the length of a reign, which of course varies from one reign to another.

May 3. The famous Irish scholar Bishop Ussher[†]
offered the following explanation of Bede's error:
Bede knew that the eclipse was near the beginning
of May of 664, but did not know the exact date.
He found by calculation that the age of the ecclesiastical moon was 1 day on 664 May 3 and hence
assigned that as the date of the eclipse. This
explanation sounds unlikely for a variety of reasons.

First, it does not agree with the principles of the lunar calendar. The first day of a
new lunar month is supposed to be the day following the first visibility of the lunar crescent
after sunset. In other words, the moon passes
conjunction with the sun on the last day of a lunar
month. Thus an eclipse can occur only on the last
day of a lunar month, and never on the first day, if
the calendar is accurate. It is possible, of course,
that Bede overlooked this point.

Second, the ecclesiastical lunar calendar was not intended to be exact. It was recognized that the times entering into the basic calculation of lunar position were rounded to a day,
and that the rounding often caused an error of 1
day and occasionally of 2 days. Bede almost surely knew that the ecclesiastical calendar could not
be relied upon to give an exact day. (The rounding errors are in addition to the slow accumula-
tion of average error that Bede probably did not
know about.)

In the third place, the explanation
does not account for Bede's other errors. Bede

[†] Bede's record is discussed under the designation
664 May 1 B,E in Section VI.3 and also in connection with the records 664 May 1a B,I and 664
May 1b B,I in Section VII.1. I have not tried to
locate Ussher's work. Many writers have mentioned
it, but MacCarthy [1901] is the only source I have
seen that identifies it; it is "Brit. Eccl. Antiq.
Wks., vi. 516" according to MacCarthy. See the
record 664 May 1b B,I in Section VII.1.

recorded eclipses in the years 538, 540, and 664, and, if Plummer is right about a "second edition",[†] he also recorded an eclipse in 733 that certainly occurred during Bede's lifetime. Dates associated with these eclipses are summarized in Table V.4.

TABLE V.4

SOLAR ECLIPSES RECORDED BY BEDE

Correct Date	Bede's Date	Date When the Age of the Ecclesiastical Moon was 1 Day
538 Feb 15	538 Feb 16	538 Feb 15
540 Jun 20	540 Jun 20	540 Jun 21
664 May 1	664 May 3	664 May 3
733 Aug 14	733 Aug 14[a]	733 Aug 16

[a]In the M-type texts. The C-type texts have 733 Aug 15.

　　　　Bede's date is correct in two cases and wrong in two cases.[‡] Bede's date agrees with the "ecclesiastical date" in only one case, and this statement is correct whether we choose the M or C texts. Of course one can argue that the date was already in the information available to Bede in every case except 664, and hence that he only had to calculate the date for 664.

[†] See the discussion of Bede [734?] in Section VI.1.
[‡] On the assumption that the M-type text represents the original. Since the record of the eclipse in the C-type has both a grammatical and a chronological error, this assumption is plausible.

In the fourth place, when the date is wrong in records other than Bede's, it rarely agrees with the first day of an ecclesiastical moon. Most errors are surely accidental slips. They may arise from omitting a stroke in a Roman numeral, or from letting one's eye stray to the wrong place when copying, or from a variety of other accidents.

In summary, Ussher's explanation does not agree with the principles of the lunar calendar, it does not agree with Bede's error in the date of the 538 eclipse, and it agrees with relatively few errors in other medieval sources. There does not seem to be any need for it. However, since it does agree with Bede's error for the 664 eclipse, we can never prove that it is not the source of that error, and it is conceivable that Bede did do just what Ussher suggested.

Note: Dr. Wayne Osborn of the Yale University Observatory has called my attention to a paper entitled "Evidence for a Supernova of A.D. 1006" by Bernard R. Goldstein in The Astronomical Journal, 70, pp. 105-114, 1965. Professor Goldstein has found several Arabic references to the apparition, and he reviews evidence from Arabic, Chinese, Japanese, and European sources relating to the event in 1006. He cites the record from Sangallenses [ca. 1056] and he also mentions a reference to a bright star from Beneventani [ca. 1130] that I had overlooked. He did not mention the record from Quedlinburgenses [ca. 1025]. Professor Goldstein suggests that there may have been two events in 1006, one in April and a different one in May. The position that he suggests was never above the horizon at Quedlinburg in 1006. See also B.R. Goldstein and Ho Peng Yoke, The 1006 Supernova in Far Eastern Sources, The Astronomical Journal, 70, pp. 748-753, 1965.

CHAPTER VI

ENGLISH RECORDS OF SOLAR ECLIPSES

1. Sources that Contain Independent Records

I have consulted 27 medieval English
sources that contain records of solar eclipses.
There are good reasons for believing that all the
records in 9 of these are secondary. The remaining 18 sources contain independent records, that
is, records that are either primary or that, although not primary, cannot be traced any closer
to their origins.

In this section I shall discuss briefly
the nature of the sources that contain independent
records. In the next section I shall discuss even
more briefly the sources I have consulted which
have only dependent records. Finally, in
Section 3 of this chapter, I shall discuss the
independent records themselves in order to determine the characteristics that will be needed in
the astronomical investigations. I shall also
discuss a few dependent records with particularly
interesting characteristics.

Anglo-Saxon Chronicle [ca. 1154]. The
Chronicle was first compiled under Alfred around
890, probably at or near Winchester. Copies of it
were made and taken to various places in England,
where they received additions; the additions record a mixture of local and general events. The
additions made after the original was compiled
are more or less contemporaneous. There are six
main texts, known as A through F; a text called
G is much less important. The texts quit at different times. The E text continues for the longest time, until 1154.

The F text is bilingual, having an Anglo-
Saxon part and a Latin part. All editions or trans-

-131-

lations of the Chronicle that I have seen deal only
with the Anglo-Saxon text, except perhaps for an
occasional note. The Latin part has been edited
separately as the Annales Domitiani Latini [Domit-
iani Latini, ca. 1057], which is discussed in the
next section. So far as I know, this is the only
place where the Latin part of the F text can be
found in print.

          Bede [734?]. This is the Historia Ec-
clesiastica Gentis Anglorum. Bede says that he
finished his History in 731. Plummer, who edited
the version cited in the references, gives reasons
for thinking that Bede may have prepared a "revised
edition" in 734. For this reason I have given the
date as "734?" rather than the usual "731".

          This work and its sources have been ana-
lyzed exhaustively, and I have little new to contribute. Bede gives records of four eclipses, of
which he must have seen one (that of 733 Aug 14)
himself. He is not primary for the other three,
but his sources for the records are unknown. It
has often been supposed that his source for the
records of the eclipses of 538 Feb 15 and 540 Jun
20 is, or is related to, the so-called Annals of
Ravenna, which are better called Fasti Vindobonenses (see Sections III.3 and XII.1). In Section
XII.2 I show that this is probably not so and that
the eclipse in Fasti Vindobonenses is instead probably that of 534 Apr 29.

          There are many texts of Bede's Historia
Ecclesiastica, and many variants, but two types
are most important. Plummer calls these the M-
type and the C-type. Plummer tentatively concluded
that the original C text may actually have been
the "revised edition" just mentioned. The writing
of Simeon of Durham (see below) furnishes a slight
bit of evidence against Plummer's conclusion.

          Bermondsey [ca. 1433]. This source is
the annals of the monastery of Bermondsey, which
was founded in 1082; Bermondsey is a southeastern

borough of London. Luard, who edited Bermondsey, said of it: "The Annals of Bermondsey differ in a marked respect from almost all the other monastick annals, in not being written contemporaneously from year to year, but being apparently compiled from documents and other chronicles existing in the monastery in or about the last year (1433) which occurs in the MS. Their value is, therefore, chiefly confined to the history of the house of Bermondsey itself."

The information in Bermondsey that is not local comes mostly from well-known sources such as William of Malmesbury [ca. 1142], Diceto [ca. 1202], and even some continental sources such as Sigebertus [ca. 1111]. Bermondsey cannot be the primary source of any of the eclipse records found in it. However, two of the records seem to be independent of any other known records and are presumably local in origin.

Burton [ca. 1262]. This source is the annals of the abbey of Burton-on-Trent in Staffordshire. The abbey was founded in 1004, and the annals seem to be almost entirely independent. Burton contains only one eclipse record that I noticed.

Diceto [ca. 1202]. This is the Ymagines Historiarum by Ralph of Diceto, which is the second part of a chronicle by Diceto. According to the editor (Stubbs), Diceto became deacon of Middlesex in 1152 and dean of S. Paul's in 1180; he held the latter position until his death in 1202 or 1203. The archdeaconry of Middlesex was considered part of the diocese of London. Hence I shall take London as the place of observation for all the eclipses recorded in Diceto.

Stubbs also says that Diceto used many written sources even for contemporary events, and that the study of his sources is quite complex. However, his eclipse records are all in his own time and seem to be independent. If they are de-

-133-

rived, they are freer paraphrases than most chroniclers take the trouble to make. Stubbs also says
that Diceto's chronology is extremely careless.
I did not check his chronology in detail, but I
noticed that his eclipse dates are exact.

Florence of Worcester [ca. 1118]. I include the continuations of Florence's Chronicle
under this citation. Florence's Chronicle proper
ends in 1118 and was written in Worcester. There
was a continuation by John of Worcester until 1140.
The chronicle was then continued at Bury S. Edmunds
by one John de Taxter until 1265, and by an unknown
monk of Bury S. Edmunds until 1295.

Florence is based upon the continental
source Marianus Scotus (see Section XI.1), and upon Bede, the Anglo-Saxon Chronicle, Asser (see
Section VII.3), and other sources. Florence struck
me as a slavish copier. For example, Asser mentioned that he had personally seen certain things,
and Florence retained these statements. In the
middle of events for the year 1043, Florence even
refers to "Marianus, the author of this chronicle
. ."

The eclipse records in Florence that can
be safely taken as independent are all in the
continuations.

Gervase [ca. 1199]. Gervase of Canterbury wrote several works. The only one with any
eclipse records that I noticed is his Chronica,
which is the work cited. I did not see any records in his Gesta Regum, which I have not cited.
Gervase also wrote a work called Mappa Mundi,
which is not cited but which was useful to me in
locating some of the religious houses involved in
this study.

Gervase was admitted as a monk to the
cathedral monastery in Canterbury in 1163. Apparently he remained a monk there until his death
in 1200 or shortly after. He was undoubtedly an

eye witness of the eclipse of 1178 Sep 13, of
which he gave an interesting and detailed description.

Matthew Paris [ca. 1250]. The story of
this source is highly confused and confusing. It
is closely connected with Wendover [ca. 1235]. I
am not at all sure that I have the story straight.

Once there was an abbey at S. Albans
(the ancient Verulamium) in Hertfordshire. Around
1200 a historian in the abbey compiled a chronicle
which he brought down to the accession of Henry II
in 1154. This chronicle was called Flores Historiarum or else it received this title shortly after.
It was compiled by a monk named Walter according
to some authorities and by Abbot John de Cella
according to others. Hewlett, the editor of Wendover, does not think that either Walter or John
was the compiler, but he has no candidate of his
own.

About 1231, Roger of Wendover resumed
the compilation of the Flores Historiarum. By
the time of his death about 1236, Roger had brought
the chronicle almost up to date. Between 1154 and
the end of his chronicle, Wendover can be considered as mostly independent. It is not contemporary except in the last part of it, but it is independent in the sense that it preserves information,
presumably of local origin, that is found in no
other known source.

Matthew Paris (so called because he presumably came from Paris) took up the Flores where
Wendover left off. He used it, along with additions of his own, to make a work called Historia
Major, which I have not consulted. Matthew then
abridged the Historia Major to make a work often
called the Historia Minor. This is the work cited
as Matthew Paris [ca. 1250]; it is also called
Historia Anglorum by some writers.

There is also an abridgement of the Historia Major called Flores Historiarum. I did not

consult this Flores and I do not know whether it
is really the same as Historia Minor or not. In
any event, this second Flores was taken to Westminster about 1265 and continued there by a man
called John Bevere or John of London to about
1325. Later it was noticed that this Flores had
something to do with an author named Matthew and
that it was kept at Westminster. Hence it became
attributed to "Matthew of Westminster", a non-
person.

Some scholars noticed that the Flores
attributed to Matthew of Westminster and the works
attributed to Matthew Paris were related. Many
thereupon assumed that Matthew of Westminster was
the source used by Matthew Paris, but H.R. Luard
showed[†] that there was no "Matthew of Westminster".
Unfortunately, the outline of this work given at
the back of many volumes in the Rerum Britannicarum
series (often called the "Rolls Series") describes
the work as "Flores Historiarum, collected by Matthew of Westminster". I have not consulted this
work.

In summary, at least of what is needed
in this study, there is a work called Flores Historiarum by Roger of Wendover; this is a chronicle prepared at S. Albans and finished about 1235.
There is a work by Matthew Paris called Historia
Minor that, acting through another work by Matthew, is effectively a continuation of Wendover
to about 1250. Since both works are from S.
Albans, we do not need to distinguish carefully
between them for astronomical purposes.

Oxenedes [ca. 1292]. The Chronica of
Johannis de Oxenedes was written at the Abbey of

[†] In his introduction to Flores Historiarum [ca.
1325]. Luard's introduction and edition of the
Flores are published in Rerum Britannicarum Medii
Aevi Scriptores, no. 95, in 3 vols., Eyre and
Spottiswoode, London, 1890.

S. Benet Holme, near Oxnead in Oxfordshire. It
contains much information, particularly about
events in the eastern part of England, that cannot be found in other known sources. It also
draws upon standard sources.

Ralph of Coggeshall [ca. 1223]. Ralph
of Coggeshall wrote his Chronicon Anglicanum at
the abbey of Coggeshall in Essex. The abbey was
founded in 1140, so Ralph is certainly not an
original authority before that date. The character of the entries changes greatly about 1191.
It is plausible that Ralph took up the compilation
of his chronicle about that date, using local
sources to bring events down to that time, and
that Ralph is largely an independent authority
from about 1191 until the end of the chronicle.

Richard of Devizes [ca. 1192]. Richard,
apparently a native of the town of Devizes, was
a monk at S. Swithun's in Winchester. His chronicle is apparently original in the sense that he
did not use previous written sources in compiling
it; however, he necessarily used information that
was not obtained by personal observation. He
probably knew first hand about the single eclipse
(1191 Jun 23) that he recorded. Some scholars
credit Richard with starting the Annals of Winchester (see the next section).

Simeon of Durham [ca. 1129]. Simeon
was a monk at Durham who had some interest in
natural phenomena. He describes the causes of
eclipses and he writes that the tides are the result of the moon acting on the water. Simeon
used much the same sources as Florence of Worcester, but with more discrimination.

The possibility has been mentioned (see
the discussion of Bede above) that Bede prepared
a revised edition of his Historia Ecclesiastica
and that the C type text may represent the revision. If Bede did prepare a revision, we are enti-

tled to a slight expectation that it would be available in the places most closely associated with Bede.
Durham is closely associated with Bede; the monastery
of Jarrow, where Bede spent most of his life, is in
the county of Durham and Bede is supposedly buried
in the cathedral at Durham. However, Simeon used an
M type text, while his contemporary Florence of Worcester used a C type text. This is not what we would
expect if the C̄ text were closely associated with
Bede, but no strong inference can be drawn from the
circumstance.

Tewkesbury [ca. 1263]. Although the
Abbey of Tewkesbury was founded about 715, the
Annals of Tewkesbury are meager until about 1200.
This may be a consequence of the fact that the
abbey suffered greatly from fires and from the
Scandinavian invasions. Tewkesbury draws heavily
from the annals of the neighboring priory of Worcester (see Worcester below), but it also preserves some independent local records.

Waverley [ca. 1291]. The Abbey of Waverley
was near Farnham, Surrey. The Annals of Waverley
tell us that it was the first Cistercian house in
England and that it was founded on 1128 Nov 24.
It was founded by monks sent from a house in Normandy who apparently brought with them some continental sources. The Annals draw heavily upon Sigebertus [ca. 1111] (see Section VIII.1) for their
early part, but they also use some of the standard
English sources. Waverley is independent and contemporaneous during much of the 13th century but
is rarely so in earlier times.

Wendover [ca. 1235]. This source is
closely associated with Matthew Paris (see above)
and it has already been discussed.

William of Malmesbury [ca. 1142]. William,
who was a monk of the Abbey of Malmesbury, wrote
two main works, De Gestis Regum Anglorum about 1125

and its sequel Historia Novella about 1142. The
first does not have any records of solar eclipses
that I noticed, although it does have some other
scientific observations. The later work is contemporaneous.

Worcester [ca. 1377]. The monastery at
Worcester dates from the 8th century but its annals do not become independent until the 13th century, except perhaps for a few entries. Worcester,
Waverley, Tewkesbury, and Winchester (see next section) are closely related. The compilers of Worcester copied from many sources and often recorded
the same event two or more times. An instance of
this will be seen in connection with the solar eclipse of 1133 Aug 2.

Wykes [ca. 1289]. The manuscript of the
Chronicon of Thomas Wykes says, in a fairly modern
hand, that it is known as the chronicle of the monastery of Salisbury. However, there is no evidence
that connects Wykes's Chronicon with Salisbury. On
the contrary, it is closely connected with the monastery of Osney in Oxfordshire. The connection is
so close that Wykes and the Annals of Osney are
published in a single volume, and I have not tried
to distinguish between them in this study. Wykes
(or Osney) is mostly independent after about 1230.

2. Some Sources with No Independent Records

Domitiani Latini [ca. 1057]. The F
text of the Anglo-Saxon Chronicle has parallel
entries in Anglo-Saxon and in Latin. All editions
of the Anglo-Saxon Chronicle that I know have only
the Anglo-Saxon part. The Latin text has been edited by Magoun under the name of Annales Domitiani
Latini and published separately from the rest of
the Chronicle. The Anglo-Saxon and Latin parts of
the F text are not mere translations of each other,
and much of Domitiani Latini is based upon an independent work compiled at Canterbury. However, its
eclipse records all come either from Bede [734?]

or from the Anglo-Saxon Chronicle (or perhaps derived from the same sources as the Chronicle but independent of it).

Dunstable [ca. 1297]. The Annals of Dunstable, down to about 1200, are taken from Diceto [ca. 1202] except for a few entries from other standard sources. Dunstable is mostly independent and contemporary from about 1200 on. It has several derivative records of eclipses. It also has records of two eclipses that cannot be identified.

Under 1186, Dunstable has: "That year in May the moon was eclipsed, and a few days later the sun experienced an eclipse." The general chronology is accurate in this part of Dunstable. There was no lunar eclipse in May any time close to this. April was the only month in which there were both types of eclipse in any year close to this one. The editor of Dunstable says that this entry is taken from Diceto. However, in the edition of Diceto that I used, the only possible solar eclipses, namely those of 1185 May 1 and 1186 Apr 21, are explicitly and correctly dated.

Under 1208 Dunstable has: "Monsters were seen in England, because† the sun and moon fought at the same time. Also a horrible eclipse appeared." The only large eclipse of the sun near this time was the eclipse of 1207 Feb 28; however, the chronology of Dunstable is quite reliable in this part of the annals, and this identification seems unlikely. It is possible that this refers to the lunar eclipse of 1208 Feb 3. Ginzel identified this as the solar eclipse of 1207 Feb 28.

Ethelwerd [ca. 975]. Ethelwerd (Aethelweard) was a great-great-grandson of Ethelred I

---

† The text has quia. I suspect a typographical error for quae, which. If monstra is read as "portents", however, quia makes grammatical sense.

(not the Unready), older brother and predecessor of King Alfred. He wrote this chronicle for the pleasure of his distant cousin Mathilda, a great-great-granddaughter of Alfred, who was married to the Holy Roman Emperor Otto I (the Great) and who therefore lived abroad. Ethelwerd recorded the eclipses of 538, 540, 664, and 733 in words almost identical with those of Bede. Under the year 879 he has: "In the same year the sun was eclipsed." This is even shorter than the entry in the Anglo-Saxon Chronicle for the eclipse of 878 Oct 29; it is possibly independent but probably is not.

Gaimar [ca. 1100]. It was once popular to write metrical histories, perhaps as a continuation of the bardic tradition. Gaimar wrote one called Lestorie des Engles. It has references to the solar eclipses of 538, 540, 664, and 733; the references are almost surely based upon Bede or upon the Anglo-Saxon Chronicle.

Henry of Huntingdon [1154]. Henry's Historia Anglorum was first "published" in 1130 and later extended to 1154. It records the eclipses of 538, 540, 664, 733, and 878, and it apparently took the records from the Anglo-Saxon Chronicle. The translation cited in the references has under the year 1133: "There was an eclipse of the sun on the 10th of August." However, the Latin edition cited does not contain this record. It is safest not to use this record.

Joannis Asserii. The annals of S. Neots were formerly ascribed to Asser. See the discussion under S. Neots below.

Osney [ca. 1347]. The Annals of the Monastery of Osney are almost identical with the Chronicon of Thomas Wykes over most of their time span, and I have not distinguished between the two sources in this study. See the discussion under Wykes in the preceding section.

Roger of Hovenden [ca. 1201]. The Chronica of Roger of Hovenden (also spelled Roger of
Howden and Rogeri de Houedene) has original material for its last few years, but the original material does not include any records of solar eclipses.
Roger has the early eclipses from Bede and the eclipse of 878 Oct 29 from Asser [ca. 893]. He has
the eclipse of 1133 Aug 2 entered under the year
1134, but, except for the date, the language used
is that found in Bede for the eclipse of 733 Aug
14. This may be a valid record made by someone who
knew Bede well, but it is safer not to use it.

Roger also has records of the eclipses
of 1185 May 1 and 1191 Jun 23. Stubbs, the editor
of the edition cited in the references, says that
these records are taken from a Chronicle[†] by Benedict of Peterborough, as is practically all the
material in Roger's Chronica for the years 1170 to
1191. Since I have many records of these eclipses,
I decided not to consult Benedict's Chronicle.
This allowed terminating this chain of references
without leaving any loose links; thus the study of
English sources could be kept to manageable length.

S. Neots [ca. 914]. The Chronicon Fani
S. Neoti contains much material found in Asser
(see Section VII.3). Perhaps this is the reason
why this chronicle was once attributed to Joannis
Asser (see above). According to Stevenson, the
editor cited in the references, the only existing
manuscript of S. Neots has on the first page, in
a hand of the 16th century: "Annals Io. Asser,
Epi. Wigorn." This notation may be by the antiquarian Archbishop Parker who, still according to
Stevenson, published an inaccurate and dishonest
edition of S. Neots in 1574.

[†] Benedict of Peterborough, Gesta Regis Henrici
Secundi Benedicti Abbatis, ca. 1192. There is an
edition by William Stubbs in Rerum Britannicarum
Medii Aevi Scriptores, no. 49, in 2 vols., Longmans,
Green, Reader, and Dyer, London, 1867.

TABLE VI.1

REFERENCES TO SOLAR ECLIPSES FOUND IN
MEDIEVAL ENGLISH SOURCES

Source	Date	Conclusions
Anglo-Saxon Chronicle	538 Feb 15	From Bede
	540 Jun 20	From Bede
	664 May 1	From Bede
	733 Aug 14	Independent
	809 Jul 16	Independent
	878 Oct 29	Independent
	1133 Aug 2	Independent
	1140 Mar 20	Independent
Bede	538 Feb 15	From continental sources[a]
	540 Jun 20	From continental sources[a]
	664 May 1	Independent
	733 Aug 14	Independent
	753 Jan 9	Independent
Bermondsey	1124 Aug 11	Independent, low reliability
	1140 Mar 20	Probably from Gervase
	1207 Feb 28	Independent
Burton	1023 Jan 24	Independent

[a] See Section XII.2.

TABLE VI.1 (Continued)

Source	Date	Conclusions
Diceto	798	Propaganda from Constantinople[b]
	1178 Sep 13	Independent
	1185 May 1	Independent
	1186 Apr 21	Independent
Domitiani Latini	538 Feb 15	From Bede
	540 Jun 20	From Bede
	664 May 1	May be independent, but did not use
	733 Aug 14	From Bede
	809 Jul 16	From ASC[c]
	878 Oct 29	From ASC[c]
Dunstable	798	Propaganda, see Diceto
	1178 Sep 13	Probably from Diceto
	1185 May 1	Probably from Diceto
	1186	Data not consistent with any eclipse
	1208	Data not consistent with any eclipse
Ethelwerd	538 - 733	Same eclipses as Bede
	878 Oct 29	May be independent, but did not use
Florence of Worcester	538 - 733	Same eclipses as Bede

[b]See Section XV.4.   [c]ASC = Anglo-Saxon Chronicle.

-144-

TABLE VI.1 (Continued)

Source	Date	Conclusions
Florence of Worcester	840 May 5	From Marianus Scotus
	878 Oct 29	From Asser
	1093 Sep 23	Not original but cannot trace; see Simeon of Durham
	1133 Aug 2	Independent
	1140 Mar 20	Independent
	1191 Jun 23	Independent
	1207 Feb 28	Independent
	1208 Feb 3	Was recorded as solar eclipse, but was lunar
	1261 Apr 1	Independent
Gaimar	538 - 733	Same eclipses as Bede
Gervase	1140 Mar 20	Probably independent
	1140 Mar 22	Obviously an error; cannot tell what is meant
	1178 Sep 13	Independent
	1185 May 1	Independent
	1186 Apr 21	Independent
	1187 Sep 4	Independent
	1194 Apr 22	Independent
Henry of Huntingdon	538 - 733	Same eclipses as Bede
	878 Oct 29	From ASC

TABLE VI.1 (Continued)

Source	Date	Conclusions
Henry of Huntingdon	1133 Aug 2	In one edition but not in another; did not use
Matthew Paris	1124 Aug 11	Independent, low reliability
	1133 Aug 2	Probably from ASC
	1140 Mar 20	Independent
	1178 Jan 8	Obviously an error; cannot tell what is meant
	1178 Sep 13	Many errors in record, did not use
	1191 Jun 23	Probably from Wend-over
	1207 Feb 28	Independent
	1230 May 14	From Wendover
	1239 Jun 3	Independent
	1241 Oct 6	Independent
Osney		Listed together with Wykes
Oxenedes	1241 Oct 6	Probably from Matthew Paris
	1255 Dec 30	Independent
	1261 Apr 1	From Florence of Worcester
Ralph of Coggeshall	1133 Aug 2	Independent of known sources
	1140 Mar 20	Independent

TABLE VI.1 (Continued)

Source	Date	Conclusions
Ralph of Coggeshall	1185 May 1	Independent
	1191 Jun 23	Not original but cannot trace; see Wendover
Richard of Devizes	1191 Jun 23	Independent
Roger of Hovenden	733 Aug 14	From Bede
	878 Oct 29	From Asser
	1133 Aug 2	Same language as for 733 eclipse; did not use
	1185 May 1	From Benedict[d]
	1191 Jun 23	From Benedict[d]
S. Neots	664 May 1	Probably not independent
	878 Oct 29	From Asser
Simeon of Durham	733 Aug 14	From Bede
	878 Oct 29	From Asser
	1093 Sep 23	Not original but cannot trace; see Florence of Worcester
Tewkesbury	1140 Mar 20	Probably not independent
	1147	Does not say whether solar or lunar; only gives year

[d]See Section VI.2.

TABLE VI.1 (Continued)

Source	Date	Conclusions
<u>Tewkesbury</u>	1191 Jun 23	Probably not independent
	1239 Jun 3	Probably independent
	1241 Oct 6	Independent
<u>Waverley</u>	937 ?	From <u>Sigebertus</u>
	1133 Aug 2	From <u>Robertus de Monte</u>
	1185 May 1	Not original but cannot trace; see <u>Margan</u>
	1191 Jun 23	Not original but cannot trace; see <u>Margan</u>
	1263 Aug 5	Independent
<u>Wendover</u>	1191 Jun 23	Not original but cannot trace; see <u>Ralph of Coggeshall</u>
	1230 May 14	Independent
<u>William of Malmesbury</u>	1133 Aug 2	Independent
	1140 Mar 20	Independent
<u>Winchester</u>	1133 Aug 2	From <u>William of Malmesbury</u>
	1140 Mar 20	Too short to use safely

TABLE VI.1 (Concluded)

Source	Date	Conclusions
Worcester	538 Feb 15	From Bede
	540 Jun 20	From Bede
	1133 Aug 2	Two reports, neither original
	1140 Mar 20	Probably independent
	1163 Jul 3	Words identical to one report of 1133 eclipse, did not use
	1178 Sep 13	Independent
	1185 May 1	Too short to use safely
	1191 Jun 23	Too short to use safely
	1239 Jun 3	Probably from Tewkesbury
	1241 Oct 6	Probably independent
Wykes	1124 Aug 11	Independent, low reliability
	1133 Aug 2	Probably from ASC
	1178 Sep 13	Too short to use safely
	1191 Jun 23	Too short to use safely
	1241 Oct 6	Probably independent
	1263 Aug 5	Probably independent
	1288 Apr 2	Independent

S. Neots has a short note about the eclipse of 664 that might have been taken from almost anywhere, and a record of the eclipse of 878
Oct 29 taken from Asser.

        Winchester [ca. 1277]. The Annals of
the Monastery of Winchester are certainly derivative down to at least the middle of the 12th century. Winchester has what is almost surely a condensation of the account of the eclipse of 1133
Aug 2 found in William of Malmesbury [ca. 1142],
including William's error in date. Winchester also has a short notice of the eclipse of 1140 Mar
20 which could be a condensation from any of several sources. The editor attributes the part of
Winchester down to about the Norman conquest to
Richard of Devizes.

3.   Discussions of the Records of Solar Eclipses

        In this section I shall discuss the
texts of the records of solar eclipses that I
take as independent, along with a few other records that present interesting features or problems. Table VI.1 contains a list of almost all
solar eclipses found in the sources that have
been consulted, whether they are considered independent or not. The table indicates the conclusion that has been reached about each record. I
did not bother to record all the cases in which
the early eclipses from Bede and from the Anglo-
Saxon Chronicle were copied.

        538 Feb 15 and 540 Jun 20. Many English
sources refer to these eclipses. It is clear that
all such references are derived from Bede [734?],
and it is almost certain that Bede derived the
records in turn from continental sources (see the
discussion in Section XII.2). So far as I know,
there are no independent English records of these
eclipses. Bede's eclipses are discussed further
under the designations 534 Apr 29 E,I, 538 Feb 15
E,I, and 540 Jun 20 E,I in Section XII.2. There

I assign the places of observation of all three
of these eclipses to be most of Italy.

        664 May 1 B,E. Reference: Bede [ca.
734?, Book III, Chapter 27]. "In the same year
664 of the Incarnation, there was an eclipse of
the sun on the third day of May about the 10th
hour of the day." This is followed by a notice
of a pestilence. This record has been discussed
in Section IV.1 of AAO. Reliability: 0.5. Standard deviation of the magnitude: 0.1. Place:
England.

        Bede and many other sources give May 3
rather than the correct date of May 1 for this
eclipse. An explanation of Bede's error is discussed
in Section V.5. Bede apparently wrote the date as
the third day of May and not as the 5th nones May.

        Domitiani Latini [ca. 1057] has, under
the year 664: "Eclipse of the sun on the 5th calends May (= Apr 27), and Archbishop Deusdedit
died." It is hard to decide whether this is independent of Bede or not. The date is different
but still wrong. The printed form of the annals
has "V Kalendas Maii"; if some other mark had been
mistaken for "V" the date would be the calends,
which is correct. However, "V Kalends" is an easy
mistake to make when "V Nonas" is intended, so
Domitiani Latini may have meant to copy Bede.

        Since the date is different, and since
the close of the annal is different, Domitiani
Latini may be independent of Bede here. In order
to be conservative, I shall assume that it is dependent.

        There are many other English references
to this eclipse that are obviously derived from
Bede.

        733 Aug 14a B,E. Reference: Anglo-
Saxon Chronicle [ca. 1154]. Most texts read something like: "In this year Aethelbald occupied

Somerton, and there was an eclipse of the sun."
I discussed this record in Section IV.1 of AAO
under the designation 733 Aug 14 B. In AAO, I
put the observation in Canterbury for reasons that
will be discussed under the next record. On the
basis of additional study, I realize that the earlier conclusions are not justified. I shall take
the place of this observation to be England. Reliability: 0.5. Standard deviation of the magnitude: 0.1.

          733 Aug 14b B,E. Reference: Bede [734?].
(This is in the "Continuation".) "Year 733, there
was an eclipse of the sun on the 19th calends September (= Aug 14) about the 3rd hour of the day,
such that almost the entire disk of the sun appeared to be covered by a horrible black shield.†

          Domitiani Latini [ca. 1057] has: "Hic
sol obscuratur, quia eclipsis facta est, et totus
orbis solis quasi nigerrimo et horrendo scuto
videbatur esse cooperatus (thus) circa III^{am} horam
diei." In AAO, I concluded that this was from a
Canterbury source. The editor (Magoun) of Domitiani Latini also concluded, on far more extensive
grounds, that this is from a Canterbury document.
When I wrote AAO, I had not discovered the "Continuation" of Bede. The similarities of wording
make it almost certain that Bede is the original
source of this record in Domitiani Latini, and

---

† There are two main variants of the text, which are
described by Plummer in the edition cited in the
references. The typical M-type text has: "Anno
DCCXXXIII, eclypsis facta est solis XVIIII. Kal.
Sep. circa horam diei tertiam, ita ut pene totus
orbis solis quasi nigerrimo et horrendo scuto
videretur esse coopertus." A typical C-type text
has "XVIII Kal." rather than "XVIIII Kal." and
"sicut" rather than "scuto"; sicut does not seem
to make sense, and "18th calends" is the wrong
date. A note by Plummer (V.II, p. 346) says that
the use of scuto implies a partial eclipse, but
I did not understand his argument.

hence that Jarrow is the original place of observation, even though Domitiani Latini as a document is from Canterbury. I shall take Jarrow as the place of observation, with a reliability of 0.5 attached to this assignment.

Domitiani Latini says "the whole disk of the sun"; for this reason I took the standard deviation of the magnitude to be 0 in AAO. Domitiani Latini dropped the significant word "pene" (almost) from the original. Bede's words suggest an eclipse that is larger than the average observed eclipse, so I shall take the standard deviation to be 0.05 rather than 0 as it was in AAO or than 0.1 as it is for the average observed eclipse.

There are many other English references to this eclipse. All are derived from Bede. Most of them are either pure M-type or pure C-type, but there are a few mixtures or other variants. It is possible, but it seems unlikely on the basis of the annal wording, that the Anglo-Saxon texts of the Chronicle are derived from Bede, and I shall take the Chronicle record (record 733 Aug 14a B,E above) to be independent.

753 Jan 9 B,E. Reference: Bede [734?]. (This is in the "Continuation".) "In the year 753, the 5th year of Eadberht's reign, there was an eclipse of the sun on the 5th† ides January (= Jan 9). Later that same year and month, which is the 9th calends February (= Jan 24), the moon endured an eclipse, covered with a horrible black shield, like the sun a little before." (The Latin text is as rough as my translation.)

There are two disturbing features of this report, besides the need to supply "5th" as

---

† The arrangement in the cited edition is " . . anno regni Eadbercti quinto, [quinto] Idus Januarias . ." I deduce from this that the editor supplied the second quinto, which could easily have been omitted in the original.

described in the footnote. The editor (Plummer)
says that the original has the year as 756, but
that he changed it to 753 because 753 was the
only year for a long time in which both a solar
and a lunar eclipse were visible in January in
England. Since the day of the month is correct
for both eclipses, provided that the year is really 753, this action of the editor seems justified.

It is more disturbing that the record
uses language clearly taken from the record 733
Aug 14b B,E above. This makes one wonder how reliable the copying of the original text was in
even the oldest manuscripts that we now have.
Since the days of the month are correct, I shall
assume that the copyist merely let his eye wander
to an earlier annal, and that the record as we
now have it is basically correct.

I shall lower the reliability to 0.3
because of all the problems with the record.
Standard deviation of the magnitude: 0.1. Place:
Jarrow.†

809 Jul 16 B,E. Reference: Anglo-Saxon Chronicle [ca. 1154]. Most texts read something like: "There was an eclipse of the sun at
the beginning of the 5th hour of the day on the
17th calends August (= Jul 16), the 2nd feria,

---

† Plummer [1899] says that the entries in the
"Continuation" to Bede's Historia Ecclesiastica
are concerned mostly with Northumbria, but that
they have been influenced by southern forms of
the Anglo-Saxon Chronicle. In other words, it
is possible that some information in the continuation is from sources outside Northumbria. However, the eclipse of 753 Jan 9 is not recorded in
any known form of the Chronicle, not even in
Domitiani Latini, which copied Bede for the eclipse of 733 Aug 14. Thus it seems safe to take
the place still to be Jarrow, which is in Northumbria. It certainly seems safe to use Jarrow
for the eclipse of 733 Aug 14, which happened
in Bede's own lifetime.

the 29th day of the moon." I used this record in
AAO. Its characteristics are: Reliability, 0.5;
standard deviation of the magnitude, 0.1; place,
England.

According to a footnote in the translation of the Chronicle cited in the references,
Domitiani Latini [ca. 1057] has "on Sunday, the
12th day of the moon"rather than what is quoted
above. When I wrote AAO, before I had located an
edition of Domitiani Latini, I assumed that the
writer of Domitiani Latini had a report of the
lunar eclipse of 809 Jul 1, which was on the 1st
feria, as well as the Anglo-Saxon text before him
and that he mixed them up accidentally. When I
found the published edition of Domitiani Latini,
which is the only existing version of the Chronicle
written in Latin, I observed that it says exactly
the same thing as the Anglo-Saxon text at this
point.

Faced with this conflict between published sources, I got a photographic copy[†] of the
relevant part of the manuscript from the British
Museum and observed that the published edition of
Domitiani Latini is correct and that the translation is in error at this point. I believe that
the problem is a typographical error. Under 806,
the Anglo-Saxon counterpart (the text that is usually called F) of Domitiani Latini has an entry
about a wonderful sign that appeared in the sun on
the 3rd calends September (= Aug 30); the other
Anglo-Saxon texts do not have this entry. Domitiani Latini does have it with the additional information that the event was on "luna XII, die Dominica,
hora IIII". I imagine that the footnote about "Sunday, the 12th day of the moon" was meant to apply
to this entry but that it was transferred to the

[†] I thank Mr. D.G. King-Hele, FRS, of the Royal
Aircraft Establishment, for arranging to get
this copy for me.

entry of 809† through a typographical error.

806 Aug 30 was indeed on the 1st feria and on the 12th day of the ecclesiastical moon. The presence of this information alone in <u>Domitiani Latini</u> does not prove it to be independent of the <u>Anglo-Saxon</u> text, because the annalist could have calculated it easily. However, "hora IIII" could not have been calculated and its presence shows that <u>Domitiani Latini</u> is independent of the Anglo-Saxon text at this point.

Several French sources have the same information about the eclipse of 809 Jul 16 as do the <u>Chronicle</u> and <u>Domitiani Latini</u>. Since the French sources in question (see Section X.3) seem to be later compilations, it is likely that one or more of them used <u>Domitiani Latini</u>. It is well known that many continental annalists used <u>Bede</u> [734?], but I do not recall seeing any discussion of the <u>Anglo-Saxon Chronicle</u> as a source for continental chroniclers.

878 Oct 29 B,E. Reference: <u>Anglo-Saxon Chronicle</u> [ca. 1154]. "And the same year there was an eclipse of the sun for one hour of the day." This record was used under the designation 878 Oct 29 B in <u>AAO</u>. Because this eclipse was used as a reference for dating in the <u>Chronicle</u>, I assumed that it was bigger than average, and gave it a standard deviation of the magnitude of only 0.03, along with a reliability of 1 and a place somewhere in England.

<u>Domitiani Latini</u> follows the other texts of the <u>Chronicle</u>. This is the last solar eclipse in the <u>Chronicle</u> except in the E text.

<u>Asser</u> [ca. 893] also has a record of the eclipse of 878 Oct 29. Because of his close asso-

---

† I am indebted to Miss J.M. Backhouse, Assistant Keeper, Dept. of Manuscripts, British Museum, for suggesting this explanation.

ciation with King Alfred, Asser is often counted
as an English source. Since the observation was
probably made in Wales, I have put Asser among the
Welsh sources in Section VII.3. Several English
sources have copied Asser, including Florence of
Worcester [ca. 1118]. The editor of Florence unfortunately identified the eclipse as that of 880
Mar 14.

Waverley [ca. 1291] records a peculiar
darkening of the sun in 937. The record is copied
from the Belgian source Sigebertus [ca. 1111] and
is discussed under the designation 934 Apr 16 E,
BN in Section VIII.2. It is doubtful that the
event is correctly reported in Sigebertus.

1023 Jan 24 B,E. Reference: Burton
[ca. 1262]. Under the year 1023 Burton has: "The
sun was darkened on the 9th calends February."
This seems to be an original entry, and the date
is correct. Reliability: 1. Standard deviation
of the magnitude: 0.1. Place: Burton-on-Trent,
Staffordshire.

1093 Sep 23a B,E and 1093 Sep 23b B,E.
References: Florence of Worcester [ca. 1118] and
Simeon of Durham [ca. 1129]. Both sources have
identical entries under 1093: "A very wonderful
sign appeared in the sun." It is odd to see identical entries in these two sources, which are presumably independent and contemporaneous at this
point in time. It is not odd to see an eclipse
referred to as a sign in medieval records, but it
is unusual to see an eclipse that is not described
as an eclipse or at least as a darkening. It is
almost certain that some words have been omitted
by accident.

Since the year is given, since other
events in the same year are dated accurately in
both sources, and since there was a large eclipse
in that year, I shall assume that this is an eclipse record, but with low reliability. If there
were a single record, a reliability of 0.2 would

be reasonable for a record that has probably been copied. Since there are two identical records, with no way to choose between them, I shall treat them as independent, with a reliability of 0.1 each, and with a standard deviation of the magnitude of 0.1. I shall take the places to be Worcester and Durham, respectively.

The Swiss source Bernoldus [1100] also refers to this eclipse as a sign.† Otherwise there is no resemblance in wording between the English record and that of Bernoldus. It is possible but unlikely that Bernoldus served as the source for Florence and Simeon.

1124 Aug 11a B,E. Reference: Bermondsey [ca. 1433]. "This year, on the 8th ides April (= Apr 6), the sun appeared like a new moon. And the same year, Pope Callixtus II died; Honorius II succeeded him." The year is given as 1124, which is correct for the death of Callixtus II and the accession of Honorius II. This report will be discussed under the report 1124 Aug 11c B,E.

1124 Aug 11b B,E. Reference: Matthew Paris [ca. 1250], who has, under the year 1124: "The sun became like a new moon." Matthew's dating is good in this part of his history. This report will be discussed along with the next one.

1124 Aug 11c B,E. Reference: Wykes [ca. 1289]. Under the year 1124, the reference has: "The sun was eclipsed to the point that it looked like a new moon." The dating is accurate in this part of the reference.

There are three English records of this

---

† See the record 1093 Sep 23c E,CE in Section IX.3. It was a sign because it was on the wrong day of the moon.

eclipse, all in sources that were prepared long after the event. Matthew Paris is the oldest of the three. However, the Bermondsey annal could not have been derived from Matthew Paris[†]; it is possible but unlikely (because a copyist rarely changes the words so much) that the Wykes entry is derived from Matthew Paris. The use of "like a new moon" arouses suspicion of a common source. Since the eclipse was apparently partial, it is reasonable to describe it in this way, but it is unlikely that three sources would choose these words and that none would use "partial" or some synonym. If there were a common source, it would have to be the one used by Bermondsey.

In the absence of a better course, I shall take the three reports to be independent, but with the lowered reliability of 0.1 assigned to each. The place for Bermondsey is London, for Matthew Paris is S. Albans, and for Wykes is Oxfordshire. Since the likely places are all close together, the error in using the sources as independent is probably small. I shall use 0.1 for the standard deviation of the magnitude because the record seems to imply a large eclipse, although not total.

Bermondsey has the date as "viii. idus Aprilis" instead of the correct date of "iii. idus Augusti". The appearance of some manuscripts makes this a plausible copying error. The editor of Bermondsey attributes this passage to "Matthew of Westminster", a mistake for Matthew Paris. The passage could not have come from the manuscript used to prepare the cited edition of Matthew Paris, since it contains detail, even though wrong, that is not to be found in the editions of Matthew. The editor of Matthew Paris attributes the passage found in that source to Wendover. The only edition of Wendover [ca. 1235] that I found starts with the accession of Henry II, so I cannot comment

---

[†] Because Bermondsey contains information not found in Matthew Paris.

on that point. The editor of Wykes attributes
the record in that source to William of Newburgh
[ca. 1198]; I did not notice a reference to any
eclipse in the only edition of William of Newburgh
that I found.

　　　　　　1133 Aug 2a B,E. Reference: Anglo-
Saxon Chronicle [ca. 1154]. On 1133 Aug 1, King
Henry I boarded ship and the next day he started
what proved to be his last channel crossing. He
died in France on 1135 Dec 1. The translation of
the Chronicle cited in the reference has, under
the year 1135: "In this year King Henry went
overseas at Lammas,† and the next day, when he
was lying asleep on board ship, the day grew dark
over all lands, and the sun became as if it were
a three-nights'-old moon, with stars about it at
midday.‡ People were very much astonished and
terrified, and said that something important would
be bound to come after this - so it did, for that
same year the king died . ."

　　　　　The record of the eclipse could not have
been strictly contemporaneous in this form, but
it must have been written within a few years after.
Hence the entry in the Chronicle was probably
written from notes. The Chronicle has no events
between 1132 and 1137 except the eclipse and the
death of Henry. Thus, although the Chronicle
clearly makes the eclipse into a magical eclipse,
it is possible that the compiler simply misread
his notes. This record will be treated as it was
in AAO. Place: Peterborough or Canterbury. Reliability: 0.5 (because of the slight magical
implication). Standard deviation of the magnitude:
0.03; an intermediate value is used because stars
were seen but a crescent was also stated.

---

† Aug 1.

‡ Totality would have been about 12^h local time.
Since Henry was still in bed, one wonders if he
were seasick.

Matthew Paris [ca. 1250] and Wykes [ca. 1289] give accounts of this eclipse that are independent of each other. Both have the correct year but describe the sun as being like the moon on its third day, as the Chronicle does. Since they are rather late authorities, they have probably combined the Chronicle with another source having the right year.

1133 Aug 2b B,E. Reference: William of Malmesbury [ca. 1142, p. 535]. William erred in a direction opposite to the Chronicle. He specified, in two different ways, that the eclipse and Henry's crossing were in 1132. After mentioning Henry's embarkation, William wrote: "That was, as I said, on the nones of August; and on the 4th feria, the elements accompanied with their sorrow the crossing of such a prince. On that very day, at the sixth hour, the sun covered his bright head with gloomy rust, as the poets are accustomed to say, frightening peoples' minds by his eclipse: and in the early morning of the sixth feria next, there was a quake in which the earth was seen to subside, and a horrible sound was heard under the earth. I saw myself the stars around the sun; and in the earthquake, the walls of the house in which I was sitting twice were lifted up and re-seated."

The eclipse date was on the 4th feria, as William said. The day of the month was the 4th nones, not the nones as he said; since two "4th's" were needed in quick succession, William may have omitted one by accident. Since 1133 Aug 5 (the nones of August) was the date of an ecclesiastical "luna 1", William may have supplied the eclipse date from lunar tables.† In any event, the error in date does not change the fact that William was almost surely an eye witness. As I did in AAO, I put this record at Malmesbury,

---

† However, if he did this, he used the year as 1133 in the calculation, and not as 1132, which is the year he specified. Also, see Section V.5.

with a standard deviation of the magnitude of 0.01.
Because William's report of the eclipse of 1140
Mar 20, which will be discussed below, indicates
that William had a tendency toward exaggeration,
I shall use a reliability of only 0.5.

Winchester [ca. 1277] gives a condensation of William's account, including his error in
the year. It has nothing to indicate an independent observation of this eclipse.

Worcester [ca. 1377] gives two accounts
of this eclipse, as if there had been two eclipses.
At a guess, one account was condensed from William
of Malmesbury and the other from Matthew Paris or
from a source that Matthew used. The editor of
Worcester said that the Worcester compiler often
repeated himself even when he was recording events
independently.

1133 Aug 2c B,E. Reference: Florence
of Worcester [ca. 1118]. This account is actually
in the first continuation by John of Worcester.
It is under the year 1132, but the words suggested
to me that he meant to put it under 1133 and misplaced it by accident. The passage is too long to
quote. Florence (or John, rather) has the king
walking about on board and watching the eclipse,
while the Anglo-Saxon Chronicle had him asleep.
John says that he heard of the eclipse from all
over England. In some places the day was only
gloomy while in others stars appeared. The account is contemporary and independent.

Reliability: 1. Place: England.
Standard deviation of the magnitude: 0.03.

1133 Aug 2d B,E. Reference: Ralph of
Coggeshall [ca. 1223]. Under the year 1133 Ralph
has: "There was an eclipse of the sun on the 4th
nones August (= Aug 2), the 27th day of the moon,
and stars appeared, and it was dark at the 6th
hour of the day." Most of the details in this
report could have been obtained from other sources,
but this is the only English report of this eclipse

which gives the complete date correctly. Hence, although the report is not contemporary, it must represent information that is independent of other known reports. We must hope that the independent information was local to the abbey of Coggeshall in Essex. Reliability: 0.5. Place: Coggeshall. Standard deviation of the magnitude: 0.01.

Waverley [ca. 1291] has taken a record of this eclipse from Robertus de Monte [ca. 1186]; see Section X.3.

1140 Mar 20a B,E. Reference: Anglo-Saxon Chronicle [ca. 1154]. The translation cited in the references has: "Thereafter during Lent the sun and the day were eclipsed about noontide[†] of the day, while men ate, so that men lighted candles to eat by; and that was XIII kalends of April; and men were greatly astonished." This sounds like a straightforward account of an eye witness, and the date is correct. Reliability: 1. Standard deviation of the magnitude: 0.05.[‡] Place: Peterborough or Canterbury.

1140 Mar 20b B,E. Reference: William of Malmesbury [ca. 1142, p. 562]. William writes concerning the year 1140: "That year in Lent, on the 13th calends April, 9th hour, 4th feria, there was an eclipse, through all England I heard. Certainly with us and our neighbors it was such a notable eclipse that men, being about to sit to tables when it happened, because it was Lent,

---

[†] The original text has nontid rather than nones; it is probable that nones, the 9th hour, is meant.

[‡] I use a smaller value than the standard because of the wording. In AAO I used 0.02 because of the detail about candles. However, after watching the eclipse of 1970 Mar 7, I think that covering 0.95 of the sun would make it dark enough indoors, where men ate, to require candles.

feared the primeval chaos; then, after the matter
was known, they went out and saw the stars around
the sun." This sounds like the account of an eye
witness at Malmesbury; I used it as such in AAO,
with a standard deviation of the magnitude of 0.01.
William sounds to me a little more astonished than
he should be, when we remember that he was also an
eye witness of the eclipse in 1133, less than 7
years before. Therefore I used a reliability of
0.5 for this account and for his account of the
1133 Aug 2 eclipse.

Winchester [ca. 1277] has, under the
year 1140: "An eclipse occurred, and stars were
seen around the sun." This is much shorter than
the entry in Winchester for the eclipse of 1133
Aug 2, which had been condensed from William of
Malmesbury. The 1140 entry is so short, and the
phraseology is so different, as to suggest that
it was not taken from William. However, the statement could have been derived from William, and the
editor of Winchester says that the latter drew
heavily from William in this part of the chronicle.
Therefore I shall not take the Winchester entry
to be independent.

1140 Mar 20c B,E. Reference: Gervase
[ca. 1199, p. 109]. Gervase has under 1139:
"This year an eclipse of the sun occurred about
the 9th hour on the 13th calends April, and many
stars appeared around the sun." This account differs from William's in having the year wrong and
in having "many stars" rather than simply "stars".
Gervase was not an eye witness, since he was not
born yet, and he might have used William. Because
of the differences, I shall assume that Gervase
took his information from a Canterbury source
independent of William, and I shall attach a reliability of 0.5 to this conclusion. Place: Canterbury. Standard deviation of the magnitude: 0.01.

Bermondsey [ca. 1433], which is a much
later source than Gervase, has the same mistake in
year, and the same words about the stars. It is
probably not derived directly from Gervase; more

probably, it and Gervase derived their information from a source that is now lost.

The record just discussed was entered under the year 1139. On p. 112 under the year 1140 Gervase has: "There was an eclipse of the sun about the 3rd hour on the 11th calends April (= Mar 22), 2nd feria." 1140 Mar 22 was on the 6th feria and the eclipse date 1140 Mar 20 was on the 4th feria, and the 3rd hour is far too early. The record says only "3rd hour" and not "3rd hour of the day", so one must keep in mind that possibility that this refers to a lunar eclipse and that "sun" was an error. However, I cannot plausibly identify this with any eclipse, solar or lunar.

1140 Mar 20d B,E. Reference: Florence of Worcester [ca. 1118]. In a continuation, under 1140, is the passage: "An eclipse of the sun occurred while the moon was in the tail of the dragon, the light shining on the head." The "tail of the dragon" is an ancient and medieval term for the descending node[†] of the moon's orbit and the "head of the dragon" is the opposite point in the heavens, namely the ascending node. The information in the passage is correct, and there was no other eclipse near this time visible in England for which the information would be correct.

Florence also referred to the anatomy of the dragon in his account of the eclipse of 1133 Aug 2. I imagine that this practice reflected a growing interest in astrology, perhaps as astronomical and astrological information began to filter into Europe from Islamic areas. The Florence chronicle with its continuations is now contemporary, and this is probably an independent observation. I shall give it a reliability of 1, observed at the city of Worcester, and with 0.1 for the standard deviation of the magnitude.

[†] That is, the point where the moon crosses the ecliptic going south.

1140 Mar 20e B,E. Reference: Matthew Paris [ca. 1250]. Matthew has, under the year 1140: "And that year a horrible and dark eclipse of the sun occurred over all England, as the heavenly bodies in the lowest positions came together in his disappearing."[†] This could be a quite free paraphrase of William's account, but it seems more plausible that it is taken from a local S. Albans source. I shall take this to be an observation made at S. Albans, with a reliability of 0.5 and a standard deviation of magnitude of 0.05 (there is some emphasis on the extent of darkening).

1140 Mar 20f B,E. Reference: Ralph of Coggeshall [ca. 1223]. Under the year 1140 Ralph has: "An eclipse of the sun took place on the 13th calends April, on the 28th of the moon." The Coggeshall annals are independent and contemporary now, according to the editor, and this sounds like an independent entry. Reliability: 1. Place: Coggeshall. Standard deviation of the magnitude: 0.1.

1140 Mar 20g B,E. Reference: Worcester [ca. 1377]. Under 1140: "There was an eclipse of the sun on S. Cuthbert's day." This terse sentence is the only record that identifies the day (correctly) as S. Cuthbert's day rather than as the 13th calends April. A monk who was thoroughly acquainted with the church calendar could have made the change automatically as he was copying, but it is more plausible that this is an independent record. It is presumably not original, but it is probably from a now-lost Worcester source. Reliability: 0.5. Standard deviation of the magnitude: 0.1.

[†] The last clause is "... ut imis corpora coelestia in suo defectu concordarent." It is hard to translate imis in this context. I assume that Matthew subscribed to a geocentric universe in which the moon and sun had the lowest positions and that imis refers to their places, but I am not happy about the resulting translation.

1163 Jul 3 B,E. Reference: <u>Worcester</u> [ca. 1377]. The reference has, under the year 1163: "Two hours before noon[†] the sun became like a third moon; accordingly it did not have light, except like the light of a 14th moon."[‡] I presume that "like a third moon" means "having the shape of a 3-day-old moon". I suppose that the sentence means that the sun had the shape of the moon 3 days after new moon and the amount of light of a full moon. However a lunar eclipse would be on or about the "14th moon", and it is also possible that the chronicler copied part of the record of a lunar eclipse by mistake.

The words in this record are identical with the words used in <u>Worcester</u> in one of the records of the eclipse of 1133 Aug 2 mentioned above; the hour of the day is reasonable for either eclipse. Something has almost surely been copied wrong. It is possible that this is a record of the eclipse of 1163 Jul 3, but it is safer to ignore it.

1178 Sep 13a B,E. Reference: <u>Diceto</u> [ca. 1202, v. 1, p. 427]. Part of the entry for 1178 says: "The sun suffered an eclipse on the ides of September." We are in Diceto's own time, and he records a number of floods and other natural events in this part of his work. The editor says of these: "Notices of the winter floods and of an eclipse of the sun .. read like personal reminiscenses." It seems reasonable to accept this as an independent observation, although it does not sound very personal. Reliability: 1. Standard deviation of the magnitude: 0.1.

According to the editor of his work, Diceto was archdeacon of Middlesex from 1152 to 1180, when he became dean of S. Paul's in London.

[†] <u>Meridiem</u>, noon in the modern sense.

[‡] I am not sure of the meaning of the part of the quotation after the semicolon. It reads: "..; <u>non ergo lumen habuit, nisi quasi lumen xiv. lunae.</u>"

Since the archdeaconry of Middlesex was part of the diocese of London, I imagine that Diceto was already in London in 1178. Hence I shall take London as the place of observation.

1178 Sep 13b B,E. Reference: <u>Gervase</u> [ca. 1199, p. 277]. Under the year 1178, <u>Gervase</u> writes: "In September, on the eve of Holy Cross Day, the 4th feria, the 27th of the moon,[†] at about the 6th hour, there was an eclipse of the sun in Kent, not total but partial." Gervase then describes how the sun appeared to have horns, first pointing toward the west, then downward,[‡] and then toward the east; the details seem to be accurate. Finally he says: "Elsewhere the eclipse was total, however, so that it looked like night at midday, and one could not see a person next to him. The eclipse was also seen in France."

The wording suggests to me that the eclipse was seen as total elsewhere in England, but not in France. If the chart of the eclipse path in <u>Oppolzer</u> [1887] is approximately right, however, the eclipse was total in southern France but nowhere in England.

Gervase does not make an explicit statement about the degree of totality, but he mentions changes in sky color, and the account reads like one of a large eclipse. I shall assume that the magnitude was 0.95 with a standard deviation of 0.05. Reliability: 1. Place: Canterbury. Path assumed to be south of England.

1178 Sep 13c B,E. Reference: <u>Worcester</u> [ca. 1377]. Under the year 1177: "Violent wind, snow, and hail; and an eclipse of the sun." In

---

[†] The date given is Sep 13, the ides. The weekday is correct, but it should be the 28th of the moon.

[‡] "As if looking at the earth" (<u>quasi in terram respicientes</u>).

the same year, Worcester records the reconcilia-
tion between the Holy Roman Emperor Frederic and
Pope Alexander, which occurred in 1177. Even so,
the eclipse must have been the one of 1178 Sep
13. The accompanying weather report makes it likely that this report is independent of the others
that have been found, even though it was probably
not written in its present form until perhaps
thirty years later. Reliability: 0.5. Place:
Worcester. Standard deviation of the magnitude:
0.1.

1178 Sep 13d B,E. Reference: Matthew
Paris [ca. 1250]. In v. 1, p. 401, under 1178,
Matthew has: "The sun suffered an eclipse on the
6th ides January (= Jan 8)." In v. 1, p. 423, under 1181, he has: "The same year, on the eve of
the Exaltation of the Holy Cross, an eclipse of
the sun occurred at the third hour." Either Matthew badly misused his sources here or his sources
were in error. I cannot identify the first record with any eclipse date, lunar or solar, by postulating a single error in writing the date. The
only likely identification I can find by postulating two errors is the penumbral eclipse of 1179
Feb 8 (6th ides February). The information in
Oppolzer [1887] indicates that the eclipse of 1179
Feb 8 would have been barely visible in England.

The second record clearly refers to the
date Sep 13. The error in the year is unreasonably
large; most records of this eclipse have the year
right. This is the only one I have found with 1181
for the year, and this suggests that the record is
independent. The error in the hour of the day is
also unreasonably large, and this error is shared
with some continental records (see, for example,
the record 1178 Sep 13a E,F in Section X.3), indicating a dependent record. Since the situation
is so confused, it is safest to give these records
a reliability of 0.

1185 May 1a B,E. Reference: Diceto [ca.
1202, v. 2, p. 37]. "The sun was eclipsed after

the 9th hour on the calends of May." (The year was 1185.) This record should be contemporary and original. Reliability: 1. Place: Diceto was certainly in London by now. Standard deviation of the magnitude: 0.1.

1185 May 1b B,E. Reference: Gervase [ca. 1199, p. 326]. Gervase says, under the year 1185: "The following month of May, on the first day of the month, there was a partial eclipse of the sun about the 7th hour." This record was contemporary and apparently original. Reliability: 1. Place: Canterbury. Magnitude: 2/3, with standard deviation of 1/6. Path to the north of Canterbury.

1185 May 1c B,E. Reference: Ralph of Coggeshall [ca. 1223]. Ralph says in 1185: "There was an earthquake on the 15th calends May (= Apr 17), and there was an eclipse of the sun." I believe that the date is meant to apply only to the earthquake and not to the eclipse. Several sources mention both the earthquake and the eclipse. The others give the date of the eclipse but not of the earthquake; Ralph is the only source I have seen that does the opposite. Thus this is an independent and apparently contemporaneous record. Reliability: 1. Place: Coggeshall. Standard deviation of the magnitude: 0.1.

1185 May 1d B,E. Reference: Waverley [ca. 1291]. Under the year 1185: "The sun, after changing and appearing eclipsed on the calends of May, changed its brightness in an unusual manner." This account is not derived from any of the sources usually listed for Waverley. Both for this eclipse and for the one of 1191 Jun 23, Margan (see Section VII.3) and Waverley seem to be closely related. I shall give a reliability of 0.2 to each account. Place for this record: Waverley. Standard deviation of the magnitude: 0.1.

1186 Apr 21a B,E. Reference: Diceto
[ca. 1202, v. 2, p. 40]. "The sun suffered an
eclipse on the 11th calends May (= Apr 21) at
the first hour of the day." This was a penumbral
eclipse. The sun was probably eclipsed at sunrise. This circumstance makes it likely that a
fairly small eclipse would be seen, and I should
perhaps assign a larger standard deviation to the
magnitude than usual. I shall not bother with
this refinement.

Reliability: 1. Place: London. Standard deviation of the magnitude: 0.1.

1186 Apr 21b B,E. Reference: Gervase
[ca. 1199, p. 334]. Under 1186 appears: "This
year a total eclipse of the moon happened on the
5th day of the month of April at the first hour
of the night. Afterwards there was a flame-col-
ored partial eclipse of the sun on the first day
of May and the first hour of the day." The information about the lunar eclipse is correct. The
information about the solar eclipse looks to be
correct except for the day of the month, which
should have been the 11th of the calends of May.
Apparently Gervase copied the date of the eclipse
of 1185 by mistake.[†] Since the account is contemporary, and since the record of this eclipse
is definitely different from the record of 1185
May 1, I shall take this as an independent contemporary record of the penumbral eclipse of 1186
Apr 21. Reliability: 1. Place: Canterbury.
Magnitude: 2/3, with standard deviation of 1/6.

1187 Sep 4 B,E. Reference: Gervase
[ca. 1199, p. 380]. Under 1187 is: "The day

[†] In the edition cited, "first day of May" occurs,
not "calends" of May. If this represents what
Gervase wrote, the error was not simply the accidental omission of "xi" before calends. It
would be necessary to see the manuscript in order to decide the point.

before the nones of September, the 6th feria and the 6th hour, on the 28th day of the moon, appeared a partial eclipse of the sun in England." It is interesting that this eclipse was recorded from as far away as Jerusalem (see Section XV.5). The feria is correct, the hour looks reasonable, and the day of the moon is correct for the ecclesiastical moon. Reliability: 1. Place: Canterbury. Magnitude: 2/3, with a standard deviation of 1/6. Path assumed north of England.

1191 Jun 23a B,E and 1191 Jun 23b B,E. References: Wendover [ca. 1235] and Ralph of Coggeshall [ca. 1223]. Under the year 1191, Wendover has: "This year, in June, on the eve of S. John the Baptist, the 1st feria, an eclipse of the sun appeared about the 6th hour of the day and lasted until the 8th hour, on the 27th of the moon, the sun being in Cancer." Ralph of Coggeshall has an entry that is identical except that he does not have "1st feria" but does have that the eclipse was partial. This circumstance makes it likely that both Wendover and Ralph used a common source that I have not found. The date and the weekday are correct, but the hours of the day look late, judging from the map in Oppolzer [1887]. The age of the moon is correct for the ecclesiastical moon.

I shall assume that these reports represent a single report which would have a reliability of 1. Since I must use these two equally, I shall give each a reliability of 0.5. The places will be S. Albans and Coggeshall, respectively. I shall assume that the original record was of a partial eclipse. Accordingly, I shall take the magnitude to be 2/3 with a standard deviation of 1/6 for each record, even though Wendover does not say that the eclipse was partial.

The record of this eclipse in Matthew Paris [ca. 1250, v. 2, p. 22] is probably taken from Wendover; at least it is from the same basic source. Tewkesbury [ca. 1263], Worcester [ca. 1377], and Wykes [ca. 1289] have brief notices of this eclipse. The information in them could be independent. I shall neglect their records.

1191 Jun 23c B,E. Reference: Waverley
[ca. 1291]. Waverley has, under the year 1191:
"There was an eclipse of the sun on the Lord's
day about the 6th hour." The Margan entry (see
Section VII.3) for this year is identical except
that it also contains "on the eve of S. John the
Baptist" after "on the Lord's day". This fact
suggests that the Waverley account was taken from
Margan. However, for the eclipse 1185 May 1, Waverley and Margan seemed to come from the same source
but neither seemed to be taken from the other. I
shall assume, but without much conviction, that
the same thing has happened here, and give a reliability of 0.2 to each report. Place for this
report: Farnham. Standard deviation of the magnitude: 0.1.

1191 Jun 23d B,E. Reference: Florence
of Worcester [ca. 1118]. The continuation made
at Bury S. Edmunds has: "The sun had an eclipse
on the 9th calends July, such that stars appeared
for 3 hours." The visibility of stars is apparently
exaggerated,[†] but the mention of stars is not found
in any other record. The record is contemporary
and probably reliable, even if we assume it to be
exaggerated. Reliability: 0.5 (not 1 because of
the possible exaggeration). Place: Bury S. Edmunds.
Standard deviation of the magnitude: 0.01.

The continuation does not give the year,
for this or for any year. However, it gives many
events in each year that can be dated independently.
It records the coronation of Henry, Holy Roman Emperor, and the conquests of Cyprus and Acre by
Richard the Lion-Hearted correctly during the year
of the eclipse.

[†] A comma inserted after "appeared" would make the
meaning reasonable. "3 hours" would then apply
to the entire eclipse.

1191 Jun 23e B,E. Reference: Richard
of Devizes [ca. 1192]. This record was used in
AAO, and there is nothing new to add; it is a
straightforward notice of the eclipse, followed
by a statement implying that eclipses have no por-
tentous significance. Reliability: 1. Place:
Winchester (where Richard lived at this time).
Standard deviation of the magnitude: 0.1 (I used
the value 0.06 in AAO, which was the conventional
value adopted there).

1194 Apr 22 B,E. Reference: Gervase
[ca. 1199, p. 527]. The reference has: "On the
10th calends of the same[†] there was a partial eclipse of the sun at the 6th hour." This contemporary record receives a reliability of 1. Place:
Canterbury. Magnitude: 2/3, with a standard deviation of 1/6. Path assumed north of England.

1207 Feb 28a B,E. Reference: Matthew
Paris [ca. 1250, v. 2, p. 113], which has under
1207: "This year the sun suffered an eclipse from
the 6th to the 9th hour. And also the moon under-
went an eclipse within the year, in sad foreboding
of things in the near future." The lunar eclipse
is presumably that of 1208 Feb 3. This report is
in a time when Matthew is usually not yet independent of Wendover [ca. 1235], but it seems to be
independent in this instance. The "foreboding"
suggests that the record was written after some of
the unpleasant events of John's reign, such as the
Interdict which was imposed in 1208, but probably
a short time after and from contemporary notes.
Reliability: 0.5. Place: S. Albans. Standard
deviation of the magnitude: 0.1.

1207 Feb 28b B,E. Reference: Florence
of Worcester [ca. 1118]. The continuation at Bury

[†] The context makes it clear that this is the 10th
calends May (= Apr 22). The reference also gives
the year as 1194.

-174-

has: "There was an eclipse of the sun on the 2nd
calends March (= Feb 28)." The year is correctly
given as the same year that saw Stephen Langton
elected Archbishop of Canterbury (the cause of
next year's Interdict) and the birth of the future
Henry III. The continuation is believed to be contemporary. Reliability: 1. Place: Bury S. Edmunds. Standard deviation of the magnitude: 0.1.

       1207 Feb 28c B,E. Reference: <u>Bermondsey</u>
[ca. 1433]. "This year an eclipse of the sun happened on the 2nd calends March about the 6th hour,
and it lasted until the 9th hour." This late record could have been the result of combining the
information in the preceding two records. If so,
it is a rather free paraphrase, and compilers of
annals did not usually take enough trouble to ab-
sorb meaning and produce distinct paraphrases. I
shall assume that this represents an independent
record taken from local Bermondsey sources, but I
shall give it a reliability of only 0.5 because it
is not original. Place: London. Standard deviation of the magnitude: 0.1.

       1230 May 14 B,E. Reference: <u>Wendover</u>
[ca. 1235]. The reference has a long passage under the year 1230 that begins: "This year happened
an eclipse of the sun, contrary to custom, immediately after sunrise, on the day before the ides
of May, in Rogation days, namely on the 3rd feria,
. . ." The passage finishes by describing the
admiration of the farmers and others who saw it,
and the return of the sun to its clarity an hour
later. The "3rd feria" is correct, and <u>Oppolzer</u>
shows the sunrise point of this eclipse in Kent.
I guess that "contrary to custom" (<u>contra morem</u>)
means that an eclipse at sunrise is unusual, but
I am not sure that this is the intended meaning.

       This is a contemporary record and re-
ceives a reliability of 1. The place is still
S. Albans. An eclipse at sunrise or sunset is
more prominent than an eclipse during the middle
of the day. For this reason, I would usually

give a sunrise or sunset eclipse a greater standard deviation than usual. In this case, the description suggests a large eclipse, and I shall use a standard deviation of 0.1, the usual value.

Matthew Paris [ca. 1250] has what is clearly a shortened form of this record.

1239 Jun 3a B,E. Reference: Matthew Paris [ca. 1250, v. 2, p. 421]. "Also within the year, on the third day of June, the sun experienced an eclipse at the 6th hour, in a sign that may be intended for the church, which like the sun enlightens mankind."† Wendover [ca. 1235] has now quit and Matthew is presumably a contemporary and independent authority. Reliability: 1. Place: S. Albans. Standard deviation of the magnitude: 0.1.

1239 Jun 3b B,E. Reference: Tewkesbury [ca. 1263] has under 1239: "The sun experienced an eclipse on the 3rd nones June (= Jun 3) about the 6th hour." Worcester [ca. 1377] has, also under 1239: "An eclipse of the sun appeared on the 3rd nones June." It is generally considered that Tewkesbury is taken from Worcester. Here the relation is clearly the other way around, unless both passages are taken from another source. Luckily Tewkesbury and Worcester are only about 25 kilometers apart, and it does not matter much which is taken as the original. I shall treat these records as a single observation with a reliability of 1, made at Tewkesbury, with a standard deviation of the magnitude of 0.1.

1241 Oct 6a B,E. Reference: Matthew Paris [ca. 1250, v. 2, p. 457], who has under the year 1241: "Also, that same year, the day before the nones of October, that is the day of S. Fidis, an eclipse of the sun occurred." The date is correct. Reliability: 1. Place: S. Albans. Stan-

---

† An editor's note says that this quotation is on an erasure except for the last clause.

dard deviation of the magnitude: 0.1.

1241 Oct 6b B,E. Reference: Tewkesbury
[ca. 1263] has under the year 1241: "On the day
of S. Fidis there was an eclipse of the sun for
two hours, not total but partial." Reliability:
1. Place: Tewkesbury. Magnitude: 2/3 with standard deviation of 1/6. Path assumed to the east
of England.

1241 Oct 6c B,E. Reference: Worcester
[ca. 1377]. The reference has under the year 1241:
"On the day of S. Fidis about the 6th hour there
was an eclipse of the sun." Since this is the only
English record that gives the hour of the day, and
correctly, it is probably independent. Reliability:
1. Place: Worcester. Standard deviation of the
magnitude: 0.1.

1241 Oct 6d B,E. Reference: Wykes [ca.
1289]. The reference says: "An eclipse of the
sun on the day of S. Fidis." This brief notice
could have been easily taken from Matthew Paris,
which is a known source for Wykes. Since Wykes
is mostly original at this time, I give this report a reliability of 0.5. Place: Osney. Standard deviation of the magnitude: 0.1.

Oxenedes [ca. 1292], which is from a
place near Wykes, has a similar brief notice of
this eclipse. Much of Oxenedes is taken from Matthew Paris. By neglecting the record in Oxenedes,
I effectively treat the records in Oxenedes and
Wykes as a single record, which may not be original.

1255 Dec 30 B,E. Reference: Oxenedes
[ca. 1292]. The reference describes how the king
spent Christmas at Winchester and then adds: ". .
and while the joys of the Nativity were still
being celebrated there was an eclipse of the sun
on the 3rd calends January (= Dec 30)." The reference gives the year as 1256, but that may be be-

cause the reference calends was the calends of
January, 1256.[†] Reliability: 1. Place: Oxnead.
Standard deviation of the magnitude: 0.1.

        1261 Apr 1 B,E. Reference: <u>Florence of
Worcester</u> [ca.1118]. The continuation <u>at Bury S.
Edmunds</u> has: "An eclipse of the sun occurred on
the calends of April, at the end of the 4th Arabic
month, 6th feria, 3rd hour of the day." The year
is not specifically identified in the continuation,
at least not in the cited edition; however, Pope
Alexander IV died and Urban IV was elected in the
same year, which is therefore 1261. 1261 Apr 1
was indeed on the 6th feria.

        <u>Oxenedes</u> [ca. 1292] has: "Year of grace
MCCLXI. An eclipse of the sun occurred on the
calends of April, Arabic month, 6th feria, 3rd
hour of the day." This differs from the account
in <u>Florence</u> in two ways: It gives the year explicitly, and it omits "at the end of the 4th"[‡] before "Arabic month" and thus makes nonsense out
of the remark. It seems certain that the account
in <u>Oxenedes</u> is taken from <u>Florence</u>, but that the
<u>Oxenedes</u> chronicler had some way to supply the
year, perhaps a different copy of <u>Florence</u> from
the one used to supply the published edition.

        I shall take the account in <u>Oxenedes</u> as
a derivative. I give the account in <u>Florence</u>,
which is believed to be a contemporary compilation
at this stage, a reliability of 1. Place: Bury
S. Edmunds. Standard deviation of the magnitude:
0.1.

[†]Or it may be that this chronicler began his year
at Christmas.

[‡]<u>Florence</u> has <u>in fine quarti mensis Arabum</u> where
<u>Oxenedes</u> has <u>mensis Arabum</u> only.

1263 Aug 5a B,E. Reference: Waverley
[ca. 1291]. "This year an eclipse of the sun happened on the day of S. Dominic the confessor at
about the 9th hour; and it lasted for the space of
a meal or longer, and it was the 27th of the moon."
This report, like the reports 1140 Mar 20a B,E
and 1140 Mar 20b B,E, reflects the custom of having the main meal of the day at the hour of nones,
the 9th hour, the middle of the afternoon. It
sounds like a personal account, suggesting that
the darkness was considerable throughout the meal.

The day of S. Dominic, the founder of
the Dominican order, is actually Aug.4. It is peculiar that this is his day, since he died on Aug
6 (of 1221).

Reliability of the record: 1. Place:
Farnham. Standard deviation of the magnitude:
0.1.

1263 Aug 5b B,E. Reference: Wykes [ca.
1289].† "That same year on the nones of August
there was an eclipse of the sun at the 6th hour."
The year is given as 1263. The "6th hour" is
much too early, but the error has the merit of
showing that this is an independent record. Reliability: 1. Place: Osney. Standard deviation
of the magnitude: 0.1.

1288 Apr 2 B,E. Reference: Wykes [ca.
1289]. The reference has, under the year 1288:
"On the day of Venus‡ in Easter week,* that is,

_____

† Occasionally Osney [ca. 1347] and Wykes are distinct, and this is an instance of it. The record
is actually in Osney.

‡ The chronicler has adopted a pagan method of giving the week day; the "day of Venus" is the 6th
feria.

* The text has in septimana Paschae. Septimana
is not the usual word for week, but I see no
other meaning here.

the 4th nones April (= Apr 2) appeared an eclipse
of the sun about the 6th hour." 1288 Mar 28 was
Easter and 1288 Apr 2 was the "day of Venus" in
the week of Easter. The eclipse would have been
greatest at about the 6th hour. This contemporary
report receives a reliability of 1. The place is
Osney, and the standard deviation of the magnitude
is 0.1.

CHAPTER VII

RECORDS OF SOLAR ECLIPSES FROM IRELAND,
SCOTLAND, WALES, AND THE ISLE OF MAN

1. Records from Ireland

          The records from England outnumber those
from Ireland, Scotland, Wales, and Man put together.
I do not know whether this is a consequence of
their histories or of my ignorance of their litera-
tures. I shall discuss the records from Ireland,
Scotland, and Wales in this chapter, devoting a
separate section to each. This procedure is convenient because there was apparently not a great
deal of interaction between annals and chronicles
in the different areas, in spite of their close
historical connections. I shall include the records from the Isle of Man in the same section with
the Scottish reports, for reasons that will be explained in Section VII.2.

          The two best known Irish annals[†] are
probably the Annals of Ulster [Ulster, ca. 1498]
and the Annals of the Kingdom of Ireland by the
Four Masters [Four Masters, 1636]. Both are late
compilations by known persons. The Ulster entry
for the year 1498 (translation by Ó Maille [1910])
contains: "Mac Maghnusa Mag Uidhir died this year,
that is Cathal Og son of Cathal . . a man full of
good qualities and knowledge in every science, ...
who planned out and compiled and collected this
book from several other books." This entry was
clearly written by one of the persons who continued Ulster for some time after 1498; the continuations are too late to interest us. In parts of
Ulster, Mac Maghnusa Mag Uidhir worked from an

[†]I thank Dr. J.A. O'Keefe for calling my attention
to the Irish annals.

earlier compilation [Ó Maílle, 1910].

Four Masters has an introductory statement by the four scholars who compiled it and who later came to be known as the "four masters", although they never applied the term to themselves. This statement gives the names of the scholars, the dates when they worked, and the sources they consulted. The leader of the four masters, Brother Michael O'Clerigh, gave the reason for compiling the annals from the ancient documents that he had collected; he feared that "they would not again be found to be put on record or commemorated to the end and termination of the world."

O'Clerigh's fears have been justified. Aside from a few fragments, only one of the ancient[†] sources used by the four masters is still known [O'Donovan, 1856], and it in translation only and not in its original language.

MacCarthy [1901], Ó Maílle [1910], and O'Donovan [1856] give reasons for believing that the sources used for Ulster and Four Masters were indeed ancient and approximately contemporaneous with the events they described.[‡] Most of their reasons involve ancient declensional forms and other archaic features of the native language (when it was used; Ulster is written in a mixture of Irish and Latin, while Four Masters is entirely in Irish). Some of their reasons involve the eclipse accounts that will be discussed below. Thus there is a good chance that Ulster and Four Masters are careful copies or paraphrases of primary sources.

[†] The four masters consulted the Annals of Ulster, which I do not count among their ancient sources. Comparison of Ulster and Four Masters proves that the four masters did not copy Ulster, but that they prepared an independent document.

[‡] At least from roughly the sixth century on. Accounts of earlier events are probably a mixture of myth and of genuine history handed down by oral tradition.

Unfortunately it is almost certain that some of the primary sources were not Irish. For example, Ulster refers to the Byzantine chronicler Marcellinus [ca. 534] (see Section XV.3) in its entry for the year 535. Thus some foreign sources were known to the compilers of Ulster. In fact, it is almost certain that the eclipse records discussed below under the designations 497 Apr 18 B, I (?) and 512 Jun 29 B,I are copied from Marcellinus.[†]

There is some question about whether the Chronicum Scotorum [Scotorum, ca. 1650] is a compilation or a copy of an earlier set of annals. The editor of Scotorum thinks that it is mostly a copy (or perhaps a translation) of the Annals of Clonmacnois that the four masters listed as one of their sources, and it may be the single surviving source mentioned by O'Donovan. However, at least as far as eclipse reports are concerned, Scotorum shows much closer affinities with Ulster than with Four Masters.

When they contain identical material, I shall treat Ulster as the earlier source and shall use Scotorum only when it helps clarify the text of Ulster. This decision is somewhat arbitrary, but the way in which the sources treat Marcellinus gives some confirmation to it.

There is a document that has appeared in printed form under the title Annals of Loch Cè.[‡] It has also been known at times, without justification in the opinion of the editor, as Annals of

---

[†] Schove [1954] refers to an "international trade in marvels" during the Middle Ages while giving a useful warning against accepting the early records uncritically. There was a trade in annals of all sorts, as we saw in Chapter I. The places of origin of many early brief notices of eclipses must be regarded as quite uncertain.

[‡] Annals of Loch Cè, edited by W.M. Hennessy, in Rerum Britannicarum Medii Aevi Scriptores, no. 54, in 2 vols., Longman and Co., London, 1871.

the Old Abbey of Inis-Macreen, an Island in Lough-
Kea, as the Continuation of the Annals of Tigher-
nach, as the Book of the O'Duigenans of Kilronan,
and as the Annals of Kilronan. The last title was
listed by the four masters as one of their sources,
but it is far from certain that the form we have
is what they used. The only eclipse record that
I noticed in Loch Cè is a record of the eclipse of
1023 Jan 24, in words identical with those in Ulster
(see the record 1023 Jan 24 B,I below). I have not
included Loch Cè in the list of sources for the present work.

In sum, I shall use Ulster and Four Masters, which seem to be careful secondary sources,
as the basic sources for Irish eclipse records.
Because they are secondary and late, I shall assign a maximum reliability of 0.5 to records from
these annals.

The annals contain two types of dating
error. Random errors, when they can be detected,
may amount to as much as 5 years. The Annals of
Ulster also contain a systematic error for dates
through 1014. Many sources used in the compilation of Ulster specified the year by giving the
week day or the age of the ecclesiastical moon
(see Section II.4) on Jan 1.[†] The compilers of
Ulster calculated the years from these data, but
they did not know all the variety of ecclesiastical moons that have existed. In calculating
the years, they systematically assigned a year
that was 1 year too early for years through 1014.
In citing dates from Ulster I shall correct this
error; the reader should keep this in mind in comparing quotations with the cited text.

If I have understood them correctly,
MacCarthy and O'Donovan feel that Ulster contains
no random errors in date. It seems to me that

[†] This practice suggests that the corresponding
sources were derived from Easter tables (see
Section III.2).

-184-

TABLE VII.1

REFERENCES TO SOLAR ECLIPSES FOUND IN
MEDIEVAL IRISH SOURCES

Source	Date	Conclusions
Four Masters	664 May 1	Probably independent; may be from Bede
	1030 Aug 31	Independent
Scotorum	497 Apr 18 ?	Same as Ulster
	512 Jun 29	Same as Ulster
	594 Jul 23 ?	Independence doubtful
	612 Aug 2 ?	Probably not an eclipse
	664 May 1	Same as Ulster except for year
	688 Jul 3	Same as Ulster except for year
	865 Jan 1	Same as Ulster
	878 Oct 29	Probably not independent
	885 Jun 16	Same as Ulster
	1023 Jan 24	Probably abridged from Ulster
	1133 Aug 2	Probably independent
Ulster	497 Apr 18 ?	Copied from Marcellinus
	512 Jun 29	Copied from Marcellinus
	591 Sep 23 ?	Probably 594 Jul 23 instead

TABLE VII.1 (Concluded)

Source	Date	Conclusions
Ulster	592 Mar 19 ?	Probably 594 Jul 23 instead
	664 May 1	Independent
	688 Jul 3	Assumed independent
	753 Jan 9	Independent
	764 Jun 4	Independent
	865 Jan 1	Assumed independent
	878 Oct 29	Assumed independent
	885 Jun 16	Assumed independent
	1023 Jan 24	Independent

specifying the year by giving the age of the moon
on Jan 1, for example, is no more a guarantee of
accuracy than specifying it by giving the year of
the Christian era. Hence I shall assume that random dating errors are present in Ulster as well
as in the other chronicles.

Table VII.1 has a list of eclipses reported in Irish sources, arranged by source.

497 Apr 18 B,I(?). References: Ulster
[ca. 1498] and Scotorum [ca. 1650]. Under the
year 496 Ulster has: "Solis defectus apparuit."†
Under the year 493 Scotorum has "Defectus solis
apparuit"; this is the same except for a trivial
change in word order. This record is best discussed along with the following one.

512 Jun 29 B,I. References: Ulster
[ca. 1498] and Scotorum [ca. 1650]. Ulster has
"Defectus solis contigit" (An eclipse of the sun
happened) under the year 512. Scotorum has the
identical entry under 510, continuing the pattern
of being earlier than Ulster.

If this eclipse and the preceding one
are Irish, there are no satisfactory identifications, although MacCarthy [1901] accepted the dates
496 Oct 22 and 512 Jun 29, respectively, without
question. We should check the dates by independent evidence. Scotorum gives the accession of
Pope Anastasius as 494, but the correct date [Mer-
cati, 1947] is 496. If the error is the same in
both annals, the date of the first eclipse should
be 495 according to Scotorum, compared with 496 in
Ulster. There is no way that I noticed to check
the other dates.

The eclipses of 496 Oct 22 and 512 Jun
29 were both small in Ireland and unlikely to have
been recorded. The nearest eclipses large enough

---

† Defectus is sometimes used to denote an eclipse.

to have been noticed are probably the penumbral
eclipses of 492 Jan 15 and 513 Nov 13; the nearest
umbral eclipses are those of 487 Nov 1 and 507 Mar
29. It is not likely that the dating errors are
this large.

Luckily there is little question about
the origin of these records. I mentioned above
that Ulster refers to Marcellinus [ca. 534].† Under the year 497 Marcellinus has "Solis defectus
apparuit." Under the year 512 he has: "His fere
temporibus solis defectus contigit."‡

Note that Ulster, Scotorum, and Marcellinus use identical words in describing the eclipses, with only trivial differences in word order; the transpositions in order could easily be
made in copying. The noun "defectus" and the verbs
"apparuit" and "contigit" are not the words most
commonly used in eclipse notices, so their identical appearances are almost certainly the result
of copying and not of accident. We know by their
direct remark that the compilers of Ulster knew
Marcellinus. Hence there seems little question
that the Irish records of these eclipses are copied from Marcellinus.

I conclude in Section XV.4 that 512
Jun 29 is the most likely identification for one
of the eclipses in Marcellinus, but that 511 Jan
15 cannot be excluded. I also show that 497 Apr
18 is the most likely identification for the other,
but that 496 Oct 22 cannot be excluded. Hence I
take the latter record to be unidentifiable, and
use the date 497 Apr 18 only for purposes of desig-

---

† Ulster says that Marcellinus brought his chronicle
to the year 535. This shows that the compilers of
Ulster had a manuscript that included the original
chronicle and the continuation to 534, but not
what the cited editor Mommsen calls "Additions to
Year 548", which are anonymous. See Section XV.3
for more details about Marcellinus.

‡ "Near these times an eclipse of the sun happened."

nating the record. 512 Jun 29 is accepted for the
former record, subject to the calculations described
in Section XV.4 and XVII.2.

Ulster seems closer to Marcellinus than
Scotorum is. Ulster but not Scotorum refers directly to Marcellinus. For the eclipse of 497 Apr
18 (?) Ulster has the identical word order as Marcellinus, although both have put "defectus" before
"solis" for 512 Jun 29. Thus if either source is
directly dependent upon the other, it is Scotorum
that depends upon Ulster.

591 Sep 23 B,I(?). References: Ulster
[ca. 1498] and Scotorum [ca. 1650]. This report
is best discussed along with the following one.

592 Mar 19 B,I(?). References: Ulster
[ca. 1498] and Scotorum [ca. 1650]. Ulster has:
"Defectio solis .i. mane tenebrorum" for an unspecified day in the year 591. Ulster also has: "Matutina tenebrosa" for an unspecified day in 592.
MacCarthy unquestioningly accepted these as records
of the two eclipses stated. Both records really
say the same thing, namely that there was darkness
or a failure of the sun in the morning. This statement does not reasonably apply to either eclipse in
Ireland.

The fact that both records say essentially the same thing raises the possibility that one†
and not two eclipses are meant. The suspicion is

---

† That is, the same eclipse may have been recorded
in two ancient sources, but under different years
because of dating error. The compilers of Ulster
could then have interpreted the records as reports
of different eclipses. The French source S. Maxentii [ca. 1140] has three records of the eclipse
of 878 Oct 29! See Section X.3. Since mane is
the stronger term (early morning as opposed to
morning without qualification), there is a mild
presumption that the original record has mane and
not matutina.

reinforced by the fact that Scotorum has but one
entry, which is for the year 590: "Defectus solis,
tenbrarum."[†]

An embarrassing number of eclipses occurred near the years 591 and 592. The eclipse of
590 Oct 4 passed through southern Norway and was
largest in Ireland at perhaps 11^h. The eclipse of
591 Mar 30 passed through northern Africa at about
sunset. The eclipse of 591 Sep 23 passed almost
through Dakar and was largest in Ireland sometime
in the morning. The eclipse of 592 Mar 19 passed
through southern Italy and was largest in Ireland
at about 11^h. The eclipse of 594 Jul 23 was probably total in Ireland shortly after sunrise. Finally, the eclipse of 596 Jan 5 crossed the Alps
near sunset.

The only two of these to which "early
morning" could properly be applied in Ireland seem
to be those of 591 Sep 23 and 594 Jul 23. It is
questionable whether the first of these would have
been seen in Ireland. I think that the most likely
explanation of the entries is that all entries refer to the eclipse of 594 Jul 23. I shall assign
zero reliability to this identification. However,
I shall calculate the circumstances of the eclipses
of 590 Oct·4, 591 Sep 23, 592 Mar 19, and 594 Jul
23 for the center of Ireland, in order to get better
estimates of the times and thus to see whether a
stronger identification can be made. The results
will be presented in Section XVII.2.

Schove [1954] said of the two entries
in Ulster: " . . it seems probable that the first
reference is a further continental import,[‡] partic-

[†] Or teibrarum in one manuscript. Both spellings
are unconventional.

[‡] This remark follows Schove's discussion of the
eclipses of 538 and 540 in Bede [734?], which
probably came from the Continent. This explains
his use of "further". See the discussion of the
records 538 Feb 15 E,I and 540 Jun 20 E,I in
Section XII.2.

ularly as a word for eclipse was not likely to have been known to the original annalists." However, the Irish church was well established by 590, and the annalists, who were apparently church people since they worked with Easter tables, should have been well acquainted with Latin. Hence they would probably know a word for eclipse, whether there was a special term for one in the native language or not. It should be noticed that the early part of the annals, at least in their present form, is in Latin. Schove concluded that the second entry in Ulster was perhaps a record of the eclipse of 594 Jul 23. He did not consider Scotorum.

The early French source "Fredegarius Scholasticus" [ca. 641] does indeed report an eclipse "a mane usque ad mediam diem" (from early morning to midday); he states explicitly that the eclipse was partial.[†] It does not seem likely to me that "Fredegarius Scholasticus" is the source used by the Irish annalists, but the possibility cannot be definitely denied. My feeling arises from the great disparity in wording; neither Irish record is merely an abbreviated form of the French record. In Section X.3 I conclude, somewhat uncertainly, that the French record refers to the eclipse of 592 Mar 19.

612 Aug 2 B,I(?). Reference: Scotorum [ca. 1650]. The entry for the year 614 says: "A star was seen at the seventh hour of the day." The entry says nothing about an eclipse, but it is conceivable that the original had "An eclipse of the sun and a star was seen . . .", and that the first part of the record was lost in copying. The nearest plausible eclipse to 614 is the eclipse of 612 Aug 2, which would have been large in Ireland around the 7th or 8th hour. However, the pattern of dating errors in this portion of Scotorum suggests that the correct year is 617 rather than 612 or 614. I consider it doubtful that this record refers to an eclipse, and I consider the eclipse to

---

[†] See the record 592 Mar 19 E,F in Section X.3.

be unidentifiable if it does.

It is possible that a star seen at the 7th hour was a super-nova, and I have listed this record in Section V.2. I have no evidence about the visibility of comets in the daytime.

664 May la B,I. Reference: Ulster [ca. 1498]. The Ulster entry for 664 contains a passage that can be translated as follows: "Darkness on the calends of May at the ninth hour, and in that same summer the sky was seen to be on fire." The time agrees well with the chart given by Oppolzer [1887]. Bede [734?] (see the record 664 May 1 B,E in Section VI.3) wrote that the eclipse was on May 3 at about the tenth hour. Hence the Ulster report is independent of Bede. The fire in the sky may be an aurora. Reliability: 0.5. Place: Ireland. Standard deviation of the magnitude: 0.1.

664 May lb B,I. Reference: Four Masters [1636]. Part of the entry for the year 664 states that there was an eclipse of the sun on the third day of May; the entry does not give an hour. This report makes the same mistake in date that Bede made, which I have already discussed in Section V.5. MacCarthy [1901] cites Ussher's explanation that is discussed there. Because Ulster does not have this error, MacCarthy concluded that the report in Ulster (the report 664 May la B,I above) was the contemporaneous report of an eye witness; it does not seem to me that a contemporaneous record is the only kind of record that can be accurate. O'Donovan [1856], without referring to the eclipse of 664 May 1 specifically, used the general accuracy of the eclipses in Ulster as a proof of the antiquity and authenticity of its sources.

The fact that Bede and the "four masters" made the same mistake in date makes one wonder if the Four Masters report is independent. Since Four Masters does not give the hour of the eclipse, as Bede does, and since the "four masters" did not list

Bede's works among their sources,[†] I shall assume
that the Four Masters report is independent. Because both arguments are weak, I shall give this
report the low reliability of 0.05. Place: Ireland. Standard deviation of the magnitude: 0.1.

           688 Jul 3 B,I. References: Ulster
[ca. 1498] and Scotorum [ca. 1650]. Ulster for
the year 688[‡] and Scotorum for the year 685 contain an identical entry that can be translated as
follows: "A part of the sun was obscured." Ulster for the year 691[‡] and Scotorum for the year
688 also contain an identical entry that can be
translated as: "The moon was turned into the color
of blood on the "natale" of S. Martin." The
"natale" of S. Martin is Nov 11.

           There was indeed a lunar eclipse on 691
Nov 11 that would have been visible in Ireland.
This entry and the record 764 Jun 4 B,I suggest
that there are random dating errors in Ulster as
well as the systematic error that has been discussed. This suggestion is strengthened by the
fact that Ulster lists the accession of the Byzantine emperor Justinian the Younger under the
year 689 (before correction for a systematic error) rather than under the correct year 685.
Since the errors in the eclipse dates and in the
date of Justinian are quite different, it is plausible that the astronomical entries and the Byzantine entries had different origins.

           The date for the lunar eclipse suggests
that 688 is the correct year for the partial solar
eclipse. The eclipse of 688 Jul 3 would have been
partial but fairly large, on the basis of Oppolzer's

[†]
This is not a strong argument, since the source
that the four masters used could have been derived from Bede.

[‡]
In this case, this is the year assigned by the
compilers before correction for the systematic
error mentioned above.

<u>Canon</u>, and it is the only possibility within 5
years. I shall accept this identification, but
with the fairly small reliability of 0.2. Place:
Ireland. Magnitude: 2/3; path to the north of
Ireland. Standard deviation of the magnitude:
1/6.

       753 Jan 9 B,I. Reference: <u>Ulster</u> [ca.
1498]. The first entry for 753 notes t<u>hat the</u> sun
was darkened, but gives no other details. The
fact that this is the first entry for the year cre-
ates a mild presumption that the eclipse happened
early in the year. I did not see any other events
near this one that could be dated independently,
so there is no way to test the accuracy of the year.
There were rather large eclipses in Ireland on 753
Jan 9 and 758 Apr 12. Since there was a large eclipse during the year stated, and since it came
early in the year as the position of the entry suggests, I shall accept the identification of 753 Jan
9. Since there was another possibility with a dating error of 5 years, I shall use the small reliability of 0.2. Place: Ireland. Standard deviation of the magnitude: 0.1.

       764 Jun 4 B,I. Reference: <u>Ulster</u> [ca.
1498]. The entry for 763 (after correc<u>ting f</u>or the
systematic error) notes that the sun was darkened
at the third hour of the day, but does not give the
day of the year. I did not notice any historical
events that would allow testing the accuracy of the
year. <u>Ulster</u> notes a lunar eclipse for 773 Dec 4.
There w<u>as a t</u>otal eclipse of the moon visible in
Ireland on that date, but this is too far away to
give good confirmation.

      However, the entry for 762 records "a
great snow and an eclipse of the moon". The reference to the great snow and the fact that this is
the first entry of the year both suggest[†] an ec-

[†] The two events could not have occurred at exactly
the same time, because one cannot see the moon
during a storm.

lipse occurring early in the year. There was an
eclipse of magnitude 0.4 on 762 Jan 15 visible in
Ireland, a total eclipse on 763 Jan 4, and an eclipse of magnitude 0.8 on 763 Dec 25.[†] There were
no other winter eclipses visible in Ireland for
several years on either side of these dates.

Hence there is a strong presumption that
the year 763 is close to being right for the solar
eclipse. The only possible eclipse that occurred
during the morning was that of 764 Jun 4; it would
have been maximum in Ireland at about the 4th hour.
Hence this is almost surely the right identification, and the lunar eclipse was thus probably that
of 763 Jan 4. This identification is strengthened
by the consideration that the lunar eclipse of 763
Jan 4 is the only one of the various possible eclipses that was total.

I shall accept the identification of 764
Jun 4 for the solar eclipse, with a reliability of
0.5. Place: Ireland. Standard deviation of the
magnitude: 0.1.

865 Jan 1 B,I. Reference: Ulster [ca.
1498]. Part of the entry for the year 865 can be
translated as: "Eclipse of the sun on the calends
of January, and an eclipse of the moon in the same
month." Both statements are correct. There was a
solar eclipse on 865 Jan 1 that was nearly total
in Ireland. There was a total lunar eclipse on
865 Jan 15 for which Oppolzer [1887] gives a conjunction time of $18^h\ 25^m$ UT; thus this eclipse
would have been visible in Ireland.

[†]
Many early annalists considered the year to begin
on Christmas. Many events occurring on or after
Dec 14 were in practice referred to the following
year, because of the nature of the Roman calendar.
For example, 763 Dec 14 could have been written
as the 19th of the calends of January of 764, and
this date could thus seem to belong to the following year. See Section I.3.

Reliability: 0.5. Place: Ireland.
Standard deviation of the magnitude: 0.1.

878 Oct 29 B,I. Reference: Ulster [ca.
1498]. Part of the entry for 878 can be translated as: "Eclipse of the moon on the ides of October (= Oct 15), 14th day of the moon, about the
third watch,[†] on the 4th feria; and an eclipse of
the sun on the 4th calends November (= Oct 29),
28th day of the moon, about the 7th hour of the
day, on the 4th feria, 15 days of the sun intervening." There was a total lunar eclipse on 878 Oct
15, the 4th feria, for which Oppolzer [1887] gives
the time $4^h 26^m$ at conjunction. The solar eclipse
of 878 Oct 29 was also on the 4th feria. According
to Oppolzer's chart, the path of this eclipse passed just north of Ireland, and the eclipse would
have been large in all Ireland at about the 7th or
8th hour of the day.

In spite of the amount of detail[‡] given,
I see no reason to doubt the validity of this report. Its wording does not suggest copying from
any other record that I have found. An even stronger

[†] MacCarthy [1901] states that this means 4:30 a.m.

[‡] MacCarthy [1901, p. civ] says: "A statement of
such minute and accurate detail proceeded from no
other than an eye witness." I do not understand
this argument. Unusually minute detail raises
the suspicion of calculation rather than observation. Thorndike [1923] gives a large number of
medieval records that contain calculations relating to eclipses. There is in fact no information in the record just quoted whose calculation
would have been beyond the capacities of the compilers of Ulster. Further, there are many instances in medieval annals in which the feria and
the age of the moon given were calculated, as we
can prove because of errors contained in them.
The records in Romualdus [ca. 1178; see Section
XII.1] and in Hydatius [ca. 468; see Chapter XIV]
furnish two of many examples.

argument is the fact that the hour is earlier than in other records of this eclipse. This is as it should be, since Ireland is farther west than the other places of observation.

Reliability: 0.5. Place: Ireland.
Standard deviation of the magnitude: 0.1.

Scotorum [ca. 1650] has a brief notice of this eclipse that cannot be used safely.

885 Jun 16 B,I. Reference: Ulster [ca. 1498]. A passage from the entry for 885 can be translated thus: "An eclipse of the sun, and stars were seen in the sky." Although there are no details that aid in identification, this almost surely refers to the eclipse of 885 Jun 16. Reliability: 0.3; lowered because of the absence of detail. Place: Ireland. Standard deviation of the magnitude: 0.01.

1023 Jan 24 B,I. Reference: Ulster [ca. 1498]. The entry for 1023 contains a passage which Hennessy, in the translation cited, renders as follows: "An eclipse of the moon on the 14th of the January moon, that is, on the 4th of the ides of January, a Thursday. An eclipse of the sun also, on the 27th of the same moon, a Thursday, at the end of a fortnight, on the 9th of the calends [of February]." This is the only eclipse record in Ulster that is in Irish rather than Latin. The dates specified are 1023 Jan 10 and 1023 Jan 24, respectively.

There was indeed an eclipse of the sun on the 5th feria, 1023 Jan 24, which was total in Ireland according to Oppolzer's Canon [1887]. According to the same source, there was a nearly total lunar eclipse on the 4th feria, 1023 Jan 9, rather than on the 5th feria, 1023 Jan 10. However, the mistake of the annalist could have arisen from uncertainty about whether the eclipse happened before or after midnight.

Reliability: 0.5. Place: Ireland.
Standard deviation of the magnitude: 0.1.

Scotorum has, under 1021: "An eclipse
of the sun at midday, and an eclipse of the moon
in the same month." This sounds as if it may have
been abridged from Ulster.

1030 Aug 31 B,I. Reference: Four Masters [1636]. According to the entry for 1030,
there was an eclipse of the sun on the day before
the calends of September, that is, on Aug 31. It
is remarkable that this is the only record I have
found, except for the questionable passage by Sigvat (see the record called 1030 Aug 31 E in AAO,
Section IV.4), of this famous eclipse which many
have tried to make coincident with the death of S.
Olaf. Lack of reference to the eclipse of 1030
Aug 31 is particularly remarkable in view of the
many records of the eclipses of 1023 Jan 24 and
of 1033 Jun 29 that have been found.[†] It may be
that the day 1030 Aug 31 was cloudy over most of
Europe but clear over Ireland. (See also Section
XIII.2).

Reliability: 0.5. Place: Ireland.
Standard deviation of the magnitude: 0.1.

1133 Aug 2 B,I. Reference: Scotorum

[†] O'Donovan [1856, p. xlix] mentions a reference
to the eclipse of 1030 Aug 31 in a manuscript
that he identifies only as Cod. Clarend. Other
notes by O'Donovan lead me to believe that Cod.
Clarend. means a manuscript owned by the British
Museum that contains some of the source material
used for Four Masters (but it is not necessarily
the source used by them). It is doubtful that
the records in Cod. Clarend. and in Four Masters
are independent, and I have not tried to learn
more about Cod. Clarend.

[ca. 1650]. The entry for the year 1133[†] notes an
eclipse of the sun at the 3rd hour of the day. The
3rd hour looks early for the eclipse of 1133 Aug 2
in Ireland, but not unreasonably so. Since the
chronology seems to be accurate, the record almost
surely refers to this eclipse. Since the hour given
does not occur in any other known record of this
eclipse, the record is almost surely independent of
other known records, although it is not original.

Reliability: 0.5. Place: Ireland.
Standard deviation of the magnitude: 0.1.

2. Records from Scotland and the Isle of Man

I have found only two sources from Scotland that contain records of solar eclipses. These
are Melrose [ca. 1275] and Pictish Chronicle [ca.
990]. I searched through the Scottish Text Society
Series without noticing any sources for eclipse records, although it is certainly possible that I missed
some. I saw an allusion to a set of Scottish annals
called Chronicon Sanctae Crucis, which are supposedly
quite ancient, but the allusion gave no clue as to
where this chronicle can be found.

The original monastery of Melrose, which
is spelled Mailros in its own annals and in other
medieval sources, was founded in the 7th century.
It suffered greatly from the Scandinavian invasions
and was abandoned in the 10th century. A new abbey,
the one where the source annals were prepared, was
founded in 1136 a few kilometers from the old site.
The site is about 60 kilometers southeast of Edinburgh.

[†] The editor of Scotorum identifies the year as
1129 and puts this date in the margin. There are
several events listed in the same year, aside
from the solar eclipse, that can be dated independently. These events indicate that the correct year is 1133. Further, 1133 is consistent
with the "Easter table" data and 1129 is not.

TABLE VII.2

REFERENCES TO SOLAR ECLIPSES FOUND IN
MEDIEVAL SCOTTISH AND MANX SOURCES

Source	Date	Conclusions
Manniae	1133 Aug 2	Almost identical with Melrose
	1185 May 1	Probably independent
Melrose	733 Aug 14	From Bede
	1133 Aug 2	Probably independent
	1140 Mar 20	Probably independent
	1178 Sep 13	Independent
	1185 May 1	Independent
	1191 Jun 23	Independent
Pictish Chronicle	885 Jun 16	Independent

The Melrose annals were started by extracting from older sources such as Simeon of Durham (see Section VI.1). Within a few years, the annals were up to date and apparently remained mostly original and contemporaneous until their close in 1275.

The Pictish Chronicle, according to its editor, was compiled during the reign of Kenneth, son of Malcolm (reigned 977 to 995), and it gives events beginning in the reign of Kenneth Macalpin (reigned for 28 years, beginning in 832 or 834, depending upon the authority). It has material found in no other source, but most of it is not contemporaneous.

The only source that I have found from the Isle of Man is Manniae [ca. 1266]. It is hard to decide under what provenance to put the chronicle from Man. During the period covered by the chronicle, Man was nominally under Norse rulers but it was often independent in effect, being governed by rulers who called themselves kings of Man and the Isles. Geographically the Isle of Man is about equidistant from the four large areas of Britain.

Since the chronicle from Man contains only two records of solar eclipses, and since one of these is derived from a record from Melrose, I shall put Manniae in with the Scottish records. Manniae shows little affiliation with the Scandinavian sources discussed in Chapter XIII. It was compiled by one unknown author about 1266. Others then made isolated entries for the years 1270, 1313, and 1316.

Table VII.2 contains a list of the solar eclipses in Scottish and Manx sources, arranged alphabetically by source.

Melrose has records of the solar eclipse of 733 Aug 14 and of the lunar eclipse of 734 Jan 24. These are from Bede [734?], probably through intermediate sources. See the record 733 Aug 14b B,E in Section VI.3.

885 Jun 16 B,SM. Reference: <u>Pictish</u>
<u>Chronicle</u> [ca. 990]. On p. 9 of the reference is
the entry: ". . and also in his[†] 9th year, in
fact on S. Cyric's day,[‡] there was an eclipse of
the sun." The eclipse is also recorded in Irish
sources, but nowhere else that I have found, and
this record seems to be independent. Reliability:
0.5 (because it is not contemporaneous). Place:
Brechin. Standard deviation of the magnitude: 0.1.

1133 Aug 2 B,SM. Reference: <u>Melrose</u>
[ca. 1275]. "In the year 1133 there was an eclipse
of the sun on the 4th nones August (= Aug 2), the
4th feria, such that at one time day seemed to be
turned into night." Although this entry cannot be
contemporaneous, it is independent of all others
known. It is one of the few British records of
this eclipse that has the correct date. The weekday is also correct. Reliability: 0.5. Place:
Melrose. Standard deviation of the magnitude:
0.05, since the record implies a large eclipse.

<u>Manniae</u> [ca. 1266] has what is clearly
an abridgement of this record. Since <u>Manniae</u> does
not have the other eclipses in <u>Melrose</u>, it is unlikely that the Manx chronicler copied <u>Melrose</u>
directly. It is more likely that the compilers of
<u>Melrose</u> and <u>Manniae</u> used the same source.

[†] "His" refers to King Grig or Ciric (spellings of
his name vary). I have not found dates for this
king, but the editor says that 885 agrees well
with his 9th year. The editor also says that
these annals originate from Brechin or Brechne,
in northeast Scotland.

[‡] It seemed suspicious to me that the eclipse occurred on the day of a saint who seemed to have
close connections with the king, judging from the
name. The editor explains this by saying, if I
understood him, that the king took the saint's
name as an adult; perhaps he was impressed by an
eclipse that occurred on the saint's day. His
day is June 16.

1140 Mar 20 B,SM. Reference: Melrose [ca. 1275]. "Year 1141. There was an eclipse of the sun on the 13th calends April (= Mar 20), and King Malcolm was born." Because there was an eclipse on 1141 Mar 10 (the 6th ides March), the editor of Melrose inferred that Malcolm was born on that date; he ignored the date given and he overlooked the detail that the path of the eclipse of 1141 Mar 10 went from Madagascar through the Indian Ocean to New Guinea. The eclipse is undoubtedly that of 1140 Mar 20. If the annal is correct in saying that Malcolm was born in the year of the eclipse, his birth year was clearly 1140.[†] Most standard sources give 1141; one gives 1142. Since Malcolm was not king at birth, the record cannot be strictly contemporaneous. It may be that the eclipse notice was written contemporaneously and the reference to Malcolm inserted later; only inspection of the original manuscript could settle this point. I shall assume that the annal is not contemporaneous.

Reliability: 0.5. Place: Melrose. Standard deviation of the magnitude: 0.1.

1178 Sep 13 B,SM. Reference: Melrose [ca. 1275]. Melrose has under the year 1178: "The sun at midday took on a pallor and was almost totally eclipsed on the ides of September. Holland flooded." I do not think that "ides of September" (= Sep 13) is intended to apply to the flood. This report seems original and contemporaneous. Reliability: 1. Place: Melrose. Standard deviation of the magnitude: 0.05, since the report seems to imply a rather large eclipse.[‡]

[†] There is little reason to assume that Malcolm was born on the day of the eclipse.

[‡] Hence it would not be reasonable to use the conventional values for a partial eclipse that I adopted in Section IV.5.

1185 May 1a B,SM. Reference: <u>Melrose</u> [ca. 1275]. Under the year 1185: "There was an eclipse of the sun on the calends of May, the 4th feria, after nones, and stars appeared." The date is given exactly and it was the 4th feria, although the hour looks slightly late. Reliability: 1. Place: Melrose. Standard deviation of the magnitude: 0.01, because of the stars.

1185 May 1b B,SM. Reference: <u>Manniae</u> [ca. 1266]. This has: "1185. An eclipse of the sun on the day of Philip and Jacob." The medieval calendar [<u>Wormald</u>, 1934] that I use lists May 1 as the day of the Apostles S. Philip and S. Jacob. The method of dating is unique and this is probably an independent record, although it cannot be contemporaneous in its present form. Reliability: 0.5. Place: Isle of Man. Standard deviation of the magnitude: 0.1.

1191 Jun 23 B,SM. Reference: <u>Melrose</u> [ca. 1275]. Under the year 1191: "There was an eclipse of the sun on the eve of S. John the Baptist, the 9th calends of July, on the Lord's day at the 6th hour." All the data in this record are accurate. Reliability: 1. Place: Melrose. Standard deviation of the magnitude: 0.1.

3. <u>Records from Wales</u>

<u>Asser</u> [ca. 893]. King Alfred, who was a great patron of learning, brought Asser, a learned Welsh monk, to his court in 884. The original idea seems to have been that Asser would spend part of each year in Alfred's court and the rest in his own institution. However Asser rather soon moved to England permanently. He became bishop of Sherborne, probably during the lifetime of Alfred, and died in 909 or 910; his death is noted in the <u>Anglo-Saxon Chronicle</u>. Asser is described as a monk of S. David's. There were many establishments in Wales named for S. David, but the most important was probably the cathedral at Mynyw or Memevia. I

shall assume that this was Asser's residence until he moved to England.

The work usually called the Life of Alfred the Great [Asser, ca. 893] is one of the main sources for the history of the time. Among other interesting items, the work in its present form contains the famous story of the burned cakes, although the editor thinks that it is a later interpolation. The story seems to have been invented by the same type of mind that invented the story of George Washington and the cherry tree.

Brut [ca. 1282]. According to ab Ithel, the editor of Brut y Tywysogion, Geoffrey of Monmouth[†] wrote that he (Geoffrey) entrusted the writing of the chronicle of the princes of Wales to Caradog of Llancarvan, and that he entrusted the writing about the English kings to Henry of Huntingdon and William of Malmesbury. I have not attempted to verify that Geoffrey wrote this, but the man who could invent the history of Britain could certainly have had the gall to arrogate to himself the responsibility for the writings of Henry and William, who were far better historians than he was. I know nothing of this Caradog. The work in question is Brut y Tywysogion or The Chronicle of the Princes [ca. 1282]. The fact that Geoffrey knew something about it, as well as other evidence, suggests that Brut was compiled about 1150 and then continued contemporaneously.

The editor of Brut thinks that it was compiled in the monasteries of either Strata Florida or Conway. Strata Florida was in the diocese of S. David's. The monastery at Conway presumably dates from 1185; earlier entries will be located at S. David's.

[†] His preface suggested to me that ab Ithel accepted the work of Geoffrey as genuine history, but perhaps I wrong him.

Cambriae [ca. 1288]. The Annales Cambriae are probably the oldest Welsh annals. There
are three main texts. The A text was apparently
compiled by one person about the year 950; the editor says that it may have been a translation into
Latin of a previously compiled Welsh text. It contains no records of solar or lunar eclipses. It
used a "Welsh era" for dating purposes; this era
is the year 445.[†]

The B text was compiled about 1286, apparently at Strata Florida in the diocese of S.
David's. The first part of B is based upon A, but
it clearly had some independent material, such as
the eclipse records. B has the only eclipse records before 1140.

The C text was compiled about 1288. The
editor says that C drew heavily from B until 1203,
after which point it is largely independent. If
this is so, C's treatment of eclipses is peculiar.
The compiler of C had no prejudice against eclipses,
since C records the solar eclipses of 1140 Mar 20,
1178 Sep 13, and 1288 Apr 2. However, C contains
none of the earlier eclipses from B; one wonders
why the compiler of C would have chosen to pass
over all of them. Perhaps C did not use B directly,
but only some of the independent material that B
used. The editor did not locate C; I shall take
it to be at the center of the western coast of
Wales.

Giraldus Cambrensis [ca. 1200]. Giraldus
was a Welshman, otherwise known as Gerald de Barri,
who went to Ireland with John's army on two occasions, and who wrote a history of John's Irish campaigns. One visit was for about a year about 1183,
and the other occasion was apparently later. Giral-

[†]The editor (ab Ithel) says that the era is 444.
However, in most of his conversions of dates to
the common era, he actually used the "Welsh era"
as 445. Further, 445 fits events better than
444.

TABLE VII.3

REFERENCES TO SOLAR ECLIPSES FOUND IN
MEDIEVAL WELSH SOURCES

Source	Date	Conclusions
Asser	878 Oct 29	Probably independent
Brut	807 Feb 11	Probably independent
	1140 Mar 20	Independent
	1185 May 1	Not original but cannot trace; see Margan
	1191 Jun 23	Probably not independent
Cambriae	447 Dec 23 ?	Possibly independent
	624 Jun 21 ?	Unlikely identification
	807 Feb 11	Probably not independent
	1140 Mar 20	Probably not independent
	1178 Sep 13	Probably independent
	1288 Apr 2	Independent
Margan	1133 Aug 2	From Wm. of Malmesbury
	1140 Mar 20	From Wm. of Malmesbury
	1178 Sep 13	Probably independent
	1185 May 1	Not original but cannot trace
	1191 Jun 23	Possibly independent

dus has no useful records of eclipses. He says of
the period near 1180: "In a period of three years
about that time, there were three eclipses of the
sun, not total however but partial." Usually we
could identify a triennium with 3 solar eclipses,
even partial ones. The eclipses referred to are
probably those of 1178 Sep 13, 1180 Jan 28, and
1181 Jul 13. However, we cannot rule out those of
1185 May 1, 1186 Apr 21, and 1187 Sep 24, which
were all recorded in British sources. Even if we
could identify the eclipses, we could not use Giraldus' statement, because we do not know whether
the eclipses were seen in Wales or Ireland.

Margan [ca. 1232]. The monastery of
Margan, in Glamorganshire, was founded in 1147
and immediately set to work to get its own annals.
For most of the early period, the Annales de Margan
are based upon the writings of William of Malmesbury (see Section VI.1). Later annals are often
independent, but the eclipse records of 1185 May 1
and 1191 Jun 23 suggest relations with Waverley
(see Section VI.1).

Table VII.3 contains a list of eclipses
reported in Welsh sources, arranged by source.

447 Dec 23 B,W(?). Reference: Cambriae
[ca. 1288]. The first entry in Annales Cambriae,
for the year 4 of the Welsh era, is: "A day as
dark as night." This corresponds to 448 if I am
correct that the year 1 corresponds to 445. The
next entry is for the year 9 (453); it records
that Leo, bishop of Rome, changed the date of Easter. The Easter of 455 was one of the dates for
which the rules of the Roman church and of the
other churches gave different results. Easter
would have been on April 24 according to the usual
rules, but the Roman church (see Section II.2)
wanted April 17. Pope Leo began arguing for April
17 as early as 452. Thus the chronology of Cambriae is fairly close even in the earliest entries.

The entry does not necessarily refer to

an eclipse. If it does refer to an eclipse, the
only reasonable identification is 447 Dec 23.
This date does not necessarily mean that the year
448 was an error according to the annalists' cus-
toms; there were many conventions that put the date
we call 447 Dec 23 into what we call the following
year. We cannot have high confidence that the observation was made in Wales, although it is plausible that such a record could have been made there.
There was probably church activity in Wales before
this time, both as a result of Christian efforts
under the Roman Empire and of missionary efforts
from Ireland. Some Welsh monasteries may have been
founded already. However, the entry is the kind
of brief notice that could easily be taken from
one place to another, and the observation may have
been made on the continent.

I shall calculate this eclipse on the
assumption that it was observed at S. David's,
but with the low reliability of 0.05. I shall use
a standard deviation of the magnitude of 0.1.

Schove [1954], in his interesting study
of early British references to eclipses, quotes
what is probably a reference to the same eclipse
in some Irish annals that I have not found. The
eclipse was also reported from Spain by Hydatius
[ca. 468]. There is little resemblance between
the records, and it is unlikely that Hydatius is
the source for the Welsh record. See the record
447 Dec 23 E,Sp in Chapter XIV.

624 Jun 21 B,W(?). Reference: Cambriae
[ca. 1288]. The notice: "The sun was obscured"
(or "darkened") occurs under the year 180, which
is presumably 624. I saw no event within 20 years
of this that could be dated independently, so there
is no way to check the accuracy of the chronology.
It is unlikely that the magnitude of the eclipse
of 624 Jun 21 exceeded about 0.5 or 0.6 in Wales.
The large eclipses nearest in time are those of
619 Mar 21 and 634 Jun 1. The penumbral eclipse of
625 Jun 10 cannot be ruled out, at least not without detailed computation. This eclipse seems to

be unidentifiable.

807 Feb 11 B,W. References: Brut [ca.
1282] and Cambriae [ca. 1288]. The cited translation of Brut has, for the year 807: "And there
was an eclipse of the sun." The same source has
under the year 810: "Eight hundred and ten was
the year of Christ when the moon turned black on
Christmas day; . ." Cambriae notes eclipses of
the sun and of the moon under the years 807 and
810 respectively. Brut is certainly closer to
the original record for the lunar eclipse and thus
it is probably closer for both records.

There was a lunar eclipse on Christmas
Day, 809 Dec 25, that was readily visible in Wales.
It is almost certain that this is the lunar eclipse
in Brut and that the year was given as 810 not in
error but in accordance with local custom. Thus
the chronology of Brut seems to be accurate for
this period of time. The solar eclipse of 807 Feb
11 did occur in the year stated in Brut. Since it
was probably the largest eclipse near that time,
it is probable that this is the correct identification of the solar eclipse in Brut, and that both
the solar and lunar eclipses were correctly recorded. However, the partial eclipse of 809 Jul 16
(but not the eclipse of 807 Feb 11) was recorded
in the Anglo-Saxon Chronicle (see the record 809
Jul 16 B,E in Section VI.3), and this fact weakens
our confidence in the identification.

I shall assume that the entries in Brut
and in Cambriae amount to a single record of the
eclipse of 807 Feb 11, with the low reliability of
0.2. Place: S. David's. Standard deviation of
the magnitude: 0.1.

878 Oct 29 B,W. Reference: Asser [ca.
893]. Asser gives, under the year 879: "This
year there was an eclipse of the sun between nones
and vespers, but closer to nones." Since Asser
did not move to Alfred's court until 884, I shall
assume that Asser made this observation while he

was still at S. David's. However, it is possible
that Asser took this notice from English records,
although it certainly did not come from the Anglo-
Saxon Chronicle (see Section VI.3). Assignment
of the record to S. David's is weakened by the
fact that this eclipse is not recorded in either
Brut or Cambriae, which both have close connections
with S. David's. Hence I shall give low reliability to the assignment of place. Reliability: 0.2.
Place: S. David's. Standard deviation of the magnitude: 0.1.

The English sources Florence of Worcester,
Roger of Hovenden, and S. Neots (see Chapter VI)
reproduced this record from Asser, including the
error of putting it in the year 879. The editor
of Florence said that it was the eclipse of 880
Mar 14, but this is impossible. The eclipse of
880 Mar 14, as seen in England, would have been
at most a small eclipse at sunset, in clear disagreement with the record.

The editor of Asser says that this report came from Fulda in Germany. The eclipse of
878 Oct 29 is indeed reported in Marianus Scotus
[ca. 1082] and Fuldenses [ca. 901], which are
from Fulda; see the discussions in Section XI.1.
The report in Asser does not seem to have anything
in common with the reports in either Marianus Scotus or Fuldenses, however. If Asser obtained his
information from a Fulda source, it must have been
from one that I have not discovered.

1133 Aug 2. Margan [ca. 1232] is the
only Welsh source I saw that records this eclipse.
The account in Margan is clearly taken from William
of Malmesbury (see the record 1133 Aug 2b B,E in
Section VI.3).

1140 Mar 20 B,W. Reference: Brut [ca.
1282]. Brut has, under the year 1137: "And there
was an eclipse of the sun on the 12th day of the
calends of April." The text is in Gaelic, and the
translation is from the cited reference. The 12th

-211-

calends April is Mar 21, not Mar 20; however, it
would have been easy to drop a stroke from the
Roman numeral xiii. The originality of the year[†]
indicates strongly that this record is independent
of any others that have been found. There is no
other plausible identification. Reliability: 0.5.
Place: S. David's. Standard deviation of the magnitude: 0.1.

The B text of Cambriae has "Sol obscurator"
under the year 1140 while the C text has "Sol patitur
eclipsim" under 1139. This strengthens the doubt already expressed about use of B by the compiler of C.
These brief notices could be independent records,
but it is safest not to use them. The record of
this eclipse in Margan is not a copy of the record
in William of Malmesbury (1140 Mar 20b B,E in Section VI.3). However, many phrases in Margan are
identical with those in William and others seem to
be close paraphrases. Hence the record in Margan
is almost surely not independent.

This is the last eclipse in the B text.

1178 Sep 13a B,W. Reference: Cambriae
[ca. 1288]. The reference has: "An eclipse of the
sun happened on the ides of September (= Sep 13)
across[‡] midday." This is in the C text only. The
source does not state the year, but the editor assigned the notice to 1180. However, the death of
Louis VII, which was in 1180, was assigned by the
editor to 1181, and the death of Pope Alexander
III was assigned to 1183 rather than to the correct year of 1181 [Mercati, 1947]. Hence the editor's assignments seem to be uniformly too late,

---

[†] Several Danish records (see Section XIII.1) give
1137 for the year of this eclipse. There is
little similarity in wording, so it is unlikely
that the records are related.

[‡] The text has contra meridiem. "Across" is not a
good translation, but other possibilities seem
even worse.

and the reference is surely to the eclipse of 1178
Sep 13. The record sounds independent, but it is
short and could be a paraphrase of some other record.
Hence I shall lower the reliability. Reliability:
0.2. Place: Western Wales. Standard deviation of
the magnitude: 0.1.

1178 Sep 13b B,W. Reference: Margan
[ca. 1232]. Margan has, under the year 1178: "There
happened a solar eclipse on the ides of September
after the 6th hour of the day." This notice also
sounds independent but could be a paraphrase. Reliability: 0.2. Place: Margan. Standard deviation of the magnitude: 0.1.

1185 May 1a B,W and 1185 May 1b B,W.
References: Brut [ca. 1282] and Margan [ca. 1232].
Both sources give the year as 1185. Brut has, in
the translation cited: "In that year, on the calends of May, the sun changed its colour, and some
said there was an eclipse of it." Margan has:
"There was an eclipse on the calends of May about
the 9th hour; the sun after the eclipse appeared
red like the color of blood in a wonderful manner."
The English source Waverley (see the record 1185
May 1d B,E in Section VI.3) also referred to a
change in the appearance of the sun after the eclipse. There could have been a physical effect of
some sort that prompted the remarks about color,
but most records of the eclipse do not mention it.
This raises a suspicion of a connection between
the records, even though the words used are different.

I assigned a reliability of only 0.2 to
the record in Waverley because of this problem.
I shall consider the Brut and Margan references
to be different records, to be assigned the designations 1185 May 1a B,W and 1185 May 1b B,W, respectively, with a reliability of 0.2 for each.
Places: S. David's for Brut and Margan for Margan.
Standard deviation of the magnitude: 0.1 for each.

1191 Jun 23 B,W. Reference: Margan
[ca. 1232]. The source has, under the year 1191:
"An eclipse of the sun happened on the Lord's day
on the eve of S. John the Baptist about the 6th
hour." The record in Waverley (1191 Jun 23c B,E
in Section VI.3) is the same except that it does
not have "on the eve of S. John the Baptist"; in
the original Latin, the words and the word order
are otherwise identical. This strengthens the suspicion already raised of a relation between the two
sources. Hence I shall give a lowered reliability
to the Margan record, as I did to the Waverley
record.

Reliability: 0.2. Place: Margan.
Standard deviation of the magnitude: 0.1.

Brut under this year has the simple notice: "And an eclipse of the sun occurred."
This notice could have been taken from many sources.

1288 Apr 2 B,W. Reference: Cambriae
[ca. 1288]. Only the C text continues to this
year, and the last entry in the text is: "On the
2nd day of April at the 9th hour there was an eclipse of the sun in the upper part of the sun and
it was seen entering the body of the sun, and the
sun was seen to have two horns lifted upward, and
it lasted thus until the hour of vespers." All
the information in this record, including the orientation of the horns, is accurate except possibly
the hours. Because the eclipse path almost crosses
the North Pole, it is hard to judge times from the
chart in Oppolzer [1887], but the hours look too
late. Further, the English report of this eclipse
(1288 Apr 2 B,E in Section VI.3) puts the eclipse
at the 6th hour.

The record is contemporaneous and reads
like an eye witness account. Reliability: 1.
Place: Western Wales. Magnitude: 2/3 with standard deviation of 1/6; path to the north of Wales.

CHAPTER VIII

RECORDS OF SOLAR ECLIPSES FROM BELGIUM AND THE NETHERLANDS

1. The Sources

I have found only one source [Egmuṅdani, ca. 1315] from The Netherlands that contains records of solar eclipses, and it is closely related to the Belgian source Sigebertus [ca. 1111]. Therefore I have included Egmundani in the same subdivision as the Belgian sources.

Belgium was part of the Roman Empire and much if not all of it has been at least officially under the Christian church from early times. In spite of this, I have found no early records of eclipses. The earliest Belgian record that is at all likely to be independent is of the eclipse of 968 Dec 22, and the first one that can be used with high confidence is of the eclipse of 1009 Mar 29.[†]

Belgium illustrates the difficulty that may arise in trying to establish a provenance or subdivision. In many ways it would be desirable to have all of Charlemagne's empire as one subdivision, but in other ways it would be desirable to have the Holy Roman Empire as one division and medieval France as another. Many Belgian sources are related to nearby French sources, and a division based upon medieval Flanders would help with these. Over a span as long as that covered in this study, there was much change in cultural and political associations, and no classification is satisfactory over the entire span. A classification based upon modern political boundaries, though arbitrary, is at least fairly definite.

[†] It is ironic that Ginzel [1882 and 1884] identified this, wrongly in my opinion, as 1010 Mar 18.

Anselmus [ca. 1048]. There are two Belgian chroniclers named Anselmus; this is the earlier one. His work is called Gesta Episcoporum Tungrensium, Traiectensium, et Leodiensium. Leodiensium is from the Latin form of the city called Liège in French and Luik in Flemish. Tongeren or Tongres is a town about 20 kilometers to the north. The bishopric was transferred from Tongres to Liège in about 930. I do not know what Traiectensium refers to, but it must be some place in the neighborhood. Anselmus was born near Cologne about the end of the 10th century. He became a canon at Liège in 1041 and later became dean of the cathedral (S. Lambert's). He was still living on 1056 Mar 3.

Anselmus wrote this work in two parts and originally carried it to 1048. Later he discovered a chronicle by Herigerus of Lobbes and substituted it for his first part. He also revised his second part and brought it down to 1056. The edition used here seems to be a mixture. The first part is that attributed to Herigerus, which runs from patristic times to about 665. It does not contain anything of interest for this work and therefore I have not cited it separately. The second part runs only to 1048 and is thus presumably from the "first edition" of Anselmus.

Anselmus [ca. 1135]. This Anselmus is later than the other one, but we know little else about him. He was a monk at two other places before he moved to Gemblacum (Gembloux) in 1099. He is known as Anselmus Gemblacensis, and he wrote one of the many continuations of the Chronica of Sigebertus Gemblacensis. See the discussion of Sigebertus [ca. 1111] below.

Blandinienses [ca. 1292]. I have not found out for sure where these annals come from; they are probably from a monastery. Ginzel [1882 and 1884] says that they are from Blandigny. Ginzel was quite conscientious about giving the coordinates of places, but he unfortunately forgot to do so for Blandigny. Perhaps he was not able to

find the place either. The editor does not give
the place, nor does any standard reference that I
have consulted. A large-scale road map lists
Blanden near Louvain. It also lists Blandain west
of Tournai almost on the French border, in the middle of medieval Flanders. Since the annals have
many references to Flemish affairs, I shall assume
that Blandain is the place.

The editor says that the annals are written in one hand down to about 1060, with occasional
insertions in hands of the 12th and 13th centuries.
In other words, the annals were compiled in their
present form about 1060, but later readers interpo-
lated interesting items that they knew about. The
annals are presumably contemporaneous and hence
independent from about 1060 on. In fact, the eclipse records seem to be independent of other known
sources from 1000 on, although they cannot be contemporaneous.

Egmundani [ca. 1315]. The Annales Egmundani were kept at the place now called Egmond aan
Zee on the North Sea coast in the province of North
Holland. The editor says that a monastery was
founded here in the 10th century that lasted until
it was ruined in civil wars about 1573. The early
part of the annals consists mostly of extracts from
Sigebertus [ca. 1111], with some local material
mixed in. From 1112 on, the annals seem to be independent. They run continuously from here through
1207, in various and presumably contemporaneous
hands of the 12th and 13th centuries. After 1207
they fell out of ordinary use. There are later
entries only for the years 1248, 1250, 1282, and
1315. We can imagine that later monks happened
upon the annals in the monastery library from time
to time and made contemporary notations. The last
entry, for 1315, is a Latin quatrain which says
that there has not been such a dying of men and
cattle from hunger since the time of Noah.

Egmundani has brief notes about solar eclipses under the years 1009 and 1023; these notices
are almost surely taken from Sigebertus. Ginzel

[1882 and 1884] took these to be records of the eclipses of 1010 Mar 18 and 1023 Jan 24. I show in
the next section that these identifications are not
safe. I believe that the correct identifications
are 1009 Mar 29 and 1018 Apr 18.

Formoselenses [ca. 1136]. In the town
of Vormezeele, near Ieper (Ypres) in West Flanders,
was a church or monastery (I am not sure which),
dedicated to Mary. The Annales Formoselenses were
compiled there about 1096, on the basis of Blandinienses [ca. 1292][†] and other sources. Someone
with a hand of the 12th century made a few insertions in this compilation. The annals were then
kept up, in various contemporaneous hands, through
1136, when they were abandoned.

Thus we may take the annals to be independent for the period 1097 - 1136, in the absence
of contradictory information. By an unfortunate
accident, Ginzel did not notice the eclipses in the
independent part of the annals. He noticed only
those in the earlier derivative part.

Fossenses [ca. 1384]. There are two towns
named Fosse in Belgium. One is about 30 kilometers
southeast of Liège. The other is about 60 kilometers west-southwest of Liège and slightly west of
Namur.

The annals called Annales Leodienses
(see Leodienses [ca. 1121]) were kept in Liège until 1121 when they were moved, for a reason unknown
to me, to the monastery of S. Foillanus in Fosse.
This information is supplied by the editor, who unfortunately forgot to mention which Fosse. Ginzel

---

[†] The dates look to be inconsistent with this conclusion of the editor of Formoselenses. I imagine
the explanation is that someone at Vormezeele borrowed the Annales Blandinienses at an early stage
and copied them, or perhaps that he went to Blandain to copy them.

[1882 and 1884] gives the coordinates of the more
westerly Fosse, but gives no basis for the choice
that I noticed. He may have given it only because
it is by far the larger one and the easier one to
find on a map. In the absence of other information
I shall follow Ginzel. Luckily the two Fosse's
have almost exactly the same latitude and differ in
longitude by only about 1°.

The annals have a number of references
to affairs of Flanders. The Fosse chosen is somewhat closer to the center of Flanders than is the
other one. This gives a slight confirmation to
the choice made. The annals resume with the year
1123, following the move, and were presumably kept
fairly regularly until their close.

The editor dates the close as 1389 rather
than 1384 as I have given it. As it happens, the
last entry in the annals affords an opportunity to
apply astronomical chronology. Since this entry
is well outside the time period of the main study,
I shall deal with it here instead of with the other
eclipse reports in the next section.

The last entry reads as follows: "1389.
On the 17th day of August the sun suffered an eclipse from the 13th hour to the 14th hour or beyond,
and this to the extent of about a third part, according to what could be judged by the spectators,
with the rest remaining in the light." Note that
the writer is using the modern style rather than
the Roman style for the day of the month. Also,
since there can be no 13th or 14th hour in the old
style, he must also be using the modern style for
giving the hour; that is, he is giving the number
of equal hours from midnight.

There was no eclipse meeting the conditions stated, or anything close to them, in 1389.
However there was an eclipse that would have been
partial in Belgium around 1 or 2 hours after noon,
according to the chart in Oppolzer [1887], on 1384
Aug 17. Thus I think it is almost certain that the
last entry should be dated 1384 rather than 1389.
In Roman numerals, the difference is the accidental

insertion of a "V", or perhaps the accidental reading of "V" as "X". Since the preceding entry was
for 1376, no disturbance of the other dates is
needed in order to make this change. I find only
one problem with this interpretation: According to
the eclipse chart, the eclipsed part of the sun
should have been about 4/5 rather than 1/3.

     Galbertus Brugensis [ca. 1128]. Charles,
son of King Knut IV of Denmark, was Count of Flanders
from 1119 to 1127, when he was murdered at Bruges.
Galbertus was one of several contemporaries of Charles who wrote biographies of him. Galbertus continues his account of affairs after the death of Charles to the middle of 1128, so that this is probably
the year in which he wrote the work.

     Gemblacensis [ca. 1148]. This is one of
many continuations of Sigebertus [ca. 1111], which
is discussed below in this section. This one was
kept at Gemblacum by three anonymous writers, judging from the hands. The entries seem to be contemporaneous and authentic.

     Lambertus Parvus [ca. 1193]. The name
Lambertus in connection with Liège is somewhat confusing, but I believe that I have matters straight.
S. Lambertus was murdered at Liège on either 705
or 706 Sep 17; Sep 17 is S. Lambert's day. The
cathedral at Liège was named S. Lambert's, presumably in honor of this Lambertus, until it was destroyed in the wars in 1794; the church of S. Paul
was then declared the cathedral. Lambertus Parvus
says that the glorious martyr Lambertus was taken
away (delatus) in 1141; this is presumably a different Lambertus. Lambertus Parvus is apparently
still another person and not a church or monastery.
He took his annals for the period down to about
1175 from sources such as Sigebertus [ca. 1111],
the earlier parts of Fossenses [ca. 1384], and
particularly S. Iacobi [ca. 1174; see below].
Lambertus Parvus can be considered an independent
source only from about 1175 until its close in 1193.

The relation between these dates and the last year
of S. Iacobi may not be an accident.

    Laubacenses [ca. 926]. The Annales Lau-
bacenses are associated with the town of Lobbes
(Laubacum in Latin), which is about 15 kilometers
southwest of Charleroi. However, the editor says
that he does not really know that the annals were
written there. There was an abbey at Laubacum
that had close connections with the cathedral at
Liège. Until 960 the bishop at Liège was always
the abbot of Laubacum. The Annales Laubacenses
can be divided into five main parts. The first
three of these run, respectively, 707-740, 741-770,
and 771-791. The editor calls these three different sections of the first part. Through 791 there
is an entry for each year, but from then on the
annals are intermittent. What the editor calls
the second part (really the fourth) runs 796-885
and what he calls the third part (really the fifth)
runs 887-926.

    Luckily the uncertainty about where the
annals were written does not concern us. They
never became independent, at least as far as eclipse data are concerned. The eclipses in the
first three parts are noted merely by "Eclypsis
solis", which is too brief to let us determine the
source. The fourth part has eclipse records taken
from the Laurissenses family of annals (see Laurissenses [ca. 829] in Section XI.1). The last part
is so closely connected with the Swiss source
Sangallensis [ca. 926; see Section IX.3] that the
editor prints the last part of Laubacenses in parallel columns with Sangallensis, and the reader
should note that the last years of the two sets
of annals are the same. Even though Laubacenses
has no independent eclipse reports, its reports
must be discussed in detail in order to establish
the fact.

    Leodienses [ca. 1121]. I have not found
out much about these annals except that they were
kept in Liège until 1121. Then they were taken to

Fosse where they became the Annales Fossenses already discussed [Fossenses, ca. 1384]. I mentioned
in connection with Laubacenses above that the bishop of Liège and the abbot of Laubacum were the same
person until 960. There is a set of annals besides
Laubacenses that was kept at Laubacum; it is called
Annales Laubienses. Leodienses and Annales Laubienses have much in common for a long time, and for
the period 939-960 they are virtually identical.
It is not necessary to use the Annales Laubienses
and I have not included it in the list of references.

The editor says that the entries are in
various hands from about 1000 on. This usually,
but not always, indicates independence and I shall
take these annals to be independent after 1000.
It is really not very important whether the place
is Lobbes or Liège; the places are separated in
longitude by little more than 1°.

It is interesting that the chronicler, at
the year 866, has inserted a list of the Byzantine
emperors from 866 "to our time, that is, the year
1000 from the Incarnation." This shows an interest
in Byzantine affairs that antedates the Crusades
by a century.

Rupertus [ca. 1095]. Rupertus' birth
date is unknown. His death occurred on either
1127 or 1135 Mar 4, depending on the source. He
was a product of the monastery of Leodiensis S.
Laurentii. He became a priest and ultimately abbot of the monastery of S. Heribertus Tuitiensis.[†]
This reference is the Chronicon S. Laurentii Leodiensis, a chronicle of the affairs of his original
monastery. Since he finished it either 32 or 40
years before his death, it must be one of his early works. He also wrote on theology, and he wrote
a life of S. Heribertus. This chronicle is not
considered to be particularly original.

[†]This is the modern Deutz near Cologne.

S. Iacobi [ca. 1174]. The Annales S.
Iacobi Leodiensis were kept at the monastery of S.
Jacob in Liège, on the island there, according to
the editor of the edition cited. The monastery
was founded by a Bishop Baldric in 1016. According to the annals it was begun on 1016 Apr 25 and
the crypt was dedicated on 1016 Sep 6. The annals
follow a rather standard pattern. One unknown individual compiled a set of annals for the year 1 -
1055; I ignored this part of the annals so far as
the main purposes of this work are concerned.
From 1056 the annals were kept by many writers in
presumably independent and contemporaneous fashion.

Sigebertus [ca. 1111]. Sigebertus Gemblacensis (Sigebert of Gembloux, ca. 1030-1112) entered the monastery at Gemblacum at an early age.
Later he moved to Metz in Lorraine but about 1071
he returned to Gemblacum and spent the rest of his
life there. He wrote on many subjects, including
the lively question of papal prerogatives, on which
he sided with the emperors against the popes. The
work involved here is his Chronica, a history of
the world from 381 to his own time. Since he
brought it down to 1111, and since he died in 1112,
he was probably keeping it reasonably up-to-date
in its later portions. It was probably considered
to be a continuation of Jerome's translation and
continuation of Eusebius' Chronicon (see Sections
III.3 and XV.3).

The Chronica contains many mistakes and
has little material not found elsewhere. Nonetheless it was highly popular in its own time. It
was widely copied and had many continuations. Two
of these (Anselmus [ca. 1135] and Gemblacensis
[ca. 1148]) are mentioned in this section, and
others will be found among the French sources in
Chapter X. Still other chroniclers and historians,
from Britain to Italy, did not copy it but used it
as a basic reference or source.

The chronology is frequently quite bad.
Some eclipses have been misplaced so badly that they
have been identified wrongly. However the mistakes

TABLE VIII.1

REFERENCES TO SOLAR ECLIPSES FOUND IN MEDIEVAL SOURCES FROM BELGIUM AND THE NETHERLANDS

Source	Date	Conclusions
Anselmus [ca. 1048]	968 Dec 22	Independent; place uncertain
Anselmus [ca. 1135]	1133 Aug 2	Original
Blandinienses	418 Jul 19	Copied from unknown source; entered under 951
	664 May 1	Not safe to use
	807 Feb 11	Copied from French source
	812 May 14	Copied from German source
	1009 Mar 29	Independent
	1023 Jan 24	Independent
	1033 Jun 29	Independent
	1037 Apr 18	Independent
	1039 Aug 22	Independent
	1093 Sep 23	Original
	1133 Aug 2	Original
	1140 Mar 20	Original
	1187 Sep 4	Original
	1207 Feb 28	Original
	1263 Aug 5	Original

TABLE VIII.1 (Continued)

Source	Date	Conclusions
Egmundani	1133 Aug 2	Original
	1147 Oct 26	Original
Formoselenses	1109 May 31	Original
	1118 May 22	Original
Fossenses	1133 Aug 2	Original
	1191 Jun 23	Original
	1241 Oct 6	Original
Galbertus	1124 Aug 11	Original
Gemblacensis	1140 Mar 20	Original
	1147 Oct 26	Original
Lambertus Parvus	1191 Jun 23	Probably original
Laubacenses	733 Aug 14	Probably not original
	760 Aug 15 ?	Cannot identify safely
	764 Jun 4	Probably not independent
	807 Feb 11	Not independent
	810 Jul 5	Not independent
	810 Nov 30	Not independent
	813 May 4 ?	Cannot identify safely
	891 Aug 8	Copied from Swiss source
Leodienses	1009 Mar 29	Original

TABLE VIII.1 (Concluded)

Source	Date	Conclusions
Rupertus	968 Dec 22	Probably not independent
S. Iacobi	1133 Aug 2	Original
	1147 Oct 26	Original
Sigebertus	592 Mar 19	Probably taken from a French source
	693 Oct 5	Copied from German source
	828 Jul 15 ?	Copied from German source; identification uncertain
	878 Oct 29	Probably from German source
	934 Apr 16 ?	Probably not an eclipse; date used for reference only
	939 Jul 19	From unknown source; identification uncertain
	1009 Mar 29	Probably from Leodienses, usually identified as 1010 Mar 18
	1018 Apr 18	Independent; usually identified as 1023 Jan 24
	1033 Jun 29	Either from Blandinienses or French source
	1039 Aug 22 ?	From French source; cannot be identified

are not always the fault of Sigebertus; sometimes
they were already present in the sources that he
used.

2. Discussions of the Records of Solar Eclipses

In this section I shall discuss the records of solar eclipses found in the sources just
described. Many of the sources are completely derivative in their early portions, and I have omitted
the records found in these portions unless they present particular problems or unless they come from
unknown sources. I have also omitted some but not
all of the later records that are dependent upon
known sources.

All the reports discussed in this section
are listed in Table VIII.1. The table includes
the source, the date of the eclipse, and a brief
statement about the conclusions regarding the report.

418 Jul 19 E,BN. Reference: Blandinienses [ca. 1292, p. 25]. This report poses two
interesting problems. First, its dating error is
533 years. This is the second largest dating error that I have found,[†] and we must explain how
such an error is possible. Second, this record
was written about two centuries before the source
from which it is apparently derived, and we must
also explain this anomaly. I have already mentioned
this record in Sections IV.3 and V.1.

Blandinienses has, under the year 951:
"The sun was eclipsed at the 3rd hour on the 14th
calends August (= Jul 19) and a star appeared
glowing (ardens) in the east until the month of
September." The Danish source Esromenses [ca. 1307]

[†] It is interesting that the largest error occurs
for this same eclipse. It is 550 years, for the
record 418 Jul 19b E,I in Section XII.2.

(see Section XIII.1) says the same except for having "calends of September" rather than "month of September".[†] However Esromenses has its entry under the year 416. I imagine that the glowing star was a comet rather than a nova and for simplicity I shall word the rest of the discussion on this assumption; no necessary part of the argument depends upon the assumption.

　　　　Identification of the eclipse is complicated by the fact that there was an umbral eclipse on 939 Jul 19 that would have been maximum in Belgium at about the 3rd hour.[‡] There was also a penumbral

---

[†] Blandinienses has: "Sol defecit hora 3. 14. Kal. Aug. et apparuit stella ab oriente ardens usque ad mensem Septembris." Esromenses differs only in having ardens before ab oriente and in having Kalendas Septembris rather than ad mensem Septembris. A phenomenon that lasted to the calends of September necessarily lasted to the month of September and vice versa, so there is no significant difference between the forms for the purpose of this entry.

[‡] This phenomenon calls attention to an eclipse series that I do not recall seeing mentioned. Oppolzer's Canon [Oppolzer, 1887] contains the members of the series in the following table:

Date (in Julian Calendar)	Julian Day Number	Time of Conjunction hours	min.	Latitude of noon point
-1145 Jul 19	1 303 046	15	15	21°
- 624 Jul 19	1 493 342	2	18	31°
- 103 Jul 19	1 683 637	9	20	38°
418 Jul 19	1 873 932	11	04	42°
939 Jul 19	2 064 227	8	53	45°
1460 Jul 18	2 254 522	5	27	50°
1981 Jul 18	2 444 817	3	53	54°

The interval between these is 190 295 days except between the first two, for which the interval is 1 day greater. All but the first are on the 6th

eclipse on 958 Jul 19 that would have been maximum in Belgium at about the same hour. There was no eclipse that fits the record in 951.

There are several independent records of the eclipse of 418 Jul 19, and several of them mention the appearance of a comet at about the same time.[†] There are several independent records of the eclipse of 939 Jul 19, and none of them mentions a comet. There are no independent records of the eclipse of 958 Jul 19, so we cannot speak with certainty about it. However, since the record in Blandinienses is almost identical with a known record of the eclipse of 418 Jul 19, we can conclude with high confidence that Blandinienses has copied a record of the latter eclipse provided that we can give a convincing explanation of the year 951.

In Section III.2 we noted that many medieval annals were kept on Easter tables. In Section II.8 we noticed the "great Easter cycle" of 532 years; after 532 years, the tables of the ecclesiastical moon and of Easter repeat in the Julian calendar. Some annals were kept on Easter tables[‡] arranged according to the "great Easter cycle". In such annals, it is difficult to know whether an event occurred in, say, the year 418 or whether it occurred 532 years later, in 950.

Therefore I conclude the following: The compiler of Blandinienses had a source that was prepared with Easter cycle data used to identify the years. This source had a record of the eclipse of 418 Jul 19 written in such a way that it

---

feria. Since the latitude of the noon point has moved only 33° during the interval of more than three millenia, the series can probably be extended outside the range of the Canon.

[†] See the records 418 Jul 19a E,I in Section XII.2 and 418 Jul 19a M,B in Section XV.4.

[‡] Or alternatively, on tables in which the only identification of the year was by means of "Easter table" data.

would be hard to tell whether the year was 418 or
419. The compiler made a mistake of one great cycle and further assigned the eclipse to the later
year of two possible ones. Thus he put the eclipse
in the year 419 + 532 = 951.

In spite of the near-identity of wording,
Esromenses cannot be the source for Blandinienses,
because the compiling of Esromenses was not begun
until after the compiling of Blandinienses had
stopped. Further, Blandinienses cannot be the
source of Esromenses because of the discrepancy in
year. The explanation must be that both used a
common source but in a different way.

I suspect that the common source came
from the Swiss monastery of S. Gall. In connection
with the Italian record 418 Jul 19a E,I in Section
XII.2, I shall point out that one copy of the record
was made at S. Gall in the 9th century. We know
from the discussion of Laubacenses in the preceding
section that there were close relations between
Belgian monasteries and S. Gall in the 10th century. I suspect that an unknown copy made at S. Gall,
differing slightly from the known copy, was taken
to Belgium , where it served as the source for this
puzzling record in Blandinienses. I shall not use
the record in Blandinienses.

592 Mar 19(?). Since I do not believe
that this is an independent report, I do not give
it a designating symbol. Sigebertus [ca. 1111]
has the following under the year 596: "The sun
was diminished from early morning to midday down
to the third part of itself."[†] "Fredegarius Scholasticus" [ca. 641] from Burgundy (see Chapter X) has
the following entry: "In the 32nd year of Guntram's
reign from early morning to midday the sun was
diminished so that hardly a third part of itself

[†] Sol a mane usque ad meridiem minoratus est usque
ad tertiam sui partem.

appeared."† It seems almost certain to me that
Sigebertus took his report from "Fredegarius". The
phrase "minoratus est" is not common in descriptions of eclipses, and the statements that a third
of the sun was not eclipsed could hardly be independent.

I have used the date 592 Mar 19 even
though Sigebertus put the eclipse in 596. I shall
discuss Fredegarius' report in more detail under
the designation 592 Mar 19 E,F in Section X.3, and
I shall give the reason for the identification
there. It is barely possible that Fredegarius'
record, perhaps by way of Sigebertus, is the origin
for the Irish records designated as 591 Sep 23 B,I
and 592 Mar 19 B,I in Section VII.1.

664 May 1 E,BN. Reference: Blandinienses [ca. 1292]. This appears under the year 664:
"In this year an eclipse of the sun happened." I
have found no other report of this eclipse with
this wording. However the annalist could easily
have compressed a longer report into these few
words, and it is not safe to use this report. Hence
I give it a reliability of 0.

693 Oct 5. Reference: Sigebertus [ca.
1111]. Sigebertus has an entry under the year 695
that is identical with an entry under the year 687
in the German source Xantenses [ca. 873]; see Section XI.2. Sigebertus is surely not original here
and I shall not give his report a designation.

Laubacenses [ca. 926] has the words "Eclipse of the sun" under the years 733, 761, and
764. These brief notes are not safe to use as
independent reports.

---

† In Section XIII: Anno xxxij. regni Guntchramni,
ita a mane usque ad mediam diem sol minoratus est,
ut tertia pars ex ipso vix appareret.

Laubacenses also has the words "Eclipse of the sun twice" under 807. It is conceivable that these words refer to the penumbral eclipse of 806 Sep 16 and to the umbral eclipse of 807 Feb 11, but I doubt it. It is possible that the annalist used the same source as Blandinienses [ca. 1292] (see below) for 807. This record has one eclipse of the sun and two of the moon in 807, and the annalist might have become confused. I think it is more likely that the annalist accidentally put down under 807 an entry that he could have found in many places for 810, which does record correctly two solar eclipses for 810. I think that this is the most likely explanation, even though Laubacenses also notes "Eclipse of the sun twice" under 811.

Blandinienses [ca. 1292] has a record of the solar eclipse of 807 Feb 11 and the lunar eclipses of 807 Feb 26 and 807 Aug 21, all in one sentence under 807. It erroneously gives the date of the last eclipse as Aug 22 (11th calends September). The sentence is identical with one found in the older French source Sithienses [ca. 823] (see Chapter X). Therefore I shall not use the report from Blandinienses.

812 May 14. Reference: Blandinienses [ca. 1292]. The source has, under 812: "An eclipse of the sun occurred on the ides of May." I believe that this is a condensation of the following entry in the German source Laurissenses [ca. 829][†]: "There was an eclipse of the sun on the ides of May after midday."

813 May 4 E,BN(?). Reference: Laubacenses [ca. 926]. The phrase "an eclipse of the sun" occurs under 813. There was an umbral eclipse on 813 May 4 that was recorded in Constantinople (see

---

[†] See Chapter XI. Laurissenses is itself probably not independent here.

Section XV.4). It would have had a magnitude of
perhaps 0.75 in Belgium at sunrise and should have
been observed there, weather permitting. In spite
of these facts, I do not believe that the phrase
refers to the eclipse of 813 May 4. I have found
no reference to this eclipse outside of Constantinople, so I think that the sky must have been overcast that morning in western Europe. However,
there are several records of the umbral eclipse of
812 May 14, and I imagine that the annalist at Laubacum put his note under the wrong year. Hence I
shall not use this record.

        Sigebertus [ca. 1111] has a brief note
about a lunar and a solar eclipse under the year
833. Xantenses [ca. 873] has almost the identical
entry. We also suspected that Sigebertus copied
Xantenses for the eclipse of 693 Oct 5. In Section
XI.2 I shall show that Xantenses paraphrased a report in Fuldenses [ca. 901] that was already too
garbled to permit identification of the eclipse.

        878 Oct 29 E,BN. Reference: Sigebertus
[ca. 1111]. Sigebertus has the following under
880: "Sol hora diei nona ita obscuratus est, ut
stellae in caelo apparerent."† The German source
Fuldenses [ca. 901] has the following under 878:
". . . sol quoque in 4. Kal. Novembris post horam
nonam ita obscuratus est per dimidiam horam, ut
stellae in coelo apparerent, . . ."‡ The entry in
Sigebertus is a paraphrase of this that can be

---

† The sun was so darkened at the ninth hour that
stars appeared in the sky.

‡ . . . the sun also on the 4th calends November
(= Oct 29) after nones was darkened for half an
hour, so that stars appeared in the sky . . .
The sentence starts with a record of the lunar
eclipse of 878 Oct 15 and ends by saying that
everyone thought that night had come (during the
solar eclipse). See the record 878 Oct 29 E,G in
Section XI.2.

accomplished by omitting words. Thus I believe
that Sigebertus took his report from Fuldenses,
or perhaps from an intermediary source, in spite
of the difference in year. Hence I give Sigebertus' report a reliability of 0.

                891 Aug 8 E,BN. Reference: Laubacenses
[ca. 926]. This is the last eclipse report in Laubacenses, which has the following under 891: "Stella
cometis. Eclipsis."[†] The Swiss source Sangallensis
[ca. 926] has, under the same year: "Stella cometis.
Eclypsis solis." The part of Laubacenses from 887
to 926 is almost identical with the corresponding
part of Sangallensis, and we could conceivably have
trouble deciding which is the original. Luckily
the word "solis" in Sangallensis that is not in
Laubacenses makes it almost certain that Sangallensis is the original. With no more information than
Laubacenses provides, the annalist at S. Gall could
not have told the kind of eclipse. Hence I give a
reliability of 0 to the report in Laubacenses.

                934 Apr 16 E,BN(?). I doubt that this
is a report of an eclipse, and I have used a date
only for purposes of designation. Under the year
937, Sigebertus has: "This year a marvel took place.
The sun was darkened in a clear sky; indeed it sent
blood-red rays through the windows of our homes.
The mountain, where later King Henry was buried,
vomited forth flames in many places." Sigebertus
has the death of Henry (the Fowler, king 919-936)
and the election of Pope Leo VII (pope, 936 Jan 3
to 939 Jul 13) in the same year, which is thus 936
rather than 937. Henry was buried in the town of
Quedlinburg in Saxony.

                The event sounds less like an eclipse
than like a volcanic eruption with a consequent
darkening of the sky by ash. If this is so, it
is surprising; I had not thought that there were
active volcanoes north of the Alps in historic

[†] The star comet. An eclipse.

times. I doubt that the event has been correctly
reported. It first appears in the German source
Widukindus [ca. 973],[t] which also lists a number
of other portents for that year. However, the
event does not appear in the annals written in
Quedlinburg [Quedlinburgenses, ca. 1025], which
are discussed in Section XI.1. It seems unlikely
that these annals would fail to report such a remarkable event happening in the immediate vicinity.
Widukindus probably misplaced an account of an
eruption that happened somewhere else.

        939 Jul 19 E,BN. Reference: Sigebertus
[ca. 1111]. Sigebertus has under 944: "The sun
suffered a horrible eclipse on the 6th feria in
the 3rd hour of the day." The only possible eclipse in 944 was the penumbral eclipse of 944 Sep
20, which was on the 6th feria. I think it would
have been maximum earlier than the 3rd hour, as
well as I can judge from Oppolzer [1887], but the
3rd hour cannot be excluded without detailed calculation. However the eclipse of 944 Sep 20 should
not have been large enough to merit the adjective
"horrible". The eclipse of 939 Jul 19 was also
on the 6th feria and should have been large in
western Europe at about the 3rd hour. 939 Jul 19
is thus a more plausible identification; Sigebertus
makes other errors of 5 years in his chronology.

        The report reads as if it is independent
of other reports of the eclipse of 939 Jul 19.
However it is certainly not original. Since Sigebertus used many sources from all parts of Europe,
we have no good way to know the point of origin.
Thus I shall give this report a reliability of 0.

        968 Dec 22a E,BN. Reference: Anselmus
[ca. 1048, p. 202]. This is the earliest eclipse
report that can be actually assigned to Belgium

---

[t] I did not find any independent reports of solar
eclipses in Widukindus, so I do not discuss it
in Section XI.1.

with any degree of plausibility. Anselmus in the 11th century is writing about the affairs of the bishops of Liège and surrounding places during the reign of Otto I, Holy Roman Emperor 936-973. Otto spent the years 966 to 972 in Italy. Anselmus writes the following: "With his army spread far and wide throughout the country, the emperor was holding the territory of Calabria when behold an unexpected eclipse of the sun struck all with great fear. And indeed I heard (or, I have heard) Bishop Wazo explaining, as he saw the sun becoming eclipsed little by little, when much light still remained, that as it developed, the day would become dark as at evening and as at night the cattle would hasten to the stables and the bees to the hives." The discussion of eclipses goes on for many more sentences.

We must first decide the significance of the first person and of the appearance of Bishop Wazo in the narrative. The only Bishop Wazo I can find connected with Liège is the one who became bishop in 1041 and who was therefore bishop at the time Anselmus was writing. Thus the "I" probably does not refer to a person who saw the eclipse of 968 Dec 22 whose account Anselmus is copying. Instead, it is probable that Anselmus and Wazo saw an eclipse together and that Wazo described what would happen when an eclipse is large enough. If so, this report is implicitly a report of two eclipses, but we cannot identify the second one.

If we accept this conclusion, it is hard to decide where the eclipse was seen. Was it seen by the emperor's party in Italy, or was it seen "meanwhile back at the ranch" and recorded in a Belgian source that Anselmus used? It is amusing that a Byzantine record of this same eclipse[†] poses the same problem, but with regard to the Byzantine emperor and Constantinople.

With the Byzantine record, I conclude from the wording that the eclipse was seen at home

---

[†] The record 968 Dec 22a M,B in Section XV.4.

in Constantinople and not with the emperor. Here
I can obtain no strong feeling. The context is
mostly concerned with affairs in Belgium, and there
are references, for example, to edicts of the emperor received from Calabria. This suggests that
most of Anselmus' material was local but does not
prove that the eclipse report was; it could have
come from Calabria along with the statement that
the emperor was occupying it.

I shall assign Liège as the place for
this observation, but with low confidence. I shall
express the low confidence by using a reliability
of only 0.05 for the record. I shall take the eclipse to be total on the basis of Anselmus' discussion. As I have said, I think that Anselmus
and Bishop Wazo saw a large or total eclipse together in the 11th century, but there is no safe way
to identify which one.

968 Dec 22b E,BN. Reference: Rupertus
[ca. 1095, p. 263]. Rupertus wrote a chronicle of
the monastery at Liège. He described the eclipse
of 968 Dec 22 thus: ". . . in the middle of the
day in the south an eclipse of the sun suddenly
occurred, . . ." He goes on to comment about the
terror that struck all the spectators.

The editor of Rupertus says that Rupertus
derived this account from the account in Anselmus
just discussed, and this may well be so. If it is,
Rupertus made a quite free paraphrase; I did not
notice any similarity in the wording. Both Rupertus
and Anselmus, however, emphasize the terror that
accompanied the eclipse. Such an emphasis is fairly uncommon. This makes it likely either that
Rupertus was inspired by Anselmus or that both
writers drew inspiration from the same source.

I incline toward the belief that both
Anselmus and Rupertus used a local Belgian (or
nearby French) source, but I can advance no strong
arguments either for or against this belief. Accordingly, I adopt it but assign the low reliability
of 0.05 to the report when used as a report from

Liège. This would be equivalent to using the reports in Anselmus and Rupertus as a single report
with a reliability of 0.1, except that we must use
0.1 rather than 0 for the standard deviation of the
magnitude in Rupertus' report.

1009 Mar 29a E,BN. Reference: Leodienses [ca. 1121]. This source has under 1009: "An
eclipse of the sun took place about the 2nd hour
of the day." The only entry nearby that I noticed
that might help with the chronology was the death
of Pope Silvester II, which was correctly put in
1003. I can find no eclipse within several years
of this entry that occurred at about the 2nd hour
except the penumbral eclipse of 1009 Mar 29, and
I see little doubt about the identification. We
are now at about the time when Leodienses is presumably independent and contemporaneous, so I give
this report a reliability of 1. Place: Liège.
Standard deviation of the magnitude: 0.1.

This report has been copied by Sigebertus
[ca. 1111]. Ginzel [1882 and 1884] noticed it
there but not in Leodienses, and he identified it
as the umbral eclipse of 1010 Mar 18. According
to Oppolzer [1887], the path of that eclipse started
at sunrise near the Galapagos Islands and ended at
sunset near Venice. It could not have occurred at
the 2nd hour anywhere in Europe. In contrast, both
the year and the hour given are correct for 1009
Mar 29.

1009 Mar 29b E,BN. Reference: Blandinienses [ca. 1292]. This source has the eclipse
under 1008, where it says: "Countess Mathilda
died. And there happened an eclipse of the sun
on the 4th calends April (= Mar 29) on the 28th
day of the current moon." Since the day stated is
Mar 29, this is surely the eclipse of 1009 Mar 29.
However, this was the 29th day of the ecclesiastical
moon.

This report and the preceding one are

independent of each other. Since this one is not
contemporaneous,[†] I give it a reliability of only
0.5. Place: Blandain. Standard deviation of the
magnitude: 0.1.

                1018 Apr 18 E,BN. Reference: Sigebertus
[ca. 1111]. Under 1023 Sigebertus writes: "In
Easter[‡] an eclipse of the sun occurred." Egmundani
[ca. 1315] has used this record, with the same year.
Ginzel takes both records to be of the eclipse of
1023 Jan 24, apparently because of the year. However, I do not think "in Easter" could possibly be
used to refer to Jan 24, so I think that either the
year is wrong or that "in Easter" is wrong.

                Sigebertus' chronology is a thorough mess
along here. He has the accession of Pope John
(XIX) in 1025 rather than 1024, so this is not too
bad. However he has the accession of Pope Benedict
(VIII) in 1017 rather than 1012 as it should be.
In 1009 he has the accession of Pope Leo; the near-
est possibilities are Leo VIII on 963 Dec 6 and
Leo IX on 1049 Feb 12.

                Thus there is a good chance that the
year is wrong, and an error of 5 years is reasonable. It is possible but much less likely that
"in Easter" is wrong. I can find no eclipse within many years that one could describe as being "in
Easter" except the eclipse of 1018 Apr 18; Easter
was on 1018 Apr 6. I shall take the place of the
observation to be Gembloux. However Sigebertus
is by no means contemporaneous here, and the record could have come from elsewhere, so I give the
record the low reliability of 0.1. Standard deviation of the magnitude: 0.1.

                1023 Jan 24 E,BN. Reference: Blandin-
ienses [ca. 1292]. "This year occurred an eclipse

[†]The source was compiled by one person to 1060.

[‡]Sigebertus has "In pascha" which I think means
"near Easter" or "in the Easter season".

of the sun on the 9th calends February (= Jan 24) at the 6th hour. Duke Godfrey died." The year is given as 1023. I have not tried to learn about Duke Godfrey. The 6th hour is reasonable on the basis of the chart in Oppolzer. We are not quite to the point where Blandinienses can be considered contemporaneous, so I give this a reliability of 0.5. Place: Blandain. Standard deviation of the magnitude: 0.1.

1033 Jun 29 E,BN. Reference: Blandinienses [ca. 1292], which has under 1033: "This year occurred an eclipse of the sun on the 3rd calends July (= Jun 29) at the 6th hour." The year given is 1033 and the eclipse should have been maximum in Belgium at about the 6th hour. Reliability: 0.5. Place: Blandain. Standard deviation of the magnitude: 0.1.

Sigebertus reports the eclipse in identical words except that he says "about meridiem" instead of "at the 6th hour". Thus Sigebertus may have used Blandinienses as his source. The editor says that Sigebertus used a source called "Bald." for this record. I believe that this is the source better called Cameracensium [ca. 1042], for reasons that I shall discuss in Section X.2. Neither edition of Cameracensium that I consulted has a record of the eclipse of 1033 Jun 29, but the editions are incomplete, and the editor's attribution may be correct.

1037 Apr 18 E,BN. Reference: Blandinienses [ca. 1292]. The reference has, under 1037: "This year occurred an eclipse of the sun on the 14th calends May (= Apr 18), on the 28th day from the new moon, at the first hour of the day." The day stated was the 28th day from the ecclesiastical new moon. "First hour" looks a little early but not enough to endanger the identification. This report can still not be considered contemporaneous. Reliability: 0.5. Place: Blandain. Standard deviation of the magnitude: 0.1.

1039 Aug 22 E,BN.   Reference: Blandinienses [ca. 1292], which has under 1039:   "This
year occured an eclipse of the sun on the 11th
calends September (= Aug 22) from the 3rd hour to
the 6th, on the 28th of the moon." The statement
is correct, although the hour looks somewhat early.
Reliability: 0.5. Place: Blandain. Standard
deviation of the magnitude: 0.1.

    Sigebertus has an eclipse record under
1039 that is identical with a passage from the
French source Cameracensium mentioned above. This
source will be cited and discussed in Chapter X.
Ginzel identified the record as 1039 Aug 22 where
it occurred in Sigebertus but as 1037 Apr 18 in
Cameracensium. I do not believe that either identification is acceptable; see Section X.3.

    1093 Sep 23 E,BN.   Reference: Blandin-
ienses [ca. 1292]. The source has: "On the 9th
calends October (= Sep 23), on the 6th feria, from
the 3rd to the 8th hour, on the 28th of the moon,
Indiction 1, epact 1, there was an eclipse of the
sun." It is interesting to note that the eclipse
records in Blandinienses from 1009 through 1039
were presumably compiled by one person and that
their wording followed a set formula. By now we
have a different person who uses quite different
wording and who seems to be a fanatic on chronology.
The entry occurs under 1093, so the date stated is
1093 Sep 23, which was on the 6th feria. From the
"3rd to the 6th hour" looks more reasonable than
what is said, but there may have been a typographical error. I have not checked the other calendrical information. Blandinienses is now contemporaneous. Reliability: 1. Place: Blandain.
Standard deviation of the magnitude: 0.1.

    1109 May 31 E,BN.   Reference: Formosel-
enses [ca. 1136], which has, under 1109:   "There
was an eclipse of the sun about the end of May."
Although this is a brief entry, I believe that it
is original. The source is being compiled contem-

poraneously here. I have found only one other
record[†] of this eclipse and I see no suggestion
that the sources are related. Reliability: 1.
Place: Vormezeele. Standard deviation of the
magnitude: 0.1.

        1118 May 22 E,BN. Reference: <u>Formosel-</u>
<u>enses</u> [ca. 1136]. Under 1118 the source has:
"There was an eclipse of the sun on the 11th calends June (= May 22)." The date is correct and
this is the only record of this eclipse that I
have found. Reliability: 1. Place: Vormezeele.
Standard deviation of the magnitude: 0.1.

        1124 Aug 11 E,BN. Reference: <u>Galbertus</u>
[ca. 1128, p. 562]. In Section 2 of his <u>life of</u>
Count Charles of Flanders, Galbertus writes: "In
the year 1124 from the Incarnation of the Lord,
in the month of August, there appeared to all the
inhabitants of the earth an eclipse in the body of
the sun about the 9th hour, . . . but not the whole
sun however was obscured, but in part, as that
cloud[‡] crossed the circle of the sun, crossing
from east to west. . ." He goes on to say that
this warns of future famine and deadly peril.

        In spite of the monitory aspect, we can
safely take this as a genuine account, apparently
by an eye witness albeit a superstitious one. It
is interesting to note that Galbertus has the wrong
sense of the direction of travel of the eclipsing
agent. It would be easy to make this mistake inadvertently in writing, however.*

[†] From <u>Hildesheimenses</u> in Chapter XI.

[‡] Galbertus did not seem to be aware of the causes
of eclipses. He twice used the word <u>nebula</u> in referring to the agent that darkened the sun.

* Or he might have watched by reflection. See Section IV.5.

Reliability: 1. Place: Belgium.
Magnitude: 2/3, with a standard deviation of 1/6;
path assumed north of Belgium.

1133 Aug 2a E,BN. Reference: <u>Anselmus</u>
[ca. 1135]. Anselmus wrote under 1133: "On the
4th nones August (= Aug 2), the 27th of the moon,
in the middle of the day there was an eclipse of
the sun for about half an hour, and stars were
seen in the sky." The date given was the 27th of
the ecclesiastical moon. This is a contemporaneous
account written while Anselmus was living at Gembloux. Reliability: 1. Place: Gembloux. Standard deviation of the magnitude: 0.01, because of
the reference to stars.

1133 Aug 2b E,BN. Reference: <u>Fossenses</u>
[ca. 1384]. This is another contemporaneous account that says: "On the fourth nones August at
midday the sun was obscured to such an extent that
stars appeared in the sky." Reliability: 1.
Place: Fosse. Standard deviation of the magnitude:
0.01.

1133 Aug 2c E,BN. Reference: <u>Blandin-
ienses</u> [ca. 1292]. This source has, under 1132:
"An eclipse of the sun occurred on the 4th nones
August." There was no entry for 1133. Reliability: 1. Place: Blandain. Standard deviation of
the magnitude: 0.1.

1133 Aug 2d E,BN. Reference: <u>S. Iacobi</u>
[ca. 1174]. This source has: "1133. On the 4th
nones August there was a horrible eclipse of the
sun about midday, with the moon appearing against
the disk of the sun, and with stars shining brightly on account of the great darkness, and men were
struck with terror." This was written at Liège
and sounds like the account of an eye witness. Reliability: 1. Place: Liège. Standard deviation
of the magnitude: 0.01.

        Lambertus Parvus [ca. 1193] copied this
but put it under 1138.

        1133 Aug 2e E,BN. Reference: Egmundani
[ca. 1315], under 1133: "In that same year on the
4th nones August, at the 6th hour of the day the
sun was darkened, not covered by clouds but eclipsed, and there was such darkness that day was turned into night, and stars with no overcast to hinder
appeared as if at night. And yet, since it was the
8th of the moon,[†] no one can consider that this
was a natural eclipse of the sun, for there cannot
be an eclipse when the moon is so many days from
it but only when it is in the ecliptic and the sun
and moon are in conjunction." The chronicler seems
to know, not only that a solar eclipse must occur
at the new moon, but also that one can occur only
when the moon and sun are near the moon's node. He
is puzzled because the (ecclesiastical) moon is
days away from the sun at the time of this eclipse.
It has not occurred to him to doubt the accuracy of
his tables.

        This was written at a time when Egmundani
seems generally to be original and contemporaneous,
and this report sounds original. I think we can
assume that a description this vivid is meant to
apply to a total eclipse. Reliability: 1. Place:
Egmond aan Zee. Standard deviation of the magnitude:
0.

        1140 Mar 20a E,BN. Reference: Gembla-
censis [ca. 1148]. This contemporaneous source
has, under 1140: "An eclipse of the sun took place
on the 4th nones April[‡] as evening was coming on

[†]
I imagine that the chronicler meant to write
"28th"; actually it was the 27th. The chronicler's
point is correct, however, for either the 27th or
28th of the moon.

[‡]
In the abbreviations common to annalists, the correct date would probably have been written XIII.
Kal. Apr. in the original manuscript. Careless
writing would change the numeral to IIII. The
change of Kal. to Non. is probably just someone's
inadvertence.

and the sky was clear and the sun was seen to be
surrounded by a hideous misty cover." In spite of
the error in date, the identification is secure.
It seems to me that the annalist is trying to describe the corona, and hence that the eclipse was
total. Reliability: 1. Place: Gembloux. Standard deviation of the magnitude: 0.

         1140 Mar 20b E,BN. Reference: Blandin-
ienses [ca. 1292]. This reads, under 1140: "An
eclipse of the sun of a kind unknown before occurred
at the time of the vernal equinox, and there was
unusual thunder on Easter eve." The phraseology
makes this sound bigger than the average eclipse,
but I do not think that we can safely infer totality. I shall use 0.02 for the standard deviation
of the magnitude. Reliability: 1. Place:
Blandain.

         1147 Oct 26a E,BN. Reference: Gembla-
censis [ca. 1148]. Under the year 1147 the reference has: "An eclipse of the sun took place
about the 3rd hour of the day, on the 7th calends
November (= Oct 26). And the sun did not regain
its whole light until it was surrounded by a misty
cover,[†] and many saw the stars prominent in the
sky." The date is correct and the hour is reasonable. It seems to me here, as with the earlier
report 1140 Mar 20a E,BN from the same source,
that the annalist is trying to describe the corona
(see Section V.1) and hence a total eclipse. There
is a problem with this interpretation for this eclipse: The eclipse of 1147 Oct 26 was annular and

---

[†] Both here and in the preceding report from this
source (1140 Mar 20a E,BN), the annalist at Gembloux used the phrase "eclyptica caligine". I do
not know whether he meant eclyptica in the sense
of a circle (around the sun), or in the sense of
"cover" that goes with the meaning of an eclipse,
or perhaps in some other sense. The possible meanings do not seem to affect the general sense of
the passage.

not total, according to Oppolzer [1887]. Until I
can make detailed calculations, I cannot tell whether
my interpretation of the passage is wrong or whether
the approximations that Oppolzer had to make caused
him to mistake a total for an annular eclipse. For
working purposes, I shall take the eclipse to be
total. Reliability of this record: 1. Place:
Gembloux. Standard deviation of the magnitude: 0.

1147 Oct 26b E,BN. Reference: S. Iacobi
[ca. 1174], which has: "1147. On the 7th calends
November, on the 1st feria,† an eclipse of the sun
from nearly the 3rd hour of the day until the full
6th hour; the reddening of the sun is to show how
much Christian blood will be shed." I imagine
that the shedding of Christian blood refers to the
disastrous Second Crusade.

I do not know whether to ascribe the "reddening" to the annalist's observation or to his obvious dramatic flair. Reddening occurs during an
eclipse when most of the disk is obscured, so that
a large fraction of the light reaching us comes
from the cool outer layers of the sun. Thus reddening suggests but does not prove a fairly large
eclipse. I shall adopt a compromise position and
use 0.05 for the standard deviation of the magnitude. Reliability: 1. Place: Liège.

1147 Oct 26c E,BN. Reference: Egmundani
[ca. 1315]. This report is prosaic by comparison
with the other reports of this eclipse. It says:
"In the year 1147 the sun was obscured from the 3rd
hour of the day to the 6th in the month of October."
Reliability: 1. Place: Egmond aan Zee. Standard
deviation of the magnitude: 0.1.

1187 Sep 4 E,BN. Reference: Blandinienses [ca. 1292]. Under 1187 appears: "An

---

†The source has "in dominica". I have assumed that
die was omitted by accident.

eclipse of the sun on the 2nd nones September
(= Sep 4). Reliability: 1. Place: Blandain.
Standard deviation of the magnitude: 0.1.

          1191 Jun 23a E,BN. Reference: Fossenses
[ca. 1384], which has under 1191: "This year occurred an eclipse of the sun on the eve of S. John
the Baptist." S. John's is Jun 24, so the date is
correct. Reliability: 1. Place: Fosse. Standard deviation of the magnitude: 0.1.

          1191 Jun 23b E,BN. Reference: Lambertus
Parvus [ca. 1193]. Lambertus writes under 1191:
"An eclipse of the sun and a flood of waters in
Liège outside the castle took place." In the same
year, Lambertus says that Richard and Henry prepared an expedition to the Holy Land. Actually
Richard (Coeur-de-Lion) and Philip started in 1190
and arrived with their expedition in 1191, but this
does not help much either to confirm or impeach the
chronology. For the next year, Lambertus writes
that the Paschal moon was seen on the 17th calends
April (= Mar 16) instead of the 14th (= Mar 19).
1192 Mar 19 was the ecclesiastical new moon whereas the new moon would have been visible in Belgium
on 1192 Mar 16, so the chronology is confirmed.
Reliability: 1. Place: Liège. Standard deviation of the magnitude: 0.1.

          1207 Feb 28 E,BN. Reference: Blandin-
ienses [ca. 1292], which says: "On the 2nd calends March (= Feb 28) an eclipse of the sun from
the 3rd hour of the day to the 9th, on the 28th
of the moon." He has this under 1206. The duration of the eclipse sounds long, but it is possible since he has rounded to the hour and since
an hour of the day would have been only about 54
minutes. Thus the duration could have been considerably less than 6 of our hours. I do not
know what convention the annalist was using for
the beginning of the year, so I do not know if
1206 is an error or not. In any case, the identification of the eclipse is secure. Reliability:

1. Place: Blandain. Standard deviation of the
magnitude: 0.1.

1241 Oct 6 E,BN. Reference: Fossenses
[ca. 1384]. The annals of Fosse say under the
year 1241: "In this year the sun went into an eclipse from midday to vespers on the octave of S.
Lambert." The hours of the day look reasonable.
I believe that few people outside of Belgium would
date an eclipse by means of S. Lambert's day, and
at that the annalist is wrong. S. Lambert's day
is Sep 17 and the octave is the 8th day thereafter,
inclusive, or Sep 24. I find no eclipse on Sep 24
for any nearby year. However, Oct 6 is the octave
of S. Michael's day (the archangel). I imagine
that the annalist meant to write S. Michael but,
because of his associations, inadvertently wrote
S. Lambert instead. Reliability: 1. Place:
Fosse. Standard deviation of the magnitude: 0.1.

1263 Aug 5 E,BN. Reference: Blandinienses [ca. 1292]. The reference has: "This
year there was an eclipse of the sun on the nones
of August (= Aug 5) at about the 10th hour" under
the year 1263. Reliability: 1. Place: Blandain.
Standard deviation of the magnitude: 0.1.

CHAPTER IX

RECORDS OF SOLAR ECLIPSES FROM
CENTRAL EUROPE

1. Records from Austria

In the term "Central Europe" I include
the areas of present-day Austria, Czechoslovakia,
and Switzerland. The annals from these three
different areas have little to do with each other,
although the records from Austria and Switzerland
have some independent relations with German and
other records. There are probably enough records
from Austria to warrant setting up a separate subdivision for them, but there are not enough from
Czechoslovakia and Switzerland. Hence I have included these areas with Austria for convenience.

I shall devote a separate section of
this chapter to each of the three areas, taking
them in the order that they occur alphabetically
in English. As usual, I shall for each area first
discuss the sources, then tabulate the eclipse
records by source, and finally discuss the individual records chronologically in order to establish the characteristics of the observations.

Admuntenses [ca. 1250]. Admont is in
central Austria roughly 60 kilometers south of
Linz. The Annales Admuntenses are closely related
to Mellicenses [ca. 1564], and more information
will be found in the discussion of that source.
Admuntenses remains almost entirely dependent upon
Mellicenses until about 1130, although its report
of the eclipse of 1093 Sep 23 seems to have an
independent origin. Admuntenses is often cited as
going to 1425. However, the only entry after 1250
is for 1425, when there were earthquakes on Aug 22
and Aug 24. Hence I think that it is more accurate
for our purposes to use 1250 as the date for cita-

tion.

Cremifanensis [ca. 1216]. This is another in the family of annals connected with Mellicenses [ca. 1564]. These seem to be independent
from about 1139 until their close. They are described as being from a monastery on the river
Krems in Austria. Unfortunately there are two
rivers named Krems in Austria. One runs into the
Danube from the north at the city of Krems about
50 kilometers upstream from Vienna. The other is
in central Austria and has a place named Kremsmünster on it. Since Cremifanensis is an inflected form of the Latinization of Kremsmünster, I
take the latter to be the place; Ginzel [1882 and
1884] reached the same conclusion.

Garstensis [ca. 1257]. This is still
another in the Mellicenses family. The only independent report of an eclipse that I noticed in it
was that of 1241 Oct 6; earlier records are derivative. Since this has no independent records before 1200 I shall not use this source; I have mentioned it only for completeness. Garsten is about
20 kilometers east of Kremsmünster.

Gottwicensis [ca. 1054]. Gottweih is a
place about 30 kilometers northeast of Melk. The
editor of Herimannus [1054][†] uses the symbols 4
and 4b to distinguish two of the many copies of
Herimannus that are known; these copies were made
at Gottweih. The differences between codices 4
and 4b and other copies exceed mere differences
in spelling or similar editorial matters. There
are items of information in these that do not appear in other copies, but that do appear in Mellicenses [ca. 1564], which is discussed below.
This makes it appear that the person or persons
who prepared the Gottweih copies of Herimannus

[†] See Chapter XI for further discussion of this
source.

-250-

had access to a local document which, if not written at Gottweih, was probably at least prepared in the vicinity. We should remember that Gottweih and Melk are about 30 kilometers apart but about 500 kilometers from Reichenau where Herimannus lived and worked.

Thus I have postulated a source for which I have invented a name and approximate date for purposes of citation. It contains those parts of the Gottweih copies of Herimannus that do not appear in other copies; it may have had other information as well. It is clear that the compiler of Mellicenses used Gottwicensis as thus defined.

There are also what are called Annales Gottwicenses. These appear in v. IX of the Monumenta Germaniae Historica, Scriptores series, but I do not list them in the references because I have not used them.

I believe that it is reasonable to consider that the contents of Gottwicensis, as I have defined it, are not part of Herimannus.

Juvavenses [ca. 975]. Juvavum is the ancient name of Salzburg. Juvavum was part of the Roman Empire, and it has early Christian tombs that date, I believe, from about the 3rd century. Christianity lost its place in the later upheavals and was formally restored by the dedication of a church there in 774.[†] These annals are far from continuous. They are written in three different hands down to 835. After the entry for 835 many pages have been cut out, to the great detriment of history as the editor Pertz says. The only surviving entry after 835 is for 975.

Pertz says that the three hands before 835 are coeval. I was not sure whether he meant

---

[†] However there was a monastery in Salzburg at an earlier date. See the discussion of S. Rudberti [ca. 1168] below.

with each other or with events. I should have
thought that he meant the latter except that three
people could not be contemporaneous with events
from the middle of the 6th century to 835. I imagine that he means that the hands are coeval with
events after they pass some of the conventional
derivative early entries and that the annals may
be contemporaneous from perhaps 800 on.

These are called "major annals" by the
editor. There are also some "minor annals" that
did not contain anything I saw of interest for this
study.

Lambacensis [ca. 1283]. This is also a
member of the Mellicenses [ca. 1564] family. However it is such a distant member that we find it
more convenient to discuss it separately.. The
source Cremifanensis [ca. 1216] is one of the continuations of Mellicenses, kept at Kremsmünster.
A copy of Cremifanensis was taken to Lambach.
There it received auctaria, that is, insertions in
the already-compiled part. In addition it was continued to the year 1348. Since there are only two
entries after 1283, which record earthquakes in
1330 and 1348, it is clear that regular use of this
set of annals stopped in 1283, and I have used this
date in the citation. The auctaria are spread
through the annals, with the last one in 1197. For
our purposes it is not necessary to distinguish the
auctaria from the continuation, and I have cited
them together under the heading Auctarium et Continuatio.

Lambach is about 20 kilometers west of
Kremsmünster.

Mellicenses [ca. 1564]. Melk is a town
about 70 kilometers west of Vienna on the super-
highway to Salzburg. There was a monastery there
which, judging from the annals, was founded in the
11th century. In 1121 [Mellicenses, ca. 1564] one
Erchenfridus was made abbot. One of his first acts
was to prepare, or to have prepared, an epitome of

the chronicle of Herimannus [1054], with a continuation to his own time. More precisely, the continuation associated with Erchenfridus[†] runs through
1123.

Through 1054 the annals are almost entirely derivative. Although this source is called an
epitome of Herimannus, the annalist did not use
Herimannus as his sole source. In particular the
annals before 1054 contain several records of eclipses that are independent of Herimannus or any
other known source; I shall return to this point
in a moment.

After 1123 the work was continued at Melk
through 1564 by many different unknown annalists.
The editor says that it is rare for the same hand
to continue for more than 5 years. The inference
is that the annals are original and contemporaneous
for the entire period 1123-1564. Few original
sources cover such a great time span.

Many annalists regarded Mellicenses as
a prime source and continued it at a number of
places. One copy was apparently made and taken to
Admont in about 1139. The editor of Admuntenses
[ca. 1250] says that the "codices Garstensis,
Vorowensis, and Admuntensis" contain almost identical continuations of Mellicenses for some time
after 1139, and I deduce that Admuntenses is the
primary one of these. It seems to be original
actually from a few years before 1139, and it is
the only one of the three named codices that I
have used. Garstensis [ca. 1257] is the only other
one of these three that I shall even bother to cite;

[†] One should perhaps use Erchenfridus as the author
and cite the annals up to here as "Erchenfridus
[1123]". However, it is not certain that Erchenfridus wrote them himself, as compared with having one of his monks write them. Further, the
most important part of the work is that for years
after 1123. Accordingly, I follow tradition and
call this Annales Mellicenses.

the editor prints it as a distinguishable source beginning with 1182.

Cremifanensis [ca. 1216] is another continuation. It too began in 1139 but seems to have been separate from the others mentioned from its beginning. One wonders why 1139 was such a favorite year for copying Mellicenses. Lambacensis, as I have already mentioned, can be regarded as a continuation of Cremifanensis.

S. Rudberti [ca. 1286], which is discussed below, is described by its editor as being identical with Mellicenses in its early parts, but the editor did not mention which version of Mellicenses. Thus it can also be viewed as a continuation of Annales Mellicenses. The two sources that I cite as Zwetlensis 102 [ca. 1170] and Zwetlense 255 [ca. 1189] are still other continuations.

There is an interesting point about the relations between Mellicenses and Herimannus. The only reports of eclipses before 968 that appear both in Mellicenses and in any copy of Herimannus are those that appear in the Gottweih copies. In other words, it appears that the originator of Mellicenses ignored the early eclipses in the copy of Herimannus that he used. Instead, it appears that he used the local source of eclipse records that the Gottweih copier of Herimannus must have used. This is the source that I have postulated and called Gottwicensis [ca. 1054].

Reicherspergenses [ca. 1279]. This is an Austrian source that is not a continuation of Mellicenses. Reichersberg is a town on the Austrian bank of the Inn River. In AAO (p. 89) I inadvertently listed it as being in Germany, although I used its correct latitude and longitude. A monastery dedicated to S. Michael was founded in Reichersberg in 1084. Soon thereafter its possessions were seized by an "antibishop". The editor says that it was restored about 1122 with the aid of the "thriving monasteries of Saxony".

The editor (W. Wattenbach) says that there are several annalistic sources associated with the monastery and that it is not easy to separate them. I believe that the following represents Wattenbach's conclusions but I refer the interested reader to his discussion in the edition cited.

The core of the annals starts with what Wattenbach calls codex A. This is a set of annals that was probably compiled between 1157 and 1167 on the basis of many sources. This codex has often been attributed to a "presbyter" named Magnus, but Wattenbach believes that this attribution is incorrect. He points out that the codex is written in many hands, not in one hand only. Expanded versions of this codex were then prepared by several unknown people.

Magnus then took one of these expanded versions and used it (as well as other sources) in preparing a chronicle from the birth of Christ to his own time. This chronicle goes to 1195. One copy of this chronicle has a continuation that records the death of Magnus in 1195. From this we may guess that Magnus had brought his chronicle up to date and that his chronological labors were interrupted only by his death. The continuation goes rather regularly through 1279, and thereafter has only a single entry for the year 1355. Thus I have taken 1279 as the "characteristic date" of Reicherspergenses. What I am calling Annales Reicherspergenses is thus a complex work. It is the combination of the original annals compiled around 1167, an expansion by an unknown, a modification by Magnus for incorporation into his chronicle, and finally continuations by unknown persons to 1279. If the source had a large amount of data, I should have divided this composite into its components; with the small amount of data that it affords, it is not worth the trouble.

The work seems independent from the first part of the 12th century to the end of Magnus' contribution in 1195. Oddly, the continuation is mostly derivative until about 1223; thereafter it seems independent until its close.

S. Rudberti [ca. 1168]. The abbey of
S. Peter in Salzburg was founded about 700 by S.
Rudbertus, by whose name it was known for a long
time. It has given us two sets of annals, of
which this is the earlier and smaller. Wattenbach,
who edits both, calls this one S. Rudberti Salisburgensis Annales Breves. It is associated with
lists of kings, bishops, and other personages and
contains 38 entries for the years 991-1168 inclusive. Several entries from 991 through 1060 are
in the same hand, proving that they were all put
down at about the same time. Other entries in
various hands are mixed in with these, representing
events that other readers thought should be included. After 1060 the entries are in various
contemporaneous hands and are presumably original.

S. Rudberti [ca. 1286]. Wattenbach
calls these Annales S. Rudberti Salisburgensis.
They are the main annals of the abbey at Salzburg,
but they are not independent until a relatively
late date. In fact, they seem to be a continuation of Mellicenses, with which they are identical
in their earlier parts that I ignored. They cannot be considered independent until about 1177.
This is unfortunate, because we should expect a
monastery this old to have annals being kept at a
much earlier date. Perhaps earlier annals were
lost somehow and replaced by copying Mellicenses.
From 1177 through their close in 1286 these can be
considered as an original source.

Zwetlensis 102 [ca. 1170]. We have to
deal with two items from the library at Zwettl in
north central Austria. The editor has supplied
their catalog numbers (this one is Codex No. 102)
which I use to distinguish them. This is basically
a variant copy of the original form of Mellicenses
plus its continuation at Melk to 1139. From then
to its close it is an independent continuation of
Mellicenses. However, even in its earlier parts it

-256-

TABLE IX.1

REFERENCES TO SOLAR ECLIPSES FOUND IN
MEDIEVAL SOURCES FROM AUSTRIA

Source	Date	Conclusions
Admuntenses	1093 Sep 23	Independent of known sources
	1133 Aug 2	Original
	1153 Jan 26	Probably original
	1241 Oct 6	Original
Cremifanensis	1187 Sep 4	Perhaps original
	1191 Jun 23	Perhaps original
Gottwicensis	760 Aug 15	Independent of known sources
	764 Jun 4	Independent of known sources
	810 Jul 5	Perhaps independent
	810 Nov 30	Probably independent
Juvavenses	807 Feb 11	Probably independent
	810 Nov 30	Probably independent
Lambacensis	1153 Jan 26	Probably not independent
	1239 Jun 3	Original
	1241 Oct 6	Original
Mellicenses	346 Jun 6	Not independent but cannot locate source

TABLE IX.1 (Continued)

Source	Date	Conclusions
Mellicenses	418 Jul 19	Not independent but cannot locate source
	664 May 1	Probably copied from Bede
	760 Aug 15	Copied from Gottwicensis
	764 Jun 4	Copied from Gottwicensis
	787 Sep 16	Probably not independent; cannot locate source
	807 Feb 11	Independent of known sources
	810 Jul 5	Copied from Gottwicensis
	810 Nov 30	Copied from Gottwicensis
	840 May 5	Probably not independent
	878 Oct 29	Probably not independent
	968 Dec 22	Copied from Herimannus
	1033 Jun 29	Probably from German sources
	1039 Aug 22	Probably from German sources
	1133 Aug 2	Original
	1187 Sep 4	Original
	1191 Jun 23	Original
Reichersperg-enses	1133 Aug 2	Probably original
	1191 Jun 23	Original

TABLE IX.1 (Concluded)

Source	Date	Conclusions
Reichersperg-enses	1241 Oct 6	Original
S. Rudberti [ca. 1168]	1133 Aug 2	Original
S. Rudberti [ca. 1286]	1187 Sep 4	Original
	1191 Jun 23	Original
	1239 Jun 3	Original
	1241 Oct 6	Original
	1263 Aug 5	Original
	1267 May 25	Original
Zwetlensis 102	1133 Aug 2	Probably not independent
Zwetlense 255	1087 Aug 1	Probably independent
	1133 Aug 2	Probably not independent

often has information not found in Mellicenses.
In fact, to me it seemed to have little in common
with Mellicenses after 1125.

Zwetlense 255 [ca. 1189]. Someone at
Zwettl wanted to continue Zwetlensis 102, which is
itself a continuation of Mellicenses.† He was not
content to add entries at the end of Zwetlensis
102, however. He prepared a separate copy and inserted numerous entries, presumably of local interest, into the body of the annals. He then exe-
cuted his continuation at the end of this expanded
copy, which is Codex No. 255 in the Zwettl library.
For the period from 1170 to 1189 this seems to be
an original document. The auctaria for earlier
years have some information that is independent of
other known sources, but they cannot be considered
as original documents.

There is still a third continuation at
Zwettl. It is much later and I did not use it.
According to Ginzel [1882 and 1884], it has records
of the eclipses of 1261 Apr 1, 1267 May 25, and
1270 Mar 23. I have not tried to verify this information.

The discussions of the eclipse reports
follow. I give the customary list of the reports,
arranged by source, in Table IX.1. As usual, I
have not listed most of the reports from early
derivative portions of annals.

346 Jun 6 E,CE. Reference: Mellicenses
[ca. 1564] which says under 346: "There was an
eclipse of the sun, and an earthquake overthrew
many cities." I have not noticed another report
of this eclipse with just this wording. However
Mellicenses is not at all an independent source for
this part of history, and there are many places

---

† Which is itself a continuation of Herimannus
[1054].

from which this information could be derived.
This report should be ignored.

418 Jul 19 E,CE. Reference: Mellicenses
[ca. 1564]. The source says under 418: "There was
an eclipse of the sun on the 14th calends August
(= Jul 19); and a comet was seen until the month
of September." No other source that I have noticed
has all this information. However, we cannot infer
the place of the observation safely, and this report is best ignored.

Mellicenses has a note of the eclipse of
664 May 1, along with some information about the
debates between the Irish and English on the "Easter question". Mellicenses almost surely got this
information from Bede [734?].

760 Aug 15 E,CE. Reference: Gottwicensis [ca. 1054]. Under 760 we find: "There was
an eclipse of the sun on the 18th calends September
(= Aug 15) at about the 9th hour." The hour looks
reasonable for the eclipse of 760 Aug 15. I have
noticed only two other independent reports of this
eclipse, one from Constantinople (Section XV.4)
and the other from Belgium (Section VIII.2). Neither could have served as the source for Gottwicensis. Since we are at a time when an independent
report from Austria is possible, and since I have
found no other source from which this report could
be derived, I shall use it with low reliability.
Reliability: 0.1. Place: Gottweih. Standard
deviation of the magnitude: 0.1.

Mellicenses copied this, but put it under
759.

764 Jun 4 E,CE. Reference: Gottwicensis [ca. 1054], which has the following under
764: "The winter was severe this year, and there
was an eclipse of the sun on the 2nd nones June
(= Jun 4) at about the 6th hour." Under 764 the

-261-

German source Laurissenses [ca. 829; see Chapter XI] has: "This year there was an eclipse of the sun on the 2nd nones June at the 6th hour." The wording could be this close by accident; some ways of wording eclipse reports are almost conventional. Nothing else that I have noticed suggests a relation between Laurissenses and Gottwicensis. Hence I shall use the report in Gottwicensis with low reliability, as before. Reliability: 0.1. Place: Gottweih. Standard deviation of the magnitude: 0.1.

Mellicenses copied this under the year 762.

787 Sep 16 E,CE. Reference: Mellicenses [ca. 1564]. This report appears under the year 785: "There was an eclipse of the sun on the 15th calends October (= Sep 17) at about the 3rd hour." There is a peculiarity about reports of this eclipse. I have found seven reports of it from western Europe that give the date, and only one gives it correctly.[†] One has Oct 1. The others give the day of the year as Sep 17. It is hard to believe that five different annalists made the same mistake, yet the reports seem mostly independent of each other. It is possible but unlikely that the day of the month was calculated. 787 Sep 17 was the last day of an ecclesiastical lunar month (see Section V.5). Only this report gives the hour as the 3rd hour.

Since I can reach no firm decision, I shall use this record with low reliability. Reliability: 0.1. Place: Melk. Standard deviation of the magnitude: 0.1.

807 Feb 11a E,CE. Reference: Juvavenses [ca. 975]. Under 807 we find: "There was an ec-

---

[†] I am speaking here only of the day of the year, not of the year itself. For further irony, the only independent Byzantine record gives Sep 9.

lipse of the sun on the 2nd ides February (= Feb
12) from the 4th to the 7th hour." We can accept
this report in spite of the error of a day in the
date. No other report gives these hours of the
day. We are now close to the point, if we are not
already there, when Juvavenses becomes a contemporaneous and original document, so I give this a
reliability of 0.5. Place: Salzburg. Standard
deviation of the magnitude: 0.1.

                807 Feb 11b E,CE. Reference: Mellicen-
ses [ca. 1564]. This report appears under 806,
continuing the tendency to be early: "There was
an eclipse of the sun on the 3rd ides February
(= Feb 11) at about the 6th hour." This is the
only report that gives either this year or this
hour, so I shall use it as independent. Since the
source is still 3 centuries away from being contemporaneous, I shall use a reliability of 0.1. Place:
Melk. Standard deviation of the magnitude: 0.1.

                810 Jul 5 E,CE. Reference: Gottwic-
ensis [ca. 1054]. This record appears under 810:
"There was an eclipse of the sun on the 7th ides
June (= Jun 7) at about the 2nd hour." The nearest alternate identification is the eclipse of 809
Jul 16, which would be wrong for the month, the
day of the month, and the hour. We have an oddity
here somewhat like that for 787 Sep 16. This is
one of many reports that are probably of the penumbral eclipse of 810 Jul 5.[†] All reports have
the date wrong, but there is little consistency in
the errors. They all give the date as some number
of days before the ides of some month, but the
number of days before, and the month named, vary
from report to report.[‡] No other report contains
the data found in this one. With some misgivings,

[†] It is odd that there was a penumbral eclipse on
810 Jun 5 also, but it was visible only in the
southern hemisphere.

[‡] The correct date is the 3rd nones July.

I take this as independent, with a reliability of
0.1. Place: Gottweih. Standard deviation of the
magnitude: 0.1.

I assume that this is not a report of the
eclipse of 809 Jul 16, because that date would be
written as the 17th calends August, and the errors
involved if this is the date are even less plausible
than those involved for 810 Jul 5.

810 Nov 30a E,CE. Reference: Juvavenses
[ca. 975]. Here we have a contrast with the reports of the eclipse of 810 Jul 5 in the same year;
all reports that state the day of the year do so
correctly. Juvavenses says under 810: "On the day
before the calends of December there was an eclipse
of the sun from the 4th to the 7th hour on the Sabbath day." The date is correct and the hours look
reasonable. The annalist used "Sabbath" in the Jewish sense, not as an equivalent to die Dominica.
Reliability: 0.5. Place: Salzburg. Standard
deviation of the magnitude: 0.1.

810 Nov 30b E,CE. Reference: Gottwicensis [ca. 1054]. This appears under 810: "In
the same year there was an eclipse of the sun on
the 2nd calends December (= Nov 30) at about the
3rd hour." The hour is different from that given
in Juvavenses and looks early, judging from the
chart in Oppolzer [1887]. However, I shall ignore
this point and use this report with a reliability
of 0.1. Place: Gottweih. Standard deviation of
the magnitude: 0.1.

Mellicenses has copied the records of
810 Jul 5 and 810 Nov 30 from Gottwicensis, but
has put them under 809. Mellicenses also has
records of the eclipses of 840 May 5 and 878 Oct
29. The information in these records could have
come from many different sources, so I shall ignore them. They do not seem to come from either
Gottwicensis or Herimannus. Mellicenses also has
a report of the eclipse of 968 Dec 22 that is from

-264-

Herimannus [1054]. It has reports of the eclipses of 1033 Jun 29 and 1039 Aug 22 that are from Suevicum [ca. 1043] rather than from Herimmanus. It is interesting that the years of eclipses in Mellicenses become correct with the last three reports mentioned.

1087 Aug 1 E,CE. Reference: Zwetlense 255 [ca. 1189]. An auctarium for the year 1088 reads: "There was an eclipse of the sun. Conrad son of Henry was made king and the country rebelled." No eclipse in 1088 looks likely, although it is barely possible that the eclipse of 1088 Jul 20 was visible as a small eclipse at sunset. The eclipse of 1087 Aug 1 is the only likely one that I find. The rulers mentioned were apparently local rulers under the Holy Roman Empire and I do not have a history detailed enough to discuss them. However the editor says in a footnote that Conrad was chosen in 1087 and hence that 1087 is the correct year. Since the astronomical and historical identifications thus confirm each other, I take this to be a record of the eclipse of 1087 Aug 1. The source is not contemporaneous but I have found no other record of this eclipse at all and it thus represents an independent source. Reliability: 0.5. Place: Zwettl. Standard deviation of the magnitude: 0.1.

1093 Sep 23a E,CE. Reference: Admuntenses [ca. 1250], which has: "1093. An expedition to Jerusalem began. There was a great mortality of men. An eclipse of the sun for 1 hour." If the "expedition" was the First Crusade, that did not start until 1096. However there were occasional earlier stirrings toward Jerusalem and this may be one of them; if so, I find no other mention of it. In any case, there is no plausible eclipse except that of 1093 Sep 23. There are many records of pestilence and famine at about this time (see Appendix IV) and the mortality was probably a consequence of them rather than of the expedition. Reliability: 0.2. Place: Admont. Standard deviation of the magnitude: 0.1.

1133 Aug 2a E,CE. Reference: <u>Mellicenses</u>
[ca. 1564]. <u>Mellicenses</u>, which is now a contemporaneous source, has under 1133: "There was a
horrible eclipse of the sun on the 4th nones August (= Aug 2) at about the 9th hour, on the 4th
feria." From the chart in <u>Oppolzer</u>, the 6th hour
looks closer; perhaps the annalist was already using "none" in reference to the hour to mean the
time that we call noon. Otherwise the information
is accurate. The adjective "horrible" suggests
that the eclipse was larger than usual, but I shall
not take it to be total. Reliability: 1. Place:
Melk. Standard deviation of the magnitude: 0.03.

1133 Aug 2b E,CE. Reference: <u>S. Rudberti</u>
[ca. 1168], which has: "1133. This year on the
4th nones August in the midday heat the sun suddenly disappeared, and shortly after it was seen as
if covered over terribly like a round bag,† and
stars moreover appeared in the sky." It is tempt-
ing to read a description of the corona into
this description, but I do not feel certain enough
of the translation to do so; I shall leave that to
an expert in medieval Latin. Whether the annalist
means that the sun was surrounded by something
round, or whether he means that it was kept from
view by something round, I believe that he is giving us an eye witness account of a total eclipse.
Reliability: 1. Place: Salzburg. Standard deviation of the magnitude: 0.

1133 Aug 2c E,CE. Reference: <u>Admuntenses</u>
[ca. 1250]. This source contains the following
passage: "1133. In this year on the 4th nones
August at about <u>meridiem</u> there was an eclipse, that
is, a disappearing, of the sun so great that stars
appeared and there was darkness over the whole

---

† I am not sure of this translation. The text has
<u>saccus cylicinus</u>. My dictionary does not list
<u>cylicinus</u>. I assume that this is an error for
<u>cyclicus</u>, which I freely translate.

earth for an hour." Although this is clearly exaggerated, and may be influenced by the account of
the darkness at the Crucifixion of Jesus in the
synoptic gospels,[†] it sounds original and reliable
to me.[‡] Reliability: 1. Place: Admont. Standard deviation of the magnitude: 0.

    1133 Aug 2d E,CE. Reference: Zwetlensis
102 [ca. 1170]. Under 1133 this source has:
"There was an eclipse of the sun on the 4th nones
August at the 9th hour." This source could be
still derivative from Mellicenses or it could be
independent at this time. This report sounds like
an abbreviation of the account 1133 Aug 2a E,CE
above, so I shall not use it.

    1133 Aug 2e E,CE. Reference: Zwetlense
255 [ca. 1189]. The editor of this codex from
Zwettl has the following: "1133. (Eclypsis -
feria 4.) luna 26." I think he means that the
words enclosed within parentheses are identical
with the corresponding passage in Mellicenses
(report 1133 Aug 2a E,CE above), which does start
with "Eclypsis" and which ends with "feria 4".
If so, the annalist at Zwettl merely copied the
Melk account and added the calculated age of the
moon. If he used the ecclesiastical moon, he erred,
because 1133 Aug 2 was on the 27th of the moon.

[†] Mark 15:33, for example.

[‡] The editor put this passage in small type, which
is a device used to mean a passage copied from
some other source. He said it is from Summa
Honorii. This means the work called Summa Totius
by the French historian Honorius Augustodunensis,
which is discussed under the citation Honorius
[ca. 1137] in Section X.2. Both Honorius and the
annalist at Admont apparently saw a most impressive solar eclipse. Aside from the fact that both
gave the date correctly, I saw little resemblance
between their accounts, and I believe that the
editor was mistaken. See the discussion of the
record 1133 Aug 2d E,F in Section X.3.

Because of the probability of copying, and not because of the error in calculation, I shall not use this report.

1133 Aug 2f E,CE. Reference: Reicher-spergenses [ca. 1279]. The earlier form of these annals (codex A) has: "1133. In this year was a great eclipse of the sun on the 4th nones August about midday, on the 27th of the moon, in the 13th year of the 19-year cycle." The statements about the moon and the year of the cycle are correct. The expanded version of the annals, the one that lies between codex A and the work of Magnus, has this copied into it, with additions saying that many stars were seen and that it was like night for about half an hour. The "expander" then wondered if this was a prodigy rather than a real eclipse since it did not occur at the new moon; he did not go to the point of explicitly doubting the accuracy of his lunar (ecclesiastical) tables.

Codex A was prepared in its present form about 25 years after the eclipse. The only information in its report that could not have been calculated with ease is the date of the eclipse and its greatness, and the annalist could have found this information in many places. However the report did not strike me as a paraphrase or condensation of any other report that I have found, so I shall take it to be from an independent source at Reichersberg.

The "expander" worked still later and from other sources. There are many sources[†] that he could have used and several medieval writers were struck by the discrepancy between the date of the eclipse and the date of the ecclesiastical new moon. I can cite no particular source that he used, but it seems plausible to me that he would have used several sources and hence that he might

[†] As I remarked in Section V.1, the eclipse of 1133 Aug 2 was probably the most widely reported eclipse in medieval times.

have written in his own words rather than copying.
Accordingly I shall ignore the contributions of
the "expander" and use only what is in codex A.[†]
For that source only, I use a reliability of 0.5.
Place: Reichersberg. Standard deviation of the
magnitude: 0.05; I use a smaller value than usual because the eclipse is said to be large.

1153 Jan 26 E,CE. Reference: <u>Admuntenses</u>
[ca. 1250]. This source has: "1153. There was an
eclipse of the sun on the 4th calends February (=
Jan 29) in the evening." "In the evening" looks
definitely late; the 9th hour of the day looks to
be closer. Perhaps the annalist was thinking of the
office of nones, made an accidental transfer in
his mind to the office of vespers, and wrote "ves-
pere" as a result. The date given is also wrong;
the correct date is the 7th calends February.
Even though <u>Admuntenses</u> is presumably an original
source by now, I shall use a reliability of 0.5
because of the errors that suggest copying.[‡]
Place: Admont. Standard deviation of the magnitude: 0.1.

<u>Lambacensis</u> [ca. 1283] says "There was
an eclipse of the sun" under 1153. Such a brief
notice is not safe to use in a source that could
still be secondary.

[†] The reader may object that I have not been consistent in my treatment of the "expander" and of the
compiler of codex A. I agree that I have not been
and that I am going only on the basis of my sub-
jective assessment of the sources. However, I
do not believe that I am committing a serious
error, since I am working in ignorance of the
astronomical consequences. Unbiased error can
be tolerated.

[‡] Some copies of <u>Admuntenses</u> definitely suggest
copying. They say: "It is said that there was
an eclipse of the sun." I do not recall seeing
such a wording anywhere else.

1187 Sep 4a E,CE. Reference: <u>Mellicenses</u>
[ca. 1564]. The first sentence under 11<u>88</u> reads:
"There was an eclipse of the sun on the 2nd nones
September (= Sep 4) at about the 9th hour; and the
Sepulchre of the Lord, besieged by pagans, is destroyed." Note the shift from a past tense to the
present tense. Again we have the use of "nona hora"
with an eclipse that probably occurred close to the
hour that we call noon. I suspect that the entry
was not meant to appear under 1188, although inspection of the manuscript would be required to settle
the question. The editor of the text has marked the
places where the handwriting changes. This sentence
is in the same hand as the entry for 1187 but in a
different hand from the rest of the 1188 entry.
Thus it is plausible that one person made the entry for 1187 and let it "spill over" to the space
for 1188. In any event, the day is given correctly,
and we may accept this as a contemporaneous record.
Reliability: 1. Place: Melk. Standard deviation
of the magnitude: 0.1.

1187 Sep 4b E,CE. Reference: <u>S. Rudberti</u>
[ca. 1286]. The annalist of <u>S. Rudbertus</u> starts
his entry for 1187 with a reference to affairs in
Palestine and then says: "In the month of September an eclipse of the sun took place." He then
mentions the death of Pope Urban III and the election of Gregory VIII in October. Thus the report
of the eclipse is surrounded by correctly dated
events, and the annals are presumably independent
by now. Hence we may take this as an original account of the eclipse of 1187 Sep 4 in spite of the
absence of detail. Reliability: 1. Place: Salzburg. Standard deviation of the magnitude: 0.1.

1187 Sep 4c E,CE. Reference: <u>Cremifan-
ensis</u> [ca. 1216]. <u>Cremifanensis</u> first records the
change of popes mentioned in the preceding report
and then says: "There was an eclipse of the sun
at midday, and stars were seen." I am mistrustful
of this record, mainly because of the mention of
stars. <u>Oppolzer</u> [1887] shows the path of this eclipse lying well to the east, passing from southern

Sweden through Poland and to the Crimea. His calculation is confirmed by the reports that I have
found. Only one report of this eclipse, namely
1187 Sep 4b E,Sc from Scandinavia, mentions stars
or otherwise indicates a large eclipse. Further,
the only other eclipse from Cremifanensis, which
is 1191 Jun 23c E,CE below, also has a suspicious
circumstance. If only this report had seemed odd,
I should have given it a reliability of 1 in spite
of my misgivings, since the source is supposed to
be contemporaneous by 1187. As it is, I shall lower the reliability to 0.5. Place: Kremsmünster.
Standard deviation of the magnitude: 0.01.

        1191 Jun 23a E,CE. Reference: Mellicen-
ses [ca. 1564]. Under 1193, this source notes the
death of Pope Clement III, the accession of Celestine III, and the coronation of Henry VI as Holy
Roman emperor by the latter. It then says: "There
was an eclipse of the sun on the 9th calends July
(= Jun 23) at about the 7th hour." I believe that
the year given is simply an accident, because all
of the information is correct for 1191. Reliability: 1. Place: Melk. Standard deviation of the
magnitude: 0.1.

        1191 Jun 23b E,CE. Reference: S. Rudberti
[ca. 1286]. This source has the information about
the coronation under the correct year, and then
has under the same year: "An eclipse of the sun
occurred on the 9th calends June about the middle
of the day." The text has "June" where it should
have had "July", but this mistake is encouraged
by the nature of the Roman calendar, and it is in
fact surprising that it did not happen more often.
Reliability: 1. Place: Salzburg. Standard deviation of the magnitude: 0.1.

        1191 Jun 23c E,CE. Reference: Cremifan-
ensis [ca. 1216]. This source has, under 1191:
"There was an eclipse of the sun on the 9th calends July at about the 9th hour." This is the report that I mentioned above in connection with the

record 1187 Sep 4c E,CE, also from Cremifanensis.
The suspicious circumstance here is that the wording of this report is identical with that of the
same eclipse (report 1191 Jun 23c E,G in Section
XI.2) from the relatively near monastery of S.
Stephani in Freising (see Section XI.1), except
that the Freising report includes the correct additional information that the eclipse was on the 1st
feria. This information could have been supplied
by calculation.

Now the reports follow a rather standard
formula for reporting eclipses, and the coincidence
in wording would not be highly significant except
for the hour. These are the only reports that put
the eclipse at "about the 9th hour", which is definitely too late. It may be that horam nonam means
the same as meridiem, but even so it is odd that
only these reports have this usage. The monastery
at Freising is much older than the one at Kremsmünster so that its annals should be taken as original if we must choose between the two places.

Accordingly I shall do as I did with the
report 1187 Sep 4c E,CE above and use a reliability
of 0.5.[†] Place: Kremsmünster. Standard deviation
of the magnitude: 0.1.

1191 Jun 23d E,CE. Reference: Reicherspergenses [ca. 1279]. Under 1191 we find: "There
was an eclipse of the sun that same year, on the
9th calends July, on the Lord's day, at the meridian hour, on the 27th of the moon, in the 14th
year of the 19-year cycle." All the information

[†]The eclipse of 1191 Jun 23 was certainly larger
at Kremsmümster than that of 1187 Sep 4. If
stars were seen at either time, it should have
been in 1191, not 1187. I was concerned above
with the possibility that the phrase about stars
was accidentally transferred from 1191 to 1187,
suggesting that Cremifanensis might not be strictly contemporaneous. See the report 1191 Jun 23c
E,G in Section XI.2 for additional considerations.

is correct, but most of it could have come from
calculation. The annals were being worked on actively at this time. Reliability: 1. Place:
Reichersberg. Standard deviation of the magnitude:
0.1.

          1239 Jun 3a E,CE. Reference: S. Rudberti
[ca. 1286]. "On the 3rd nones June (= Jun 3) there
was an eclipse of the sun that lasted for 2 hours."
The duration probably refers to the interval from
the first to the last observed contact. Reliability: 1. Place: Salzburg. Standard deviation of
the magnitude: 0.1.

          1239 Jun 3b E,CE. Reference: Lambacensis
[ca. 1283], which has: "1239. There was an eclipse of the sun on the 3rd nones June at about the
9th hour." This contemporaneous report receives a
reliability of 1. Place: Lambach. Standard deviation of the magnitude: 0.1.

          1241 Oct 6a E,CE. Reference: S. Rudberti
[ca. 1286]. This source has a long entry under
1241 that is correct where I can test it. Part of
it reads: "There was such an eclipse of the sun on
the Lord's day in the octave of S. Michael's that
the heavens were darkened and stars could be seen
in the sky." The "octave of S. Michael's" is Oct
6, which was on the 1st feria in 1241. Reliability:
1. Place: Salzburg. Standard deviation of the
magnitude: 0.01, because of the stars.

          1241 Oct 6b E,CE. Reference: Admunten-
ses [ca. 1250]. This has: "1241. There was an
eclipse of the sun, and the darkness was so heavy
that stars appeared in the sky, about the 9th hour."
Reliability: 1. Place: Admont. Standard deviation of the magnitude: 0.01.

          1241 Oct 6c E,CE. Reference: Lambacensis
[ca. 1283]. This has: "1241. There was darkness

over all the earth, and stars appeared, on the day
before the nones of October,[†] about the 9th hour."
Later, under the same year, the entry for 1241 refers to the Mongol invasion of Hungary and says
that 80,000 people were killed. I think the annalist is trying to describe a total eclipse. Reliability: 1. Place: Lambach. Standard deviation
of the magnitude: 0.

       1241 Oct 6d E,CE. Reference: Reicher-
spergenses [ca. 1279]. This is a contemporaneous
entry in the continuation after the work of Magnus: "In the year 1241 on the day before the nones
of October the sun, while in its full brillance after midday, was suddenly covered over with a not-
small blackness in such a way that no part of it
could be seen, and stars were seen just as if it
were night, almost for 4 hours."[‡] I think that
the duration refers to the entire eclipse and not
just to the time that stars were visible. I also
think that the annalist is trying to describe a
total eclipse. Reliability: 1. Place: Reichersberg. Standard deviation of the magnitude: 0.

       1263 Aug 5 E,CE. Reference: S. Rudberti
[ca. 1286]. Under 1263 this has: "There was a
partial eclipse of the sun and two of the moon."
The solar eclipse of 1263 Aug 5 was partial in
Salzburg, and the partial lunar eclipses of 1263
Feb 24 and 1263 Aug 20 were at times that would
make them visible in Salzburg, weather permitting.

[†] The nones of October is Oct 7, so the eclipse is
correctly dated.

[‡] After the entry for 1241 the annalist has given
us an almost-rhyming couplet in what Sayers
[1928] would have called "rather doggy Latin".
I venture to give a free (but better rhymed)
translation into equally doggy English:

     In year forty, twelve hundred, plus one,
     On 6 October eclipse of the sun.

Reliability: 1. Place: Salzburg. Magnitude: 2/3 with standard deviation of 1/6. Path assumed to the east of Salzburg.

1267 May 25 E,CE. Reference: S. Rudberti [ca. 1286]. "1267. An earthquake occurred about the 3rd hour of the day, and an eclipse of the sun followed thereafter." The eclipse would have been a maximum at perhaps the 5th hour, and thus it could have followed an earthquake at the 3rd hour. However, I do not think that it is safe to infer that the earthquake and the eclipse were necessarily on the same day. The eclipse could have been on a later day, but probably not long after the earthquake. Reliability: 1. Place: Salzburg. Standard deviation of the magnitude: 0.1.

2. Records from Czechoslovakia

Cosmas [ca. 1125]. Cosmas Pragensis was a prominent citizen of Prague and was dean of the cathedral there for some time. Some reference that I forgot to note called him "the Herodotus of Bohemia". I believe that this is merely a flowery way of saying that he is the first known Bohemian historian and that it is not meant as an assessment of the quality of his work. A note inserted at the end of his chronicle says that he died on 1125 Oct 21. Not far before the end of the chronicle, Cosmas refers to himself as an octogenarian, therefore he was born about 1045. His work was popular and found many continuators.

Gradicenses [ca. 1145]. These annals are from the monastery of S. Stephan at Hradisch,[†] in Moravia. The monastery was founded in 1077 but these annals were probably not started until about 1130. They are all in one hand. The first entry is for 894, but through 999 the material is from

---

[†] Or Uherské Hradiste.

the source that the editor calls "Ekkehard".[†] In
later years the annalist used other sources such
as Cosmas [ca. 1125]. Some original material begins to appear with the founding of the monastery,
but the annals do not become fully independent until about 1130.

In 1145 civil disturbances caused the annals to be taken to the monastery of S. Laurentius
in Opatowitz, a short distance to the northwest,
where they were continued for a few years. The
continuation at Opatowitz does not contain any information needed for this work, and it will not be
used.

Pragenses [ca. 1220]. These annals are
written on tables of the indiction, and lists of
popes and emperors, in a single hand of the early
13th century. Except for some local material, they
seem to be entirely derived from known sources. I
have referred to them only in order to point out
what may be an error made by Ginzel [1882 and 1884].
See the record designated only as "1186?" below.

Pragensium [ca. 1283]. The editor called
this the "second continuation of Cosmas"; see the
discussions of Cosmas [ca. 1125] above and of
Wissegradensis [ca. 1142] below. It is a late continuation and has no eclipses before 1200. I made
an exception and included it in this study, however, because it has a rarity among medieval European records, namely a record of a possible super-
nova (see Section V.2).

Sasavensis [ca. 1170]. This is a continuation of Cosmas by an anonymous monk in the monastery at Sazava, a short distance south-east of

---

[†] This probably means the Chronicon Universale of
Frutolf [1103], which was formerly attributed to
Ekkehardus Uraugiensis (Ekkehard von Aura).
See Section XI.1.

TABLE IX.2

REFERENCES TO SOLAR ECLIPSES FOUND IN
MEDIEVAL SOURCES FROM CZECHOSLOVAKIA

Source	Date	Conclusions
Cosmas	1093 Sep 23	Probably original
	1124 Aug 11	Original
Gradicenses	1133 Aug 2	Probably original
Pragenses	939 Jul 19	Not independent; cannot locate source
	1133 Aug 2	Not independent; cannot locate source
	1186 ?	Cannot identify
Pragensium	1241 Oct 6	Original
	1255 Dec 30	Original
Sasavensis	990 Oct 21	Copied from German source
	1133 Aug 2	Probably independent
Wissegradensis	1130 Oct 4 ?	Event is probably not an eclipse
	1133 Aug 2	Original

Prague on the Sazava River. The monastery was
founded in the middle of the 11th century. This
continuation of Cosmas is independent of the line
of continuations that starts with Wissegradensis
[ca. 1142] below, and the monk inserted some early
material from other sources. He only brought the
continuation down to 1162. On the basis of
certain incidental remarks, the editor concludes
that he was still writing it at least as late as
1170 (in other words, he was running at least 8
years behind time), and I have used this date accordingly. The editor also thinks that the author
had lived many years at Sazava.

Wissegradensis [ca. 1142]. The editor
calls this the "first continuation of Cosmas". It
is independent of Sasavensis above. Pragensium
[ca. 1283], which the editor called the second
continuation, is the successor to Wissegradensis.
According to the editor, all we know of the author
of this continuation is that he was a canon of the
monastery of Wissegrad. Cosmas [ca. 1125] says
that the monastery was founded by King Wratislaus
in 1070. Wissegrad, which is also spelled Wissegrada
and Bizzenrad in the Latin sources, is spelled Vyse-
hrad[†] in the modern English sources that I have consulted. Wissegrad is the name of a hill in Prague,
as well as of a castle and of a district of the
city. Wissegradensis seems to be a contemporaneous
and original document.

The reports found in these sources are
listed in Table IX.2. It is interesting that only
three of the eclipses found in Czechoslovakian
sources are also found in Austrian ones.

Under both 939 and 1133 Pragenses [ca. 1220]
has the entry: "The sun was seen diminished (min-
utus)." These occurrences are probably meant to
be the eclipses of 939 Jul 19 and 1133 Aug 2. There
is no way to tell the source from which the compiler
of Pragenses, working around 1220, got these records,

[†] I am indebted to Mr. Kaye Weedon of Blommenholm,
Norway, for pointing out this identification.

and it is not safe to use them. There is no particular reason to assume that they are local to
Prague.

Under 990 Sasavensis [ca. 1170] has a
record of the eclipse of 990 Oct 21 that has the
same words, slightly rearranged, as the German
record 990 Oct 21a E,G in Section XI.2.

1093 Sep 23b E,CE. Reference: Cosmas
[ca. 1125]. Cosmas wrote in Book II, Chapter 51:
"In the same year there was an eclipse of the sun
on the 12th calends October (= Sep 20), on the 6th
feria, after midday." The year is not explicitly
stated but the context clearly puts the record in
1093. The day of the month is wrong but the day
of the week is right, showing that the data were
not calculated and that the error is accidental.
Since Cosmas was about 50 years old at the time,
this may well be an original observation. Reliability: 1. Place: Prague. Standard deviation
of the magnitude: 0.1.

1124 Aug 11 E,CE. Reference: Cosmas
[ca. 1125]. Cosmas wrote in Book III, Chapter 57
of his chronicle: "And that same year on the 3rd
ides August (= Aug 11) at the 11th hour of the day
was an eclipse of the sun, and later there was a
great pestilence of cows, sheep, and pigs; many
bees died and there was an exceeding lack of honey."
The hour of the day is clearly wrong; I imagine
that "VI" has been misread as "XI". This is a
contemporaneous record. Reliability: 1. Place:
Prague. Standard deviation of the magnitude: 0.1.

1130 Oct 4(?). Reference: Wissegradensis [ca. 1142] has the following entry under
1130: "This year on the 8th ides October (Oct 8)
some monster like a serpent was seen flying over
Bohemia and many other places for a short time
about sunset. After that many saw another very
bright sign in the morning hour." The eclipse of
1130 Oct 4 might have been visible in Bohemia at

sunrise; it would take detailed calculation to answer the question. However, the sign seems to be an excess of light and not an absence of it. Thus I doubt that this is a report of the eclipse, in spite of the near-coincidence of date and time. I shall not use this as an eclipse report.

1133 Aug 2g E,CE. Reference: Sasavensis [ca. 1170], which has, under 1133: "This year there was an eclipse of the sun in the middle of the day." In spite of its brevity it seems safe to take this as an independent report, and the editor apparently had the same impression. However, I do not feel safe in taking it as original. Reliability: 0.5. Place: Sazava. Standard deviation of the magnitude: 0.1.

1133 Aug 2h E,CE. Reference: Wissegradensis [ca. 1142]. Under 1133 this source has: "On the 4th nones August (= Aug 2) an eclipse of the sun appeared in a wonderful way; gradually diminishing, it grew so small that it became a corona like that of a crescent moon in the southern part, that later turned to the east, subsequently to the west, and then it returned to its original condition." This is a contemporaneous account. I think that it describes an eclipse that was almost but never quite total. By the "crescent moon" I shall guess that he means a moon that has become easily visible and I shall guess that this means that about 0.03 of the surface is illuminated.[†] To be more specific, I shall take 0.97 to be the magnitude at maximum eclipse, and I shall take 0.03 as the standard deviation of this estimate. The chart in Oppolzer is not accurate enough to let us decide whether the path of the eclipse went to the north or the south of the observer. I should not like to guess whether the annalist meant that the southern part was the part that was dark or illuminated. Hence I must do what I do not like to do: I shall decide

---

[†] This is in the range of illumination for a moon between 1 and 2 days old.

this point on the basis of the calculations to be
made, provided that they allow a clear decision.
Reliability of the record: 1. Place: Prague.
If the calculations do not allow a clear decision
about the direction to the path, I shall treat
this as an ordinary report with a standard deviation of 0.03.

1133 Aug 2i E,CE. Reference: Gradicenses [ca. 1145]. This says under 1133: "On the
4th nones August there was an eclipse of the sun
at the 10th hour of the day." This report comes
during the short time, roughly 1130 to 1145, that
these annals were independent. Even within this
period, however, the annalist could have included
an interesting report from elsewhere. However,
the stated hour of the day, which does not look
accurate, occurs in no other report that I have
found, so this report is probably original. Reliability: 1. Place: Hradisch. Standard deviation of the magnitude: 0.1.

1186(?). Reference: Pragenses [ca. 1220].
These annals have under 1186: "An eclipse of the
sun on the fifth feria; there was mortality of men."
I presume that the last part of the entry refers to
a pestilence. The entry for 1187 refers to the
capture of Jerusalem, thus it seems likely that
1186 was the year meant by the annalist. The only
eclipse I can find near this time on the 5th feria
was that of 1187 Mar 12, and it was visible only
in the South Pacific. It is possible that the
feria is wrong, but this is unlikely because the
Latin text has the number written out, as I have
given it in the translation.

Ginzel [1882 and 1884] took this to be
the eclipse of 1187 Sep 4, which was on the 6th
feria. If this identification is correct, both
the numerical data and the historical context of
the report are wrong. Further, one cannot justi-
fy 1187 Sep 4 by ignoring the data and using the
nearest eclipse. 1185 May 1 was as near in time
and probably as large an eclipse at Prague; inci-

dentally, it was on the 4th feria.

It is clear that the eclipse cannot be identified confidently. It is conceivable that the entry represents someone's attempt to predict an eclipse.

1241 Oct 6e E,CE. Reference: Pragensium [ca. 1283]. Under 1241 this contemporaneous source has "On the 3rd nones October (= Oct 5) after midday there was such an eclipse of the sun that people called it twilight." The word used in the text is crepusculum. I believe that the use of this word does not forbid totality but I believe also that it does not imply totality. Use of 0.02 for the standard deviation of the magnitude seems reasonable. Reliability: 1. Place: Prague.

1255 Dec 30 E,CE. Reference: Pragensium [ca. 1283], which has, under 1255: "On the 3rd calends January (= Dec 30) the sun suffered an eclipse." Reliability: 1. Place: Prague. Standard deviation of the magnitude: 0.1.

3. Records from Switzerland

Bernoldus [1100]. Bernoldus, so the editor tells us, was born in the German diocese of Konstanz and became a monk at S. Blasien, about 75 kilometers west of the city of Konstanz. While there, he began this chronicle, in 1072 or 1073. In 1091 he moved to the monastery of S. Salvatoris in Schaffhausen, about halfway between S. Blasien and Konstanz, taking his chronicle with him. He died in Schaffhausen on 1100 Sep 16.

Bernoldus was apparently interested in astronomy. After a necrology of kings and bishops, he enters on a discussion of calendrical

problems, and refers to Herimannus Contractus[†]
(1013-1054), the noted astronomer, inventor, and
chronicler from nearby Reichenau. Under the year
1091 in his chronicle, Bernoldus wrote in praise
of Wilhelm, abbot of Hirsaugiensis, who invented
a natural horologium on the pattern of the celestial hemisphere. It is not possible to tell from
this just what Wilhelm invented, except that it
was a device for reading the time. It could have
been a hemispherical sun dial. It could also have
been a clockwork mechanism that might, for example,
have shown the motions of the solar system. Such
a device should not have been beyond the technology
of the time; after all, this was almost 3 centuries
after Charlegmagne had a striking clock [Fuldenses,
ca. 901].[‡] Wilhelm also showed how to make studies
of the equinoxes and solstices and the condition
of the heavens. He uses the term certis experi-
mentis in connection with the equinoxes and solstices and this, I think, implies more than mere
calculation.

Bernoldus' Chronicon should certainly
be an original source for the period 1070-1100;
the editor in fact took it to be original from
1055 on.

Engelbergenses [ca. 1175]. The monastery at Engelberg, or Mons Angelorum, was founded
in central Switzerland in 1120. A few years later
its current abbot Frowinus was given the task of
restoring the monastery of S. Blasien, in the
Black Forest, that had fallen into neglect and

[†] Also known as Hermannus Augiensis, Herman the
Lame, and Herman of Reichenau. His chronicle
[Herimannus, 1054] is one of the sources used
in Chapter XI.

[‡] It is true that Charlemagne's clock was not made
in Europe, but that it was sent to him by the
Caliph Harun-al-Rashid of Thousand-and-One Nights
fame. In the intervening centuries, there was
ample time for Europeans to learn how to make
such devices, if indeed they had to learn.

decay. One result of Frowinus' reforms was the start of a set of annals at S. Blasien. In 1147 these annals were transferred to Engelberg; if a reason was given for this transfer, I did not notice it. The annals were maintained continuously at Engelberg until 1175. A set of annals found in a different codex runs from 1178 to 1546. This set can be regarded as a continuation of the first set, and contains nothing that will be used in this study. For our purposes, Annales Engelbergenses[†] means the set of annals kept at Engelberg from 1147 to 1175.[‡]

      Ratpertus [ca. 883]. Ratpertus became a monk of S. Gall about 850. He wrote a history about S. Gall, both the man and the monastery, down to 883, so that this is probably the year of his death. His history was continued to 1329. Amazingly, the continuations have no data useful for this study, or at least none that I noticed. The history of Ratpertus is probably an original document from about 850 to its close, and earlier material may be independent of other known sources. See the discussion of Sangallenses [ca. 1056] below.

      Sangallensis [ca. 926]. See the discussion of Sangallenses [ca. 1056] below. This source is closely related to the Belgian source cited as Laubacenses [ca. 926] in Chapter VIII.

      Sangallenses [ca. 1056]. About 614, a man from Ireland named Gall, Cellach, or Caillech came to live in what is now northeastern Switzerland. Gall is variously described as a hermit and a missionary.* Others came to live at the same place,

---

[†] This is a Latinization of a Germanicization of a Latin name.

[‡] The part of the annals kept at S. Blasien forms the source called S. Blasii in Chapter XI.

* It is perhaps possible to be both, but being a hermit might impair one's effectiveness as a missionary.

which gradually became a place of pilgrimage. In
the middle of the 8th century, the monastery of S.
Gall was founded here.[†] The monastery became celebrated as a center of learning and possesses one
of the greatest of medieval libraries.

Many annals, chronicles, and histories
have been prepared at S. Gall. Ratpertus [ca. 883]
is one example. The source Sangallensis [ca. 926]
is another. The latter is part of the complex of
annals that the editor Pertz calls Annales Alamannici; he calls it a continuation of the original
series of "German annals", and it has some original
material for several decades before its close.
Finally there is the source that I am citing as
Sangallenses [ca. 1056], which the editor calls
Annales Sangallenses Maiores.

What Pertz calls Annales Alamannici were
developed jointly at S. Gall, Reichenau in the Bodensee (Lake Constance), and a place that I have not
located whose name is Murbach or something like that.
From their beginning in 709 down through 918 the
Annales Sangallenses Maiores are drawn from this
set. From 918 to their close in 1056 these annals
are apparently an original source.

Annales Sangallenses Maiores have been
called Annales Sangallenses Breves; it is interesting that some editors find them "short" while some
find them "major". They have often been attributed
to one Hepidannus, apparently in error. The annals,
particularly when attributed to Hepidannus, are
sometimes described as going to 1044 rather than
1056. It is correct that the regular keeping of
the annals stopped with 1044, and the only entry
after 1044 records the death of the Emperor Henry
in 1056.

The German source cited as Augienses

[†] This summary is drawn from the standard sources.
Little of this history is found in the annals of
S. Gall that I saw, and I do not know the source
of it.

TABLE IX.3

REFERENCES TO SOLAR ECLIPSES FOUND IN
MEDIEVAL SOURCES FROM SWITZERLAND

Source	Date	Conclusions
Bernoldus	1093 Sep 23	Original
Engelbergenses	1147 Oct 26	Two different accounts; one probably original, other probably brought from Germany
Ratpertus	840 May 5	Probably from German source
Sangallensis [ca. 926]	891 Aug 8	Probably original
Sangallenses [ca. 1056]	840 May 5	Probably from German source
	939 Jul 19	Original
	968 Dec 22	Original

[ca. 954] in Chapter XI is also part of the Alamannici family. There are other Swiss sources that I have not mentioned, but I did not notice any data useful for this study in them.

The eclipse records discussed below are tabulated in Table IX.3.

840 May 5 E,CE. Reference: Ratpertus [ca. 883]. After noting that Emperor Louis (I, le Débonnaire) died on 840 Jun 20, Ratpertus wrote: "That same year was an eclipse of the sun on the 3rd nones May (= May 5) on the eve of the Ascension of the Lord." Ascension Day was on May 6 in 840. The German source Augienses [ca. 954] has: "An eclipse of the sun on the 3rd nones May, between the 8th and the 9th hours, on the eve of the Ascension of the Lord. Louis died." Both Ratpertus and Augienses became independent sources at about the same time, but after this eclipse. From their dates alone, we would not be able to tell which of these copied the other. However, Augienses has the hour of the day, correctly so far as I can judge, and Ratpertus does not. Therefore we may assume that Ratpertus is secondary to Augienses. It is true that Ratpertus has the date of Louis' death and Augienses does not, but Ratpertus could have learned that from many sources. I shall give Ratpertus' report a reliability of 0.

The report of this eclipse in Sangallenses [ca. 1056] is almost surely copied from Augienses, with the omission of the phrase about Ascension Day.

891 Aug 8 E,CE. Reference: Sangallensis [ca. 926]. This source has, under 891: "The star comet. Eclipse of the sun." I have already discussed this report in connection with the Belgian report 891 Aug 8 E,BN and decided that this is the original one of the two. Sangallensis was probably an independent source in 891, but I dislike giving full reliability to such a brief notice when there is a possible question about independence. Reliabil-

-287-

ity: 0.5. Place: S. Gall. Standard deviation
of the magnitude: 0.1.

        939 Jul 19 E,CE. Reference: <u>Sangallen-</u>
ses [ca. 1056]. Under 939 these annals have:
"There was an eclipse of the sun about the 3rd hour
of the day on the 14th calends August (= Jul 19)
in the 4th year of Otto on the 6th feria, the 29th
of the moon." All the information in this record
is correct, and the annals were being kept contemporaneously. Reliability: 1. Place: S. Gall.
Standard deviation of the magnitude: 0.1.

        968 Dec 22 E,CE. Reference: <u>Sangallen-</u>
ses [ca. 1056]. I do not know why the editor assigned the date 968 to this record and then put 970
in parentheses after it: "This year there was an
eclipse of the sun on the 11th calends January (=
Dec 22) on the 28th of the moon at the 3rd hour of
the day." The information is correct and the report is contemporaneous. Reliability: 1. Place:
S. Gall. Standard deviation of the magnitude:
0.1.

        1093 Sep 23c E,CE. Reference: <u>Bernoldus</u>
[1100, p. 457]. Bernoldus wrote, under 1093: "A
sign occurred in the sun on the 9th calends October
(= Sep 23) before <u>meridiem</u>, in that a kind of circle appeared in it and it shone very darkly in a
clear sky. But some considered it to be rather an
eclipse than a sign, mainly because the moon was
at its 28th day. For indeed the illustrious cal-
culator Herimannus wrote that an eclipse happened
on the 27th of the moon in the year 1033." The eclipse was on the 28th day of the ecclesiastical
moon, as Bernoldus writes. <u>Herimannus</u> [1054]† reported the eclipse of 1033 Jun 29 in his chronicle,
but he did not say that it was the 27th of the moon.
However, he may have said so somewhere else, or
Bernoldus may have calculated that fact. He may
merely have meant that Herimannus reported the ec-

_____
† See the report 1033 Jun 29b E,G in Section XI.2.

lipse and that it was on the 27th day of the moon.
Bernoldus' account seems original and contemporaneous.

The eclipse of 1093 Sep 23 was annular.
The statement that a circle appeared in the sun
and that it "shone darkly" sounds as if Bernoldus
was within the umbral region. However, I shall
not assume that the report is this definite; I
believe that one could use such words about a large
but nonetheless partial eclipse. Reliability: 1.
Place: Schaffhausen. Standard deviation of the
magnitude: 0.01.

1147 Oct 26 E,CE.   Reference: Engelbergenses [ca. 1175]. The second sentence of the entry
for 1147 reads: "Also there was an eclipse of the
sun on the 7th calends November (= Oct 26) before
midday on the Lord's day." 1147 Oct 26 was on the
1st feria as stated, and the eclipse was before
midday. The next sentence records the death of a
Duke Frederick, and the sentence after that says:
"In that same year there was an eclipse of the sun
on the 7th calends November at the 3rd hour of the
day on the 28th of the moon." The age of the moon
is correct.

1147 is the year in which the first annals at Engelberg appear, and we may speculate that
this is the year in which they were transferred
from S. Blasien. It is possible that two annalists
at Engelberg put down different records of the eclipse. I think it more likely, however, that the
eclipse was recorded at S. Blasien in the first of
the two records above, that the annals were then
taken to Engelberg, and that an annalist there recorded the eclipse again, not noticing that it was
already recorded. Accordingly, I shall assign the
first of the two records to be a record made at S.
Blasien[†] and the second to be one made at Engelberg.
Because of doubt about the place, however, I shall

_____

[†]This record will therefore appear in Section XI.2
as the report 1147 Oct 26f   E,G.

-289-

use a reliability of 0.5 rather than 1. Place:
Engelberg. Standard deviation of the magnitude:
0.1.

CHAPTER X

FRENCH RECORDS OF SOLAR ECLIPSES

1. Editions of the Sources

There have been many editions and translations of various individual works written in the Middle Ages in what is now France. There have also been many large collections. Which of these editions would be most useful in research depends upon the nature and purpose of the research. It is beyond the scope of this work to attempt an evaluation of the various editions. I only wish to point out 3 large collections that seem to combine value for the purposes of this work with reasonable availability.

The first of the three is the large collection prepared under the auspices of the Société de l'Histoire de France. It contains about 500 volumes prepared at times from the first part of the 19th century to the present. However, I have not made much use of this series because a short inspection of the titles indicated that it dealt mostly with years subsequent to those of interest in this study. I used it mostly for its collection of Angevin chronicles.† It could be quite useful in an extension of this work.

The second collection is that usually denoted by the name of Dom Martin Bouquet (1685-1754), a Benedictine monk. The collection contains 24 volumes. Lest the reader should think this to be a small collection compared with the first, let me point out that these are folio volumes of about 1000 pages. Each volume contains probably as much

---

† See the citation for the reference Rainaldus [ca. 1152], for example.

material as ten or more volumes of ordinary size.
Bouquet himself lived to complete only the first
8 volumes, which appeared between 1738 and 1752.
Later volumes have a variety of editors and have
appeared over an extended time; the 24th volume is
dated 1904. The editors of the first several volumes after the 8th are identified only as Benedictine monks of the Congregation of S. Maur, the
Congregation to which Bouquet himself belonged.
They perhaps chose to remain anonymous in order
that the series could stand as a fitting memorial
to Bouquet's scholarship.

In spite of its impressive scholarship,
I found the series hard to use. The critical commentary needed to evaluate the texts is given in
the prefaces to the volumes, separated from the
textual material itself, with at best awkward means
of matching commentary with text. The editors presented in each volume a large collection of material
dealing with a restricted period of time. This
means that many texts are divided among several volumes, and, more inconveniently, the critical commentary is also divided.

The editors also seemed to follow a particular principle in selecting the material that
they edited. Apparently their interest was that
of presenting documentation useful in preparing
histories on the national and international scale.
This interest caused them to have a sometimes-ex-
pressed objection to documents of only local interest and to prefer the sources with extensive interests. Unfortunately, for our purposes the latter are the less useful type of source; we need
sources that are local to an identifiable geographical place.

The third and most useful collection for
this work is the series called Monumenta Germaniae
Historica that has been used extensively in the
preceding two chapters. Materials useful in studying medieval German history come from all of Europe.
I believe that the series contains documents from
almost everywhere in Europe west of Russia, including a small amount from the Byzantine Empire.

The series is divided into several main parts that are identified by a term placed after Monumenta Germaniae Historica, and each part contains many volumes. I did not notice any material useful for this work in the parts called Leges and Necrologia. The most important part for this work is the part called Monumenta Germaniae Historica, Scriptores, of which there are 30 volumes. The parts called Auctorum Antiquissimorum and Scriptores Rerum Langobardicarum et Italicarum, Saec. VI-IX, also contain useful information.

This series is extremely valuable and quite easy to use. With few exceptions, each source text is presented in a single volume, and the critical commentary is placed immediately in front of the text. The editors recognized that there are many needs in research and they included many local as well as general sources. Further, with few exceptions, if they presented a text they presented all of it. The editors of the Bouquet series, in contrast, often omitted portions of a text, apparently on the basis that it held no interest to the student. If the first part of some text was a mere copy of another, the editors of Scriptores often omitted it. Otherwise, they rarely omitted anything from a source document, at least so far as I noticed.

The famous historian Georg Heinrich Pertz (1795-1876), of Hannover, was the first editor of Monumenta Germaniae Historica. He edited the first 12 volumes of the Scriptores part, between the years 1826 and 1856. This included both the overall responsibility for the volumes as well as the detailed editing of most of the individual items. Some of the other volumes were also edited by Pertz and some were edited by Georg Waitz.

2. The Sources

In this section I shall describe briefly the sources I have found that contain independent records of solar eclipses. I have examined so many sources in which I found no independent records

that it would not be feasible even to list them.
However I shall mention a few obvious sources in
order to let the reader know that I have not overlooked them.

Ademarus [ca. 1028]. Ademarus is sometimes called Ademarus Cabannensis, after the town
of Chabanais between Limoges and Angoulême. He
was born about 988 of a noble Aquitanian family
and was educated in the monastery in Angoulême.
Later he lived for some time at Limoges and wrote
a history of the abbey there up to 1025. It is
known that he moved from Limoges back to Angoulême
at some time, and we may guess that it was after
1025, since the history relative to Limoges stops
then. All else that we know about him is that he
made a pilgrimage to Jerusalem and may have died
there. He died about 1030.

He wrote a history of the Franks in three
books. The editor said that the first two were entirely derivative and he did not publish them. The
source being cited here is only the third book,
which is regarded as a valuable original source
document for the period that it covers.

Adonis [ca. 869]. Adonis was Archbishop
of Vienne from 860 to his death in 875. He wrote a
chronicle from early times to 869. A longer chronicle has been ascribed to him, but the parts after
869 were added by someone else. I found no independent records of eclipses in his chronicle.

Aquicense [ca. 1166]. One of the many
copies of the chronicle by the Belgian writer Sigebertus [ca. 1111][†] was taken to Anchin, near Douai.
There someone unknown inserted additional material
into the main body of the chronicle and then continued it to 1166. The auctarium and continuation

---

[†] See Chapter VIII.

are apparently from local sources, and the continuation is presumably contemporaneous.

The editor of Aquicense mentions still another extension of the work of Sigebertus that he calls Affligemense. Much of the latter is a copy of Aquicense, and I do not cite it separately.

Aquicinctina [ca. 1237]. This is a continuation of Aquicense, which is itself a continuation of Sigebertus. Its account of the eclipse of 1153 Jan 26 is copied from Aquicense, while its account of the eclipse of 1178 Sep 13 is independent. I did not try to identify more closely the time when Aquicinctina became an independent source. Aquicinctina was compiled at the monastery of S. Salvatoris on the island of Aquicinctus. The monastery [Aquicense, ca. 1166] was founded in 1079. I have not been able to locate the island, but it must be in the vicinity of Douai.

Bellovacense [ca. 1163]. Still another copy of Sigebertus [ca. 1111] was taken to Beauvais, where it received enlargement and continuation. The added material contains an account of the eclipse of 1147 Oct 26 that may be independent.

Besuenses [ca. 1174]. The editor said that these annals are from the monastery of Besuense, called Blaise today, near Dijon in the village of "Hattoariorum"; I have not tried to guess the nominative case of this name. I do not find Blaise, and I cannot find any place to match "Hattoariorum" on my maps of France. However I find Beze about 25 kilometers northeast of Dijon and I believe that this must be the place. Ginzel [1882 and 1884] reaches the same conclusion.

These annals are written on the margins of Easter tables. Before about 880 they are derived from the Spanish writer Isidorus (see Chapter XIV), the French source Lugdunenses [ca. 841; see

below],[†] and the German source that the editor of
Besuenses calls Annales Alamannici.[‡] For later
years there is some question about the relation between Besuenses and the source S. Benigni [ca. 1214]
discussed below. Pertz (the editor of Besuenses)
says that an earlier editor asserted that Besuenses
is derived in part from S. Benigni, which was written in Dijon. If I understand Pertz correctly, he
believes just the opposite. I shall accept Pertz's
conclusion. Luckily, the places are so close together that the correct answer is unimportant for our
purposes.

        Burburgensis [ca. 1164]. This is still
another continuation of Sigebertus [ca. 1111], this
one being kept at Bourbourg. It seems to be contemporaneous.

        Cameracensium [ca. 1042]. This is from
the town in northern France whose modern name is
Cambrai. The town dates from at least Roman times,
and it had a bishop in the 5th century. This work
is a history of the bishopric. In its earlier parts
it is derived from sources such as Gregory of Tours
[ca. 592] and other standard works such as the biographies of Charlemagne. It seemed to me to be independent from about 950 on, although I noticed no
statement by the editor about the matter. It is
apparently the work of one person.

        This source has been designated by sever-
al names, and there has been some controversy about
the identity of the author. In volumes X and XI
of the Bouquet series there is a work identified

[†] However, I think the borrowing from Lugdunenses
was incidental to borrowing that monastery's copy
of Bede's De Temporum Ratione [Bede, 725]. See
Chapter II, and see the discussion of the record
878 Oct 29c E,F in the next section.

[‡] See the discussion of the source cited as Augienses [ca. 954] in the next chapter.

as Chronicon Cameracensium et Atrebatensi,[†] auctore Baldericus. I have not checked this work in its entirety but I have checked several places and found it to be identical with Cameracensium. The editor of Cameracensium says that a Baldericus has been suggested as the author, which tends to confirm the identity of the works. If they are not the same, there has certainly been extensive copying. I shall assume that they are the same.

The anonymous editors of v. X of the Bouquet series say that the author was Balderic, Bishop of Noyon, not far from Arras. The anonymous editors of v. XI say that the author was not Bishop Noyon but Balderic, "Chantre de Térouanne", another town in the area. Bethmann, the editor of Cameracensium, says, as I have remarked, that Bishop Baldericus has been suggested, but that the author is actually an unknown canon at Cambrai acting under the instructions of Bishop Gerardus I. To confuse matters still more, this source is one of the sources used by Sigebertus [ca. 1111], which is also edited by Bethmann. When he is acting as the editor of Sigebertus, Bethmann identifies this work as Baldericus, whom he denies as the author elsewhere. I shall assume that the author is unknown.

This still does not end the confusion. Ginzel [1882 and 1884] did not realize that the sources are the same. He identified an eclipse report in Cameracensium as 1037 Apr 18. He identified the identical report[‡] in Chronicon Cameracensium et Atrebatensi as 1039 Aug 22. Unfortunately, the eclipse in question cannot be identified at all; the data given in the record are not consistent with any eclipse.

[†]This equals the modern Arras. The bishopric was originally at Arras. It was transferred to Cambrai in the 5th century and it was transferred back to Arras about 1100.

[‡] Whether the works are identical or not, the eclipse reports are.

Clarius [ca. 1124]. Clarius was a writer
in Sens who prepared a chronicle up to the year
1124. A continuation to 1180 is cited separately
as S. Petri [ca. 1180], after the monastery where
Clarius was a monk. The editors of volume 12 of
the Bouquet series say that the chronicle of Clarius is one of the best works of its period. If
the parts of it that concern us are typical, this
judgment seems unwarranted to me; "careless" is a
more suitable adjective. See the discussion following the report 1044 Nov 22c E,F in the next
section.

Elnonenses Maiores [ca. 1224]. Ginzel
gives a town S. Amand in northern Belgium as the
place where these annals were compiled. It is
clear from the annals themselves that they have to
do with an abbey founded by S. Amand, whose dates
are given as 571-661. The town S. Amand-les-Eaux,
about 40 kilometers southeast of Lille, is built
around an abbey founded by S. Amand in 647, and it
is at the junction of the Elnon with the Scarpe.
Thus I believe that this, and not the Belgian town,
is the correct place. If I am correct, this is
probably the only error that Ginzel made in his
places. I do not know why the annals are named
for the stream rather than for the abbey.

The annals are written in a variety of
hands from the late 11th century on, so they are
probably an independent source for the 12th century.

Elnonenses Minores [ca. 1061]. As I
have implied by the name, these are a smaller set
of annals kept at the same place as the last source.
The editor does not say much about these except that
someone obviously copied them carelessly from an
older source. However, they do contain two reports
of eclipses that are independent of any other sources
I have found.

Engolismenses [ca. 870]. There are two
sets of annals called Annales Engolismenses in the

Monumenta Germaniae Historica, Scriptores. One
set, edited by Pertz in v. IV, does not have any
records of eclipses. In v. XVI Pertz says that,
subsequent to the appearance of v. IV, L.C. Bethmann found the present set of annals in the Vatican library and copied them for him. There are
three independent parts to the annals that Bethmann found. The original part covers the years 815
to 870. Then there is a continuation for 886-930
and another continuation, almost as brief, for
940-991. Only the first part has any information
useful for this study, and the citation refers
only to it. The place is Angoulême.

Flaviniacenses et Lausonenses [ca. 985].
There seem to be two towns named Flavigny in France.
The one that is on the road map is not in the en-
cyclopedias that I have consulted, and vice versa.
The one we want is the one that is not on the map.
According to Ginzel, it is about 60 kilometers
northwest of Dijon and about 5 kilometers southeast of the probable site of Alesia, where the last
significant battle of Caesar's Gallic Wars was
fought. A monastery was established at Flavigny
in the 8th century. The annals are in one hand
down to 800. Later parts are in various hands and
are probably original. About 840 the annals were
taken to, and continued at, a place that the editor calls Lausona. Since the only eclipses took
place before the transfer, I have not tried to
identify Lausona.

Floriacenses [ca. 1060]. The town whose
Latin name was Floriacum is the modern S. Benoit-
sur-Loire, about 40 kilometers upriver from Orléans.
The monastery there, founded in the 7th century,
had considerable importance during the Middle Ages.
The annals are written on a 9th century copy of a
table of the "great Easter cycle" for 532-1063.
The editor says that the entries before 854 are all
copied from various sources, but he does not name
them. From 854 on, the entries seem to be original.

Folcwinus [ca. 962]. Folcwinus was a
member of an illustrious family descended from
Charles Martel, and the editor of his work gives
his family tree. He was born about 935, and was
ordained at the abbey of S. Bertini Sithienses[†]
in 961. Apparently one of the first things he did
after ordination was to write this history of the
abbey of S. Bertini. Some years later Folcwinus
moved to the abbey at Lobbes in Belgium, where he
wrote a history of the abbacy of Lobbes. At some
time he also wrote a biography of his great uncle
S. Folcwinus of Thérouanne. He became abbot of
Lobbes and died there in 990.

Because he wrote about both S. Bertini
and Lobbes, Folcwinus was long thought to be two
different people.

"Fredegarius Scholasticus" [ca. 641]. A
Burgundian chronicle written about the middle of
the 7th century has traditionally been ascribed to
a person called Fredegarius Scholasticus. The editor says: "We do not know why this work has been
attributed to Fredegarius, since there is no manuscript that carries this name." Since the work
cannot be associated with a specific place nor with
a particular person, I shall call it by the name
"Fredegarius Scholasticus" for want of an alternative, but I shall put the name in quotation marks
to indicate that it is not to be taken literally.

The unknown author names some sources that
he used, which include Eusebius Pamphili [ca. 325]
and Hydatius [ca. 468].[‡] In addition, he either
used an active imagination or he used works that he
did not name and that are now lost. He shows a
reasonable acquaintance with recent Byzantine affairs. There are several continuations of this

---

[†] For more information about this abbey, including
where it is, see the discussion of the source
Sithienses [ca. 823] below.
[‡] See Chapter XIV and Section XV.3 for discussions
of these sources.

chronicle that have no information needed here and
that are not cited.

Burgundy, at the time of this chronicle,
lay in parts of modern France and Switzerland. It
extended roughly from Besançon to Vienne north-
south and from Autun to Fribourg (Switzerland) east-
west. I shall use the region centered at 46°.5N,
5°.5E, with a radius of 1° of a great circle, as
being equivalent to Burgundy.

<u>Gaufredus</u> [ca. 1182]. Gaufredus was a
member of a well-connected family of Limousin. In
connection with an event in 1151, he describes himself as a little boy in school; hence we may guess
that he was born about 1140. He was ordained in
1167 and became prior of the monastery at Vigeois[†]
(= Vosiensis) in 1177. He wrote this chronicle of
the monastery about 1182. The editors say that
his work would be excellent if it were better or-
dered, but they complain that Gaufredus hardly mentions events outside the region of Limousin. This
may be a disadvantage for their purposes, but it
is an advantage for ours.

<u>Gregory of Tours</u> [ca. 592]. This name is
so well known that I use the English rather than
the Latin form of it.[‡] Gregory belonged to an illustrious senatorial family that had already contributed several important members to the early
church. He was born on 538 Nov 30 and became bishop
of Tours in 573. In that position he played a
prominent part in both the religious and political
affairs of the time. He died on 594 Nov 17.

This source, which is his <u>Ecclesiastical</u>

[†] Vigeois is a town about 50 kilometers southeast
of Limoges.
[‡] Since the name is so well known, it seems ironic
that it is not his correct name, which is Georgius
Florentius.

History of the Franks in Ten Books, is his best
known work. It is one of our main sources for the
history of the early Merovingian line of Frankish
kings. It should be noted that Gregory, like "Fredegarius", kept up with the affairs of the Byzantine
Empire.

Unfortunately, we cannot acquit S. Gregory
of the charge of being superstitious. He explains
[Gregory of Tours, ca. 580] how we can tell from the
shape of a comet whether it portends danger to the
king or the kingdom. In each book of his history
after he comes reasonably close to his own time, he
devotes one chapter to signs such as comets or eclipses, and he also usually gives us at least one
miracle per book. However, we must agree that his
superstitious concern has been valuable to us because it is the reason he recorded eclipses.

Honorius [ca. 1137]. Honorius Augustodunensis wrote two works, Summa Totius Mundi and
Imago Mundi. The one used here is the first, whose
title means The Totality of the Whole World or something like that. Such grandiose titles were popular
in his time. Honorius used many of the standard annals and chronicles in writing his own. Otherwise,
we know of him only that he lived in Autun, as his
"place name" indicates, and that he flourished in
the early 12th century. I have ignored the reports
of eclipses before his own time. His work was
fairly popular and had several extensions. I did
not notice any information in them that is useful
for this study.

Hugo Flaviniacensis [ca. 1102]. Hugo, of
an illustrious family, was born in 1065 and became
a monk at S. Vanne in Verdun in his youth. In 1097
he became abbot of the monastery of Flavigny. This
fact accounts for his name and also explains why
his chronicle has sometimes been called a chronicle
of Flavigny; it is really a chronicle of Verdun.
The editors say that the part before 919 contains
almost as many errors as words, and that from 919

to 966 it is a copy of the chronicle of Frodoard[†] but with little accuracy. As for the rest, they say, it has little order, the dates are confused, there are anachronisms, and there are still other sins against the historical truth. They say that the work nonetheless has some value because it contains material not found elsewhere. I could not help wondering whether such material is not found elsewhere because it never happened.

If Hugo's treatment of eclipses is typical, his work merits the comments of the editors. His reports are but careless copies of the reports in Rodulfus [ca. 1044], in which he has managed to compress three eclipses into two. In consequence of Hugo's chronicle and of the many chronicles for which it served as a source, we have a large collection of eclipse reports that have been wrongly identified. Clarius [ca. 1124] in turn made a careless copy of Hugo's careless copies, creating still more problems.

Hugo Floriacensis [ca. 1137]. Hugo wrote many historical works, fragments of which are published in the Monumenta Germaniae series. None of these fragments has anything useful for this study, and I cite them only to spare the reader the trouble of a search. The editors of this series usually published all parts of a work that they thought might have any value at all, so it is doubtful that any of Hugo's work will be useful.

Laudunense [ca. 1145]. This is still another of the many enlargements and continuations of Sigebertus [ca. 1111]. This one was made at Laon and is probably original for the time it covers.

Lemovicenses [ca. 1060]. This is a

[†] I found no material in Frodoard's work that is useful for this study and have not cited it.

sparse set of annals kept at the monastery of S.
Martialis at Limoges. The editor says that the
early part is copied but that the part after 867
was compiled at Limoges. He also says that the
annals are written on a table of the 19-year cycle.
One might interpret this to mean a table of the
ecclesiastical moon (see Section II.4), but I do
not see how annals covering a period of several
centuries can be kept on a table that repeats every 19 years. Hence I imagine that the editor
means a standard Easter table, which is based upon
the 19-year lunar cycle.

       Lugdunenses [ca. 841]. These annals
were written in the margins of the copy of Bede's
De Temporum Ratione[†] that was formerly owned by
the cathedral at Lyon. There are only seven entries, for the years 769, 782, 792, 804, 816, 840,
and 841. The editor says that there were once
some others that have faded and that cannot be
restored by any chemical treatment. I found these
annals, or at least the first five entries, rather
personal and touching. They gave me the impression
that someone studying Bede was inspired to record a
few events of his own life. The first entry says:
"769. In this year I was born." The second says:
"782. In this year I came from Spain into Gaul."
However, the entries for 840 and 841 are fairly
long formal entries of an impersonal nature, and
they read to me as if they had been written by a
different person. Perhaps a study of the handwriting could tell us how many authors are involved.
The number of references to Lyon, and the almost
complete lack of reference to other places, makes
it almost certain that all the annals were written
at Lyon. However, the book was probably taken to
several places, perhaps to be copied, and it seems
likely to me that it was in Beze about 878.[‡]

[†] This is the reference Bede [725] discussed in
Chapter II.

[‡] See the discussion of the record 878 Oct 29c E,F
in the next section. Copying, if it took place,
was probably not for the sake of the annals but
because the original book was by Bede.

Mettenses [ca. 905]. These annals were apparently compiled by a single author who was a monk of S. Arnoul at Metz. He copied almost word-for-word from many sources, including the work of Gregory of Tours and "Fredegarius Scholasticus". I do not believe that it has independent information that is useful for this study. I cite it because it furnishes an extreme example of the difficulty of working with Bouquet's format. As the reader can see by consulting the references, this source is divided among 5 different volumes; such a procedure might simplify an intensive study of a limited portion of history if one could take the sources at face value. However, an important part of historical study is a study of the value of the sources, and this is difficult if a source is fragmented.

Ginzel [1882 and 1884] says that Mettenses has a record of the eclipse of 840 May 5. I cannot verify this from the cited edition because the annals for 831 through 840 are omitted, apparently because the editor thought that their information would not interest anyone. Since it is doubtful that the eclipse record is independent, I have not tried to find it in some other edition.

Mortui-Maris [ca. 1180]. Since Mortemer is the French equivalent of Mortui-Maris, I imagine that this chronicle is from the Norman village of Mortemer, the seat of the Mortimer family that is famous in English history. This is one of what the editors call "thirteen little chronicles" that "pose no difficulties except the few mentioned in footnotes." The editors tell us nothing about the time coverage of the chronicle, whether it is all in one hand, whether they have printed all of it, and other "unimportant" matters.

This may be a continuation of Sigebertus [ca. 1111]; it is in a favorable time and place to be one. The first entry, marked 1113, says that Sigebertus, a man incomparably skilled in omni

scientia litterarum, died on Oct 5 of that year.[†]
The next entry says that he died in 1114 "as we
count" the years; at this time there were still
several usages for numbering the years. Thus the
annals, at least in the edited form, begin where
Sigebertus' chronicle leaves off.

Mosomagenses [ca. 1371]. The editor and
Ginzel [1882 and 1884] both give the modern name of
this place as Mousson, and Ginzel gives a set of
coordinates. My map gives a village called Pont-
à-Mousson near the specified point, so this is
probably the place. It is near Metz in Lorraine.
These annals are from an abbey there. They are
written by many people, and there are numerous gaps
in them. This suggests that they were kept by
various people on no assigned basis but merely as
the individuals were moved to keep them. Perhaps
they were forgotten at times and then discovered
again. Most of the entries before 1187 are brief,
and most of the entries after 1208 are long, but
still scattered in time. There are only 22 entries
for the years 1226 to 1371. The editors describe
the annals as going to 1452, but there is only the
single brief entry for 1452 after 1371. Hence I
have used 1371 as being a more significant date for
the close of the annals.

Some of the words cannot be read with
surety. The editor has suggested words for these
places and has put his guesses in italics.

Nithardus [ca. 843]. Nithardus (795?-
844) was the natural son of Bertha, daughter of
Charlemagne, and of S. Angilbert, an abbot and a
confidant of Charlemagne. Apparently saints were
allowed more latitude then than they would be now.[‡]

[†] The standard sources say that Sigebertus died on
1112 Oct 5.

[‡] And so, perhaps, were Charlemagne's daughters.
A story goes that his daughter Emma was enter-
taining a lover in her rooms one night when a
snowfall made it impossible for him to leave
without making telltale footprints. Emma thereupon carried her lover back to his quarters in

The editor describes Nithardus as "spirit, counseler, and right arm" of Charles the Bald (Charles II, first king (843-877) of what may be called France). Nithardus' history is an important source for the history of the last part of the reign of Louis I and for the fratricidal wars that followed his death. Nithardus died on probably 844 Jun 14, and the last event in his history is the lunar eclipse, followed by a heavy snowfall, of 843 Mar 19, which he put on Mar 20.

Nivernenses [ca. 1188]. These annals are from Nevers, in central France. A church in Nevers [Nivernenses, ca. 1188] was dedicated in 858 in honor of God, the Savior, Mary, and S. Peter. The annals are written between lines of Easter tables in various hands belonging to various centuries. This characteristic suggests originality, but the suggestion is not necessarily valid in Nivernenses. Many entries were inserted at times much later than the events, by persons using Engolismenses [ca. 870] and other Aquitanian sources. The editor has studied the hands carefully in order to distinguish contemporaneous records from later insertions.

Praemonstratensis [ca. 1155]. This is a continuation of Sigebertus [ca. 1111] from a place that the editor describes as being in the diocese of Laon or Reims. I have found no other clue to the place. The institution involved was probably the original house of the Premonstratensian order. The document should be original. I shall use Laon as the place.

Rainaldus [ca. 1152]. Rainaldus is described as an archdeacon of the cathedral of S. Maurice at Angers in Anjou. Much of his chronicle

---

order that he should leave no incriminating marks. The story may be doubted. It is told of more than one person in medieval times, and there is no contemporary evidence that Charlemagne had a daughter Emma [Holland, 1910].

is practically a copy of earlier works, and it is
an independent source only from about 1050 to its
close. Even so, the editors believe that it is
probably the oldest Angevin chronicle. It was used
extensively by later chroniclers.

Robertus Autissiodorensis [ca. 1211].
Robert is described as "Robertus Canonicus S.
Mariani Autissiodorensis". There are several possible Robert's. His chronicle, including the records
of solar eclipses, is largely derivative, and I did
not spend much effort on it. The place is Auxerre.

Robertus de Monte [ca. 1186]. This Robert
is sometimes called Robert de Torigni. His appel-
lation "de Monte" comes from the fact that he was
abbot of Mont S. Michel from 1154 until his death,
probably in 1186. This work is called a chronicle
but it is also one of the many continuations of
Sigebertus. Robert's office put him in a good position to be well acquainted with Anglo-Norman affairs, and his work is described as valuable for
affairs from 1154 to 1170. Otherwise it has little
useful material.

Rodulfus [ca. 1044]. This work is one
of the few which, in my opinion, are better edited
in the Bouquet series than in the Monumenta Germaniae Historica series. The former gives the entire
work, and in one volume, while the latter gives
only part of it. Rodulfus was born in Burgundy,
was educated in a monastery, and lived in several
monasteries including S. Benigni Divionensi (at
Dijon). In either 1001 or 1002 [Rodulfus, ca. 1044,
Book III, Chapter 3] he moved to Cluny and apparently
spent the rest of his life there. He began his history at Dijon but wrote the parts of it that concern
us at Cluny.

The editors of the Bouquet series do not
seem to like Rodulfus. They describe his wanderings
by saying that he deserted the monastery where he had
dedicated himself to God, and that he quitted seve-

ral others before he went to Cluny. His work, they
say, is unformed and poorly digested, and he suffers
from many sins unbecoming to an historian. In spite
of these flaws, they give him considerable space because he has material not found elsewhere. This
seems an odd basis for importance. If his work has
little value where it can be tested, why should it
have value where it cannot be? Perhaps the very
sins they object to are the reasons that he has information "not found elsewhere".

Rodulfus is the first contemporaneous
writer I have found who says that the year 1000 was
accompanied by many signs.

Rotomagensi [ca. 1174]. This, like Mortui-
Maris [ca. 1180], is one of "thirteen little chronicles" that pose "no difficulties" needing editorial
comment. At least as edited, this chronicle starts
in 1074 and thus it is not a continuation of Sigebertus [ca. 1111]. The entries are relatively
sparse. The place is Rouen.

S. Albini [ca. 1357]. This work is denoted by "chronicles" in the plural rather than the
singular. The library of S. Aubin in Angers once
had at least seven manuscripts containing chronicles.
This work is the combination of all seven, with the
combining being done by editors; they said that all
the chronicles were variants of a single one. The
editors thought that the entries were contemporaneous
from about 1000 on.

S. Andreae [1133]. This chronicle, in three
books, was written by an unknown monk at the monastery of S. Andreas in Cambrai. The last few lines
tell us that it was written in 1133. There is
nothing else to say about it.

S. Benedicti [ca. 1110]. The editors
entitled this a "fragment of Frankish history" found
in an ancient manuscript from the monastery at S.

Benoit-sur-Loire. Since their title does not seem
to be a useful one for citing, I have invented the
title used here. The editors did not give much information about the fragment. They say that it is
an open question whether it should be attributed to
Hugo Floriacensis or not. At least this tells us
that the work was apparently written in the first
part of the 12th century, and that it is hence at
least approximately contemporaneous; otherwise there
could be no question about attributing it to Hugo.[†]
It is also written by a single author who, according to the editors, does a poor job with his chronology.

        S. Benigni [ca. 1214]. There are two
copies of these annals. One that extends to 1285
is, if I understood the editor, entirely in hands of
the 13th century and hence it is not original in
the parts that concern us. One can regard it perhaps as a copy and later continuation of the other
version. The second, which is the one being cited,
is more extensive for the period before 1214. It
was compiled by several people, perhaps around or
in 1125. One hand goes to about 1063, and it is
possible that someone made an initial compilation
then. It is also possible that someone around 1125
got as far as 1063 before something stopped his
labors, and that others then brought the annals up
to date and then kept them that way to about 1214.
It is interesting that S. Benigni contains a copy
of the entries from Ludgunenses [ca. 841] for 840
and 841. This work was written in Dijon. The parts
after 1125 are probably original. Those between
the middle of the 11th century and 1125 may be, but
they must be used with caution if at all.

        S. Denis [ca. 1150]. This is from the
famous church north of Paris. It has no data for
this work, but it is interesting for two reasons:
It is in French rather than Latin, and it contains

[†] See the discussion of Hugo Floriacensis [ca. 1137]
above.

two contradictory accounts of the expedition of
Charlemagne into Spain in 778.[†]

        S. Florentii [ca. 1236]. This chronicle
is from the abbey of S. Florent de Saumur. The
editors did not give the part before 700 because it
offers "no interest for our national history." They
did not give any opinion that I saw about the originality of later parts. It has two conflicting reports of the eclipse of 878 Oct 29, neither under
the right year, so it was probably not original for
the 9th century. Some of its parts for the 12th
century are probably original.

        S. Maxentii [ca. 1140]. This chronicle
was compiled in the 12th century in the monastery
of Maillezais near La Rochelle. It has no original
information useful for this work that I noticed.
It was compiled, mostly by copying, from S. Albini
[ca. 1357], Rainaldus [ca. 1152], Vindocinense
[ca. 1251], and S. Florentii [ca. 1236],[‡] as well
as from now-lost works. He seems to have copied
entire sections at once in some places, with no
attention to the contents or to chronology. As a
result the material is often out of chronological
order and the same event is often given more than
once. The compiler has two reports of the eclipse
of 1033 Jun 29, both under 1042, along with a demise
of the French crown that took place in 1031. He
has three reports of the eclipse of 878 Oct 29,
under the years 875, 876, and 877.

        S. Petri [ca. 1180]. This is the contin-
uation of Clarius [ca. 1124] mentioned earlier.
The two sources appear as a single set of annals in
the manuscript. There are no entries between 1124

[†]
See the record 779 Aug 16 E,F in the next section.

[‡]
The reader may notice that S. Maxentii closes
sooner than any source I have mentioned from which
the compiler copied. This presumably means that
he copied from early stages of these works.

and 1128, and all authorities that I have consulted agree that the author changed at this point. There are two copies of the continuation.

S. Sergii [ca. 1215]. This is a fairly late chronicle from the abbey of S. Sergius near Angers. Much of it is taken from Rainaldi and S. Albini, but it has some original material.

S. Vincentii [ca. 1280]. This is also a rather late chronicle that becomes contemporaneous after about 1154. It was prepared in Metz and used Belgian sources as well as older sources from Metz for its early information.

Senonensis [ca. 1203]. S. Columba, virgin, was martyred in 266 [Senonensis, ca. 1203], and these are annals kept in a church in Sens dedicated to her on 853 Jul 22. After 868 the annals seem to be contemporaneous and original. Unfortunately the annalists did not seem to be interested by solar eclipses; the only ones I noticed were in the part before 868.

Sithienses [ca. 823]. In either the 7th or 8th century (the sources disagree) a monastery dedicated to S. Bertini was founded on an island called Sithiu in the River Aa, at the town of S. Omer in northern France. It flourished in the 8th century but was harassed by the Norsemen in the 9th century and declined in importance. It was later restored and Folcwinus (see the discussion of Folcwinus [ca. 962]) was one of its later abbots.

These annals have much in common with the earlier part of the German annals Fuldenses [ca. 901]. There has been a considerable controversy about which of the two is the original. The editor of Sithienses feels strongly that Fuldenses is original, saying of Sithienses: "I . . hold these annals to be nothing except excerpts from Annales Fuldenses." The German annals Laurissenses [ca. 829] also come into the

question. All three sources have identical records of several eclipses and all sources are of
about the same antiquity.

Fuldenses and Laurissenses come from
places quite close together, and the uncertainty
about them is almost negligible for our purposes.
Sithiu, however, lies about 7° to the west. Luckily, the eclipses involved are old enough that an
uncertainty of 7° in longitude is only uncomfort-
able and not vital.

The weight of editorial opinion[†] is that
Sithienses is the secondary source, perhaps because
Sithienses shows more interest in Saxon affairs than
we would expect in a place so far west. Hence I
shall take Fuldenses or Laurissenses as primary when
either shares material with Sithienses. However,
Sithienses has some material that does not appear
in the others. I shall take such material as original with Sithienses.

Turoldus [ca. 1100]. This is one of the
12th-century versions of the Song of Roland. The
text names Turoldus as the author, but we do not
know whether he was the composer or merely a scribe
who copied or translated an older version. I used
Turoldus in AAO because it furnishes a "textbook"
example of a magical eclipse.

Vedastini [ca. 900]. These annals are
from the monastery of S. Vedasti (S. Vaast) in
Arras. The monastery is an old one but I have not
found a specific date for its founding. The editor
says that the annals contain nothing before 877
that is not found in other annals, and he does not
publish those parts. The annals seem to be original
for the short period from 877 to their close.

Vindocinense [ca. 1251]. The name refers

[†] See the discussion of Laurissenses in Section XI.2.

to the modern place of Vendôme. The annals have
been known as Chronicon Andegavense, but the editor
thinks they would be more appropriately named after
Évière. The abbey of Évière was a dependency of the
abbey of Vendôme; it was founded about 1056 at Angers. The annals to about this time were compiled
by a monk at Évière who worked mainly by copying
Rainaldus. The annals were then continued at Vendôme,
and the part after 1060 has "more originality", as
the editors say.

                Weissemburgenses Minores [ca. 846]. The
Monumenta Germaniae Historica series contains two
works both called Annales Weissemburgenses. It would
be possible to distinguish them by their last years,
but it is better to use different names if there is
a significant reason for the difference. This set
is much smaller than the other; it contains only 15
entries between the years 763 and 846 inclusive.
Hence I shall designate this set as Annales Weissemburgenses Minores, using Maiores for the set that
will be discussed next. The entries seem to be the
contemporaneous work of more than one person. They
are written in a work kept in the monastery at Wissembourg, in Alsace.

                Weissemburgenses Maiores [ca. 1075]. This
set of annals, which I am calling the major set, is
closely related to the German sets of annals cited
as Hildesheimenses [ca. 1137], Quedlinburgenses [ca.
1025], and Lambertus [ca. 1077] in the next chapter.
They are so close that the editor prints all four
sets in parallel columns down through 973. More
information about the annals through this point will
be found in the next chapter. Weissemburgenses
Maiores has a small amount of information for years
before 973 that is not found in the others. This
information was probably inserted after a copy of
the early annals was brought to Wissembourg and
hence it is probably not contemporaneous even when
it is independent of other known sources. The annals
are presumably original after 973. The only entries
after 1075 are for 1087 and 1147, so I have used
1075 rather than 1147 for the date.

TABLE X.1

REFERENCES TO SOLAR ECLIPSES FOUND IN
MEDIEVAL FRENCH SOURCES

Source	Date	Conclusions
Ademarus	840 May 5	Too short to be safely used
	1023 Jan 24	Place somewhat uncertain
Adonis	807 Feb 11	Copied from German source
	810 Jul 5	Copied from German source
	810 Nov 30	Copied from German source
Aquicense	1153 Jan 26	Original
Aquicinctina	1153 Jan 26	Copied Aquicense
	1178 Sep 13	Original
	1191 Jun 23	Original
Bellovacense	1147 Oct 26	Original
Besuenses	878 Oct 29	Probably an original report in imitative language
	961 May 17	Probably not independent
	1033 Jun 29	Probably has a common origin with Rodulfus

TABLE X.1 (Continued)

Source	Date	Conclusions
Burburgensis	1133 Aug 2	Probably original
	1140 Mar 20	Perhaps original
Cameracensium	1039 ?	An unidentifiable record
Clarius	1033, 1039, and 1044	An account carelessly assembled from three different records
Elnonenses Maiores	1185 May 1 ?	Cannot identify; could be 1186 Apr 21
	1191 Jun 23	Original
Elnonenses Minores	878 Oct 29	Not independent but cannot trace
	1033 Jun 29	Original
Engolismenses	840 May 5	Original
Flaviniacenses et Lausonenses	764 Jun 4	Probably independent
	840 May 5	Not safe to use
Floriacenses	878 Oct 29	Original
	968 Dec 22	Original
Folcwinus	878 Oct 29	Independent
	891 Aug 8	Independent

TABLE X.1 (Continued)

Source	Date	Conclusions
"Fredegarius Schol- asticus"	592 Mar 19	Independent; identi- fication slightly questionable
	603 Aug 12	Probably independent
Gaufredus	1178 Sep 13	Original
Gregory of Tours	563 Oct 3	Probably original
	590 Oct 4	Original
Honorius	1133 Aug 2	Original
Hugo Flaviniac- ensis	1033 and 1039	Ran two reports togeth- er in copying
	1044 Nov 22	Careless copy of Rodulfus
Hugo Floriacensis	840 May 5	Copied Senonensis
	878 Oct 29	Copied Floriacenses
Laudunense	1133 Aug 2	Cannot tell whether this or Praemonstra- tensis is the original
Lemovicenses	1037 Apr 18	Original; identifica- tion slightly question- able
Lugdunenses	840 May 5	Original
Mettenses	878 Oct 29	Copied from German source

TABLE X.1 (Continued)

Source	Date	Conclusions
Mortui-Maris	1133 Aug 2	Perhaps independent; not enough information to judge
	1147 Oct 26	Perhaps independent; not enough information to judge
Mosomagenses	1005 ?	Unidentifiable
	1039 Aug 22	Original
Nithardus	841 Oct 18	Original
Nivernenses	840 May 5	Copied from Engolismenses
	1033 Jun 29	Original
	1044 Nov 22	Original
	1147 Oct 26	Original
Praemonstratensis	1133 Aug 2	Cannot tell whether this or Laudunense is original
	1147 Oct 26	Original
Rainaldus	809 Jul 16	Copied from Anglo-Saxon Chronicle
	878 Oct 29	Shared with other Angevin annals
Robertus Autissiodorensis	1133 Aug 2	Copied Praemonstratensis
	1178 Sep 13	Probably copied from S. Petri

TABLE X.1 (Continued)

Source	Date	Conclusions
Robertus de Monte	1133 Aug 2	Copied from Rotomagensi
	1153 Jan 26	Original
Rodulfus	1033 Jun 29	Probably common origin with Besuenses
	1039 Aug 22	Original
	1044 Nov 22	Original
Rotomagensi	1133 Aug 2	Original
S. Albini	1095 ?	Probably not an eclipse
	1191 Jun 23	Original
	1207 Feb 28	Original; slight question about identification
S. Andreae	1133 Aug 2	Original
S. Benedicti	1037 Apr 18	Independent
	1044 Nov 22	Independent
S. Benigni	1033 Jun 29	Possibly independent
	1178 Sep 13	Original
S. Florentii	809 Jul 16	Copied from Anglo-Saxon Chronicle
	878 Oct 29	Shared with other Angevin annals
	1178 Sep 13	Possibly independent

TABLE X.1 (Continued)

Source	Date	Conclusions
S. Florentii	1185 May 1	Probably independent
S. Maxentii	840 May 5	Careless copy, probably of Engolismenses
	878 Oct 29	Copied three accounts
	1033 Jun 29	Copied two accounts
S. Petri	1133 Aug 2	Not safe to use
	1178 Sep 13	Seems copied
S. Sergii	1178 Sep 13	Original
S. Vincentii	1191 Jun 23	Probably original
Senonensis	809 Jul 16	Copied from Anglo-Saxon Chronicle
	840 May 5	Probably independent
Sithienses	787 Sep 16	Probably copied from German source
	807 Feb 11	Perhaps independent
	810 Jul 5	Copied from German source
	810 Nov 30	Copied from German source
	818 Jul 7	Copied from German source
Turoldus	779 Aug 16 ?	Classic example of a magical eclipse

TABLE X.1 (Concluded)

Source	Date	Conclusions
Vedastini	878 Oct 29	Original
Vindocinense	809 Jul 16	Copied from Anglo-Saxon Chronicle
	878 Oct 29	Shared with other Angevin annals
	1191 Jun 23	Original
Weissemburgenses Minores	840 May 5	Original
Weissemburgenses Maiores	1033 Jun 29	Original

3. Discussions of the Records of Solar Eclipses

The usual list of the records, arranged by source, is given in Table X.1. As usual, I have not listed all appearances of a much-copied record.

563 Oct 3 E,F. Reference: <u>Gregory of Tours</u> [ca. 592], which says in Chapter 4.31: ". . on the calends of October the sun was so obscured that not a fourth part of it remained shining, but it appeared hideous and discolored like some sort of bag." This appears in the chapter that Gregory devoted to signs near this time, and the year is not given exactly. However there does not seem to be any question about the identification of the eclipse, even though the day of the year is wrong by 2. The eclipse was in Gregory's time, but this does not sound like a strictly contemporaneous account and I shall give it a reliability of only 0.5. Since he says "not a fourth part", I shall take the magnitude to be 0.8 (leaving 1/5 uneclipsed), with a standard deviation of 0.1. Place: Tours, path assumed north of Tours.

590 Oct 4 E,F. Reference: <u>Gregory of Tours</u> [ca. 592], who writes in Chapter 10.23: "The sun suffered an eclipse during the 8th month and its light diminished until the part remaining in light had horns like a 5th moon." I believe that "8th month" means October in its original literal meaning and that "like a 5th moon" means like the moon when it is 5 days old. A moon that is 5 days old has about 0.25 of its surface illuminated,[†] so I shall take the magnitude of the eclipse to be 0.75 of the area. This is equivalent to a magnitude of about 0.8 as defined in Section IV.5.[‡] I shall use this as the magnitude, with a

---

[†] See, for example, the Ephemeris for Physical Observations of the Moon in any recent issue of the <u>American Ephemeris and Nautical Almanac</u>.

[‡] See, for example, Section V.1 in <u>AAO</u>.

standard deviation of 0.1. Place: Tours. There
is no question about the identification of the
eclipse, and the account was written within a short
time of the eclipse. Hence I shall use a reliability of 1. The path is assumed northeast of Tours.

592 Mar 19 E,F. Reference: "Fredegarius
Scholasticus" [ca. 641; Section XIII]: "In the
32nd year of the reign of Guntram, the sun was eclipsed from morning to midday, so that hardly a
third of it was seen". Guntram or Gontran was king
of Burgundy from 561 to 593, and 592 was thus the
32nd year of his reign. Further, Oppolzer [1887]
shows that the eclipse of 592 Mar 19 should have
been in the morning in Burgundy. In spite of this
agreement we cannot regard the identification as
secure. The eclipse of 594 Jul 23 was also in the
morning and it should have had about the same magnitude in Burgundy as the eclipse of 592 Mar 19. The
chronology definitely excludes the eclipse of 594
Jul 23 if we can accept it as accurate, but mistakes
of 2 years in the date of an event are unfortunately
rather common. Therefore I shall assume that the
probability is 0.5 that the identification of 592
Mar 19 is correct. Since the account is not quite
contemporaneous, I shall further lower the reliability to 0.25. Place: 7th century Burgundy, as
defined in the preceding section. Magnitude: 2/3
with a standard deviation of 0.1; path south of
Burgundy.

603 Aug 12 E,F. Reference: "Fredegarius
Scholasticus" [ca. 641; Section XXIV]. "Fredegarius"
puts this eclipse in the year after Phocas killed
Mauricius and assumed the Byzantine throne, which
happened in 602. The record says: "In that year
the sun was eclipsed." I have found no earlier
record of this eclipse and the eclipse was within
less than 40 years of the compilation of the chronicle. Thus there is a good chance that the record
is independent and local. Reliability: 0.5. Place:
7th century Burgundy. Standard deviation of the
magnitude: 0.1.

764 Jun 4 E,F. Reference: Flaviniacenses et Lausonenses [ca. 985]. This source has,
under 764: "The sun suffered an eclipse on the
2nd feria on the 2nd nones Jun (= Jun 4)." The
date and the week day are correct. This is in the
part of the annals that were apparently compiled
around 800, so this is not contemporaneous. It is
the only account that gives the week day, but the
week day can be calculated and hence it does not
prove independence. Since the eclipse is reasonably
close in time to the time of compilation, there is
a moderate probability that the record is independent, and I shall give it a reliability of 0.5.
Place: Flavigny. Standard deviation of the magnitude: 0.1.

779 Aug 16 E,F. Reference: Turoldus
[ca. 1100]. Turoldus' 12th-century version of the
Roland legend makes a magical eclipse occur at the
death of Roland. I discussed this account at some
length in AAO. I did not advance it seriously as
a record of an eclipse; instead I used it pedagog-
ically. I do not believe that anyone would take it
seriously, but it has as much claim to validity as
a number of eclipse reports that have been used
seriously in calculations of the lunar motion and
in astronomical chronology. It is instructive to
compare this record with some that have been used.
It served its purpose in AAO and I shall not use it
in this study.

In fact, there is no basis in contemporaneous records for a major battle at Roncevaux.
S. Denis [ca. 1150] shows this point nicely. S.
Denis has two incompatible accounts of the action.
In the earlier account in Book I of the deeds of
Charlemagne, it says that he returned from the
Spanish campaign "safe and sound, except for a
little mischief on crossing the mountains produced
by the malice of the Gascons"; this leaves no room
for the major action of later legend. The account
that was written later recounts much of the legend
of Roland, but it does not have an eclipse at his
death. Thus the eclipse is not even a firmly established part of a legend.

787 Sep 16 E,F. Reference: Sithienses
[ca. 823]. Sithienses says under 787: "There was
an eclipse of the sun on the 15th calends October
(= Sep 17). Charles came with his army from Rome
to Capua." Fuldenses [ca. 901] reports the eclipse[†]
in identical words, including the mistake in date.
Fuldenses then says that Charles went to Benevento
rather than to Capua and makes no mention of Rome.
This shows that Sithienses is something more than
a series of excerpts from Fuldenses; see the discussion of Sithienses in the preceding section.

The identity in wording makes it almost
certain that the eclipse reports are not independent.
Since the weight of opinion favors Fuldenses as the
original, I shall not use the report in Sithienses.

807 Feb 11 E,F. Reference: Sithienses
[ca. 823]. Under 807 we find: "There was an eclipse of the sun on the 3rd ides February (= Feb 11)
and an eclipse of the moon on the 4th calends March
(= Feb 26). Another eclipse of the moon on the 11th
calends September (= Aug 22)." This is a correct
record of the solar eclipse of 807 Feb 11 and the
lunar eclipses of 807 Feb 26 and 807 Aug 21, except
that the latter date is the 12th rather than the
11th calends September. Fuldenses does not have a
report of these eclipses. Laurissenses [ca. 829]
has a report in quite different words. Thus the
record in Sithienses seems to be independent and
contemporary. However Sithienses is still copying
annals from other places in 818, so the record may
not be original in spite of appearances. Since it
is independent of other known records, I give it a
reliability of 0.5. Place: S. Omer. Standard
deviation of the magnitude: 0.1.

The Belgian source Blandinienses discussed
in Chapter VIII copied the report from Sithienses,
while Adonis [ca. 869] copied Laurissenses.

[†] This report is discussed in connection with the
record 787 Sep 16b E,G in Section XI.2.

Rainaldus [ca. 1152], S. Florentii [ca. 1236], Senonensis [ca. 1203], and Vindocinense [ca. 1251] have all copied the record of the eclipse of 809 Jul 16 from the Anglo-Saxon Chronicle[†] or from its Latin counterpart Annales Domitiani Latini, either directly or though intermediary sources.

Adonis and Sithienses have the same reports of the eclipses of 810 Jul 5 and 810 Nov 30 that we find in Laurissenses in the next chapter. Sithienses also has the report of 818 Jul 7 that we find in Laurissenses.

840 May 5a E,F. Reference: Senonensis [ca. 1203]. This says under 840: "Emperor Louis died, and there was an eclipse of the sun on the 4th feria before the Ascension of the Lord at the 9th hour on the 2nd nones May."[‡] The week day is correct and the hour is plausible, although it seems somewhat early. Louis died on 840 Jun 20. This record is before the time when Senonensis is considered to become original, but I cannot find the hour and the date given in any earlier source. Hence I shall take this report to be independent but not contemporaneous and give it a reliability of 0.5. Place: Sens. Standard deviation of the magnitude: 0.1.

Hugo Floriacensis [ca. 1137] has copied this report, as has the Italian source Romualdus discussed in Chapter XII.

840 May 5b E,F. Reference: Lugdunenses [ca. 841]. The first sentence in the entry for 840 reads: "An eclipse of the sun happened during

---

[†] See the record 809 Jul 16 B,E in Section VI.3.

[‡] The date given equals May 6 rather than May 5. Ascension Day in 840 was on May 6, and the date of Ascension Day was probably written inadvertently in place of the eclipse date.

the days of the litanies[†] on the 3rd nones May
(= May 5), the 4th feria, about the 8th hour of
the day, and it lasted for about half an hour, and
it became so dark that stars could be seen clearly
in the sky." This is a clear and accurate statement of a large eclipse, I think, and it is original, judging from all appearances. Reliability:
1. Place: Lyon. Standard deviation of the magnitude 0.01.

840 May 5c E,F. Reference: Weissemburgenses Minores [ca. 846]. The entry for 840
says: "Emperor Louis died. There was darkness
on the 3rd nones May (= May 5) at about the 7th
hour of the day." A number of sources have passages similar to this one. However, the circumstances surrounding the source make it likely that
this is an original and contemporaneous entry, and
no other source gives this hour. It is probable
that the similarity in wording of several entries
results from the fact that this is one of the simplest ways of wording the information. Reliability:
1. Place: Wissembourg. Standard deviation of the
magnitude: 0.1.

840 May 5d E,F. Reference: Engolismenses
[ca. 870]. The entry for 840 reads: "On the 3rd
nones May, the 4th feria, at the 8th hour, an eclipse of the sun was made, and on the 12th calends
July (= Jun 20) the emperor Louis ended his hours."
All the information in this account is correct and
it seems to be original and contemporaneous. Reliability: 1. Place: Angoulême. Standard deviation of the magnitude: 0.1.

Nivernenses [ca. 1188] has copied this
report almost verbatim. Many other sources have
reports rather like this one or the one before,
and the safe course is not to use them. S. Maxentii

[†] In diebus laetaniarum. The three days of the
Lesser Litany are the three days before Ascension
Day. In 840 these three days were May 3, 4, and 5.

[ca. 1140] has copied so carelessly that he makes
it sound as if the eclipse were on the day Louis
died, which he makes "ii. kalends junii" (= May 31)
instead of "xii. kalends julii".

        841 Oct 18 E,F. Reference: Nithardus
[ca. 843]. Nithardus wrote in Chapter II.10:
"As I was writing this, being halted upon the Loire
near sanctum Fludualdum, an eclipse of the sun
happened in Scorpio at the first hour, on the third
feria, on the 15th calends November (= Oct 18)."
The information in this account is correct. Unfortunately no one seems to know where sanctum Fludualdum was. The editor suggests "S. Claude above
Blois". Since the Loire runs almost straight from
Orléans to Tours, since Blois is halfway between the
two, and since the stretch between them is a reasonably large fraction of the Loire, I shall take Orléans and Tours as two possible places, thus automatically including places between and nearby. Reliability: 1. Standard deviation of the magnitude:
0.1.

        878 Oct 29a E,F. References: S. Maxentii
[ca. 1140], Rainaldus [ca. 1152], S. Florentii [ca.
1236], and Vindocinense [ca. 1251]. Under the year
875 S. Maxentii has: "There was an amazing (mirabilis)
eclipse of the sun at the 9th hour." Under 876 it
has: "A comet was seen in the month of July, and
an eclipse of the sun on the 5th calends November
(= Oct 28) was amazing." Under 877 it has: ". .
and an eclipse of the sun was amazing at the 9th
hour." It is amazing that the annalist did not notice that he was copying three accounts of the same
eclipse and putting them under different years.
There is a possible excuse for not noticing that
the middle record goes with the other two. The
other reports differ only by a trifling change in
word order, and it is hard to see how the annalist
would list them as separate events. I have noticed
several places where an annalist has the same event
twice, but this is the only place I have noticed an
event given thrice.

The treatment in S. Maxentii does not end
the confusion. S. Florentii has the 876 record
from S. Maxentii under 874,† and it has the 875
record under 877. Several Angevin sources, including Rainaldus and Vindocinense, have the record of the comet in July and the eclipse on "Oct
28" under a variety of years.

It seems safe to conclude that none of
these records is original. However, the record
of a comet in July and the eclipse on Oct 28 occurs nowhere that I have seen outside of Anjou,
so it perhaps represents a record that originated
there. Hence I shall take all the Anjou sources
as representing a single record to which I shall
give a reliability of 0.2. Place: Anjou. Standard deviation of the magnitude: 0.1. I shall
take Anjou to be the region centered on Angers
with a radius of 1° of a great circle.

878 Oct 29b E,F. Reference: Vedastini
[ca. 900]. Under 878 these annals have: ". . .
and in that same month there was an eclipse of the
sun at the 8th hour of the day, in the 12th indiction." The context identifies the month as October.
This seems to be a contemporaneous record. Reliability: 1. Place: Arras. Standard deviation
of the magnitude: 0.1.

878 Oct 29c E,F. Reference: Besuenses
[ca. 1174]. These annals have the following under
878: "On the third calends November (= Oct 30)
an eclipse of the sun happened on the 4th feria,
about the 8th hour of the day, and it lasted for
about half an hour, and it became so dark that
stars could be seen clearly in the sky." Except
for the statement of the date, this record is
identical with the record 840 May 5b E,F from
Lugdunenses [ca. 841].

---

† Even the editor joined in. He put a footnote at
this point saying that the correct year may be
973!

We can guess that the following is the explanation of the near-identity of the records of two eclipses that happened more than 38 years apart: In the discussion of Lugdunenses in the preceding section it was pointed out that the Annales Lugdunenses were written in the margin of a copy of Bede's work on reckoning time. This was a popular work and we can guess that the monastery at Beze borrowed the copy from Lyon in or about 878 in order to make a further copy. Someone at Beze saw the marginal note about the eclipse of 840 May 5 near the time of the eclipse of 878 Oct 29 and was struck by the resemblance between the data for the two eclipses. He thereupon recorded the eclipse of 878 Oct 29 by copying the record of 840 May 5 in all respects except the date.

If this is correct, it is amusing that the feria and the hour, which were copied, are actually correct for 878 Oct 29 while the date, which was original at Beze, is wrong by one day.

I shall take this to be an original record of the eclipse of 878 Oct 29. Reliability: 1. Place: Beze. Standard deviation of the magnitude: 0.01.

This record also appears in S. Benigni [ca. 1214], so it does not help in deciding upon the relation between S. Benigni and Besuenses.

878 Oct 29d E,F. Reference: Floriacenses [ca. 1060], which has: "In that year on the ides of October (= Oct 15) was an eclipse of the moon, since it was the 14th,[†] and following was an eclipse of the sun on the 4th calends November (= Oct 29), the 28th of the moon, and so both stars were eclipsed within 15 days." All the data in this record are correct, including the ages of the (ecclesiastical) moon. This is a contemporaneous account. Reliability: 1. Place: S. Benoit-sur-Loire. Standard deviation of the magnitude: 0.1.

[†] That is, the 14th day of the moon.

Hugo Floriacensis [ca. 1137] copied this
record.

        878 Oct 29e E,F. Reference: Elnonenses
Minores [ca. 1061]. These annals have: "875. On
the 4th calends November was an eclipse of the sun
after the 9th hour of the day that was so great
that stars appeared." This report cannot be original but it is independent of any report I have
found, so I shall use it. Reliability: 0.5. Place:
S. Amand-les-Eaux. Standard deviation of the magnitude: 0.01.

        878 Oct 29f E,F. Reference: Folcwinus
[ca. 962]. In Section 88 Folcwinus writes: "This
year the monastery of S. Peter and S. Bertin for
the first time, as others have been before it, was
burned by the Normans on the 5th calends August
(= Jul 28); and there was an eclipse of the sun on
the 4th calends November, the 4th feria, at the 9th
hour; and there was a mortality of men and cattle."
Other events put in the same year with these make
it clear that the year is 878. I don't believe
that Folcwinus intended to make the mortality (probably from famine or pestilence) a consequence of
the eclipse. This report seems to be independent
and local although it cannot be original. Reliability: 0.5. Place: S. Omer. Standard deviation
of the magnitude: 0.1.

        One may speculate that the Normans did
not cause calamitous damage. The annals continue
with no obvious interruption and with no mention
of rebuilding after the burning.

        The record of this eclipse in Mettenses
[ca. 905] is a copy of the German record 878 Oct
29d E,G in the next chapter.

        891 Aug 8 E,F. Reference: Folcwinus
[ca. 962]. Section 96 is headed "Of an eclipse of
the sun, and of the coming of the Normans." It
starts: "Three years later there was an eclipse

of the sun on the 18th calends September (= Aug 15)
at the 2nd hour; and a great drought in the months
of May, June, and July; and the star comet appeared."
Later he tells us that the Normans came on May 2;
and that they blinded 12 men and injured others.
The hour given for the eclipse looks slightly early
but possible, and the date is wrong by exactly a
week. The context makes it clear that the year is
891. Reliability: 0.5, since the report is still
not contemporaneous. Place: S. Omer. Standard
deviation of the magnitude: 0.1.

Besuenses [ca. 1174] has a brief note
under 961: "There was an eclipse of the sun." The
chronology is fairly accurate and it is safe to
conclude that the eclipse was that of 961 May 17.
However there is some question about the originality
of Besuenses here, and this brief note could have
come from several places. Hence I shall not use
this report.

968 Dec 22 E,F. Reference: Floriacenses
[ca. 1060]. The following series of sentences is
given under 956 in the edition cited: "This year
on the 4th nones September (= Sep 2) the moon was
turned to blood. That same year in June a wonderful sign appeared in the sky, namely a great dragon
but without a head. Later was the death of Hugo
the great prince of the Franks, Burgundians, Britons,
and Normans. There was an eclipse of the sun on the
11th calends January (= Dec 22), and stars appeared
from the 1st hour to the 3rd." The next entry is
put under 969 or 970; I was not sure which.

This may well be a case in which one
could learn something useful from an inspection of
the manuscript. I think it rather doubtful that
the annalist meant to put all these events under
the same year. The moon's turning to blood
was probably the lunar eclipse of 955 Sep 4. A
dragon is not an astronomical phenomenon and we
cannot calculate when one might have appeared.
The Hugo mentioned is probably Hugo (or Hugh) the
Great, also known as Hugo the White, the father of
Hugh Capet; Hugo died in 956. The solar eclipse

of 968 Dec 22 was the only visible eclipse (in
France, that is) on Dec 22 within decades of this
time.

Thus what probably happened was that
events through the death of Hugo were recorded in
or near 956. There was then a gap in the annals,
probably because no one was interested in keeping
them, until the eclipse of 968 Dec 22, after which
they were resumed. Either the annalist or the
editor failed to make clear the long gap in time.
I conclude that the eclipse record is original and
contemporaneous. Reliability: 1. Place: S.
Benoit-sur-Loire. Standard deviation of the magnitude: 0.01.[†]

Under the year 1005 the source Mosomagenses [ca. 1371] has: "Decimo Kal. Septembris
sol tenebratus est."[‡] The underlined letters are
ones that the editor could not read but that he
has supplied on the basis of the context. I cannot
find any eclipse that this could be. However,
Mosomagenses has a record of the eclipse of 1039
Aug 22 in almost identical words. Perhaps the
annalist put the entry under 1005 by mistake, then
realized his mistake and tried to erase it, thus
making the entry hard to read.

1023 Jan 24 E,F. Reference: Ademarus
[ca. 1028]. Ademarus wrote in Chapter LXII: "In
those days, in the month of January, about the 6th
hour of the day an eclipse of the sun took place
for an hour; . ." The context makes it clear that
this is the eclipse of 1023 Jan 24. The main problem concerns the place of the observation. Ademarus
finished his history after he moved from Limoges to
Angoulême about 1025, but the eclipse took place

[†]
I think that "from the 1st hour to the 3rd" is
meant to apply to the duration of the significant eclipse and not to the time that stars were
visible.

[‡]
On the 10th calends September (= Aug 23) the sun
was darkened.

before his move. I do not see any way to decide
whether he took with him, when he moved, a note of
what would have been a personal observation of
the eclipse, or whether he used a record that he
found in Angoulême. Since he almost surely saw
the eclipse himself, I shall assume that the observation came from Limoges and that he had either
notes or a memory of it when he wrote. Because of
the uncertainty I shall not give the record full
reliability even though it is contemporaneous. Reliability: 0.5. Place: Limoges. Standard deviation of the magnitude: 0.1.

           1033 Jun 29a E,F. References: Besuenses
[ca. 1174] and Rodulfus [ca. 1044]. Under 1033
Besuenses has: "On the 3rd calends July (= Jun 29),
that is, on the day of the solemnities of the apostles Peter and Paul, on the 6th feria, the 28th of
the moon, there was an eclipse or failure of the
sun from the 7th hour of the day to the 9th hour,
as had never been known to be so horrible. For the
form of the sun was seen to be in the image of the
moon on the 4th [day?] from its rekindling. It
repeatedly took on a sapphire color." The report
later mentions that stars shone before and behind
the sun. In Chapter IV.9 Rodulfus has: "In the
year 1000 from the Passion of our Lord,[†] on the
3rd calends July, on the 6th feria, the 28th of the
moon, there was an eclipse or failure of the sun
from the 6th to the 8th hour of that day, excessively terrible. And indeed the sun was made a
sapphire color, showing in its upper part the form
of the moon on the 4th [day?] from its relighting."
In these translations I have been rather careful to
use identical words where the originals were identical and to use different words where the originals
differ. "[day?]" has been supplied by me.

        Although the information in these accounts
differs somewhat, it is hard to believe that they
are independent. The hours differ, but that could

[†] Rodulfus puts the Crucifixion in the year that we
call 33, so this is 1033.

be the result of a copying error. Besuenses
mentions stars while Rodulfus does not. However
a moon on its 4th day should have about 1/8 of its
surface illuminated. If this much of the sun were
uneclipsed, it is highly doubtful that even Venus
could be seen. (See Section IV.5.) Thus I suspect
that the mention of stars in the report of Besuenses
is the product of imagination[†] rather than observation.

I think it almost certain that one of
these reports represents an original observation.
The question of which one is original concerns us
only in the matter of assigning the place of observation. I suspect that Rodulfus is the original
and that the annalist of Besuenses has copied Rodulfus, perhaps in combination with other reports. If
this were correct, we should take Cluny to be the
place; otherwise we should take Beze. Luckily the
two places are so close together that it matters
little which one we use. As a compromise, I shall
use Dijon, which is between the two. Reliability:
1. Magnitude: 7/8 eclipsed, with a standard deviation of 0.06 (about half of the amount that was
stated to be illuminated). The chart in Oppolzer
is not sufficiently detailed to let us decide the
position of the observer with respect to the central
path. If it is correct that the upper half of the
sun remained bright, the path was south of the
observer. Instead of relying upon this interpretation I shall calculate both possibilities and make
a choice if the calculations provide a clear-cut
difference. If not, I shall take the eclipse to
be central, with a standard deviation of 0.1 for
the magnitude.

"Sapphire" seems an unlikely term to apply
to the sun during a partial eclipse. Since Besuenses refers to stars, it may be that the eclipse was
total and that "sapphire" is an attempt to describe
the corona. On the basis of present knowledge, it

[†] The annalist at Beze could easily have gotten
the idea of stars from the record 878 Oct 29c
E,F above.

would not be possible to exclude this interpretation, even though it is not the one I am adopting.

        1033 Jun 29b E,F. Reference: <u>Weissem-</u>
<u>burgenses Maiores</u> [ca. 1075]. This has <u>under 1033</u>:
<u>"The oratory of S</u>. Peter in Wissembourg was dedicated
by Reginbold, bishop of Spirensus,[†] and an eclipse
of the sun occurred on the nativity[‡] of the apostles."
This may represent a genuine coincidence. June 29
is the day of the apostles Peter and Paul, and it
would be a plausible choice for the day when an ora-
tory to one of them would be dedicated. An eclipse
on that day would strike the spectators powerfully.
However the text does not say explicitly that the
dedication and eclipse were on the same day.

        Regardless of this point, we have an orig-
inal report of the eclipse of 1033 Jun 29. Reliability: 1. Place: Wissembourg. Standard deviation
of the magnitude: 0.1.

        1033 Jun 29c E,F. Reference: <u>S. Benigni</u>
[ca. 1214]. This source has the entry: <u>"1033. In</u>
this year there was an eclipse of the sun on the
feast day of S. Peter and Paul, 6th feria, in the
middle hours, and a star was clearly seen." This
entry is in a part of the annals that was mainly
compiled from other sources. However I have not
found any other record that is a plausible origin
for this one, so it may represent an original record at Dijon. I shall use it as an independent
record. Reliability: 0.2. Place: Dijon. Standard deviation of the magnitude: 0.01.

[†] The text has <u>Spirensi</u> in the genitive case. I
have guessed <u>this</u> nominative for the place; I do
not know where it is.

[‡] From <u>nativitas</u>, Late Latin. In this context, the
word <u>does not</u> mean the day of birth but the day
dedicated to the saint.

1033 Jun 29d E,F. Reference: Elnonenses
Minores [ca. 1061]. This source has the entry:
"1033. An eclipse of the sun was on the 3rd calends
July (= Jun 29), the 27th of the moon, about the
8th hour. His face looked green at first and then
yellow (croceus), and then those colors were affected
(colores movebantur) as gold or silver is usually
affected when they are cleaned by being dipped in
lead. Thereupon the clothes and faces of men looked
as if they were yellow. Then his light returned
from the west, like a moon in its first night and
then its second and third and fourth, and finally
all its light was restored."

The editor says that these annals were
carelessly copied. I see no carelessness in this
record, although the hour seems late. It seems to
be an original, accurate, and striking account. The
date and the feria are correct, and the age of the
moon is correct for the ecclesiastical moon. It is
not clear what the statements about the colors mean,
perhaps because I have never dipped gold or silver
into molten lead. This sounds like an original record and I have not found its counterpart elsewhere.
Reliability: 1. Place: S. Amand-les-Eaux. Standard deviation of the magnitude: 0; I do not believe that the phases of the recovery from the eclipse would read as they do unless the eclipse was
total.

It is interesting that both "green" and
"sapphire" have been used in describing this eclipse.
It is rather unusual to mention any colors at all,
so there may have been something remarkable about
the appearance of this eclipse. It was annular
rather than total, according to the calculations of
Oppolzer.

1033 Jun 29e E,F. Reference: Nivernenses
[ca. 1188]. This source has: "1033. In this year
on the 3rd calends July on the nativity of apostles
Peter and Paul was an eclipse of the sun, in fact
between the 3rd hour and midday." The editor took
this to be an original entry. I have not found
these hours in any other record, and this confirms
the editor's opinion. Reliability: 1. Place:

Nevers.   Standard deviation of the magnitude: 0.1.

S. Maxentii [ca. 1140] has only two records
of this eclipse.   It dates both of them 1042 Jun 30.
I cannot identify a specific origin for either, but
each could have been obtained from several places by
slight changes in wording.   Hence I shall not use
either report.

1037 Apr 18a E,F.   Reference: S. Benedicti
[ca. 1110].   This source has: "In the year 1037
of the Incarnation, in the 1st hour of the day after
the octave of Easter, the sun lost its radiance, and
it appeared in the form that the moon should have on
its 2nd day; and about the 3rd hour like the moon
on the 5th day, and a little later on the 8th."   The
octave of Easter in 1037 was on Apr 17, so the date
is accurate, and the hours are reasonable.   At greatest eclipse the sun was like the moon on its second
day, when about 0.03 of the area is illuminated.   I
shall take this as equivalent to a magnitude of
about 0.98 of the diameter.†   Reliability: 0.5;
the record cannot be contemporaneous.   Place: S.
Benoit-sur-Loire.   Magnitude: 0.98 with a standard
deviation of 0.01.   It is not possible to tell the
relation of the place to the central path from
Oppolzer [1887] for an eclipse so near totality.
Hence I shall try both possibilities.

1037 Apr 18b E,F.   Reference: Lemovicenses
[ca. 1060].   This has under 1029:   "On the 13th
calends May (= Apr 19) there was an eclipse of the
sun from the 4th hour almost to the 6th and it was
not as usual, but first the sun appeared ◐ and
later ◐ and still later ◑ ."   One should
not try to scale the magnitude from the drawings,
I think; I also think it is safe to take this as
the record of an eclipse that was nearly total.   I
shall use 0.95 as the magnitude, with a standard
deviation of 0.03.   The entries in Lemovicenses

---

† See Figure V.2 in AAO.

are sparse and almost entirely local, so the place
can be safely taken as Limoges. The only problem
concerns the identification.

Since the entries are local and sparse,
I found no nearby events by which I could test the
accuracy of the chronology. The nearest event I
could use for a test was the lunar eclipse of 1009
Oct 6, which was put under 1010.

Ginzel [1882 and 1884] took this as the
eclipse of 1033 Jun 29, but it seems to me that
this ignores the clear statement of the day. The
day was probably written "xiii. Kal. Mai." Jun 29
would be written "iii. Kal. Iul." I do not think
that this is a plausible recording error. However
the eclipse date of Apr 18 would be correctly written
"xiiii. Kal. Mai." and the accidental omission of
one i in a string of four i's does seem like a plausible error. Hence I think that the record almost
surely refers to the eclipse of 1037 Apr 18. Because
of the uncertainty I shall lower the reliability to
0.5.

1039 Aug 22a E,F. Reference: Mosomagenses
[ca. 1371]. The entry for 1039 reads in part: "11.
Kal. Septembris sol contenebratus est."† The words
are almost identical with some entered under 1005,
as I mentioned above. The date given is 1039 Aug
22, and this is a contemporaneous record. Reliability:
1. Place: Pont-à-Mousson. Standard deviation of
the magnitude: 0.1.

1039 Aug 22b E,F. Reference: Rodulfus
[ca. 1044, p. 69]. This record reads: "Also 4
years later there was an eclipse of the sun on the
11th calends September (= Aug 22), feria 4, hour 6,
and, as it always happens, on the 28th of the moon."

---

† The underlining indicates a word that the editor
has "restored". It is most likely that the manuscript has "xi" rather than "11" for the numeral;
it is possible that the numeral is written out.

It would be difficult to tell exactly what year is meant but it is certainly close to 1039. The information is correct for the eclipse of 1039 Aug 22, including the age of the moon, but it is not correct that a solar eclipse is always on the 28th of the moon. The eclipse of 1033 Jun 29 was on the 27th of the ecclesiastical moon. Reliability: 1. Place: Cluny. Standard deviation of the magnitude: 0.1.

1039 Aug 22c E,F(?) Reference: Cameracensium [ca. 1042]. In Chapter III.55 this has: "... there was an eclipse of the sun on the day before the ides of May (= May 14); and the 2nd nones June (= Jun 4), so it is said, he[†] died." The death fixes the year intended as 1039, and makes 1039 May 14 the date stated for the eclipse. The only eclipse to which I can match this record even by allowing two errors is the small lunar eclipse of 1044 May 14. Even that possibility is doubtful. Oppolzer [1887] puts the end of this eclipse at 18^h 40^m Universal Time, which would be about 18^h 50^m local time. This time, if it is correct, is well before moonrise at Cambrai on that date.

I believe that one must take this as a record of an eclipse that cannot be identified.

This report has caused a lot of trouble. Ginzel [1882 and 1884] took the eclipse to be 1037 Apr 18, presumably because it was closest to the right time of year. However, it takes 3 errors[‡] to make this identification acceptable and with 3 errors we can turn the date into 1039 Aug 22 and many other possibilities. Sigebertus [ca. 1111] copied this record and many other sources copied Sigebertus, with a consequent large amount of confusion. As I mentioned after the discussion of the record 1039 Aug 22 E,BN in Section VIII.2, Ginzel did not realize that Sigebertus had copied from Cameracensium and he identified the identical record in

---

[†] "He" is the Holy Roman emperor Conrad II, who died on 1039 Jun 4.

[‡] Apr 18 would be 14th calends May.

Sigebertus as 1039 Aug 22. The appearance of the
record in sources that copied Sigebertus has
caused still more confusion.

       1044 Nov 22a E,F. Reference: Nivernenses
[ca. 1188], which has: "1044. In this year, on
the 10th calends December (= Nov 22), the nativity
of S. Cecilia, was an eclipse of the sun, to wit at
the 3rd hour." The editor takes this to be an original entry and I see no reason to doubt it. Reliability: 1. Place: Nevers. Standard deviation
of the magnitude: 0.1.

       1044 Nov 22b E,F. Reference: S. Benedicti
[ca. 1110], which reads: "In the year 1044 of the
Incarnation of the Word, on the 6th ides December
(= Dec 8), at the 8th hour of the night, feria 5,
14th of the moon, an eclipse of the moon took place
between the Hyades and the Pleiades. In the same
year and month, and the same feria, at the 2nd hour
of the day, on the 10th calends of the same month,
a solar eclipse happened." The annalist became
mixed up in writing the month. The lunar eclipse
was on 1044 Nov 8, that is, the 6th ides November.
The solar eclipse was also in November, but the
date should have been written 10th calends December.
Aside from this slip, all the data are correct.
However, the record is not contemporaneous. Reliability: 0.5. Place: S. Benoit-sur-Loire.
Standard deviation of the magnitude: 0.1.

       1044 Nov 22c E,F. Reference: Rodulfus
[ca. 1044]. This is in Chapter V.3, which is
headed "About the third eclipse of the sun". The
chapter starts: "In the aforesaid month of November,
on the 10th calends December (= Nov 22) at the 3rd
hour of that day, there occurred the 3rd solar eclipse of our time, on exactly the 28th of the moon,
. . Also in those days, as Archipraesul[†] Wido of
Reims reports, we learn that the star called

[†]Praesul usually means a public dancer. By an odd
transfer, it means a patron or a high official in
some contexts.

Phosphorus or Lucifer repeatedly in the evening
was seen to be agitated up and down ..." Venus
has been known both as Phosphorus and Lucifer; I
cannot account for its unusual behavior. So far
as the eclipse is concerned, this is an original
and contemporaneous record. Reliability: 1.
Place: Cluny. Standard deviation of the magnitude: 0.1.

   I am now in a position to comment on the
records of solar eclipses in Hugo Flaviniacensis
[ca. 1102] and Clarius [ca. 1124], whose compounded
carelessness has caused considerable confusion.
Hugo started by making a compressed version of
Rodulfus' report of the eclipse of 1033 Jun 29 and
putting it under 1039. His eye must have wandered
from one report to the other as he wrote, and he
omitted all of the eclipse record of 1039 Aug 22
except the year. He also wrote a compressed version
of Rodulfus's report of the eclipse of 1044 Nov 22
in these words: "At that time, in November, on
the 28th of the moon, there was an eclipse of the
sun; and the star called Lucifer was seen to be
agitated up and down in the evening ..."

   Many later writers copied the record that
Hugo put under 1039, but which really applies to
1033 Jun 29. This has given rise to a large number
of wrong records. Ginzel [1882 and 1884] and the
editors involved have almost all taken the records
to be of the eclipse of 1039 Aug 22 rather than 1033
Jun 29.

   Clarius then carelessly used Hugo's
already wrong chronicle, and in doing so he achieved
some sort of a record. He took the first part of
the account that Hugo put under 1039, which is
really an account of the eclipse of 1033 Jun 29,
combined it with the last part of the 1044 record,
making it appear that the peculiar behavior of
Venus was at the same time as the eclipse, and put
the resulting mixture under 1039. Thus he has
created a single "account" of an eclipse under the
year 1039 that is made up of parts of reports of
the eclipses of 1033 Jun 29, 1039 Aug 22, and 1044

Nov 22. I believe that this is a record amount of confusion, and it has caused editors of medieval manuscripts to make many errors.

S. Albini [ca. 1357] records that the sun appeared "cerulean" on a day that is either 1095 Apr 11 or 12, and that the moon looked the same on the following night. Neither date is the date of either a solar or lunar eclipse. The dating is buttressed by several circumstances and seems to be accurate. Therefore I imagine that the appearances were the result of some kind of atmospheric phenomenon and not eclipses.

1133 Aug 2a E,F. Reference: Burburgensis [ca. 1164]. Under 1133 this source has: "The sun was obscured on the 4th nones August (= Aug 2) at the 6th hour, such that stars appeared in the sky." When an eclipse is reported as many times as this one, it becomes hard to judge the independence of a short record. This record is not identical with any other but, not surprisingly, it resembles several others. In particular, it resembles the German record 1133 Aug 2b E,G in the next chapter. The resemblances could easily be the result of chance, I believe, and I further believe that Burburgensis is considered to be an independent source at this time. Hence I shall give this record a reliability of 1, but it would not be surprising if further study showed it to be derivative. Place: Bourbourg. Standard deviation of the magnitude: 0.01.

1133 Aug 2b E,F. References: Praemonstratensis [ca. 1155] and Laudunense [ca. 1145]. Praemonstratensis has, under 1132: "There was an eclipse of the sun on the 4th nones August, feria 6, at the 6th hour." Laudunense has identical words except for having the 5th hour; it has the entry under the correct year. Since both places are in the diocese of Laon, we do not need to decide which is original.[†]

[†] Since Laudunense has the year right, there is a mild presumption in its favor.

I shall take this as equivalent to a single report from Laon. Reliability: 1. Standard deviation of the magnitude: 0.1.

The report of this eclipse in <u>Robertus Autissiodorensis</u> [ca. 1211] seems to be a copy of the one in <u>Praemonstratensis</u>.

1133 Aug 2c E,F. Reference: <u>S. Andreae</u> [1133]. This is almost the last thing in the chronicle, and the author tells us a few lines farther on that it was written in 1133: "Three months later, on the 4th nones of the following August (= Aug 2), feria 4, 6th hour of the day, the sun was suddenly obscured and there was darkness on all the earth... that lasted for almost half an hour." Later he remarks that this must have been a prodigy rather than an astronomical eclipse, because it was only the 27th of the moon, when an eclipse cannot happen. This reads to me like an eye-witness account written almost immediately after the event. Although the writer does not mention any specific indicators, it reads to me as if the eclipse was quite large if not total, and I shall take the standard deviation of the magnitude to be 0.01. Reliability: 1. Place: Cambrai.

1133 Aug 2d E,F. Reference: <u>Honorius</u> [ca. 1137, p. 131]. Honorius devotes a section in this part of his work to each reign of a Holy Roman emperor. In the section on Lothar the Saxon (emperor 1125-1137) we find: "There was an eclipse of the sun on the 4th nones August for almost an hour, of a kind that has not been seen for a thousand years. And indeed the whole sky became dark as if it were night, and stars could be seen in almost all the sky. Later when the sun left the darkness it first appeared in the manner of a star, then like a new moon, and finally took on its own proper form." The day given is Aug 2, so there is no question about the identification. Reliability: 1. Place: Autun. Standard deviation of the magnitude: 0.

Comparison of the returning sun to a star sounds like an observation of the "diamond ring" effect. If so, this is the earliest mention of that effect that I have noticed.

1133 Aug 2e E,F. Reference: <u>Rotomagensi</u> [ca. 1174], which has: "MCXXXIII. An eclipse of the sun on the 4th calends[†] August about the 3rd hour. And that same day King Henry of the English crossed to Normandy, not to return except when dead." Reliability: 1. Place: Rouen. Standard deviation of the magnitude: 0.1.

This report, including the remark about Henry, was copied by <u>Robertus de Monte</u> [ca. 1186].

1133 Aug 2f E,F. Reference: <u>Mortui-Maris</u> [ca. 1180]. This has the entry: "MCXXXII. An eclipse of the sun on the 4th nones August." Since the editor supplied none of the needed critical apparatus for this source, we cannot give it high reliability. I suspect that the record of the eclipse of 1147 Oct 26 found in this source is copied. Hence I shall give this a reliability of 0.2. Place: Normandy, which I shall take as equivalent to Rouen. Standard deviation of the magnitude: 0.1.

1133 Aug 2g E,F. Reference: <u>S. Petri</u> [ca. 1180]. This has the entry: "MCXXXIII. [The sun suffered an eclipse] Stephen the Cistercian abbot died, Rainard Abbot IV succeeded." I have reproduced the punctuation of the original. I do not know whether Rainard was the 4th abbot or the 4th Rainard to become abbot. In no place that I saw did the anonymous editors explain why they used brackets. For safety, I shall not use this record.

1140 Mar 20 E,F. Reference: <u>Burburgensis</u>

[†]This is clearly an accidental error. It should read "nones".

[ca. 1164]. This has under 1139: "There was an eclipse of the sun on the 13th calends April (= Mar 20), at the hour of vespers." A fairly short record like this one can easily be copied. However this one reads somewhat differently from any other record I have found, and the source should be contemporaneous. I shall compromise on the reliability. Reliability: 0.5. Place: Bourbourg. Standard deviation of the magnitude: 0.1.

1147 Oct 26a E,F. Reference: Bellovacense [ca. 1163]. Under 1147 this source has: "There was an eclipse of the sun, 5th calends November (= Oct 28), 3rd hour, the Lord's day." The weekday is correct for the date of the eclipse and the hour looks reasonable. Since the weekday is right and the day of the month is wrong, the data were not calculated. Hence the report is probably original in spite of the error. Reliability: 1. Place: Beauvais. Standard deviation of the magnitude: 0.1.

Mortui-Maris [ca. 1180] has this identical record, including the error in date. There is no sure way to know which is original. I have taken Bellovacense to be the original, since it closes earlier. Luckily it makes little difference to us which is original since the places are so close together.

1147 Oct 26b E,F. Reference: Praemonstratensis [ca. 1155]. This says under 1147: "There was an eclipse of the sun on the 8th† calends November, the Lord's day, from the 2nd hour to the 5th hour." Reliability: 1. Place: Laon. Standard deviation of the magnitude: 0.1.

1147 Oct 26c E,F. Reference: Nivernenses [ca. 1188]. This says under 1147: "In this year was an eclipse of the sun on the feast of the dedication of the church of S. Cyric Martyr." Note that

---

† The 7th would be correct.

this says "on the feast of the dedication of the church" and not on the feast day of the saint himself. Nivernenses for 1058 says that the dedication in question took place on 1058 Oct 26. I do not know whether this is the same S. Cyric as the one in the account 885 Jun 16 B,SM in Section VII.2 or not. The day of that saint is Jun 16 rather than Oct 26. This does not prove that the two are different since the dedication of the church in honor of the saint would not necessarily be on the saint's day even though we would expect it to be. I see no serious question about the identification of the eclipse, and the way of stating the date makes it almost certain that the report is original. Reliability: 1. Place: Nevers. Standard deviation of the magnitude: 0.1.

1153 Jan 26a E,F. Reference: Aquicense [ca. 1166]. This has under 1153: "There was an eclipse of the sun on the 7th calends February (= Jan 26) on the 2nd feria." These data are correct. Reliability: 1. Place: Douai.† Standard deviation of the magnitude: 0.1.

Aquicinctina [ca. 1237] has copied this record.

1153 Jan 26b E,F. Reference: Robertus de Monte [ca. 1186]. This says under 1152: "There was an eclipse of the sun on the 7th calends February about the 8th hour, on the 27th of the moon." The hour is plausible. The number of the year is wrong by 1. The other data are correct. This report seems to be independent of any other that I have found and since we are in Robertus' own time I take this to be original. Reliability: 1. Place: Mont S. Michel. Standard deviation of the magnitude: 0.1.

---

† I use Douai because I cannot find coordinates for Anchin, and Anchin is known to be close to Douai.

1178 Sep 13a E,F. References: S. Florentii [ca. 1236] and S. Petri [ca. 1180]. S.
Florentii has under 1178: "Ides of September (=
Sep 13), an eclipse of the sun at the 3rd hour of
the day." S. Petri has, also under 1178: "An
eclipse of the sun appeared on the eve of the Holy
Cross about the 3rd hour." Holy Cross Day is Sep
14 and its eve is Sep 13, the ides of September.
Thus both sources give the day correctly and in
ways that would be equally familiar to one who used
the church calendar extensively. The hour given by
both for the eclipse is oddly early, and this makes
me suspect copying. The editors put the entry from
S. Petri in brackets. For this reason I shall not
use the report from S. Petri. I shall use the report in S. Florentii but, because of the suspicious
circumstances, I shall lower the reliability to 0.5.
Place: Saumur. Standard deviation of the magnitude:
0.1.

The reports of this eclipse in Robertus
Autissiodorensis [ca. 1211] and in the English
source Matthew Paris [ca. 1250] are almost identical with the report in S. Petri. Hence I do not
use either of these.

1178 Sep 13b E,F. Reference: Aquicinctina [ca. 1237], which has under 1178: "There was
an eclipse of the sun on the ides of September,
before the 6th hour." This hour is more reasonable
than the one given in the preceding report; it is
perhaps slightly late. The place is close to Douai
but I do not have its exact coordinates; hence I
shall use Douai. Reliability: 1. Standard deviation of the magnitude: 0.1.

1178 Sep 13c E,F. Reference: S. Sergii
[ca. 1215], which has, under 1178: "The sun was so
darkened that it did not shine except as the moon
usually does, on the day of S. Maurilius." The day
stated is Sep 12. The error and the odd way of
giving the day make me feel that this report is
original, even though this source is often a copy.
This seems to be a partial eclipse, but one that
is rather large. I shall take 0.95 as the magnitude,

with 0.05 as the standard deviation. The chart in
Oppolzer [1887] shows the path to be south of the
place, which is Angers. Reliability: 1.

          1178 Sep 13d E,F. Reference: S. Benigni
[ca. 1214]. Under 1178 this has: "And also in
this year the sun was darkened on the ides of September (= Sep 13) about the hour of nones." We cannot tell whether "nones" means the 9th hour or the
6th hour; the chart of the eclipse makes the latter
more likely. Reliability: 1. Place: Dijon.
Standard deviation of the magnitude: 0.1.

          1178 Sep 13e E,F. Reference: Gaufredus
[ca. 1182]. This eyewitness account reads: "Year
of the Incarnation of our Lord MCLXXVIII. 4th feria,
ides of September, 28th of the moon, on a clear day
about the 5th hour the sun suffered an eclipse; its
sphere began to be covered from the east until it
was like a moon on its 2nd or 3rd. The star Venus
was seen to the north. After the 6th [hour?] in
the order that it was darkened its brightness returned until the sun shown full again. Then we
could see each other's faces, . . ." I have supplied "[hour?]". Reliability: 1. Place: Vigeois.
A moon on "its 2nd or 3rd" night would have about
0.04 of its area exposed on the average. I believe
that this conflicts with the statement that Venus
could be seen.[†] Hence I shall ignore the indication
of partiality, and I shall use this as an ordinary
report with a standard deviation of 0.02.

          1185 May la E,F. Reference: S. Florentii
[ca. 1236]. Part of the entry for 1185 reads: "..
and there was an eclipse of the sun on the calends
of May, 9th hour." The hour looks reasonable if
"9th hour" has its original meaning. S. Florentii
can still not be given full reliability. Reliability: 0.5. Place: Saumur. Standard deviation of

---

[†] See the discussion of the visibility of Venus in
Section IV.5.

the magnitude: 0.1.

1185 May lb E,F(?) Reference: Elnonenses
Maiores [ca. 1224].   This source has an entry under
1184 as follows: "In this year an eclipse of the
sun happened about the 9th hour on the octave of
Easter."   There is no entry for 1185.   In 1185 the
octave of Easter was Apr 28, so the dating is poor
both as to year and day if 1185 is the identification;
however, the eclipse was at about the 9th hour.   In
1186 the octave of Easter was Apr 20 and there was
a penumbral eclipse on 1186 Apr 21.   This is closer
to the correct day of the year, but the eclipse
should have been in the morning.   I conclude that
this eclipse cannot be identified safely.

1191 Jun 23a E,F.   Reference: Aquicinctina [ca. 1237].   This says under 1191:   "There
was an eclipse of the sun on the 9th calends July
(= Jun 23), on the Lord's day, at the 6th hour, on
the eve of John the Baptist,[†] but its light remained
in the northern part about to the extent of a moon
on the 3rd day."   Reliability: 1.   Place: Douai.
The magnitude indicated is about 0.94.   If the
northern part remained illuminated the central path
should have been south of the observer.   Oppolzer's
chart [Oppolzer, 1887] shows the path to be a few
degrees north of Douai.   I believe that it is possible for Oppolzer's approximations to cause this
much error.   Hence I shall start by assuming the
path to be to the south, but I shall alter the
assumption if the calculations show this to be
seriously in error.   I shall use 0.03 for the standard deviation of the magnitude.

1191 Jun 23b E,F.   Reference: S. Albini
[ca. 1357].   This says under 1191:   "On the eve of
John the Baptist was an eclipse of the sun."   A
brief note like this one could easily be derived.

[†] The day of John the Baptist is Jun 24.

However, the editors of S. Albini seem to think that the entries in this part are genuine, and I shall follow them since there is no evidence to the contrary. Reliability: 1. Place: Angers. Standard deviation of the magnitude: 0.1.

1191 Jun 23c E,F. Reference: Vindocinense [ca. 1251]. This source says under 1191: "This year, as the two kings† were crossing to Jerusalem against the gentiles (presumably the Moslems), on the eve of John the Baptist, on the 27th of the moon, there was an eclipse of the sun." The age of the moon is correct. This should be an independent report. Reliability: 1. Place: Vendôme. Standard deviation of the magnitude: 0.1.

1191 Jun 23d E,F. Reference: Elnonenses Maiores [ca. 1224], which says under 1191: "This year there was a wonderful eclipse of the sun on the 9th calends July (= Jun 23) at the middle of the day." Since the eclipse was "wonderful" it was probably larger than the average observed eclipse. Hence I shall take the standard deviation of the magnitude to be 0.06 rather than 0.1. Reliability: 1. Place: S. Amand-les-Eaux.

1191 Jun 23e E,F. Reference: S. Vincentii [ca. 1280]. This source has a brief note under 1191: "On the 9th calends July was an eclipse of the sun." A brief record like this one is often unsafe to use. However Pertz feels that this source is original from about 1154 on, so I shall give this a reliability of 1. Place: Metz. Standard deviation of the magnitude: 0.1.

1207 Feb 28 E,F. Reference: S. Albini [ca. 1357]. This says under 1207: "This year was an eclipse of the sun, about the 3rd hour, on the

---

† This is a reference to the kings of England and France and to the Third Crusade.

11th calends March." The hour looks early. The
date given is Feb 19. In the manuscript the date
would have been written as "xi. kal. marcii" or
the equivalent. I imagine that a poorly made "ii"
was misread as "xi"; the 2nd calends March is
Feb 28. The only other possibility that I can find
is the penumbral eclipse of 1208 Feb 17 (= 13th calends March). This eclipse, however, should have occurred late in the day if it occurred at all before
sunset in France, so it is not compatible with the
hour. Hence I shall take 1207 Feb 28 as the identification and lower the reliability. Reliability:
0.5. Place: Angers. Standard deviation of the
magnitude: 0.1.

CHAPTER XI

RECORDS OF SOLAR ECLIPSES FROM GERMANY

1. The Sources

I have found considerably more records
of solar eclipses in sources from Germany than from
any other geographical subdivision. This does not
necessarily mean that there are more records from
Germany. Instead the fact is a tribute to the
diligence of the editors of the Monumenta Germaniae
Historica series. Almost all of the editions of
German sources used in this study are from that
series. Many of the sources are also edited in the
Bouquet series (see Section X.1), but the Monumenta
edition is almost always easier to use than the
Bouquet edition.

There is a series that I do not use but
whose editor will interest some readers. This is
a series called Scriptores Rerum Brunsvicensium, in
3 volumes, published by Nicolas Foerster in Hannover
from 1707 to 1711. All the items in this series
that are used in this study are also in the Monumenta
Germaniae Historica series. I chose the latter editions because they are more than a century later
and provide more extensive critical commentary. The
editor is Gottfried Wilhelm, freiherr von Leibniz
(1646-1716), whose name is often but apparently
wrongly spelled Leibnitz in English. Leibniz was
primarily a lawyer, philosopher, and historian, and
he served as a privy councilor to the Elector of
Brandenburg for some time. In his late twenties
he became interested in mathematics and, after a
short period of essentially independent study, he
invented the calculus. Shortly thereafter his
interests returned mainly to their former fields,
although we may judge that he never entirely lost
interest in mathematics.

In this section I shall discuss briefly
the sources that contain material useful for this

study as well as a few others with interesting
features. In the next section I shall discuss
the eclipse records in chronological order.

Albertus [1256]. Albertus became the
abbot of S. Maria Stadensis in 1232, as he tells
us himself. Stade is a few kilometers west of
Hamburg. There is a tradition that this was the
oldest town of the Saxons and that it was founded
in -320. We know little else about Albertus. He
attaches a list of popes to his annals that stops
with Urban IV (1261-1264), so Albertus probably
died sometime during Urban's papacy. His annals
have material that is otherwise unknown from about
1100 on. One interesting feature of his work is
that he gives a "travel guide", complete with a
table of distances, for traveling from Stade to
Rome, and he gives information about travel in the
Holy Land. He also describes what seems to be a
nova in 1245.

Albertus' work has sometimes been called
Annales Stadenses.

Aquenses [ca. 1196]. Aquenses and
Aquisgranum are old forms of the name of Aachen.
The Annales Aquenses were kept by the canons of
the church of S. Maria Aquisgranensis. Some of
the material in them is almost identical with that
in Rodenses [ca. 1157], to be discussed below.
There is also a set of annals that the editor calls
Annales S. Petri et Aquenses that is a conflation
of Aquenses and Erphesfurdensis [ca. 1182], which
will also be discussed below. I have not cited S.
Petri et Aquenses as a separate source. Confusion
between Aquenses and Rodenses does not concern us
much because the respective places are close together.
Aachen and Erfurt are separated by about 5° in longitude and we must be careful to distinguish their
contributions.

Augienses [ca. 954]. These are from the
abbey on the island that is now called Reichenau,

in what we call Lake Constance in English. This
Benedictine abbey was founded in 724 and was a
famous seat of learning in the Middle Ages. The
editor divides the rather brief Annales Augienses
into two parts. The part through 858 is entirely
derivative. The editor uses "Genuina" to describe
the remaining part, meaning that these annals
originated in the abbey. Part of Augienses is
identical with Weingartenses [ca. 936] below.

    Augustani [ca. 1104]. These annals were
composed at the cathedral in Augusta, which is the
modern Augsburg. They were written by several
authors. Up to about 1040 each annal is rather
brief. From then to about 1065 the entries gradually
increase in length and from 1065 on they are fairly
long. Entries after 1065, and perhaps after 1040,
are probably original. Earlier ones are sometimes
independent of other known sources.

    Augustani Minores [ca. 1321]. The editor
called these Annales Augustani Minores. They are
actually a continuation of the Chronographia by
Heimo [1135] discussed below. As such, it seems
to me that it would be better to call them a continuation of Heimo made at Augsburg. A copy of
Heimo was taken to Augsburg soon after 1135, and
it was continued with moderate regularity to 1321.
The editor describes the annals as going to 1457.
I have preferred to use 1321 as the date because
the only entry after 1321 is the one for 1457.
This entry looks odd to the modern eye. It says
that Ladizlaus, king of Hungary and Bohemia, became intoxicated in Prague. I presume that it
means that the king was killed by poison.

    Bertholdus [ca. 1138]. Bertholdus had an
active career in monastic work and finally became
abbot at the monastery in Zwiefalten in 1139. The
abbey there was founded in 1089 [Bertholdus, ca.
1138]. Shortly before he became abbot, Bertholdus
wrote this account of the building of the monastery
and continued it as a chronicle of local affairs to

1138. This Bertholdus is apparently not the same
as the Bertholdus who was a continuator of Herimannus Contractus (see Herimannus [1054] below).

  Brunwilarenses [ca. 1179]. The editor of
these annals gave an extensive analysis of the hands
that appear. In the 11th century several hands are
involved but, oddly, they are interwoven and not
consecutive. For example, one person wrote the annals for 1000, 1021, 1034, 1036, and 1039 while another wrote those for 1024, 1028, 1030, 1049, 1050,
and 1051. The editor did not discuss the entries
in the 12th century, so I presume that they were
contemporaneous. In his preoccupation with the
handwritings, the editor forgot to tell us where
the annals were written. I do not know whether they
came from an abbey, a cathedral, or from some other
type of institution. Luckily, Ginzel [1882 and 1884]
says that they come from the place called Brauweiler,
for which he gives coordinates.

  Colonienses [ca. 1238]. Cologne or Köln
is an old city, having been founded by Rome about
the year 50. In spite of its age, the only annals
I have found from it are relatively late. The church
and monastery of S. Pantaleon there was founded in
964. These annals started as a compilation made
there about 1100, and they contain little independent
material before that date. After that date they
were maintained contemporaneously and have much important historical material. The editor calls these
Annales Colonienses Maximi. Since they are closely
associated with S. Pantaleon, it seems better to me
to call them Annales S. Pantaleonis Coloniensis or
some such name, but I have not ventured to change
the name used by the editor.

  Corbeienses [ca. 1148]. These annals are
from the town that is usually spelled either Corvei
or Corvey; it is on the River Weser in Westphalia.
The monastery of Corvei was founded [Corbeienses,
ca. 1148] in 822. The annals start as a textbook
example of Easter table annals (see Section III.2).

The tables are an extensive set with 8 columns and with room for about 5 or 6 words at the end of the line for each year. The annals keep this character, with a few exceptions, to 1145, but the last 4 years occupy 10 pages. The annals seem to be original from about 840 on.

Corbeienses Notae [ca. 1241]. A codex from Corbei contains a register of abbots and brothers of the monastery, beginning with the founding on 822 Aug 25, 2nd feria,[†] by a group who had lived there for 7 years, and continuing to 1146 Oct 8. The editor has published this register. At the end of the register he adds two notes that, he says, he has excerpted from folio 132 of the codex. It was not clear to me whether he meant that he had excerpted these notes from surrounding but different material or whether there were many notes of which he is publishing only two. Whichever he meant, Corbeienses Notae as used here means the two notes published after the register of the monastery.

The notes are the accounts of solar ec- lipses designated as 1133 Aug 2g E,G and 1241 Oct 6c E,G in the next section. In my opinion, they furnish strong evidence against the statement of MacCarthy [1901, p. civ] that minute detail in an account shows that the account is that of an eye- witness.[‡]

Einhard [ca. 820]. The known work of this writer does not contain any data needed for this study. However he and his work come up in the eval- uation of other sources and it is convenient to in- sert a discussion of him here. Einhard's name ap- pears in Latin forms variously as Einhartus, Ain- hardus, Heinhardus, Agenardus, Eginhardus, and Eginhartus and it appears in other languages in the forms got by leaving off us from these. "Einhard"

[†] The feria is correct.

[‡] See the discussion of the record 878 Oct 29 B,I in Section VII.1.

seems to be the most common form in recent writing.

Einhard was a friend and biographer of Charlemagne. He was born about 770 and died in 840. He is linked with one of Charlemagne's daughters in the romantic legend that I recounted in connection with Nithardus in Section X.2. One may doubt both that specific legend as well as the connection in general. After Charlemagne's death in 814 Einhard continued to serve his son Louis I (the Pious) until about 830. Einhard served as abbot of several monasteries.

The Annales Fuldenses [Fuldenses, ca. 901] and the Annales Laurissenses [Laurissenses, ca. 829] have often been attributed to Einhard, and the citation Einhard that appears in many critical discussions of medieval histories usually means one of these sets of annals, and not the biography of Charlemagne that is safely attributed to him. In connection with the record 807 Feb 11 E,G in the next section I shall give some evidence that suggests, at least to me, that Einhard did not write either set of annals.

There is a recurring legend that Charlemagne was exceedingly tall. This legend probably comes from a misreading of Einhard's Vita Karoli Magni. I shall join many others who have tried to correct this point. What Einhard wrote is that Charles' height equalled 7 of his own feet (underlining mine). This statement is true of many people who are particularly tall.

Erphesfurdensis [ca. 1182]. Erfurt is a town in Saxony that, by tradition, was founded in the 6th century. It was made a bishopric in about 740 but its see was later merged with that of Mainz. In spite of the age of the establishment, the annals do not become original until about 1125. There are variant versions of the annals. There is also a set of annals that the editor calls S. Petri et Aquenses (see the discussion of Aquenses [ca. 1196] above), which someone prepared by fusing Erphesfurdensis and Aquenses.

-358-

Frutolf [1103]. Frutolf was a prior of
Michelsberg Cloister near Bamberg. Early in 1103
he completed his Chronicon Universale; we may presume that he intended to keep it up to date but he
died on 1103 Jan 17. The abbot Ekkehardus Uraugiensis (Ekkehard von Aura) revised it several times
and continued it to 1125. The fact that the work
is mostly due to Frutolf was unknown until recently,
and earlier editors, including the editors of the
Monumenta Germaniae Historica, refer to the work
as Ekkehard.

Ekkehard was connected with the monastery
at Wurzburg. There is a Chronicon Wirziburgense[†]
that is also attributed to Ekkehard, I presume correctly. I inspected Chronicon Wirziburgense and
found nothing in it of value to this study.

The information in Chronicon Universale
that is useful for this work occurs for years before
1103. Because of the dating, I have assumed that
the information in question was written by Frutolf
rather than by Ekkehard, and I have accordingly
cited only the former. However, since Ekkehard revised Frutolf's chronicle in addition to continuing it,
it is possible that some information before 1103 is
actually due to him. If so, Wurzburg rather than
Bamberg may be the place of observation. Luckily,
confusion between Bamberg and Wurzburg is not important here because the places are only about 60
kilometers apart. The danger is that Ekkehard may
have inserted some material that he obtained from
somewhere other than either Bamberg or Wurzburg.

Fuldenses [ca. 901]. Fulda is in west
central Germany about 100 kilometers northeast of
Frankfurt-am-Main. As Fuldenses tells us, the
monastery of Fulda was founded by S. Boniface in

[†] This is different from Annales Wirziburgenses.
See the discussion of Wirziburgenses [ca.1101]
below. The Chronicon Wirziburgense is edited by
G. Waitz in the same volume as Frutolf's Chronicon
Universale.

744.

The editor Pertz divides Annales Fuldenses into five parts, with these dates and authors: 714-838, Enhardus; 838-863, Rudolfus; 863-882, uncertain; 882-887, anonymous; 882-901,† Bawarus. Pertz says that Enhardus is different from Einhard. Since Einhard is often stated to be the author of the first part, I am suspicious. I see three possibilities: (a) Einhard wrote part of the annals. (b) Enhardus, a different person, wrote part of them. (c) Neither wrote any part of them, but Einhard's name has been linked to them by legend. We can doubt (a) strongly. The most plausible is (c), but I do not know any rigorous reason to eliminate (b).

One standard source (New Catholic Encyclopedia) says that the Annales Fuldenses were actually written in Mainz. Since I have not found the evidence for this, I shall use Fulda as the place. Luckily, Mainz and Fulda are only about 100 kilometers apart, so the question is not serious.

Sithienses [ca. 823], which was discussed in Section X.2, and Fuldenses have much in common. There has been considerable controversy about which is original for the parts that they share. Scholarly opinion inclines toward Fuldenses as the original, so far as I can judge, and I have accepted this judgment. Both are also closely connected with Laurissenses [ca. 829], which will be discussed below.

Gundechar [ca. 1074]. Gundechar was born on 1019 Aug 10 and died 1075 Aug 2. He was designated bishop (as he tells us) of Eichstet on 1057 Aug 20 and was consecrated on 1057 Dec 27. He wrote an account of the bishopric through 1074, shortly before he died. This work itself does not concern us. However, bound with his work is a table that the editor apparently considered as a

---

†This part is in parallel with the fourth part through 887.

continuation. It is a list of the Holy Roman
emperors from 1105 through 1162, and in the middle
of the list is an account of the eclipse of 1133
Aug 2. Although this account is not really connected with Gundechar, there is no way provided
to cite it except by using his name.

Halesbrunnenses [ca. 1241]. The Annales
Halesbrunnenses are described by the editor Pertz
as short annals from two codices connected with
the monastery of Heilsbronn.[†] It was not clear to
me whether the codices contained variant versions
of the same annals or whether they contained different entries that Pertz merged. The difference
is not important for us. The annals contain entries
for 26 years from 1098 to 1178 inclusive, plus a
single entry for 1241.

Halesbrunnenses Notae [ca. 1133]. Judging
from his citations, I believe that this is the
source that Ginzel [1882 and 1884] called Annales
Halesbrunnenses, rather than the source just discussed. If so, this was a simple slip. Pertz
calls this source Halesbrunnenses Notae. The Notae
consist of 3 entries made on the last page of a
copy of what Pertz calls "Ekkehardus" (see Frutolf
[1103] above). The entries are for the years 1117,
which records an earthquake, for 1133, which records a solar eclipse, and for 1338. Only the first
two concern us, so I have used the date 1133 in
citation.

Heimo [1135]. Heimo was a monk of S.
Michael's at Bamberg who wrote a set of chronological tables with annalistic entries, from the beginning of the world to his own time. He writes
that he has prepared the table in 1135, notes that
it will be 460 years to the end of the 3rd great

[†]
Heilsbronn is a small place slightly west of
Nürnberg and should not be confused with the
much larger Heilbronn.

Easter cycle, and leaves a note to the posterity
of that distant date. He also leaves a note that
reads: "And thou, most select brother Burcard,
do not fail to pray for me." Heimo's work enjoyed
some popularity and had several continuations, including Augustani Minores [ca. 1321] already discussed.

Herbipolenses [ca. 1215]. The word used
in the citation is from a form of a name for Würzburg that was used only in medieval times. The
work is a continuation of Ekkehard's continuation
of Frutolf [1103]. The annals run from 1125 to
1158. There is then a gap to 1202, with extensive
entries for 1202, 1203, and 1204, and a single
paragraph for 1215. The annals contain an account
of the Second Crusade that is long enough to suggest
that the unknown writer went on the crusade. I
did not notice any use of the first person, but
several choices of word or phrase suggested personal
experience. The possibility that the annalist took
part in the Crusade clouds the interpretation of
the eclipse of 1147 Oct 26; see the discussion of
the record 1147 Oct 26a E,G in the next section.

Herimannus [1054]. The author of this
work is known as Herimannus or Hermannus Augiensis,
Hermannus of Reichenau, and Herimannus Contractus.
The last form, which means Herman the Lame, is
perhaps the most common. He is an appealing and
sympathetic figure. He was born on 1013 Jul 18 of
a noble German family, and the editor gives his
family tree for 5 generations back. He was crippled
by illness in early childhood and never walked again.
Perhaps because his illness barred him from many
pursuits, he became a learned and influential scholar
in the fields of mathematics, astronomy, music, and
history. He constructed clocks[†] and astronomical

[†] The word commonly used is horologium, which could
mean a sundial, a water clock, or a mechanical
clock as we know it.

instruments. He wrote descriptions of the astrolabe, and Europeans for some time believed him to be the inventor of it.

His chronicle is probably his best known work. It is a chronicle from the birth of Christ to his own time that he was keeping up to date when he died. In its earlier parts Herimannus drew from Bede [734?], Eusebius Pamphili [ca. 325],[†] Isidorus [ca. 624] with its continuations, Fuldenses [ca. 901], Laurissenses [ca. 829], and other sources. It is one of our main sources for Herimannus' own time. Herimannus died on 1054 Sep 24.

Herimannus' chronicle was a popular work that was copied and continued at many places. One of the main continuations was by his friend and disciple Bertholdus [Bertholdus, 1080]. Bertholdus describes the accomplishments of Herimannus at some length; we learn for example that he made an extensive study of lunar eclipses. Since Bertholdus' continuation does not contain any data that I noticed that is useful for this study, I have not discussed his work separately.

The Swiss writer Bernoldus [1100] was also an admirer of Herimannus and refers to his work on calendrical problems and to his experiments concerning the equinoxes and solstices.

Unfortunately it is necessary to express a caution: Discovery is often attributed to one whose only role was transmission.

Hildesheimenses [ca. 1137]. These annals are closely connected with the German sources Quedlinburgenses [ca. 1025] and Lambertus [ca. 1077].[‡]

[†] We may guess that he used Jerome's translation and continuation rather than the original. See Section XV.3.

[‡] This Lambertus is distinct from the Belgian Lambertus Parvus [ca. 1193].

and with the French source <u>Weissemburgenses Maiores</u>
[ca. 1075]; it is necessary <u>to discuss all four</u>
sources together. The connections are so close
that the editor prints them in parallel columns
over much of their time span.

The sources have a common origin in what
the editor calls <u>Annales Herolfesfeldenses</u>, which
apparently no longer survive as a separate document.
These are from a monastery at Hersfeld[†] that was
founded in the middle of the 8th century. They are
derived from known sources through 829. From 829
through 973 the editor calls them <u>genuina</u>. At the
year 974 the editor has marked across the parallel
annals that the annals of Hersfeld extend up to
this point.

In or about 973 the <u>Annales Herolfesfeld-</u>
<u>enses</u> were copied, presumably in order to continue
them elsewhere. The final result is that we have
the four sets of annals that have been enumerated.[‡]
One copy was taken to Hildesheim, about 30 kilometers
southeast of Hannover. There the annals were continued to about 1100. Then they were taken to Paderborn and continued to 1137; the Paderborn part has
not received a separate name for some reason. Since
even the parts before 973 contain some information
not found in the parallel annals, I have assumed
that such information is derived from local sources.
The editor believes that he can distinguish the
later writers and their politics.

<u>Lambertus</u> [ca. 1077]. The author of
these annals is usually known as Lambertus

[†] Hersfeld is about 40 kilometers north of Fulda.
It is not the same as Harsefeld, which is near
Hamburg.

[‡] It does not follow that four copies were made simultaneously; a second copy could have been made from
the first, and so on. One could undoubtedly settle
the sequence of copying by making a suitable study
of the texts, but I have not tried to study them
from this standpoint.

Aschafnaburgensis but occasionally as Lambertus Herolfesfeldensis or some variant spelling. Lambertus made one of the continuations of the Annales Herolfesfeldenses that have just been discussed. I did not see any independent information in it that is useful for this study. I gather from the standard sources that modern critics think rather poorly of his work. One of them (the New Catholic Encyclopedia) says that he is "more famous for his form than for his credibility."

Laureshamenses [ca. 803]. These are from a monastery in the village usually called Lorsch; it is on the Rhine near Worms. The monastery at Lorsch gave rise to many sets of annals, of which only two concern us. Both were edited by Pertz in the same volume of Monumenta Germaniae Historica. Fortunately, from the standpoint of clarity in citation, he used different spellings of the Latin form of Lorsch in the titles of the two. The other spelling is Laurissenses.

The annals tell us that the monastery was founded in 764. Before that date the annals could not be independent and I spent little time in studying them. From about 768 on they have many local references and give the appearance of originality.

Laurissenses [ca. 829]. This is the other set of annals from Lorsch that concern us. Their analysis is somewhat complex. They are among the annals that are attributed to Einhard.[†] In fact, Pertz edits them under the title Annales Einhardi and the citation Einhard when used by Pertz seems always to mean these annals. The annals that Pertz calls Annales Laurissenses are a different set; I examined them and found nothing of interest for this study. There are also references in some places to the Annales Laurissenses Maiores. As well as I can make out, these are the

---

[†] See the discussion of Einhard [ca. 820] above.

combination of what Pertz calls <u>Annales Laurissen-</u><u>ses</u> and <u>Annales Einhardi</u>.

The annals that Bouquet calls <u>Annales</u> <u>Loiselianos</u> [<u>Loiselianos</u>, ca. 814] and <u>Annales</u> <u>Tilliani</u> [ca. 808] must be discussed in conjunc-tion with <u>Laurissenses</u>. Bouquet has named these annals after two early editors Loisel and Tillet, about whom I can discover nothing more. The manu-script called <u>Loiselianos</u> is (or at least was in 1744) in the library of S. Germain-des-Prés in Paris, but that does not prove that the manuscript was written there. I did not notice any statement about where the manuscript called <u>Tilliani</u> can be found.

Bouquet says that <u>Loiselianos</u> was followed (for the time that it covers) by <u>Regino</u> and by many other annalists, including the "Annalist of S. Ber-tin" and the "Author of the Annals of Mets".[†] Bou-quet also says that <u>Loiselianos</u> was followed, from 801 to 813, by Einhard or by the author of the an-nals that bear his name. This is a reference to the annals that are being called <u>Laurissenses</u> here; this sentence suggests that Bouquet did not neces-sarily accept Einhard as their author.

Bouquet says still more. He says that <u>Loiselianos</u> itself is largely copied from <u>Tilliani</u>. Bouquet's statements, if correct, would mean that we could not use the reports of eclipses in <u>Lauris-</u><u>senses</u> because they would be derived in the first instance from <u>Loiselianos</u>, of which we do not know the geographical provenance, and they would be de-rived at one remove from <u>Tilliani</u>, about which we know even less.

However, I do not believe that Bouquet's statements must necessarily be accepted. There is

----

[†] "Regino" refers to the author of the source <u>Regino</u> [ca. 906] that is discussed later in this section. "Mets" refers to the source <u>Mettenses</u> [ca. 905] discussed in Section X.2. "S. Bertin" probably but not certainly refers to the source that I desig-nated as <u>Sithienses</u> [ca. 823] in the same section.

a long passage that will be designated as the record
807 Feb 11 E,G in the next section. This passage
appears under the correct year in Laurissenses and
Loiselianos and under a wrong year in Tilliani. It
seems unlikely that the compilers of Laurissenses
and Loiselianos would have made an error in copying
Tilliani that would bring the year back to the correct number. It is unlikely that the compilers
would have corrected the year by independent research, because people who do this much research
are likely to write in their own words.

My suspicion is that later research has
shown that Loiselianos and Tilliani are variants
of Laurissenses. Pertz says that Tilliani "agrees"
with Laurissenses in large part, and I believe
that what Pertz calls Codex 7 of Laurissenses is
Bouquet's Loiselianos. I shall assume that records of eclipses found in Laurissenses originated in Lorsch, except for records also found in
Fuldenses.† I shall share material found in both
Laurissenses and Fuldenses equally between the two
places. Since Lorsch and Fulda are only about 100
kilometers apart, lack of decision between them does
not matter much for our purposes.

I shall also use the record 807 Feb 11 E,G
to suggest that Einhard did not write either Laurissenses or Fuldenses.

Since Laurissenses deals mostly with royal
affairs, many writers say that it could not have
come from Lorsch or any other monastery. I do not
understand the argument; medieval monasteries often
had an interest in royal affairs. Perhaps, because
of its interests, I should assign the eclipses in
Laurissenses to a region rather than to Lorsch.
Since none of the eclipses will be taken as total,
I shall not adopt this refinement.

---

† The reader may remember from the last chapter that
Sithienses has much in common with Laurissenses,
as Bouquet's words imply. I took Sithienses as
secondary to Fuldenses or Laurissenses when they
have common material.

Magdeburgenses [ca. 1188]. These annals
are divided into two distinct parts. The first
part runs through 1188. After that they were not
used for almost 3 centuries. Then someone found
them and added entries for the years 1453-1460. It
is reasonable to treat the later entries as a separate document, and the present citation refers
only to the first and larger part. The first part
was apparently written entirely by one person.
Nothing definite is known of this person but, judging from his interests as shown by his writing, he
was a monk of S. Johannis Magdeburgensis. The annals begin with the Annunciation. About 950 they
begin to incorporate independent local material.
From about 1130 on they seem to be substantially
original.

Marbacenses [ca. 1375]. These annals
are probably from Marbach, just north of Stuttgart.
They contain brief notices of the eclipses of 1133
Aug 2 and 1191 Jun 23. Ginzel [1882 and 1884]
said these were taken from the Annales Argentinenses†
but he was misled by some marginal notes of the
editor that apply to something else. I shall not
use the eclipse notices from Marbacenses because
they could easily have been extracted from any one
of several sources.

It is interesting that the annals say,
under 1185, that the astronomer known as John of
Toledo has predicted that in about September of
the following year all the planets will be in the
same house. This will lead to many unpleasant
happenings, including the coming of the anti-Christ.
This entry is an example of the recrudescence of
astrology, from which earlier centuries seem to
have been largely free.

Marianus Scotus [ca. 1082]. Marianus
was born in Ireland in 1028. He moved to Cologne

---

† I did not find anything in Argentinenses that is
useful for this work and therefore I have not
cited it.

about 1056, to Fulda about 2 years later, and then
in a few years to Mainz where he spent the rest of
his life. He died on either 1082 or 1083 Dec 22.
His Chronicon claims to be a universal chronicle
from the Creation to 1082. It was a popular work
for some time and was used by Florence of Worcester
[ca. 1118] and other British writers. It does not
contain any information that is usable in this study,
but some discussion will be required in the next
section in order to establish that some of his eclipse
descriptions are unusable.

His last entry is interesting. It records
that the Paschal new moon in 1082 appeared on April
2 although it should not have appeared until April
4 (see Section II.4). This entry is correct. 1082
Apr 4 is the date of the 'ecclesiastical" new moon.
I calculate that the "Naval Observatory" moon was
new at about midday on 1082 Apr 2 and hence it would
have first been visible after sunrise that evening.

Two anonymous continuations are printed
along with the Chronicon. One of them records the
eclipse of 1133 Aug 2. I have counted this among
the records in Marianus Scotus since there is no
information about the author.

Monacensis [ca. 1322]. This is the Munich
member of the family that includes Mellicenses [ca.
1564] and many other Austrian annals. The editor
says that Monacensis has little value except for
some information about Richard Coeur-de-Lion, but
he did not have our interests. It contains two
eclipse reports within our time period that seem
to be original.

Naufragia Ratisbonensia [ca. 1358].
Ratisbon is the former English name, and the root
of the Latin name, of the city that is now called
Regensburg. The monastery of S. Emmerammus was
founded there in about 650. About 25 kilometers
away was another early monastery called Weltenburg.
These monasteries have yielded a few fragments
that the editor Jaffé has edited under the heading

Annales et Notae S. Emmerammi Ratisbonenses et
Weltenburgenses. The fragmentary notes hardly seem
to deserve being called annals, especially since
there is a different set of annals called Annales
Ratisbonenses.[†] In his prefatory notes, Jaffé
refers to these as "minor wreckage" (naufragia) of
history, and Naufragia Ratisbonensia sounds like
an appropriate designation to use. The fragment
that is being designated consists of annalistic
entries for the years 1036 to 1046 and then for
1048, 1074, 1241, and 1358. Two of the entries
are original reports of solar eclipses. The ten
consecutive annals were written at Regensberg and
the "continuations" at Weltenburg, according to
the editor. Weltenburg can be found on a large
map. It is on the Danube a few kilometers above
Kelheim.

　　　　Necrologici Fuldenses [ca. 1064]. The
basic part of this work is a list of deaths of
various people year by year, along with some records of other interesting events. Thus they form
a type of annals that has no apparent connection
with Easter table annals. The entries begin in
779 with the death of Sturmus, the first abbot of
Fulda. Some of the people are fully identified,
but most are not. We can speculate that the ones
who are not were monks at Fulda. If so, the monastery was large because the deaths average about
10 per year. The nature of the entries suggests
that they were prepared by one person down to about
900 and were then kept up to date reasonably well
by various people. An analysis of the handwritings
would probably confirm or deny this speculation.

　　　　Neresheimenses [ca. 1296]. The manuscript
of these annals was already lost when the edition
in Monumenta Germaniae Historica was prepared, and
the editor worked from an earlier printed edition.
The Annales Neresheimenses borrow heavily in their

[†] I found nothing to use in these and do not cite
them.

earlier parts from <u>Annales Elwangenses</u>.† Elwangenses in turn was derived from the <u>Herolfesfeldenses</u> family; see the discussion of <u>Hildesheimenses</u> [ca. 1137] above. <u>Neresheimenses</u> seems original from about 1150 on. There are three continuations that carry the annals almost continuously down to 1572, but these are outside the time period of this study.

<u>Pegavienses</u> [ca. 1227].

Pegau is perhaps 20 kilometers southwest of Leipzig. The monastery there, according to these annals, was dedicated on 1096 Jul 26, and the town subsequently grew up around ᴛᴴᴱ ᴍonastery. The editor believes that the annals were first compiled through 1149 in about that year. There are then three continuations that carry the annals through 1227. Continuations prepared in this way are usually original. In this case the annals do not become fully independent until about 1176, although they have some original earlier parts. Through 1176 the annals incorporate passages taken almost verbatim from <u>Magdeburgenses</u> and other sources. After 1176 the annals are in various contemporaneous hands and seem to be original.

<u>Pruveningenses</u> [ca. 1298].

These annals are from a monastery in the diocese of Regensburg. The entries are written in the margins of Easter tables, with reference signs to indicate the year involved. The editor says that the signs are confused and that it is often difficult to make out the year intended. Many entries were made by a hand of the late 13th century, including all of the notices of eclipses. Thus the eclipse reports have low reliability at best. Nonetheless I have included <u>Pruveningenses</u> among the sources used in this work because it contains the only record I have found that may be of the eclipse of 779 Aug 16.

---

† I have not cited <u>Annales Elwangenses</u> because they did not contain any information useful for our purposes.

Quedlinburgenses [ca. 1025]. This is
one of the sources that originated with the Annales
Herolfesfeldenses, which were discussed above in
connection with Hildesheimenses [ca. 1137]. Quedlinburg was the site of a nunnery that enjoyed much
royal favor. It was considered as a house for the
daughters of noble Saxon families. It owned extensive properties and was responsible directly to the
pope. Much of the history of the town concerns the
struggles of the townspeople to be independent of
the rule of the abbesses of the nunnery. The annals
often borrow from the Laurissenses family where they
do not derive from Herolfesfeldenses.

Regino [ca. 906]. As we learn from his
chronicle, Regino became abbot of the monastery of
Prum in 892. In 899 he moved to Trier where he
also became abbot, and remained so until his death
in 915. The editor says that his date can still be
read on his tomb at Trier, but he does not give it.
Regino's chronicle runs from the birth of Christ to
his own time. At 817 Regino inserts a note that
what has gone before is what he found in a certain
little book, and all that he has done is to correct
the grammar and to add a few things learned from
his elders. "Borrowing" from an 'old book' is a
literary device that was already hoary in Regino's
time.[†] What follows, Regino then says, is what he
has learned from his studies and from the narrations
of others. We can tell that he has used Laurissenses
and Fuldenses or sources related to them.

Rodenses [ca. 1157]. Ginzel [1882 and
1884] calls this place Klosterrath but does not
give its coordinates. Pertz calls it Klosterrad
and says it is near Herzogenrath north of Aachen.
I shall use the coordinates of Herzogenrath. The
monastery from which the annals came was founded
early in the 12th century. Rather unusually, the
annalist did not borrow extensively from earlier

[†] This agrees with Bouquet's statement that Regino
used what Bouquet calls Annales Loiselianos. See
the discussion of Laurissenses [ca. 829] above.

annals but started about the time of the foundation.
The annals cover only about 50 years and are in only
two hands. They served as a source for Aquenses
[ca. 1196].

Rosenveldenses [ca. 1131]. These annals
are from a monastery in Harsefeld; according to the
annals, the monastery was founded in either 1101 or
1102. After perhaps 1105 the annals may be original.
The annals break off after saying "Anno Domini 1131"
without any corresponding annal. We should need to
inspect the manuscript in order to know what this
signifies.

Ryenses [ca. 1288]. The monastery of Rye
is just on the German side of the present border
between Germany and Denmark. By my rules, Ryenses
should count as a German source. Since its associations and interest are dominantly with Danish
affairs, I make an exception and put Ryenses among
the Danish sources in Section XIII.1.

S. Blasii [ca. 1147]. I have discussed
this source in conjunction with the Swiss source
Engelbergenses [ca. 1175] in Section IX.3. These
annals cannot be original before about 1120, although they may incorporate some otherwise unknown
material. In 1147 they were transferred to the
monastery of Engelberg in Switzerland.

S. Disibodi [ca. 1200]. A place called
Disibodenberg apparently grew up around a monastery
dedicated to S. Disibodi (I am not sure of the
nominative case of his name) that was founded in
the middle of the 12th century. It is close to
Bad Kreuznach. The annals were first compiled
shortly after the founding of the monastery, from
Marianus Scotus [ca. 1082], Wirziburgenses [ca.
1101], and Erphesfurdensis [ca. 1182], perhaps
along with other sources. They seem to have been
kept regularly only between 1155 and 1168. From
then on there are only a few short entries until

-373-

the annals stop entirely in 1200.

The annals furnish some information about prices of books in the middle ages. Someone has written on the inside of the cover: "This book is worth 6 pounds according to Master Wirnher, but I do not believe it. I have bought other chronicles for 5 pounds." "Master Wirnher" is probably the Wirnher who was abbot of S. Disibodi from 1294 to 1305, according to the editor.

S. Maximini [ca. 987]. These annals are from the monastery of S. Maximinus in Trier. They are written on a table of the second "great Easter cycle" that ran from 532 to 1063. The notices are in one hand until 840 and in various hands from then on.

S. Stephani [ca. 1448]. The town of Freising near Munich possibly dates to Roman times. A monastery named for S. Stephan, which is often called Weihenstephan, was founded there about 725. After some disaster, of which I have not discovered the nature, it was restored in 1021 and it survived until 1803. In more recent times it has been made into a model farm and a brewery. The annals are written in various hands from the time of the res- toration onward. Someone in the 15th century in- serted some entries for earlier years that were taken directly from Mellicenses, which is discussed in Section IX.1; the editor omitted these from his edition.

S. Trudperti [ca. 1246]. These late annals were compiled toward the end of the 13th century in the monastery of S. Trudpertus about 15 kilometers south of Freiburg in the Black Forest. The editor lists Regino [ca. 906], Herimannus [1054], Engelbergenses [ca. 1175], and Zwifaltenses as being among the sources used. I am not sure what he means by Zwifaltenses, but I do not believe that it is the source called Zwifaltenses below. In spite of their late origin, the annals contain

an account of the eclipse of 1191 Jun 23 that I
have found nowhere else.

Scheftlarienses Maiores [ca. 1247]. The
monastery that Ginzel [1882 and 1884] spells Scheftlarn is on the Isar south of Munich, and Ginzel
gives its coordinates. It was founded in 780 [De
Fundatione Scheftlariensi, ca. 1341] but it was
destroyed by the "most impious duke Arnolf of Bavaria"
in a year that is not stated. A new monastery was
established in 1160. The annals must belong entirely
to the new establishment, because they are all in
one hand through 1162. Later entries are in various
hands.

Scheftlarienses Minores [ca. 1272]. These
are from the same place. They are rather odd. The
first entry is for 814 and reads: "King Charles the
Great ruled."[†] After that the only entries are for
the period 1215-1272. Thus the Minores lie outside
the time chosen for this study. I have used them
because they show that quite independent records
can be kept in the same monastery at the same time;
see the record 1241 Oct 6i E,G in the next section.

Suevicum [ca. 1043]. This chronicle was
apparently written by one person. It was written
about the same time as Herimannus [1054], it deals
with much the same geographic area as Herimannus,
and sometimes it has been attributed to Herimannus.
I gather that the editor did not believe that Herimannus wrote this chronicle, and the records of
solar eclipses tend to confirm this conclusion. The
eclipse of 968 Dec 22 is recorded in the same words
in Suevicum and Herimannus. The eclipse of 1033
Jun 29 is recorded in different words in the two
sources, and the eclipse of 1039 Aug 22 is in
Suevicum but not in Herimannus. This suggests
that Herimannus and the unknown author of Suevicum

[†] We may doubt that the annalist meant to write
this. Charlemagne died on 814 Jan 28.

used some of the same sources for records before their own times, but that in their own times they wrote independently.

This chronicle is concerned with the district later called Swabia. It consisted approximately of the areas now called Baden and Württemberg. It should be sufficiently accurate for present purposes to assume that the observations came from Stuttgart.

Theodorus [ca. 1182]. This source is usually called Annales Palidenses, with no author's name attached. However, it was written by a known person, and it seems appropriate to use his name. He was Theodorus, a monk of the monastery at Pohlde. The work was written to be a continuation of Eusebius [ca. 325], and Theodorus relies heavily upon the Spanish source Hydatius [ca. 468] for its earlier parts. After the entry for 486 Theodorus wrote across the annals: "Bishop Hydatius to here; thenceforth Theodorus wrote the annals."[†] Theodorus used many other sources for the later part. About 1140 the annals change character and it is plausible that Theodorus is an essentially original source thereafter.

There is also an anonymous continuation. It consists mostly of fragmentary lists of popes, kings, and emperors, along with about 20 annalistic entries for years from 1182 through 1421. I have not distinguished the scanty continuation from the original in citation.

Thietmarus [ca. 1018]. Thietmarus (also spelled Thietmar or Dietmar) was descended on both sides from noble or prominent Saxon families. The editor gives a long biography of Thietmarus for the sake of those who are interested. Thietmarus himself gives us (Book VI, Chapter 27) the main fact needed

---

[†] Theodorus must have used a continuation of Hydatius that I have not consulted; see Chapter XIV.

here: he was bishop of Merseburg from 1009 to 1018.
This rather long chronicle is mainly a history of
the affairs of the kings of Saxony from about 890
to his time.

Trevirensis [ca. 967]. Regino [Regino,
ca. 906] was abbot of Trier. There someone unknown
continued his chronicle from his death in 915 down
to 967. Before 950 the entries are short except
for 939. We may guess that the unknown began to
keep the continuation in 950 and that he made only
a brief summary of earlier years, except that 939
interested him for some reason. We do not need to
know whether the guess is correct or not, and no
stress should be put upon it.

Vita Hludowici Imperatoris [ca. 840].
This Louis is the son of Charlemagne and he was
the second and last ruler of Charlemagne's empire.
His policies caused the division of the empire after
his death in 840, with unpleasant consequences for
later European history. The copy of the Vita that
I used has been marked "attributed to Astronomus"
by the cataloguing department of the library, but
I know of no reason to believe that such an attribution is correct.

This work is not associated with a specific geographic place. Since it is a biography
of Louis, I shall assume that the place of an observation is the place where Louis was at the time.†
This is a weak assumption and I shall attach a low
reliability to records used with it.

Weingartenses [ca. 936]. Weingarten is
a town about 20 kilometers east and north of Karlsruhe. I have not been able to learn anything about
it but there was presumably a monastery there. The
editor of the annals says that they are written in

---

† This is not necessarily the place where Louis
dwelled.

the margins of a 10th century copy of De Temporum
Ratione [Bede, 725]. Many parts of them are identical with Augienses [ca. 954]. Since both sources
are of almost the same age, I see no way to assign
priority. Therefore I shall divide the reliability
of a report that appears in both sources equally between them. Weingarten and Reichenau (where Augienses was written) are about 150 kilometers apart.
Thus the difference between the places is appreciable but not serious for our purposes.

Wirziburgenses [ca. 1101]. These annals
have sometimes been attributed to a person named
Baluzius, apparently with little justification.
They were written in the 12th century. Thus they
cannot be original and I saw nothing in them that
was not derived from known sources. The main sources
were Hildesheimenses, Frutolf, and Herimannus; Wirziburgenses is almost identical with Hildesheimenses
from 1040 on. I mention these annals only because
they are often mentioned in editions of other sources,
and because they help us restore the correct text of
the record 1093 Sep 23a E,G in the next section.

Wormatienses [ca. 1295]. These annals
are an unreliable source by their very nature. At
their head someone in the 15th century wrote that
he had taken them from an old book that was almost
illegible because of its age. The editor says that
the annals abound in chronological and other errors
that he has corrected in footnotes wherever he could.
Thus the 19th-century editor has reproduced the
manuscript that he found as faithfully as he could,
but we have no way to know how well the editor
in the 15th century did his job. So far as I can
see, the only reason for assigning the annals to
Worms is the number of references to Worms. It
seemed to me that there were about the same number
of references to Mainz as to Worms and that it would
be difficult to make a choice. Since Mainz and
Worms are fairly close together, the question of
origin is not important for this study.

TABLE XI.1

REFERENCES TO SOLAR ECLIPSES FOUND IN
MEDIEVAL GERMAN SOURCES

Source	Date	Conclusions
Albertus	1140 Mar 20	Probably copied S. Blasii
	1241 Oct 6	Original
	1245 Jul 25	Original
Aquenses	1133 Aug 2	Taken from Rodenses
	1147 Oct 26	Original
Augienses	840 May 5	Common report with Weingartenses
	878 Oct 29	Common report with Weingartenses
Augustani	1033 Jun 29	Not original but cannot trace
	1093 Sep 23	Original
	1098 Dec 25	Original
Augustani Minores	1187 Sep 4	Original
	1241 Oct 6	Original
Bertholdus	1093 Sep 23	Independent of known sources
Brunwilarenses	1133 Aug 2	Independent of known sources

TABLE XI.1 (Continued)

Source	Date	Conclusions
Brunwilarenses, cont'd	1147 Oct 26	Original
Colonienses	1147 Oct 26	Original
	1153 Jan 26	Original
	1187 Sep 4	Probably original; may have come from Liège
	1191 Jun 23	Original
	1207 Feb 28	Original
	1230 May 14	Original
	1232 Oct 15	Original
Corbeienses	664 May 1	Probably from Bede
	878 Oct 29	Too short to use safely
	891 Aug 8	Probably from Sang-allensis
	939 Jul 19	Original
Corbeienses Notae	1133 Aug 2	Original
	1241 Oct 6	Original
Erphesfurdensis	1133 Aug 2	Probably not independent
	1148 ? ?	Cannot identify with any eclipse
	1153 Jan 26	Original
Frutolf	1093 Sep 23	Original

TABLE XI.1 (Continued)

Source	Date	Conclusions
Fuldenses	787 Sep 16	Probably not original
	812 May 14	Common source with Laurissenses
	817 Feb 5	Recorded as solar, actually lunar
	818 Jul 7	Common source with Laurissenses
	832 ? ?	Cannot identify with any eclipse
	840 May 5	Original
	878 Oct 29	Original
Gundechar	1133 Aug 2	Probably original; not by Gundechar but bound with his work
Halesbrunnenses	1133 Aug 2	May be original
	1241 Oct 6	Original
Halesbrunnenses Notae	1133 Aug 2	Original
Heimo	1133 Aug 2	May be original
Herbipolenses	1133 Aug 2	Original
	1147 Oct 26	Original record; may have been observed in Asia Minor

TABLE XI.1 (Continued)

Source	Date	Conclusions
Herimannus	787 Sep 16	Probably common source with Quedlinburgenses
	968 Dec 22	Independent of known sources; common source with Suevicum
	1033 Jun 29	Original
Hildesheimenses	878 Oct 29	Not safe to use
	990 Oct 21	Original
	1033 Jun 29	Original
	1093 Sep 23	Original; printed edition probably contains an error
	1124 Aug 11	Original
	1133 Aug 2	Original
Laureshamenses	787 Sep 16	Original
Laurissenses	764 Jun 4	Probably a local observation
	807 Feb 11	Probably a local observation
	810 Jul 5	Probably a local observation
	810 Nov 30	Probably a local observation
	812 May 14	Common source with Fuldenses
	818 Jul 7	Common source with Fuldenses

TABLE XI.1 (Continued)

Source	Date	Conclusions
Loiselianos	-----------	Probably a variant copy of Laurissenses
Magdeburgenses	1147 Oct 26	Original
Marianus Scotus	664 May 1	Probably from Bede
	840 May 5	Copied from Augienses or Weingartenses
	878 Oct 29	Copied from Augienses
	1023 Jan 24 ?	Cannot be identified safely
Monacensis	1187 Sep 4	Original
	1191 Jun 23	Original
Naufragia Ratisbonensia	1044 Nov 22	Original
	1241 Oct 6	Original
Necrologici Fuldenses	1039 Aug 22	Original
	1044 Nov 22	Original
Neresheimenses	1153 Jan 26	Probably not independent
	1239 Jun 3	Original
	1241 Oct 6	Original
	1249 ? ?	Cannot identify
Pegavienses	1133 Aug 2	Probably a local record
	1147 Oct 26	May be from Asia Minor

TABLE XI.1 (Continued)

Source	Date	Conclusions
Pegavienses, cont'd	1187 Sep 4	Original
Pruveningenses	779 Aug 16	Unique record but surely not original
Quedlinburgenses	733 Aug 14	Not original but cannot trace
	787 Sep 16	Probably a local record
	812 May 14	Copied Fuldenses
	818 Jul 7	Copied Fuldenses
	990 Oct 21	Probably original
	1009 Mar 29 ?	Probably not an eclipse
	1014 ? ?	Probably not an eclipse
Regino	878 Oct 29	Original
Rodenses	1133 Aug 2	Original
	1147 Oct 26	May be from Asia Minor
Rosenveldenses	1124 Aug 11	Original
S. Blasii	1093 Sep 23	Probably taken from Frutolf
	1098 Dec 25	Not independent
	1133 Aug 2	Probably not independent
	1140 Mar 20	Original
	1147 Oct 26	Probably a local record

TABLE XI.1 (Continued)

Source	Date	Conclusions
S. Disibodi	1133 Aug 2	May be independent
S. Maximini	809 Jul 16	From Anglo-Saxon Chronicle through Senonensis
	840 May 5	Derived but cannot trace
	961 May 17	Original
S. Stephani	1133 Aug 2	Related to Rodenses
	1153 Jan 26	Original
	1191 Jun 23	Probably original
	1207 Feb 28	Original
	1239 Jun 3	Original
	1241 Oct 6	Original
	1261 Apr 1	Identification slightly doubtful
	1263 Aug 5	Original
S. Trudperti	1191 Jun 23	Probably a local record
	1241 Oct 6	Not safe to use
Scheftlarienses Minores	1241 Oct 6	Original
	1267 May 25	Original
Scheftlarienses Maiores	1187 Sep 4	Original
	1207 Feb 28	Original
	1241 Oct 6	Probably a local record

TABLE XI.1 (Continued)

Source	Date	Conclusions
Suevicum	961 May 17	May not be an eclipse
	968 Dec 22	Common source with Herimannus
	1033 Jun 29	Original
	1039 Aug 22	May not be original
Theodorus	1147 Oct 26	May be from Asia Minor
	1153 Jan 26	Probably not independent
	1187 Sep 4	Original
	1241 Oct 6	Original
Thietmarus	990 Oct 21	Probably copied Quedlinburgenses
	1018 Apr 18	Original
Trevirenses	961 May 17	Original
Vita Hludowici Imperatoris	840 May 5	Original; place of observation uncertain
Weingartenses	840 May 5	Common report with Augienses
	878 Oct 29	Common report with Augienses
Wirziburgenses	1033 Jun 29	Copied from Suevicum
	1093 Sep 23	Probably copied from Hildesheimenses; helps us restore the text

TABLE XI.1 (Concluded)

Source	Date	Conclusions
Wormatienses	1191 Jun 23	Probably original
	1241 Oct 6	Probably original
Xantenses	693 Oct 5	May be from a Byzantine source
	810 Jul 5	Copied from Laurissenses
	810 Nov 30	Copied from Laurissenses
	833 ? ?	Probably an erroneous copy of an erroneous entry in Fuldenses
	840 May 5	Original
Zwifaltenses	1093 Sep 23	Original
	1133 Aug 2	Probably original
	1153 Jan 26	Original
	1207 Feb 28	Original

Xantenses [ca. 873]. I did not learn
much about these annals. They are in the Tiberius
portion of the Cottonian collection. The annals
change character about 830. Until that point they
look to be short extracts from other annals. After
that each entry is moderately extensive. It is
possible that one person began them about 830, using
other annals to bring his record up to date, and
maintained them until his death in 873.

Zwifaltenses [ca. 1221]. Zwiefalten is
about 100 kilometers southeast of Stuttgart. The
monastery at Zwiefalten had two main sets of annals,
according to the editor, who calls them the "major"
and the "minor" annals. He says that the minor
annals extend to 1221. However, everything that
the editor presents is labelled "major" except for
the short first section that stops in 1098. The
annals as presented are in many independent but
overlapping sections, and one of the main sections
ends in 1221. I do not know whether the year is a
coincidence or whether this section is in the minor
annals and I have made a mistake in understanding
the editor. In any case the annals seem reliable
and original. They are in many different hands,
and the editor has indicated each change of hand.
Many of the entries in the late part of the 11th
century are in rhyme.

One section of the annals continues to
1503, but I did not use it.

2. Discussions of the Records of Solar Eclipses

Table XI.1 contains the usual list of
records arranged by source, along with a brief
comment on each record.

Marianus Scotus [ca. 1082] and Corbeienses
[ca. 1148] have short notes about the eclipse of
664 May 1. They contain the same data as the report
in Bede [734?], including the error in date. It is
almost certain that the reports are derived indirectly
from Bede even though the words do not agree exactly.

693 Oct 5 E,G. Reference: Xantenses
[ca. 873]. This has, under 687: "There was an
eclipse of the sun at the 3rd hour of the day,
such that some stars clearly appeared." This
report is suspiciously like the report 693 Oct 5a
M,B, in the Byzantine source Theophanes [ca. 813],
that is discussed in Section XV.4. If one were to
translate Theophanes' record into Latin, I believe
he would write something very like the record in
Xantenses. Further, Theophanes' count of the years
from the Incarnation differs by 8 from our count,
and Theophanes does not specify the year precisely
even in his own system. Thus a western writer who
was not aware of the hazards in using Theophanes'
dates could well assign the year 687.

There is nothing inherently implausible
in the idea that a German annalist in the middle
of the 9th century used a Byzantine source. The
west never lost touch with Byzantine affairs, and
Charlemagne and the Byzantine emperors were in
active diplomatic contact half a century earlier.
Greek was actively studied at Charlemagne's court,
Theophanes was a major contemporary Byzantine
chronicler, and it is easily imaginable that a copy
of Theophanes' writing, or of some of his sources,
found its way to the Frankish court and thence to
Xanten.

However, Theophanes gives the day of the
year and the day of the week correctly, while
Xantenses omits both items. Enough clichés are
used in eclipse records that the resemblances
could have happened by chance. Further, Xantenses
does not report any of the other eclipses found in
Theophanes. The most serious problem is that the
hour is the same in both records, and it is improb-
able that the hour would be reported as the same
in places as far apart as Xanten and Constantinople.

It is almost certain that the report in
Xantenses is neither original nor local. I see no
way to reach a firm decision about whether Theophanes
is the source for Xantenses. The conservative action
is to ignore the record in Xantenses.

The Belgian source Sigebertus [ca. 1111] has the same record as Xantenses but under the year 695. I did not use his record either.

Quedlinburgenses [ca. 1025] notes "There was an eclipse of the sun" under 733. This brief note is not safe to use even though we cannot identify specifically the source from which it was derived.

764 Jun 4 E,G. Reference: Laurissenses [ca. 829]. Under 764 this source has: "In that same year was an eclipse of the sun on the day before the nones of June† at the 6th hour." The hour looks reasonable. This record does not occur in Loiselianos [ca. 814] nor in Fuldenses [ca. 901]. Therefore I shall assume that it is local to Lorsch, but I shall use a reliability of only 0.5. Standard deviation of the magnitude: 0.1.

779 Aug 16 E,G. Reference: Pruveningenses [ca. 1298]. Pruveningenses contains the following passage: "770. An eclipse of the sun on the 17th calends September (= Aug 16) in the 20th part of Leo, at about the 9th hour (nones)." The hero Roland, according to legend, was killed on 778 Aug 15, and there was a magical eclipse of the sun at his death according to the form of the legend in Turoldus [ca. 1100]. While I put little faith in the notion, I cannot absolutely deny the possibility that the eclipse of 779 Aug 16, which was almost on the anniversary of the battle, was assimilated to the occasion. Whether this is so or not, the eclipse of 779 Aug 16 did occur and it should have been recorded, weather permitting. The record in Pruveningenses is the only one I have found that could apply to 779 Aug 16.

---

† Therefore the date given for the eclipse is 764 Jun 4.

The editor of Pruveningenses[†] points out that it is difficult to tell the year which applies to an annalistic entry, and we cannot pay much attention to the year stated. There is no sure identification of the eclipse. The penumbral eclipse of 770 Aug 25 would have been a maximum at about midday. This would fit the later meaning of nones,[‡] but the date would be the 8th calends rather than the 17th calends. 779 Aug 16 is the right day of the month, but the hour should have been about the 3rd. The eclipse of 760 Aug 15 would have been maximum at about the 9th hour, but it was a small eclipse and on the wrong day. I conclude that the eclipse cannot be identified.

The "20th part of Leo" does not help. According to the medieval church calendar the sun would be in that position on about Aug 7, which is the 7th ides August. It is more plausible that "20th" was written in error for "28th" than that "17th calends September" was written in error for "7th ides August", whether the numbers are written out or expressed in Roman numerals.

787 Sep 16a E,G. Reference: Laureshamenses [ca. 803]. This set of annals from Lorsch has under 787: "There was an eclipse of the sun at the 2nd hour on the 16th calends October (= Sep 16) on the Lord's day." The hour looks reasonable and the calendrical data are correct. This record must be independent because it is the only one out of 11 that I have found that has the correct date. We are also at a time when the record should be original. Reliability: 1. Place: Lorsch. Standard deviation of the magnitude: 0.1.

[†] See the discussion of this source in the preceding section.

[‡] Nones probably still meant the 9th hour in 770, but the actual record was not written until the 13th century. By that time, nones could have the meaning of our noon.

787 Sep 16b E,G. Reference: Quedlinburgenses [ca. 1025]. Under 786 this has: "There was
an eclipse of the sun on the 15th calends October
(= Sep 17) from the 1st hour of the day until the
5th." The hours given here are probably those for
almost the whole duration of the eclipse. "15th"
(xv.) in Roman numerals could have been obtained
from the correct xvi. by accidental omission of a
vertical stroke. This is the only source in the
Herolfesfeldenses family[†] that contains these data.
Therefore I shall assume that this report represents
local Quedlinburg information, although it cannot
be original. Reliability: 0.5. Place: Quedlinburg.
Standard deviation of the magnitude: 0.1.

Fuldenses [ca. 901], Pruveningenses [ca.
1298], and Herimannus [1054] record this eclipse
on the date that Quedlinburgenses gives, but with no
accompanying detail. A plausible explanation is
that all used the same source as Quedlinburgenses
but omitted most of the detail.

807 Feb 11 E,G. Reference: Loiselianos
[ca. 814]. This source has a long and interesting
entry which I translate as follows:

And the number of the years
was changed into[‡]

DCCCVII

In the preceding year, 4th nones September
(= Sep 2), was an eclipse of the moon: the sun
was then in the 16th degree of Virgo. The moon
therefore was in the 16th degree of Pisces.
Moreover in this year on the 2nd calends
February (= Jan 31) the moon was at the 17th

[†] See the discussion of Hildesheimenses [ca. 1137]
in the preceding section.

[‡] This was a standard formula in some sources. I
doubt that such sources were written on Easter
tables.

(day) when the star Jupiter was seen to pass through it. And on the 3rd ides February (= Feb 11) was an eclipse of the sun at midday, both stars being in the 25th degree of Aquarius. Again on the 4th calends March (= Feb 26) was an eclipse of the moon, and flames appeared that night of an amazing brightness, and the sun stood in the 11th degree of Pisces and the moon in the 11th degree of Virgo. And, still more, the star Mercury on the 16th calends April (= Mar 17) was seen in the sun like a small black spot, a little above the center of that very body, and it was seen by us for 8 days, but when it first entered, and when it left, clouds kept us from observing. And again in August, on the 11th calends September (= Aug 22) was an eclipse of the moon at the 3rd hour of the night, the sun's position being in the 5th degree of Virgo and the moon in the 5th degree of Pisces. And thus from September of the preceding year to September of the present year the moon was obscured thrice and the sun once.

The flames that appeared on Feb 26 were probably auroral; they seem to have been on the same night as the eclipse. The appearance of "Mercury" in the sun cannot be a transit of Mercury; a transit of Mercury cannot last for 8 days. As many writers have observed, this must be the oldest known European record of a sunspot. The passage of Jupiter "through" the moon is a plausible event that one can test by calculation, but I have not done so.

Oppolzer [1887] lists the following eclipses in Greenwich Time:

1. A lunar eclipse, 806 Sep 1 at $22^h\ 34^m$.

2. A solar eclipse, 807 Feb 11 at $10^h\ 43^m$, easily visible in Germany.

3. A lunar eclipse, 807 Feb 26 at $2^h\ 49^m$.

4. A lunar eclipse, 807 Aug 21 at $22^h\ 53^m$.

The last lunar eclipse was not quite total; the
other lunar eclipses were total. The dates given
for the eclipses are correct, except that for the
first and last there could be a reasonable question
in Germany as to which day to assign them to; they
would have been close to midnight there. The positions of the sun and moon in the zodiac are acceptable if the positions are taken from the medieval
church calendar; they are in fact only ways of
giving an approximate date and are probably not
observations.

One can imagine the interest that this
set of events would arouse in a group of people
who, like Charlemagne's court, were studying astronomy.

This long passage appears in Tilliani [ca.
808] under the year 808. Thus it seems unlikely to
me that Tilliani is the original source. The passage
also occurs in Laurissenses [ca. 829] but not in Fuldenses [ca. 901]. Thus, in accordance with the decisions reached in the discussion of Laurissenses in
the preceding section, we may take Lorsch to be the
place. The record of the solar eclipse is contemporary and may receive a reliability of 1. Standard
deviation of the magnitude: 0.1.

This passage, or excerpts from it, have
been copied into too many sources to mention.

Einhard [ca. 820] in his Vita Karoli Magni
wrote that there were many signs before the death
of Charles, including: "For 3 successive years near
the end of his life[†] the sun and the moon were frequently eclipsed, and a black spot was seen in the
sun for the space of 7 days." It seems unlikely to
me that the same person wrote this and the long
passage in Loiselianos. The solar eclipses of 810
Jul 5, 810 Nov 30, and 812 May 14, and the lunar
eclipses of 809 Dec 25, 810 Jun 20, and 810 Dec 14
are briefly noted in Loiselianos and/or Laurissenses.
These, even together with the eclipses in 806 and

_____

[†] Charlemagne died 814 Jan 28.

807 already listed, do not make up eclipses in 3
successive years, and the last eclipse was more
than 1½ years before Charlemagne died. Still, we
may perhaps allow a biographer a little latitude.

The biographer and the annalist have
different numbers of days that the spot in the sun
was visible; we may attribute that to scribal error.
The biographer also does not mention that the spot
was Mercury, and he does not mention the other
interesting phenomena. To be sure, people may
write different ways for different purposes, and
consistency is not a universal human characteristic.
After making these allowances, I still doubt that
Einhard the biographer of Charlemagne was the same
person as the annalist of Lorsch.

Fuldenses [ca. 901] does not have any
astronomical events in 807. An event in 807 that
did strike the annalist at Fulda as important was
that "Aaron the king of the Persians"† sent Charlemagne a wonderful clock that struck the hours. It
seems less likely that Einhard wrote Fuldenses than
that he wrote Laurissenses.

S. Maximini [ca. 987] has an account of
the eclipse of 809 Jul 16 that is identical with
the one in the French source Senonensis [ca. 1203].
It is clearly taken from the account of the same
eclipse in the Anglo-Saxon Chronicle [ca. 1154].

810 Jul 5 E,G. Reference: Laurissenses
[ca. 829]. The text reads, near the end of the
entry for 810: "In this year the sun and the moon
were twice eclipsed, the sun on the 7th (or 8th)‡
ides June (or July)‡ and the 2nd calends of December

---

† This is Harun al-Rashid (764? - 809) of Thousand-
and-One Nights fame. Harun also had diplomatic
relations with China.

‡ Different copies have different numbers for the
day and also different months. The combination
7th ides June is probably the most common, but
I did not try to make an exact count.

(= Nov 30), and the moon on 11th calends July
(= Jun 21) and 18th calends January (= Dec 15).
As I have noted, different copies have different
dates for the first solar eclipse, but none has
the right date, which is the 3rd nones July. The
date of the second solar eclipse is correct. The
annalist probably began his day at sunset, because
the lunar eclipses were both shortly after sunset,
on 810 Jun 20 and 810 Dec 14 respectively.

This entry is in Loiselianos [ca. 814],
which Bouquet took as original, but it is not in
Fuldenses. Thus I take Lorsch to be the place.
Reliability: 1. Standard deviation of the magnitude: 0.1.

The report appears in various other German
and French annals with almost all combinations of
error for the date of the first solar eclipse.

810 Nov 30 E,G. Reference: Laurissenses
[ca. 829]. This record appears in the same passage
as the preceding record. It is interesting that
all copies have the correct date for this eclipse
and a wrong date for the preceding one. Reliability:
1. Place: Lorsch. Standard deviation of the magnitude: 0.1.

812 May 14a E,G and 812 May 14b E,G.
References: Laurissenses [ca. 829] and Fuldenses
[ca. 901], respectively. Laurissenses has: "In
this year was an eclipse of the sun after midday
on the ides of May." Fuldenses has the same except
that it omits the day. Both give the year as 812.
The correct day is the 2nd ides May. Since Fuldenses
does not have the day, it cannot be the original for
Laurissenses. Since the records have the same words,
they are surely not independent. I shall give each
record a reliability of 0.5; this is equivalent to
saying that they form essentially one record. Places:
Lorsch and Fulda. Standard deviation of the magnitude:
0.1. See the discussion of the next record.

Quedlinburgenses [ca. 1025] alone of the Herolfesfeldenses family has a record of this eclipse. It has copied Fuldenses.

Fuldenses [p. 356] says that there was a solar eclipse on 817 Feb 5. This is a simple slip. There was a lunar eclipse then.

818 Jul 7a E,G and 818 Jul 7b E,G. References: Laurissenses [ca. 829] and Fuldenses [ca. 901]. Both say the same thing under 818: "An eclipse of the sun occurred on the 8th ides July." The correct day is the nones of July. Here we have a dating error similar to that of 810 Jul 5. A day that should be written in relation to the nones is written in relation to the ides. It is as if the annalist thought that Jul 5 and not Jul 7 was the nones of July. I shall share this record between Fulda and Lorsch, as I did with the records of 812 May 14. Reliability: 0.5 for each. Places: Lorsch and Fulda. Standard deviation of the magnitude: 0.1.

The reader may raise an objection at this point. He may say that the record of 812 May 14 proves that Fuldenses is not original there, and hence that it probably is not original for 818 either. I would agree with this objection if the eclipses were all that we had. Other material, however, shows that Fuldenses is often independent of Laurissenses.

Quedlinburgenses [ca. 1025] and Vita Hludowici Imperatoris [ca. 840] have copied this record.

Under 832 Fuldenses [ca. 901] has: "The sun was eclipsed on the 5th nones May (= May 3) and the moon on the 2nd nones June (= Jun 4)." It is clear that the record is in error because a solar and a lunar eclipse could not have this interval. I have not been able to find any combination of two errors that would account for this record. Xantenses [ca. 873, Appendix] says that the sun and moon were eclipsed in 833. This is almost surely an

erroneous borrowing from the already serious error
in Fuldenses.

       840 May 5a E,G and 840 May 5b E,G.
References: Augienses [ca. 954] and Weingartenses
[ca. 936]. Augienses has, under 840: "An eclipse
of the sun on the 3rd nones May (= May 5), between
the 8th and the 9th hour, on the eve of the Ascension of Our Lord. Louis died." Weingartenses has
the same, except that it inverts the order of the
sentence and adds Louis' title of emperor after his
name. These records make it seem that Louis died
on the day of the eclipse, but that is not correct.
Louis (the subject of Vita Hludowici Imperatoris)
died on 840 Jun 20.

       Ascension Day in 840 was May 6. Thus
the date is stated in two distinct but equally
correct ways.

       Neither source can be original in 840,
but this record seems independent of all others
that I have found. There is no sure way to assign
priority, so I shall use the records equally,
giving a reliability of 0.2 to each. Places:
Reichenau and Weingarten. Standard deviation of
the magnitude: 0.1.

       The record of this eclipse in Marianus
Scotus [ca. 1082] is a copy of this one.

       840 May 5c E,G. Reference: Fuldenses
[ca. 901]. Fuldenses has, under 840: "Moreover
in this year on the eve of the Ascension of the
Lord, that is, the 4th ides May (= May 12), there
was an eclipse of the sun about the 7th or 8th
hour of the day so great that stars near the darkness of the sun could be seen, and the color of
things in the earth was changed."

       Here I think that the error in date
strongly suggests originality. The correct date
is "3rd nones May" and it is unlikely that a
copier would change this into "4th ides May". The

error is exactly 1 week. I imagine that the annalist
was refreshing his memory by consulting a calendar,
that he had a calendar arranged by weeks, and that
he simply read off the wrong Wednesday.[†] Reliability: 1. Place: Fulda. Standard deviation of the
magnitude: 0.01; since he still refers to color,
it is doubtful that the eclipse was total.

        840 May 5d E,G. Reference: <u>Xantenses</u>
[ca. 873], which has under 840: "And the 3rd of
May, that is, the 3rd of the Rogation days,[‡] there
was an eclipse of the sun at the 9th hour, and stars
were seen clear in the sky just as at night." The
annalist has "3rd of May", which is wrong as well
as being an unusual way of writing a date; he probably omitted "nones" from "3rd nones May" by accident. This sounds like an original description of
a total eclipse. Reliability: 1. Place: Xanten.
Standard deviation of the magnitude: 0.

        840 May 5e E,G. Reference: <u>Vita Hludowici</u>
<u>Imperatoris</u> [ca. 840, Chapter 62, p. 646]. This
says: "In that time an unusual eclipse of the sun
occurred on the 3rd day of the Major Litany[*]; as
much as its light receded darkness prevailed until
it differed in no way from true night. For indeed
the regular collection of stars was seen, and no
star was dimmed by the light of the sun; ..." The
passage continues at some length describing the
phases of the sun as its light was restored. This
sounds like the original description of an eye witness. I shall take the standard deviation of the
magnitude to be 0. There is a problem about the
place.

[†] 840 May 5 was the 4th feria.

[‡] The Rogation days are the 3 days preceding Ascension Day. Hence they were May 3, 4, and 5 in
840. See the next record.

[*] The annalist meant the Lesser Litany. The days
of the Lesser Litany are the same as the Rogation
days.

I decided in the preceding section to take
the place to be where Louis was at the time. At
the time he died, Louis was returning from Bavaria,
where he had been to suppress a revolt. He was
presumably on his way to Aachen when he died on 840
Jun 20 on an island in the Rhine near Ingelheim,
a few kilometers from Mainz. It is doubtful that
he would have been there since May 5, the day of the
eclipse, so he was probably in Bavaria on May 5.
Ingolstadt is about in the center of Bavaria, and
every point in Bavaria is within about 1° of it, so
that it can be taken as the place. In order to
compensate for the uncertainty in the place of the
observation, I shall lower the reliability to 0.5.

S. Maximini [ca. 987] has a short notice
of this eclipse that could have been derived from
many sources. I shall not use it.

878 Oct 29a E,G and 878 Oct 29b E,G.
References: Augienses [ca. 954] and Weingartenses
[ca. 936]. Under 878, Augienses has: "An eclipse
of the sun on the 4th calends November (= Oct 29)",
while Weingartenses has: "Pope John came to France,
and there was an eclipse of the sun on the 4th calends November." Pertz, who edited both annals,
took the former to be original and the latter to
be a copy. He may be correct, but I prefer to take
both as equal. Reliability: 0.5 for each. Places:
Reichenau and Weingarten respectively. Standard
deviation of the magnitude: 0.1.

878 Oct 29c E,G. Reference: Fuldenses
[ca. 901]. After identifying the month as October
of 878, this says: "There was an eclipse of the
moon on the ides of that month (= Oct 15), at the
last hour of the night; the sun also on the 4th
calends November was eclipsed after the 9th hour
for about half an hour so that stars appeared in
the sky, and everyone thought that night had come."
Oppolzer [1887] lists a total lunar eclipse at
4h 26m on 878 Oct 15. This sounds like the account
of an eye witness. Reliability: 1. Place: Fulda.

Standard deviation of the magnitude: 0.

          878 Oct 29d E,G.   Reference: Regino
[ca. 906, p. 590].   Regino writes the following:
"In the year 878 of the Incarnation of our Lord,
there was a great eclipse of the moon in the month
of October, on the 15th day.   And that same month
an eclipse of the sun happened, on the 29th day
about the 9th hour."   It was unusual to give dates
in this fashion at the time this was written.   It
is possible that Regino derived his information
from Fuldenses but I doubt it.   If he used Fuldenses
or any other source, he paraphrased with unusual
freedom; the language in Regino is quite different
from that in other sources.   We are now in Regino's
own time and this could readily be a personal observation.   If so, it was probably made in Prum.   Reliability: 1.   Standard deviation of the magnitude:
0.1.

          Marianus Scotus [ca. 1082] has the same
words as Augienses.   Corbeienses [ca. 1148] and
Hildesheimenses [ca. 1137] have brief but different
notices of this eclipse.   Both occur in portions
of the annals that could be original, but it is
safer to ignore them.

          891 Aug 8 E,G.   Reference: Corbeienses
[ca. 1148].   Under 891 this source has: "Comet
and eclipse of the sun."   If I understood the editor,
he believes that this is in an original part of the
annals.   However this entry is almost identical
with the record 891 Aug 8 E,CE from the Swiss source
Sangallensis [ca. 926] that is discussed in Section
IX.3.   Hence I shall give the report in Corbeienses
a reliability of 0.

          939 Jul 19 E,G.   Reference: Corbeienses
[ca. 1148].   This has under 939:   "The sun was
diminished[†] at the 3rd hour of the day on the 4th
_____
[†]Inminutus.

                        -401-

ides July." The day given is Jul 12, which is
exactly 1 week too early. This report seems to be
independent and it is in a part of the annals that
the editor takes as original. The choice of words
implies an eclipse that was not total. Reliability:
1. Place: Corvey. Magnitude: 2/3 with standard
deviation of 1/6; path south of Germany.

        961 May 17a E,G. Reference: Trevirensis
[ca. 967], which has under 961: "An eclipse of the
sun on the 16th calends June." The day given is
961 May 17, and I have found no other record with
this information. Trevirensis should be original
here. Reliability: 1. Place: Trier. Standard
deviation of the magnitude: 0.1.

        961 May 17b E,G. Reference: S. Maximini
[ca. 987]. This contemporaneous source has, under
961: "Otto the younger was made king, and there
was an eclipse of the sun." This is Otto II, who
became king of the Germans in 961 and Holy Roman
emperor in 973. Reliability: 1. Place: Trier.
Standard deviation of the magnitude: 0.1.

        Under 961 Suevicum [ca. 1043] says that
a sign appeared in the sun. The chronology of the
source is good, and this may be a note about the
eclipse of 961 May 17. However there are other
kinds of sign, and it is safest not to use this.

        968 Dec 22 E,G. References: Herimannus
[1054] and Suevicum [ca. 1043]. Both sources note
that there was an eclipse of the sun on the 11th
calends December in 968. The date written is an
error for 11th calends January (= Dec 22); the error
is of a sort that is encouraged by the nature of
the Roman calendar, which requires that more than
half the dates in a month be written using the
name of the following month. The error is more likely
to be made in first writing down a date than in
copying a date already written correctly, I believe.
Thus the error was probably in the original record.
Neither Herimannus nor Suevicum is likely to be
original here, so I shall assume that both sources

have copied an older record that is independent of
others known. Reliability: 0.5. Place: Reichenau.
Standard deviation of the magnitude: 0.1.

    990 Oct 21a E,G. References: Quedlinburg-
enses [ca. 1025] and Thietmarus [ca. 1018, p. 772].
In Book I, Chapter 10, Thietmarus writes: "In the
year 989 of the Incarnation of the Lord the sun was
eclipsed on the 12th calends November (= Oct 21) at
the 5th hour of the day." Quedlinburgenses gives
the same information about the eclipse, but under
the correct year. If one copied the other, it is
Thietmarus that is the copy. The way the year is
given in Thietmarus makes it unlikely that a copier
would make an error in the year, but it would be
easy to mistake the year intended in Quedlinburgenses.
Further, Quedlinburgenses has by now diverged from
the underlying Annales Herolfesfeldenses and should
be independent. On the other hand, Thietmarus was
only 14 at the time of the eclipse and it is on the
face of it unlikely that he made this as an original
entry. I shall assume that Quedlinburgenses is the
original. Since Quedlinburg and Merseburg are only
about 70 kilometers apart, an error in the place
has little consequence for us. Reliability: 1.
Place: Quedlinburg. Standard deviation of the
magnitude: 0.1.

    990 Oct 21b E,G. Reference: Hildesheim-
enses [ca. 1137]. Hildesheimenses has also diverged
from Annales Herolfesfeldenses by now. It has, under
990: "And that year was an eclipse of the sun, which
not long after was followed by a mortality of men
and beasts, on the 12th calends November." I think
the date is meant to apply to the eclipse and not to
the mortality. This should be a contemporaneous
record. Reliability: 1. Place: Hildesheim. Standard deviation of the magnitude: 0.1.

    1009 Mar 29(?). Quedlinburgenses [ca.
1025] says that, in 1009, on the 3rd calends May
(= Apr 29), on the 6th feria, the 1st of the moon,
the sun was covered by a terrible colored cloud and

that it stayed covered for 2 days, regaining its proper light only on the 3rd day. This is certainly not a description of an eclipse as it stands, but it may be an exaggerated account or it may be a confusion between an eclipse and some other event.

There was no eclipse on 1009 Apr 29, which was the first of the ecclesiastical moon but not the astronomical one. However there was a penumbral eclipse on 1009 Mar 29, and two Belgian records of it are given in Section VIII.2. Now 1009 Mar 29 was the 3rd feria and the 29th of the moon, while 1009 Apr 29 was the 6th feria and the 1st of the moon as stated. However, if Apr 29 was written by mistake for Mar 29, the feria and the age of the moon could have been calculated from the wrong date.

However, writing Apr 29 where Mar 29 is meant is a likely error only in our calendar and not in the Roman calendar. Mar 29 is the 4th calends April while Apr 29 is the 3rd calends May. Thus two mistakes are needed to create the error.

In sum, I believe that this is not a record of the eclipse of 1009 Mar 29 in spite of the near-coincidence of date.

Quedlinburgenses also says that the sun and moon and other stars gave sad signs in 1014. No eclipses, either lunar or solar, should have been seen in Germany that year.

1018 Apr 18 E,G. Reference: Thietmarus [ca. 1018, p. 863]. In Book VIII, Chapter 5, at a time that is in the spring of 1018: "In those days the sun before its setting appeared to several to be halved in a wondrous fashion." The eclipse of 1018 Apr 18 should have been about half total at sunset in Saxony. Reliability: 1. Place: Merseburg. Magnitude at sunset: 0.5 with a standard deviation of 0.1.

The last statement needs comment. This record differs in principle from all other statements about solar eclipses that have been used in this

work. It is not a statement that implies anything
about the greatest amount of eclipse at the observing site. It is instead a statement about the
amount of obscuration at an accurately known time,
namely sunset. In using this record it will be
necessary to deal specifically with the eclipse at
sunset and not necessarily with the maximum eclipse.

1023 Jan 24(?). Marianus Scotus [ca. 1082]
says that there was an eclipse of the sun at the 9th
hour in the springtime in 1023. The only possible
eclipse I can find near this time is the eclipse of
1023 Jan 24, which would have been maximum at about
12^h in the winter. It is possible that "9th hour"
means nones or noon in its modern meaning. It is
also possible that "springtime" (verno tempore)
meant a warm spell that happened to come in January.
It seems safest to conclude that the eclipse is out
of place and cannot be identified. Even if it could
be, the record could not be used safely. The editor
says that the entry is not in Marianus' original
text but has been added in the margin in a different
hand. Hence we have no idea where the record was
made.

1033 Jun 29a E,G. Reference: Hildesheimenses [ca. 1137]. This has under 1033: "An eclipse
of the sun occurred on the 3rd calends July (= Jun
29), 6th feria, on the natale[†] of S. Peter Apostle,
at the 6th hour, as the emperor with the first people
of the reign was entering Merseburg." This appears
to be contemporaneous. Reliability: 1. Place:
Hildesheim.[‡] Standard deviation of the magnitude: 0.1.

[†] Natale has no counterpart in English that I can
find. It is the day a person became a saint, in
contrast to nativity, when a person became a person.
Natale and nativitas are often used interchangeably
in medieval sources.

[‡] It may be that the actual observation was made at
Merseburg rather than at Hildesheim. The difference in magnitude at the two places is probably
no more than about 0.01, which is negligible compared with the standard deviation.

1033 Jun 29b E,G. Reference: Herimannus [1054]. Herimannus wrote under 1033: "There was an eclipse of the sun on the 3rd calends July (= Jun 29) about the 7th hour of the day." The editor printed this in the type that he reserved for original material. The date in relation to Herimannus' dates makes its originality plausible. So does the fact that "7th hour" occurs in no source that Herimannus is likely to have used. Reliability: 1. Place: Reichenau. Standard deviation of the magnitude: 0.1.

1033 Jun 29c E,G. Reference: Suevicum [ca. 1043]. This source has under 1033: "An eclipse of the sun on the 3rd calends July at almost the 6th hour." It is possible that a person could write this when he meant to copy the preceding record, and vice versa. However, I see no reason to doubt that this is an original record. Reliability: 1. Place: Stuttgart (as the approximate center of Swabia). Standard deviation of the magnitude: 0.1.

The editor of Wirziburgenses [ca. 1101] thought that the Annales Wirziburgenses were taken either from Frutolf [1103] (which he called "Ekkehardus") or from Herimannus [1054]. The record of the eclipse of 1033 Jun 29 in Wirziburgenses, however, is taken from neither. It is taken from Suevicum.

Augustani [ca. 1104] notes that there was an eclipse of the sun on 1033 Jun 29 but gives no details. We can feel sure that this report is derivative even though we cannot specify the source of the information; it could have come from many places.

1039 Aug 22a E,G. Reference: Suevicum [ca. 1043]. This source has, under 1039: "An eclipse of the sun on the 11th calends September." The date is correct. The source should be contemporaneous. Since it is not safe to use such a brief notice with confidence, I shall use a reliability of 0.2. Place: Stuttgart. Standard deviation of the magnitude: 0.1.

This, and not <u>Herimannus</u> [1054], is the probable source of the record of this eclipse in the Austrian source <u>Mellicenses</u> discussed in Section IX.1.

1039 Aug 22b E,G. Reference: <u>Necrologici Fuldenses</u> [ca. 1064]. This has under 1039: "On the 11th calends September the sun suffered an eclipse at the 7th hour." Reliability: 1. Place: Fulda. Standard deviation of the magnitude: 0.1.

1044 Nov 22a E,G. Reference: <u>Necrologici Fuldenses</u> [ca. 1064]. This has under 1044: "There was an eclipse of the sun on the 10th calends December (= Nov 22) from the 1st hour to the 9th." An hour of the day in Fulda on the day given is about 45 of our minutes, so the duration stated for the eclipse is about 6 of our hours. This still seems long, but that is not sufficient reason to doubt the reliability of the record. Reliability: 1. Place: Fulda. Standard deviation of the magnitude: 0.1.

1044 Nov 22b E,G. Reference: <u>Naufragia Ratisbonensia</u> [ca. 1358]. This says: "An eclipse of the sun from the 2nd hour of the day to about the 8th hour on the 10th calends December; and its splendor was not seen from the 10th calends December to the 8th ides December." The year is 1044. The duration is more reasonable than the value recorded at Fulda. The last part of the sentence probably records a cloudy spell that started after Nov 22 and lasted for two weeks, until Dec 6. Reliability: 1. Place: Regensburg. Standard deviation of the magnitude: 0.1.

1093 Sep 23a E,G. References: <u>Wirziburgenses</u> [ca. 1101] and <u>Hildesheimenses</u> [ca. 1137, p. 106]. Here the exact language will be helpful. <u>Wirziburgenses</u> has under 1093: "Eclypsis solis

-407-

facta est 3. hora diei, et draco visus est."[†]
Hildesheimenses has: "5. Kal. Aug. circa vesperam
eclypsis solis facta est 3. hora diei, et draco
visus est." The reader should note that the records
are identical except for the underlined words, which
mean: "On the 5th calends August (= Jul 28) about
evening". These words contradict the rest of the
passage. The date of the eclipse was the 9th calends
October, and circa vesperam contradicts 3. hora diei.

The underlined words were printed in italics
in the cited edition. This meant that the words had
become illegible and that the editor had "restored"
them with the aid of other annals. We must also
remember that Wirziburgenses is often identical with
Hildesheimenses, and we should note that it has only
the words in Hildesheimenses that are not underlined.
This means almost surely that there are no missing
words in Hildesheimenses to be restored. In other
words, I think that the words "5 kal. Aug. circa
vesperam" were meant to be inserted somewhere else
and that the editor put them in this passage by
accident. If so, the words in Wirziburgenses constitute the original record, but the record itself
came from Hildesheim.

It is imaginable that there was an unusually large prominence visible during the eclipse and
that reminded someone of the flame from a dragon.
This interpretation would imply a total eclipse. I
shall not push the word that far, and I shall make
the usual assumption about the magnitude. Reliability: 1. Place: Hildesheim. Standard deviation
of the magnitude: 0.1.

1093 Sep 23b E,G. Reference: Augustani
[ca. 1104]. This says under 1093: "A rainy autumn.
An eclipse of the sun in midday in Libra, on the
8th calends October." The sun is in Libra on Sep

[†] "There was an eclipse of the sun at the 3rd hour
of the day, and a dragon was seen." I assume that
draco does not refer to the constellation of that
name.

23, which is the 9th rather than the 8th calends.
This should be an original record. Reliability:
1. Place: Augsburg. Standard deviation of the
magnitude: 0.1.

1093 Sep 23c E,G. Reference: Bertholdus
[ca. 1138, p. 111]. In Section 28 of his work
Bertholdus writes: "In his first year, on the 10th
calends October (= Sep 22) the sun was eclipsed for
almost 3 hours; indeed about the middle of the day
the sun began to grow dark in a black or hyacinth
spot and to lose color in a horrible appearance;
and the next year followed a great pestilence of
men." Note that the day of the month is wrong by
1. "His" refers in the context to Abbot Nogger who
became abbot at about this time; I did not find an
explicit statement of his date. Thus the eclipse
is surely that of 1093 Sep 23, even though the day
stated is Sep 22. This is almost but not quite a
contemporaneous record. Reliability: 0.5. Place:
Zwiefalten. Standard deviation of the magnitude:
0.1.

1093 Sep 23d E,G. Reference: Zwifaltenses
[ca. 1221]. This is one of many annalistic entries
that are in rhyme in this part of the annals, so
I give it in the same form that the editor did:

1093.   Eclypsis solis, undenis facta Kalendis
        Octobris mensis, mansit ferme tribus horis.

This means: "An eclipse of the sun, that occurred
on the 11th calends of the month October, lasted
nearly 3 hours." Bertholdus may have gotten his
"almost 3 hours" from this record, but he has a
different date and other information that is not in
it. This record should be contemporaneous. Thus
it is quite likely that there were two independent
records of this eclipse at Zwiefalten. Reliability:
1. Place: Zwiefalten. Standard deviation of the
magnitude: 0.1.

1093 Sep 23e E,G. Reference: Frutolf [1103, p. 207]. Frutolf writes under 1093: "There was an eclipse of the sun on the 9th calends October (= Sep 23) at the 3rd hour, and a great mortality followed." This is reminiscent of the record 990 Oct 21b E,G above, but I see no reason to suspect any relation between the two records. The record seems to be independent of other known records and it is well within Frutolf's own time. Therefore it is likely to be original. Reliability: 1. Place: Bamberg. Standard deviation of the magnitude: 0.1.

S. Blasii [ca. 1147] uses slightly more words than Frutolf to say the same thing, except that it gives the date as "xi. Kal." rather than "ix. Kal." This could easily be a copying error. The editor of S. Blasii seemed to think that its record was original, but the history of the monastery makes this rather unlikely. I think it is more likely that a prolix annalist at S. Blasien used Frutolf's chronicle, and that he accidentally changed "ix" to "xi". Thus I shall not use the record in S. Blasii.

1098 Dec 25a E,G. Reference: Augustani [ca. 1104]. This entry occurs under 1098: "An eclipse of the sun on the 8th calends January (= Dec 25) after midday, about the 28th of the moon, or perhaps the 29th; because the moon this year is being lighted sooner than that which is usual, not according to calculation." In other words, the new moon is not following the ecclesiastical tables but is appearing sooner than they say. The reader may remember from the preceding section that Marianus Scotus [ca. 1082] called attention to the fact that the Paschal moon in 1082 was 2 days early. This is a contemporaneous record. Reliability: 1. Place: Augsburg. Standard deviation of the magnitude: 0.1.

We should note that the annalist does not think it necessary to explain that a solar eclipse can occur only at new moon.

1098 Dec 25b E,G. Reference: S. Blasii
[ca. 1147], which says under 1098: "Also on the
8th calends January was an eclipse of the sun."
This is still too early for us to expect original
records at S. Blasii, either in this source or
any other local source. Augsburg is not too far
away and it is possible that the compiler at S.
Blasien used the source Augustani. I shall give
this record a reliability of 0.

1124 Aug 11a E,G. Reference: Hildesheimenses [ca. 1137]. This says under 1124: "There
was an eclipse of the moon on the calends February...
There was an eclipse of the sun on the 3rd ides
August." This entry records correctly the lunar
eclipse of 1124 Feb 1 and the solar eclipse of 1124
Aug 11. Reliability: 1. Place: Paderborn. Standard deviation of the magnitude: 0.1.

1124 Aug 11b E,G. Reference: Rosenveldenses [ca. 1131], which says: "1124. There was
an eclipse of the sun at about the 6th hour, 3rd
ides August." The hour looks reasonable. Reliability: 1. Place: Harsefeld. Standard deviation of
the magnitude: 0.1.

1133 Aug 2a E,G. Reference: Marianus
Scotus [ca. 1082]. This is actually the last
entry in the first continuation, and it is the only
entry after 1106: "1155. In this year on the 4th
nones August (= Aug 2) about the ninth hour of the
day there was an eclipse of the sun so great that
stars could be seen in the sky." The hour is written
using the word nones; since this hour is unreasonably
late, nones probably has its modern meaning. The
year is not an error. Marianus had original ideas
in the matter of an era, but I believe that no one
followed him in the matter except his continuator.
This seems to be an original entry but unfortunately
there is no way to use it safely. The codex in
which it is found is in the Vatican and we do not
know where the entry was made. The prime possibilities are Mainz and Rome, and they are so far apart

that it is pointless to use this record.

1133 Aug 2b E,G. Reference: Hildesheimenses [ca. 1137]. Part of the entry for 1133 tells
us that a solar halo was seen around the church in
Paderborn on Jun 29. This is a strong indication,
if any were needed, that this part of the annals
was written in Paderborn rather than in Hildesheim.
The entry then goes on to say: "There was an eclipse
of the sun on the 4th nones August about the 6th hour
to such an extent that stars appeared in the sky."
Reliability: 1. Place: Paderborn. Standard deviation of the magnitude: 0.01.

1133 Aug 2c E,G. Reference: Gundechar
[ca. 1074, p. 251]. As I pointed out above, this
record is not written by Gundechar but is merely
bound with his work: "1132. An eclipse of the
sun occurred on the 4th nones August (= Aug 2) at
the 9th hour." Here the hour is written with the
numeral in the printed edition. This record points
out a trouble with this eclipse. There are so many
records of it that it is hard to tell whether a
short record is independent or not. The circumstances
suggest that this record is original, but to be safe
I shall lower the reliability to 0.5. Place: Eichstet. Standard deviation of the magnitude: 0.1.

1133 Aug 2d E,G. Reference: Zwifaltenses
[ca. 1221]. This is under 1133. It is in the same
general section of the annals as the record 1093
Sep 23d E,G above, but the entries are no longer in
rhyme: "An eclipse of the sun on the 3rd nones
August." Note that the date given is Aug 3. The
annals seem to be original for many years both
before and after this entry, so I shall use it in
spite of its brevity, but with misgivings expressed
by a lowered reliability. Reliability: 0.5. Place:
Zwiefalten. Standard deviation of the magnitude: 0.1.

1133 Aug 2e E,G. Reference: Rodenses
[ca. 1157], which has: "In the year 1133 of the

Incarnation of the Lord, there was darkness over
the whole earth because of an obscuring of the sun
about midday, for almost an hour, on the 17th of
the moon, on the 4th nones August. For stars
appeared just as at night, and the birds hastened
from the sky, and the ground became wet with dew,
and men were struck with great dread, fearing that
the last day had come." The annalist goes on to
say (and this may have been the cause of the fear)
that the darkness had some cause other than an
eclipse, because the sun can be eclipsed only about
the 30th of the moon, when the sun and moon are in
line.

Unfortunately the annalist erred twice
in his calculation of the age of the moon in order
to get the 17th. He must have calculated for 1132
rather than for 1133, or have made an equivalent
error some other way.[†] Even so, he made another
error, because the age of the moon on 1132 Aug 2
was 16, not 17. The correct age on 1133 Aug 2 is
27 days.[‡] This is far enough from 30 that he would
probably have reached the same conclusion even if
he had done the calculation correctly.

This report is surely original, and the
eclipse certainly sounds total. I do not believe
that I have seen the appearance of dew observed
before, in either old or modern literature on
eclipses. Reliability: 1. Place: Herzogenrath,
since I do not have coordinates for Klosterrath.
Standard deviation of the magnitude: 0.

Aquenses [ca. 1196] and others of the
related annals have used this record by omitting
most parts of it. They show the origin by retain-
ing the remark about the dew.

S. Stephani [ca. 1448] says: "1133. This
year, the 4th nones August, 17th of the moon, about

[†] One starts the calculation by dividing the year
by 19 and adding 1 to the remainder. Forgetting
to add 1 is equivalent to using 1132 rather than
1133.

[‡] Perhaps he wrote "17th" in error for "27th".

noon on a clear day there was such darkness that
stars appeared." This has the same double error
as Rodenses in the age of the moon, although any
error is rare. This also says "darkness" rather
than "eclipse", and such usage is rare. I have
not noticed any suggestion that S. Stephani and
Rodenses are related, and indeed they are almost
on opposite sides of Germany. Further, the appearances suggest that S. Stephani is an original
source at this time. However, I dislike to accept
the relations between the accounts as coincidence
and I shall not use the record in S. Stephani.

   1133 Aug 2f E,G. Reference: Heimo
[1135, p. 3]. The copies of the entry for 1133
differ somewhat. One has: "There was an eclipse
of the sun on the 4th nones August (Aug 2) on the
16th of the moon." The other has: "This year
there was darkness on the 3rd nones August (= Aug
3)." These are so different that it is unlikely
that they simply represent different copies. It
seems more likely that someone did some chronic-
ling independently of Heimo. We again have an
error in the age of the moon, but at least only
a single error. The circumstances are suspicious,
but we are in Heimo's own time and he must have
seen the eclipse. That does not mean that he
recorded it, of course. I shall lower the reliability to 0.5 because of the suspicious circumstances. Place: Bamberg. Standard deviation
of the magnitude: 0.1.

   1133 Aug 2g E,G. Reference: Corbeienses
Notae [ca. 1241, p. 277], which has: "In the year
1133 of the Incarnation of the Lord, Indiction 11,
epact 12, year 13 of the lunar cycle, 4th nones
August, 27th of the moon, in the middle of the
day occurred an eclipse of the sun so dark that
it was almost like night and many stars could be
seen. In the time of Master Folcmar abbot there
was an eclipse of the sun before the day of the

inventio[†] of S. Stephen protomartyr." The last sentence is not another account, as one might first think. The monastery was dedicated to S. Stephen, so the timing of the eclipse was interesting to its monks. Folcmar was the current abbot, and he was probably not displeased to have such a significant event recorded under his name. This is probably an original record. Reliability: 1. Place: Corvei. Standard deviation of the magnitude: 0.01.

See the record 1241 Oct 6c E,G below for further discussion of this record.

1133 Aug 2h E,G. Reference: S. Disibodi [ca. 1200]. This source has little original information, and the accounts of eclipses before and after this one are so obviously copied that I have not even bothered to mention them. The account of the eclipse of 1133 Aug 2 says: "In this year on the 4th nones August, 27th of the moon, the sun was obscured at the 7th hour of the day, and there was such darkness that stars appeared in the sky." This resembles other accounts, but I have found none just like it and especially I have found none with this hour. Therefore I shall use it but with a reliability of only 0.5. Place: Disibodenberg. Standard deviation of the magnitude: 0.01.

1133 Aug 2i E,G. Reference: Halesbrunnenses [ca. 1241]. Under 1133 this has: "There was an eclipse of the sun on the 4th nones August at the hour of nones." I use nones rather than a translation because its meaning is uncertain. This record is presumably original but, because such a short notice could easily be copied, I shall give it a reliability of only 0.5. Place: Heilsbronn. Standard deviation of the magnitude: 0.1.

[†] The inventio is the day when something significant connected with a saint, perhaps his tomb, was discovered. The inventio of S. Stephen is different from "S. Stephen's Day" in the usual sense, which is Dec 26. His inventio is Aug 3.

Erphesfurdensis [ca. 1182] has an entry
that is identical except for the omission of the
hour. It could be original but for safety I
shall not use it.

        1133 Aug 2j E,G. Reference: Halesbrunn-
enses Notae [ca. 1133, p. 13]. One of the notes says:
"In the year 1133 of the Incarnation of the Lord,
the 13th year of the 19-year cycle, when Lothair
the pious emperor was holding the government of the
realm, on the 4th nones August, the 4th feria, as
the day was verging toward nones, the sun in a moment was made black as pitch, day was changed into
night, many stars were seen, things on earth looked
as they do at night, and the streams of water were
brought to a stop." I do not understand the last
part unless it means that one could not see the
motion of the water. This seems to be an original
and contemporaneous record of a total eclipse. Reliability: 1. Place: Heilsbronn. Standard deviation of the magnitude: 0.

        One should notice the great difference
between the two records made at Heilsbronn.

        1133 Aug 2k E,G. Reference: Herbipol-
enses [ca. 1215]. This says: "This year there
was an eclipse of the sun on the 4th nones August
in the middle of the day, with such darkness that
stars could be seen everywhere as in the middle
of the night." Reliability: 1. Place: Würzburg.
Standard deviation of the magnitude: 0.

        1133 Aug 2ℓ E,G. Reference: Pegavienses
[ca. 1227]. This has: "There was an eclipse of
the sun on the 4th nones August, in the middle of
the 8th hour." The editor believes that this was
not written until about 1148, and he says that
this passage is from 'Ann. Erph!' It is not from
what he called Annales Erphordenses, a source in
which I found nothing useful. However, the part
down to the comma is found in what he calls S.
Petri Erphesfurdensis, cited here as Erphesfurdensis

[ca. 1182], as I mentioned in connection with the record 1133 Aug 2i E,G above. The phrase "in the middle of the 8th hour" is unique to this record, whether with "8th" or some other hour, and I can find no other record that uses the 8th hour. Thus this may represent a local record even though the annals as a whole are not independent. Reliability: 0.5. Place: Pegau. Standard deviation of the magnitude: 0.1.

1133 Aug 2m E,G. Reference: Brunwilar-enses [ca. 1179]. This says: "1133. At the 6th hour of the day the day was obscured, and stars appeared." The statement that the day rather than the sun was obscured is unusual, although the annal-ist may have meant to write "sun" but repeated "day" by inadvertence. The annals are presumably independent here, so I shall use this record in spite of its brevity, but with lowered reliability. Reli-ability: 0.5. Place: Brauweiler. Standard devia-tion of the magnitude: 0.1.

S. Blasii [ca. 1147] records: "1133. On the 4th nones August an eclipse of the sun, about the 6th hour." I have used other records that are equally brief, but only when there were strong reasons for believing that the annals in question were original. Here the independence of the annals in 1133 is open to question, so I shall not use this record. Pruveningenses [ca. 1298] merely notes an eclipse of the sun in 1133, with no details. I shall ignore this record.

1140 Mar 20 E,G. Reference: S. Blasii [ca. 1147]. The annals of S. Blasien record: "1140. In this year an eclipse of the sun occurred on the 13th calends April (= Mar 20) about the 10th hour, on the 28th of the moon." Albertus [ca. 1256] has an almost identical entry. Since Albertus did not write until the middle of the next century, we may conclude that his entry is the derivative one. Otherwise, no source gives this hour, and 1140 is among the few years when S. Blasii is likely to be original. Reliability: 1. Place: S. Blasien.

Standard deviation of the magnitude: 0.1.

     1147 Oct 26a E,G. Reference: <u>Herbipol-</u>
<u>enses</u> [ca. 1215]. Several German records of this
eclipse exhibit the same symptom. I have selected
<u>Herbipolenses</u> for particular reference because it
has the longest account, as follows: "There was
an eclipse of the sun in the year of this expedition,
namely 1147, on the 7th calends November (= Oct 26),
the 6th hour, the 28th of the moon, on the Lord's
Day in the feast of S. Amand." The date, the age of
the moon, and the day of the week are correct; I
do not know when S. Amand's day comes.

     As I mentioned in the discussion of
<u>Herbipolenses</u> in the preceding section, the annals
contain a strong suggestion that the annalist took
part in the Second Crusade, which is the expedition
referred to. The eclipse of 1147 Oct 26 took place
on a day that is moderately important in the history
of the Crusade, and it is recorded in at least one
account of the Second Crusade.[†] Thus the observation, although found in the annals of a German
monastery, may have been made in the Near East.

     The suspicion that the observation comes
from the Orient is strengthened by the hour. The
6th hour is definitely late for maximum eclipse in
Germany. <u>Oppolzer</u> [1887] shows the noon point to
be almost exactly at Athens, and if the observer
was with the main army he was farther east where
the time would have been later. Most sources in
western Europe give times well before the 6th hour,
and only one or two give a time as late as the 6th
hour even for the end of the eclipse.

     Because of the combination of the hour
and the likelihood that the writer was in Asia
Minor, I think that the observation was probably
made in Asia Minor. Thus if I were to use this
record, I should include it with the Byzantine

[†] The record 1147 Oct 26 M,B in Section XV.4, written
by <u>Odon</u> (de Deuil) [1148].

observations in Section XV.4. However I feel
that one cannot decide upon the place with enough
surety to let one use the record.

Theodorus [ca. 1182], Pegavienses [ca. 1227],
and Rodenses [ca. 1157] all give the 6th hour. Pegavienses is the only one of these in which the context strongly suggests the Crusade. Nonetheless the
circumstances are such that any of the sources could
have obtained the record of the observation from
Herbipolenses or from some source definitely from the
East. Thus I shall ignore all these records.

1147 Oct 26b E,G. Reference: Magdeburgenses
[ca. 1188]. Under 1147 this source writes: "In
this year on the 5th calends November (= Oct 28) an
eclipse of the sun about the middle of the day covered the earth with a horrible gloom,[†] to the point
that the circle was seen in the shape of a sickle,
which signified the shedding of the blood of human
kind." I shall guess that comparison of the sun's
circle with a sickle meant that 1/10 of the diameter
was left bright. The definite indication that the
eclipse was partial agrees with the use of caligo.
The record says "about the middle of the day", but
this is a vague phrase and does not, I think, mean
that the eclipse was seen in the Near East.

Reliability: 1. Place: Magdeburg. Magnitude: 0.9 with a standard deviation of 0.05;
path south and west of Magdeburg.

1147 Oct 26c E,G. Reference: Aquenses
[ca. 1196]. Aquenses says under 1147: "On the
6th calends November (= Oct 27) was an eclipse
of the sun from the 3rd hour to the 6th." In
regard to the record 1147 Oct 26a E,G above, note
that one could get the 6th hour from this only for

[†]The text uses the word caligo, which means fog or
vapor. It is also used for darkness, but I believe
that it usually implies less darkness than tenebrae,
for example, at least when used carefully.

the end of the eclipse. The editor of Aquenses
specifies 1147 as one of the years for which Aquenses has been "brought together" with Rodenses [ca.
1157]. The record of this eclipse in Rodenses,
found on p. 719, reads: "The sun was obscured on
the 7th calends November, the Lord's Day, about
the 6th hour." The reader should note that the
data disagree completely except for the year. The
record from Aquenses is independent of any other
that I have found and I shall use it. Reliability:
1. Place: Aachen. Standard deviation of the magnitude: 0.1.

       1147 Oct 26d E,G. Reference: Brunwilar-
enses [ca. 1179], which has: "1147 year, 7th calends November, the Lord's Day, happened an eclipse
of the sun from the 3rd hour, and it lasted until
after the 6th; in which disappearing it stood fixed
and unmoving almost an entire hour, as marked by
the horologe." I shall not try to guess what sort
of timepiece was used to make the last observation.
This source is apparently contemporaneous in the
12th century. Reliability: 1. Place: Brauweiler.
Standard deviation of the magnitude: 0.1.

       1147 Oct 26e E,G. Reference: Colonienses
[ca. 1238, p. 761]. This says under 1147: "There
was an eclipse of the sun on the 7th calends November near the hour of the day that was almost the
4th." This is an original record, I believe. Reliability: 1. Place: Cologne. Standard deviation
of the magnitude: 0.1.

       1147 Oct 26f E,G. Reference: S. Blasii
[ca. 1147]. I have already discussed this record
in connection with the record 1147 Oct 26 E,CE in
Section IX.3. The reader may remember from Section
XI.1 that the Annales S. Blasii were transferred
from S. Blasien to Engelberg in Switzerland in 1147.
The entry for 1147 is the first entry in what the
editor calls Engelbergenses as opposed to S. Blasii.
Since the entry contains two independent records of
the eclipse, I speculated that the first entry was

made in S. Blasien before the transfer and that
only the second entry was made in Engelberg. Because of doubt about the place, I shall use a reliability of only 0.5. Place: S. Blasien. Standard deviation of the magnitude: 0.1.

       1148 (?). Reference: Erphesfurdensis
[ca. 1182]. Erphesfurdensis has an entry under
1148 that I cannot understand. It says that there
was a sign near the sun on the 10th calends March
(= Feb 20) and gives a drawing of the sign like
this:

This looks remarkably like an attempt to render an
almost-eclipsed sun with 3 bright stars near it.
However there was no eclipse on 1148 Feb 20 and I
can find no plausible error that would change the
date into an eclipse date. The chronology of the
source seems good, and it is striking that the
date is the date of a nearly-new real moon as opposed to an ecclesiastical one. The drawing could
conceivably be the new moon with 3 stars nearby.
If this were so, one would not expect the annalist
to call it a sign; he was surely well acquainted
with the moon. The sign may be some sort of reflection or refraction phenomenon.

       1153 Jan 26a E,G. Reference: Zwifaltenses
[ca. 1221]. As we noted in the preceding section,
the Annales Zwifaltenses consist of a number of
independent but often overlapping sections. Two
sections record an eclipse under 1153. One section
says simply: "An eclipse of the sun." Another
says: "On the 7th calends February (= Jan 26)
there was an eclipse of the sun about the 6th hour
lasting almost until evening." No other record
gives these hours, and this seems to be a contemporaneous account. I shall use only the second
record. Reliability: 1. Place: Zwiefalten.
Standard deviation of the magnitude: 0.1.

       1153 Jan 26b E,G. Reference: S. Stephani

[ca. 1448]. This says: "1153. In this year was
an eclipse of the sun on the 7th calends February
after midday, although the moon was 27 according
to the lunar computation." The eclipse should have
been after midday, and 1153 Jan 26 was the 27th of
the ecclesiastical moon. Reliability: 1. Place:
Freising. Standard deviation of the magnitude: 0.1.

1153 Jan 26c E,G. Reference: Erphesfurdensis [ca. 1182]. Again we have a sign and a drawing: "A sign appeared in the sun on the 7th calends
February. ⌣ " It is interesting that both entries
from Erphesfurdensis that include drawings have been
copied into the conflated source called S. Petri et
Aquenses but not into the main Annales Aquenses[†]
themselves. This indicates strongly that the observation comes from Erfurt and not from Aachen. It
also suggests strongly that the same annalist wrote
the entries for 1148 and 1153, but not the earlier
entry for 1133.[‡] Whatever may have caused the confusion leading to the entry for 1148, this is surely
a record that the eclipse of 1153 Jan 26 was partial.
One may guess that 1/10 of the sun's diameter remained visible. Reliability: 1. Place: Erfurt.
Magnitude: 0.9 with a standard deviation of 0.05;
path south and east of Erfurt.

1153 Jan 26d E,G. Reference: Colonienses
[ca. 1238, p. 765]. This says under 1153: "On the
7th calends February (= Jan 26) there was an eclipse
of the sun around midday." Such a short note could
be based upon other sources. Since Colonienses should
be a primary source at this time, however, I shall
use this record. Reliability: 1. Place: Cologne.
Standard deviation of the magnitude: 0.1.

Theodorus [ca. 1182] and Neresheimenses
[ca. 1296] have records of this eclipse with about

[†] See the discussions of Aquenses and Erphesfurdensis
in the preceding section.
[‡] See the discussion of the record 1133 Aug 2i E,G
above.

the same amount of information as Colonienses.
Since neither of them is supposed to be original
at this time, I shall not use them even though I
use Colonienses.

1187 Sep 4a E,G. Reference: Augustani
Minores [ca. 1321]. This says: "1187. An eclipse
of the sun from the 6th hour to the 9th hour took
place." These annals were maintained continuously
from 1135 to 1321, so this should be original.
Reliability: 1. Place: Augsburg. Standard deviation of the magnitude: 0.1.

1187 Sep 4b E,G. Reference: Monacensis
[ca. 1322]. This continuation of some annals from
Salzburg has: "1187. There was an eclipse of the
sun on the day before the nones of September (that
is, on Sep 4) on the day of Venus[†] for about 4
hours." Reliability: 1. Place: Munich. Standard
deviation of the magnitude: 0.1.

1187 Sep 4c E,G. Reference: Theodorus
[ca. 1182]. As the dates show, this is in a continuation rather than in Theodorus' original annals:
"1187. An eclipse of the sun." The circumstances
indicate that this is an original entry. Reliability:
1. Place: Pohlde. Standard deviation of the magnitude: 0.1.

1187 Sep 4d E,G. Reference: Pegavienses
[ca. 1227]. This has, under 1187: "An eclipse of
the sun at the middle of the day on the 3rd nones
September (= Sep 3)." No other source has the day
wrong. The editor believes that these annals have
been fully original for about a decade. Reliability:
1. Place: Pegau. Standard deviation of the magnitude: 0.1.

[†] The Roman equivalent of our Freya's day, or Friday.
It is unusual for a clerical writer to use the
pagan names of the days.

1187 Sep 4e E,G. Reference: <u>Colonienses</u> [ca. 1238, p. 793]. This has, under 1187: "An eclipse of the sun about the feast of S. Lambert from the 6th hour until the 8th." The only feast day of S. Lambert that I find is Sep 17, which is not very close to Sep 4. However, I can find no other likely identification. The penumbral eclipse of 1186 Sep 14 was visible in the northern hemisphere, but it would have been around sunset in Cologne if it were visible at all there. The lunar eclipse of 1187 Sep 19 was not visible in Germany. Perhaps S. Lambert was particularly important to the annalist,[†] so two weeks away was still close to his day. The preoccupation with S. Lambert suggests a source from Liège,[‡] but I have found no Belgian source from which this record could be derived. Even if it should have come from Liège, we shall make no great geographical error in assigning it to Cologne. Reliability: 1. Place: Cologne. Standard deviation of the magnitude: 0.1.

1187 Sep 4f E,G. Reference: <u>Scheftlar- ienses Maiores</u> [ca. 1247]. These annals have the entry: "1187. There was an eclipse of the sun on the 2nd nones September." The first compilation of these annals was finished in 1162 and they are now kept up to date. Reliability: 1. Place: Scheftlarn. Standard deviation of the magnitude: 0.1.

1191 Jun 23a E,G. Reference: <u>Zwifaltenses</u> [ca. 1221]. In the middle of a rhyming section[*] dealing with other matters there is the following: "1190. On the 9th calends July (= Jun 23) there

---

[†] If so, it was not because of his church, which was dedicated to S. Pantaleonis. Maybe the annalist was named Lambert.

[‡] See the discussion of <u>Lambertus Parvus</u> [ca. 1193] in Section VIII.1, for example.

[*] Rhyming may have been a local tradition. Some entries made a century earlier were in rhyme. See the record 1093 Sep 23d E,G above.

was an eclipse of the sun near the 6th hour of the
day." The editor says that this note was written
in the same hand that wrote the entry for 1194.
The mixture of rhyming and non-rhyming entries,
together with the error in the year, suggests that
this entry may have been added at a later time.
If so, the time could not have been much later, so
I shall keep the reliability at 0.5. Place: Zwiefalten. Standard deviation of the magnitude: 0.1.

        1191 Jun 23b E,G. Reference: <u>Monacensis</u>
[ca. 1322]. This source says under 1191: "There
was an eclipse of the sun on the 7th calends of
July (= Jun 25) near midday, and all the air was
made purple.†" This is in a continuation that is
presumably up-to-date, and the record is unique
for this eclipse. I shall give it a reliability
of 1 in spite of the error in the day. Place:
Munich. Standard deviation of the magnitude: 0.05,
because of the wording.

        1191 Jun 23c E,G. Reference: <u>S. Stephani</u>
[ca. 1448]. These annals have: "1191. There was
an eclipse of the sun on the 9th calends July, the
Lord's day, about the hour <u>nones</u>. On the same day
as the eclipse a great <u>battle of</u> the Christians
with the pagans began near Acre, and, with many
from both sides being killed, victory fell to the
Christians." The siege of Acre had been going on
for a long time. It was finally captured by an
army under Richard on 1191 Jul 12. The final action
in the siege may have begun on Jun 23. When we
consider the rate at which news travelled, it is
clear that the entry could not have been written
until a few months after the eclipse. However this
fact by itself would not cause us to question the
basic reliability of the record.

---

† <u>Purpureus</u>, the word used, is also used sometimes
<u>to indicate</u> any dark color. Regardless of the
exact shade intended, the word implies a large
eclipse.

The eclipse is recorded in almost identical words in Cremifanensis, in the record 1191 Jun 23c E,CE in Section IX.1. Since there were other suspicious circumstances, I did not give full reliability; since I could not definitely prove copying, I kept the reliability as high as 0.5 because Cremifanensis is supposed to be an original source in 1191.

There is another troubling circumstance that I did not mention in Section IX.1. The eclipse of 1191 Jun 23 should have had a magnitude of around 0.8 in Acre at about the 9th hour, and there is a good chance that it would have been reported from there. If so, it would probably have been reported in much the same words, including the account of the battle, that occur in S. Stephani. It is unfortunate that the word nones is ambiguous, being able to mean either "noon" or the 9th hour. Noon would be about the correct hour in Germany, as other German records show. Since the words used can be interpreted in a way that makes a correct record, and since the source is supposed to be contemporaneous, I shall use this record. I shall lower the reliability to 0.5 because of the element of doubt. Place: Freising. Standard deviation of the magnitude: 0.1.

1191 Jun 23d E,G. Reference: Colonienses [ca. 1238, p. 801]. This says under 1191: "An eclipse of the sun in the month of June, on the 30th of the moon." It was actually the 27th of the ecclesiastical moon. Either the annalist made a mistake in calculating the moon, or he believed a statement that an eclipse had to be at the new moon, or he had accurate lunar tables. The last is certainly a possibility by 1191. Reliability: 1. Place: Cologne. Standard deviation of the magnitude: 0.1.

1191 Jun 23e E,G. Reference: S. Trudperti [ca. 1246]. These annals say under 1191: "This year on the eve of S. John the Baptist[†] on the Lord's

---

[†] Hence on 1191 Jun 23.

-426-

day the sun, at the 6th hour of the day in clear
weather, was seen diminished as if covered by a
bandage lessening its rays." These annals are not
supposed to have originated until the next century,
but this account differs from any other that I have
found, and I shall take it to represent a local
source. Reliability: 0.5. Place: S. Trudperti.
Magnitude: 2/3 with a standard deviation of 1/6;
path assumed to the north.

        1191 Jun 23f E,G. Reference: <u>Wormatienses</u>
[ca. 1295]. These annals say: "1191. On the eve
of S. John the Baptist there was a total[†] eclipse
of the sun." Because of the question about meaning,
as discussed in the footnote, and because of the
editor's remarks about the legibility of the manuscript, I shall use a reliability of 0.5. Place:
Worms. Standard deviation of the magnitude: 0.

        <u>Pruveningenses</u> [ca. 1298], which was not
compiled until the next century, has a brief notice
about this eclipse that could have been derived from
any one of many sources. I shall not use it.

        1207 Feb 28a E,G. Reference: <u>Zwifaltenses</u>
[ca. 1221]. These contemporaneous annals have the
entry: "1207. On the 2nd calends March (= Feb 28)
there was an eclipse of the sun from the 6th hour
of the day until the 9th." Reliability: 1. Place:
Zwiefalten. Standard deviation of the magnitude:
0.1.

        1207 Feb 28b E,G. Reference: <u>S. Stephani</u>
[ca. 1448]. These annals should also be contemporaneous.

[†] The word used is <u>generalis</u>. Several different
words are used in <u>medieval</u> annals to indicate a
total eclipse. In other cases, the word used is
clearly contrasted with a word meaning partial,
and the meaning is certain. Here the annalist
could mean an eclipse that was seen over a wide
area, but I think that "total" is a more plausible
meaning.

They say: "1207. In this year there was an eclipse of the sun on the day before the calends of March near the middle of the day." Reliability: 1. Place: Freising. Standard deviation of the magnitude: 0.1.

1207 Feb 28c E,G. Reference: Colonienses [ca. 1238, p. 822]. This source says under 1206: "There was an eclipse of the sun on the 2nd calends March at the 10th hour of the day." Since the same annals put the eclipse of 1153 Jan 26 in the year 1153, it is unlikely that the annalist began his year at Mar 1; in other words, he has the year wrong. The hour is odd. The path went to the northeast, and the eclipse should have been at the 10th hour somewhere around the White Sea. Hence it is unlikely that the report came from somewhere to the east, and the errors are probably simple slips. Reliability: 1. Place: Cologne. Standard deviation of the magnitude: 0.1.

1207 Feb 28d E,G. Reference: Scheftlarienses Maiores [ca. 1247]. These say: "1207. There was an eclipse of the sun on the 2nd calends March." Since the annals appear to be contemporaneous, I shall accept this record in spite of its brevity. Reliability: 1. Place: Scheftlarn. Standard deviation of the magnitude: 0.1.

1230 May 14 E,G. Reference: Colonienses [ca. 1238, p. 842]. These annals have the following entry under 1231: "There was a partial eclipse of the sun at about sunrise on the 2nd ides June (= Jun 12)." The editor assigned the date 1231 Jun 12 to this entry. Apparently he merely took the entry as correct without consulting a table of eclipses; there was no eclipse on that date. "June" is apparently a slip for "May". The 2nd ides May equals May 14. Oppolzer [1887] shows the sunrise point of the eclipse of 1230 May 14 in the English Channel. From there the path went almost over the North Pole. Thus the eclipse of 1230 May 14 would have been partial in Cologne at sunrise as stated, and it seems to me that the identification is certain

in spite of the errors in the record. Reliability:
1. Place: Cologne. Magnitude: 2/3 with a standard deviation of 1/6; path west of Cologne.

        1232 Oct 15 E,G. Reference: Colonienses
[ca. 1238, p. 843]. This source notes another
partial eclipse in 1232: "A partial eclipse of the
sun after midday was seen this year, but not very
notable, however." This can only be the eclipse of
1232 Oct 15, whose path went across the northern
part of Africa. Since the annalist says that it
was a small eclipse, I shall use a magnitude of
only 1/3, with a standard deviation of 1/6. Reliability: 1. Place: Cologne.

        1239 Jun 3a E,G. Reference: Neresheim-
enses [ca. 1296]. These annals have the following:
"1239. A prodigy was seen near the sun on the 3rd
nones June at the 9th hour. The sky turned to a
kind of blood-red and the air was darkened." The
date stated is 1239 Jun 3, and the eclipse of that
date would have been maximum in the afternoon in
Germany. Thus it seems certain that this is a record of the eclipse, in spite of the wording. The
same writer two years later does use the word
eclipsis for 1241 Oct 6 (in the record 1241 Oct 6a
E,G below), so it is probable that he did not call
this an eclipse because it did not look like one.
The best guess I can make is that the sky was overcast in the southern portion so that the sun could
not be seen directly but the general region of the
peculiar effect in the sky could be seen. In other
words, this may be a record of an eclipse occurring
in cloudy weather. The implication is that the
darkening was not total and hence it is probable
that the eclipse was partial. Reliability: 1.
Place: Neresheim. Magnitude: 2/3 with standard
deviation of 1/6; path south of Germany.

        1239 Jun 3b E,G. Reference: S. Stephani
[ca. 1448]. These have: "1239. This year was an
eclipse of the sun on the 3rd nones June (= Jun 3)
about the middle of the day." Reliability: 1.

Place: Freising. Standard deviation of the magnitude: 0.1.

1241 Oct 6a E,G. Reference: Neresheimenses [ca. 1296]. These annals say: "1241.
There was a very great eclipse of the sun on the
2nd nones October (= Oct 6) and very great darkness
at the time of midday." This record seems to be
original, although it would be more satisfying if
the manuscript still existed so that it could be
studied. Reliability: 1. Place: Neresheim.
Standard deviation of the magnitude: 0.02; super-
latives are used but no specific indicators of a
total eclipse are mentioned.

1241 Oct 6b E,G. Reference: Augustani
Minores [ca. 1321]. This source has: "1241. In
this year there was darkness on the nones of October
(= Oct 7), on the Lord's day after midday, Fidis
virgin." My church calendar does not list S. Fidis,
but many accounts of this eclipse put it on her
feast day, so her day is probably Oct 6. The day
of the eclipse, rather than Oct 7, was on the 1st
feria, so it is probable that "2nd" was omitted
before "nones" by inadvertence. Reliability: 1.
Place: Augsburg. Standard deviation of the magnitude: 0.1.

1241 Oct 6c E,G. Reference: Corbeienses
Notae [ca. 1241, p. 277]. The second of the two
notes in this source says: "In the year 1241 of
the Incarnation of our Lord, on the 2nd nones
October, the 27th of the moon, indiction __, epact
17, concurrent__, clave[†] 39, solar cycle 18 and
lunar cycle 7, an eclipse of the sun was seen, as
the masters of Paris predicted, thus about the 10th
hour on a clear day, with no clouds visible, darkness sprang up around the sun in the manner of
night, and many stars were seen." The editor says

[†]This is probably the ablative of clavis, a key.
I have no idea what is meant.

that the blanks are blank spaces in the manuscript.
It is clear that the annalist intended to supply
many items of calendrical data and left spaces for
them. He filled in some of the blanks, but forgot
to or was prevented from filling in two of them.
This shows clearly, I think, that information of
this sort may have been supplied at any time and
that its presence does not indicate an eye witness.[†]

The record 1133 Aug 2g E,G is the other
eclipse record from Corbeienses Notae, and it too
has considerable calendrical detail. This raises
the possibility that the same person wrote both
accounts. I believe that this is not the case,
for two reasons. First, the quantities given in
the two records are not the same; several quantities, of which the concurrent is an example, are
mentioned in the 1241 record but not the 1133 record. If a person after 1241 decided to put both
eclipses in the record, I think it is probable
that he would have kept more parallelism in the
accounts. It is plausible, however, that the presence of so much detail in the 1133 record inspired
the 1241 annalist, who had the 1133 record before
him, to calculate even more calendrical quantities.
Second, the 1133 annalist emphasized the name of the
current abbot, and it does not seem likely that an
annalist a century later would have preserved the
name of the abbot in 1133 but not the one in his
own time.

The year of the lunar cycle and the age
of the moon are correct. I have not checked the
other information. It is interesting that predictions of solar eclipses were being made in increasing number by 1241 and that the prediction of this
eclipse had travelled all the way from Paris to
Corvei.

Reliability: 1. Place: Corvei. Standard deviation of the magnitude: 0.

[†] See the discussion of the record 878 Oct 29 B,I
in Section VII.1, where an earlier editor asserted
that the presence of such calculable detail indicated an eye witness.

1241 Oct 6d E,G. Reference: S. Stephani
[ca. 1448]. These annals contain the entry: "1241.
On the 2nd nones October was an eclipse of the sun,
such that stars appeared plentifully in the sky in
the middle of the day." This is the record that I
designated as 1241 Oct 6n E in AAO. Reliability:
1. Place: Freising. Standard deviation of the
magnitude: 0.01.

        1241 Oct 6e E,G. Reference: Halesbrunn-
enses [ca. 1241]. The entry for 1241 is the only
entry in these annals after 1178. Part of it says:
"On the 2nd nones October there was an eclipse of
the sun about the time of vespers after nones for
some interval." Reliability: 1. Place: Heilsbronn. Standard deviation of the magnitude: 0.1.

        1241 Oct 6f E,G. Reference: Albertus
[1256]. Albertus wrote under the year 1241: "An
eclipse of the sun on the octave of S. Michael,[†]
namely the 2nd nones October, a little after midday on the Lord's day, with stars appearing and
with the sun entirely hidden from our view. And
so clear was the sky that not a cloud was to be
seen." This is a clear description of a total
eclipse that Albertus witnessed himself. Reliability: 1. Place: Stade. Standard deviation
of the magnitude: 0.

        1241 Oct 6g E,G. Reference: Naufragia
Ratisbonensia [ca. 1358]. One of the fragments
preserved from the church of S. Emmerammus reads:
"In the year 1241 of the Incarnation of the Lord,
on the 2nd nones October, on the feast of S. Fidis
virgin, on the Lord's day, around the 9th hour the
sun began to disappear by an eclipse so great that,
with its rays looking greenish or a little purple[‡]

[†] S. Michael's is Sep 29 and the octave of it is
Oct 6.

[‡] The text has vidolaceus. I assume that this is
an error for violaceus.

while vanishing, darkness covered the entire earth[†]
for two hours or more." Reliability: 1. Place:
Weltenburg. Standard deviation of the magnitude:
0. I think that the observer is indicating a total
eclipse even though he gives no confirmatory detail.

1241 Oct 6h E,G. Reference: Scheftlarienses Maiores [ca. 1247]. This source says under
1240: "This year there were two eclipses of the
sun, and the moon was turned to blood; the first
about the solstice on the day of S. Vitus, the
second in the autumn on the 4th nones October (=
Oct 4), such that for half an hour it was dark in
the manner of night for the whole earth; . . ."
S. Vitus' day is Jun 15 and the editor rightly observed that there was no eclipse on either 1240 Jun
15 or 1241 Jun 15. In spite of this, the basic
facts are correct but the writer was certainly careless. The only lunar eclipse in either 1240 or
1241 that was visible in Germany was that of 1241
Apr 27, and S. Vitalis' day is Apr 28.[‡] This is
fairly close to May 8, which is often called the
beginning of summer in medieval calendars. We may
speculate that the annalist saw a note something
like this: "There were two eclipses; the sun was
eclipsed and the moon was turned to blood. The
second was near the beginning of summer, on the day
of S. Vitalis, and the first in the autumn..." If to
him the beginning of the summer meant the solstice,
not May 8, he would have changed S. Vitalis to S.
Vitus thinking S. Vitalis was an error, and he would
have written what we see after mixing up "first" and
"second". Thus the annalist was probably working
from notes that he had not made, and we must not
give his record full reliability. He probably does
mean to indicate a total eclipse. Reliability: 0.5.
Place: Scheftlarn. Standard deviation of the magnitude: 0.

[†] Orbem. This word may refer to the solar disk.
[‡] There is a S. Vitalis whose day is Oct 24. The
one whose day is Apr 28 is S. Vitalis martyr.

1241 Oct 6i E,G. Reference: Scheftlarienses Minores [ca. 1272]. This record is quite
different from the preceding one, although both
were made at the same place. This one says: "A.D.
1241, the sun was obscured in midday, and there was
night for a long hour, and stars were seen." This
should be a contemporaneous account of a total
eclipse. Reliability: 1. Place: Scheftlarn.
Standard deviation of the magnitude: 0.

1241 Oct 6j E,G. Reference: S. Trudperti
[ca. 1246]. This source has the short notice: "1242.
On the 2nd nones October was an eclipse of the sun."
Since the last year in the annals is 1246, we might
expect this to be contemporaneous. However, the
editor says that the annals were compiled toward the
end of the century and this appears near the end
only because the last annals have been lost. Thus
this is probably not an independent record and I
shall not use it.

1241 Oct 6k E,G. Reference: Wormatienses
[ca. 1295]. This has the brief entry: "1241,
there was a total eclipse of the sun." Since this
manuscript had already been restored in the 15th
century, we should accept this record cautiously.
Reliability: 0.5. Place: Worms. Standard deviation of the magnitude: 0.

1241 Oct 6ℓ E,G. Reference: Theodorus
[ca. 1182]. This record is actually in an anonymous
continuation: "1241. An eclipse of the sun on the
Lord's day after Michael's (day)." The dating is
awkward but correct. S. Michael's day (Sep 29) in
1241 was on the 1st feria and Oct 6 was the next
following 1st feria. Circumstances indicate that
the record is contemporaneous. Reliability: 1.
Place: Pohlde. Standard deviation of the magnitude:
0.1.

1245 Jul 25 E,G. Reference: Albertus
[1256]. Albertus has given us the only record of

-434-

this eclipse that I have found. It occurs under
1245: "An eclipse of the sun on Jacob's day,[†]
1st hour, 3rd feria, 27th of the moon; and the
next lunation was so much in error that full moon
occurred on the 12th.[‡]" All details in this record are correct, including the ages of the (ecclesiastical) moon at the eclipse and the following
full moon. Reliability: 1. Place: Stade. Standard deviation of the magnitude: 0.1.

        This account comes in the middle of a
description of what seems to be a nova. See Section
V.2.

        Neresheimenses [ca. 1296] records a solar
eclipse on 1249 Aug 16 (17 calends September). There
was no such eclipse. I have not been able to think
of any plausible set of errors that would change
this into any eclipse date, whether lunar or solar.

        1261 Apr 1 E,G. Reference: S. Stephani
[ca. 1448]. This has a brief note under 1261:
"There was an eclipse of the sun." The only plausible identifications are 1261 Apr 1 and 1263 Aug 5.
Since 1263 Aug 5 is separately recorded below, the
first date is almost surely right. Since annalists
have been known to record the same eclipse twice,
I shall use a reliability of only 0.5. Place:
Freising. Standard deviation of the magnitude:
0.1.

        1263 Aug 5 E,G. Reference: S. Stephani
[ca. 1448]. This entry occurs under 1264: "This
year there was an eclipse of the sun on Oswald's
day." Oswald's day is Aug 5, so the identification
of the eclipse is certain. This is the only eclipse
in S. Stephani for which the year is wrong. It is
interesting that both this record and the preceding
one have the unusual spelling eglipsis. Reliability:

[†] Jul 25.

[‡] That is, on the 12th day of the following moon.

1. Place: Freising. Standard deviation of the magnitude: 0.1.

1267 May 25 E,G. Reference: Scheftlarienses Minores [ca. 1272], which has under 1267: "Also the sun was obscured on the eve of Ascension." Ascension Day was 1267 May 26, so the dating is correct. Reliability: 1. Place: Scheftlarn. Standard deviation of the magnitude: 0.1.

It is interesting that three of the last five entries in Augustani Minores [ca. 1321], including the isolated entry for 1457 that I did not use in the citation, record solar eclipses. The eclipses are those of 1310 Jan 31, 1312 Jul 5, and 1321 Jun 26. All three records seem to be reliable, but I do not use them because they are outside the time interval adopted for this work.

CHAPTER XII

RECORDS OF SOLAR ECLIPSES FROM ITALY

1. The Sources

Records from Italy cover the time since
the end of classical antiquity essentially without
interruption. In spite of this I have found relatively few Italian records in the time span chosen
for this study. This situation is probably a consequence of the interests of those who have edited
the medieval Italian sources. The largest collection of Italian sources is Rerum Italicarum Scriptores, edited by L. A. Muratori in the first part
of the 18th century but revised in great part in
this century. I must confess that I did not study
this collection thoroughly. I inspected about ten
volumes with considerable care but found that they
dealt almost entirely with material later than the
closing date chosen for this study. This tended
to discourage careful study of the many remaining
volumes. It is of course possible that the collection contains many more records within the chosen
time span than I have found.

I shall describe the sources briefly in
this section and the records of eclipses in the
next section. It is desirable to begin by describing
the Italian method of writing dates.

The Roman method of writing days of the
months, that is, by counting backward from the calends, nones, or ides, has been almost the only
method of writing dates that we have encountered
in the records from western Europe. We expect that
it would also be the method used in Italy, its natural home, but a different method is used about as
often.

In this method, a day in the first half of
a month was specified by writing a number followed by

intrante mense followed by the name of the month. A
day in the last half of the month was specified by
writing a number followed by stante (or occasionally adstante or exeunte) mense followed by the name of the
month; the number written was that obtained by
counting backward from the last day of the month,
counting the last day as 1. For example, the day
that we write Jul 4 would be 4 (or 4th) intrante
mense Iulii and the day that we write Dec 25 would
be 7 (or 7th) stante mense Decembris.[†]

Note that the number of the stante mense
is not the same as the number before the calends in
the Roman system. The latter number is greater by
1, since it is counted from the 1st of the following
month.

This method at least had the merit of
naming a day of the month with the use of the month
in which the day occurred rather than with the
following month. In the Roman system more than half
the days in August, for example, had to be named by
using September; Aug 14 is the 19th calends September
in the Roman system.

It must not be supposed that the intrante-
stante method was used universally. Intrante mense
was often omitted, leaving a date apparently written
in just the way we do. Sometimes the days were
counted straight through the month just as we do.
Sometimes the intrante-stante method and the Roman
method were used in the same source and sometimes
they were even mixed within the same sentence.

Agnellus [ca. 841]. Agnellus wrote a
history of the bishops of Ravenna. In the form of
the work that we have, the history is preceded by
some songs about Agnellus and his writing; the songs
were probably written by someone else. All that we
know about Agnellus is what the songs tell. He was

[†] Understanding, of course, that the numeral was
written in Roman numerals or words, or Italian words
in the case of some late sources.

born about 805 of a rich and noble family of Ravenna
and was dedicated to the service of the cathedral at
an early age. He was ordained about 828. We may
presume that he died about 841, since that is when
his history ends.

Parts of Agnellus have been edited as part
of Consularia Italica. See the discussion of Fasti
Vindobonenses [ca. 576] below.

Andreas Bergomatis [ca. 877]. Andreas
refers to himself as a "presbyter" in the second
chapter of his history. Otherwise I believe that
we know nothing of the author except that he lived
in Bergamo. On the basis of its organization we
should call this a history rather than a chronicle.
It is apparently intended as a continuation of
Paulus Diaconus [ca. 787].

Arnulfus [ca. 1077]. Arnulfus is another
writer about whom we know only what he says incidentally about himself in his writing. In referring
to Archbishop Arnulfus, who was archbishop of Milan
from 970 to 974, the writer says that his great-
grandfather was the brother of that Arnulfus. This
work is a history of the archbishops of Milan from
925 to 1077.

Bede [734?]. Bede's history is discussed
in Section VI.1. I list him here because two of the
records of solar eclipses in his work probably came
from Italy.

Beneventani [ca. 1130]. These are the
annals of the monastery of S. Sophia in Benevento,
near Naples. There are three copies. The editor
printed two of the copies as a single document,
noting where their readings differed. He printed
the third in parallel columns. It seemed to me
that it was so different that it deserved to be
called a separate source. For our purposes, however,
it can be ignored, because the observations that it

contains are abridgements of what appears in the other copies.

Bolognetti [ca. 1420]. This is an anonymous chronicle written in Italian rather than Latin. It is one of four closely related chronicles. 1 saw no independent information in it that is useful for this study and I mention it only because of its relation to the others. See the discussion of Varignana [ca. 1425] below for more information.

Casinates [ca. 1042]. I did not notice any statement by the editor about the place where the Annales Casinates were written; perhaps he thought the place was obvious. The content of the annals shows an origin in southern Italy, and Casinum is the ancient name of Monte Cassino. Thus Monte Cassino is probably the place. Ginzel [1882 and 1884] also took Monte Cassino to be the place of origin. The annals consist of only 14 entries scattered between the years 914 and 1042 inclusive. They are written in a copy of what the editor calls the "Bede cycle". The form of the annals suggests that they are original.

Cavenses [ca. 1315]. These annals are from the monastery of The Most Holy Trinity of Cava or La Cava; it is near Salerno. They are written on a set of Easter tables that goes from 534 to 1538. They were apparently compiled at one time for years through 976. Then there is a gap to 1033. From 1033 on they are in a variety of hands and are presumably original.

Dandulus [ca. 1340]. Andrea Dandulus (Dandolo in the Italian spelling) was a member of one of the most important families in Venice, and one of the most unusual families that I have read about. The first doge of Venice was elected in 697 and a Dandulus was one of the electors. Silvestro Dandolo was one of the leaders in the revolution of 1848 that made Venice independent of Austria and

eventually took it into Italy. In between, the
family contributed many leaders, including four
doges. Few families have such a long span of continued prominence and public service; the family
may still be prominent for all I know.

Andreas (ca. 1307 - 1354) was the last
member of the family to be doge, which he was from
1343 to 1354. His chronicle is one of the main
sources for Venetian history, particularly for the
period covered by its last part. His chronology is
often poor, but much of his bad chronology is ac-
counted for by his extensive use of Sigebertus [ca.
1111] as a source document.

Farfenses [ca. 1099]. These annals are
based upon a catalogue of popes and abbots, the
latter being presumably the abbots of the monastery
at Farfa. Abbot Hugo was made abbot three different
times, in 998, 1014, and 1036, as we learn from the
annals; we do not learn what kind of politics caused
this frequent turnover. A large part of the annals
were written under his intermittent abbacy. The
remaining parts were drawn up in or shortly before
1099, but the two parts overlap in time. The editor
has distinguished the parts by the type used.

Farfa can be found on large-scale maps.
It is about 40 kilometers northeast of Rome.

Fasti Vindobonenses [ca. 576]. Vindobona
is the ancient name of Vienna. The source has been
named for Vienna only because two of the main manuscript copies are found in the library there that
was once the imperial library. Naming this source
after Vienna continues a long tradition of giving
it a misleading title.

Fasti Vindobonenses is one of a group of
annals that have traditionally been published together, although they show no particular relation-
ship. During most of their published history the
group has been known as Annals of Ravenna, and the
components have usually not been distinguished in

citation. The editor of the edition cited points out
that Annals of Ravenna is much too restrictive a
title; various ones of them could have come from
many different places in Italy. He calls the group
Consularia Italica.

The editor of Consularia Italica included
excerpts from Agnellus [ca. 841; see dicussion above]
as part of the Consularia. This was the only one of
the group that seems to me to have a particular
association with Ravenna.

Fasti and Consularia are terms applied to
Roman lists of consuls or other officers, or to
annals based upon such lists.[†] While the editor
improved matters by using Italica rather than Ravenna
for the geographical part of the name, it seems un-
desirable to me to use Consularia for the set, since
one of the constituent parts by Agnellus [ca. 841]
is called "Book of the Archbishops of Ravenna" in
translation. It seems preferable to give the components separate designations.

There are three manuscripts of the Fasti
Vindobonenses, continuing the name used by the
editor. Two are found in Vienna, as already noted;
the third comes from the library of S. Gall. The
three copies are basically of the same set of annals,
but each has material (and gaps in its material) that
is unique to it. When he gets down to the details of
the annals, the editor distinguishes all three by
separate titles. I shall use Fasti Vindobonenses to
mean the entire set of 3 manuscripts, and shall
indicate an individual manuscript by a note when
necessary. The two manuscripts in Vienna date from
the 15th century in their present form, while the
S. Gall manuscript is 9th-century.

The entries in Fasti Vindobonenses could
have come from anywhere in Italy. However, most of
the population was north of the latitude of Naples,
so I shall assume that the entries came from the corresponding part of Italy. I shall use the phrase "most of

---

[†] See the discussion in Section III.3.

Italy" to mean this part.

Liudprandus [ca. 969]. Liudprandus lived
and did most of his writing in Cremona. However,
the only solar eclipse that I found reported in his
writing was observed when he was on a diplomatic
mission to Constantinople. Thus I put him among
the Byzantine sources in Chapter XV.

"Lupus Barensis" [ca. 1102]. This chronicle
was published in 1626 under the name of Lupus. The
editor of the present edition does not know why this
name was used, because it occurs nowhere in connection
with the manuscript. He used the name Lupus, and I
shall do so, because of tradition. Nothing is known
of the actual author, but we may judge that he lived
in or near Bari because of the nature of the material
in the chronicle. I shall use quotation marks to
indicate the lack of validity of using his name.

Paschale Campanum [ca. 585]. This set of
annals is based upon a combination of an Easter table
and a consular list. It was not clear to me why the
annals were specifically associated with Campania.
There are only two entries that I noted which are
specifically associated with Campania. Since they
were both eruptions of Vesuvius, the association is
not strong; a person anywhere in Italy might be interested in such events. However, I shall use Campania
as the place, on the assumption that the editor had
reasons for the name. More specifically, since
Campania is relatively small, it will suffice to use
Naples.

The main part of the source runs from 464
through 585. The annalistic entries are fairly
frequent through 512 but are absent thereafter. There
is a continuation that consists only of an Easter
table through 613, plus a single annal. The writer
apparently used an estimate of the Crucifixion as an
era; his number for the year is smaller by 27 than
the number that we would assign.

Paschale Campanum is edited as part of the Consularia Italica. See the discussion of Fasti Vindobonenses above.

Paulus Diaconus [ca. 787]. Paulus was a member of a family that had been prominent in Lombard affairs for 2 centuries. His father was named Warnefridus, and Paulus lived in the monastery of Monte Cassino for many years. Thus he is sometimes known as Paulus Warnefridi and sometimes as Paulus Casinensis. He was well educated and served at the Lombard court for some time before he entered monastic life. He attracted Charlemagne's attention and lived at his court for some time. His history of the Lombard people, which covers the history of the Lombards from 568 to 747, is perhaps his best known work.

I did not notice any independent material relevant to this study in Paulus' history. I include it in this study partly because it is a famous source that the reader may expect to find mentioned and partly in order to correct an error of 15 years that the editors have made in a minor part of Paulus' chronology.

Procopius [ca. 554]. Procopius is a Byzantine writer. I include him among the Italian sources because his single record of an eclipse, if that is what it is, was made in Italy. Procopius served the famous Byzantine general Belisarius on all his campaigns in a capacity somewhat like that of a private secretary. Procopius wrote the history of the campaigns in three books. In the course of his books, he shows a moderate amount of learning For example, he knows that a long wall was built across the north of Britain well before his time, and he gives us one of the early descriptions of Thule: It is north of Britain and ten times larger; the sun does not set there for 40 days in summer nor rise for 40 days in winter. This seems to indicate some knowledge of the northern part of Scandinavia, but one must not read too much into that.

Rampona [ca. 1425]. This is another of
the 4 related chronicles that will be discussed
under Varignana [ca. 1425] below. Rampona is in
Latin in its early parts but gradually mixes in
material in Italian, and from about 1320 on it is
almost all in Italian.

Romani [ca. 1049]. This is not a set of
annals in the usual sense, but the editor called it
Annales Romani for want of a better term. A codex
in the Vatican library contains 5 folios, not successive, that describe various contemporaneous
events in hands of the 11th and 12th centuries.
One of the folios contains a record of events in
Rome beginning with a sedition in 1044 and closing
with the accession of Pope Leo on 1049 Feb 12. The
present citation is intended to refer to this folio
only; the other folios do not seem to be part of
the same original work.

Romualdus [ca. 1178]. Romualdus is often
called "Salernitanus" but his correct name was apparently Romualdo Guarna in its Italian form. He
studied medicine but later entered the church and
then politics. He became archbishop of Salerno in
December, 1154, and was closely associated with the
regency under William II of Sicily (1153-1189, king
1166-1189). We do not know the date of his birth,
but he died on 1181 Apr 1. Romualdus wrote this as
a universal chronicle in the manner of Eusebius
[ca. 325]. It is probably an independent authority
from about 1150 to its close. Romualdus thought
rather highly of his work. Near the end of the
events for 1177, Romualdus tells us that we can
have complete confidence in what has been written
because Romualdo II, archbishop of Salerno, wrote
it, and "let ye know that his testimony is true."
He may have once intended to stop here, but the
chronicle continues for about another year.

His work furnishes proof, if any is needed,†

---

† See the record 878 Oct 29 B,I in Section VII.1.

-445-

that great detail in a record does not indicate
originality, if the detail is the sort that can
be calculated. For example, Romualdus [p. 208]
puts a lunar eclipse on 1118 Dec 11, the 4th feria.
Now 1118 Dec 11 was indeed the 4th feria but it was
not the date of a lunar eclipse. The eclipse was on
1117 Dec 11, which was the 3rd feria. Therefore the
feria is not part of the original record but has
been calculated later from the wrong date. I have
referred to this matter several times because it
seems important. It is easy to assume that detail
means accuracy, and at least one editor has made
the mistake of postulating that assumption as a
principle and acting on it.

Varignana [ca. 1425]. Varignana is one of
four related chronicles from Bologna. The others
are Bolognetti [ca. 1420], Rampona [ca. 1425], and
Villola [ca. 1376]. The editor presented all four
in parallel with the composite title Corpus Chronicon Bononiensium.[†] All four chronicles are late
compilations, of the late 14th or early 15th century,
each being made by one person or by a few persons.
They are heavily interdependent, either because
some copied from the others or because they used
the same sources such as Dandulus [ca. 1340]. I
believe that both practices were followed to some
extent. For example, Rampona, Villola, and Bolognetti give the same event in 1350 in almost identical
words, but Varignana gives a differently worded description of the same event.

Villola stops sooner than the others, in
1376. Bolognetti continues to 1420, while Varignana
and Rampona continue slightly farther, to 1425.
Varignana and Rampona are longer than the others
with regard to the time covered, but they are also
longer (that is, they give more information) even
within the period covered by all. I have put the
discussion under Varignana because it exhibits a
greater degree of independence than the others.
Rampona perhaps has the most independence in the

[†] Bononia is an old name for Bologna.

TABLE XII.1

REFERENCES TO SOLAR ECLIPSES FOUND IN
MEDIEVAL ITALIAN SOURCES

Source	Date	Conclusions
Agnellus	840 May 5	Original
Andreas Bergomatis	840 May 5	Probably independent
Arnulfus	1033 Jun 29	Probably independent
Bede	538 Feb 15	Probably used Italian source
	540 Jun 20	Probably used Italian source
Beneventani	939 Jul 19	Not safe to use
	968 Dec 22	Probably independent
	1033 Jun 29	Probably independent
	1067 Feb 16	Probably independent
Bolognetti	----------	Almost same as Varignana for solar eclipses
Casinates	939 Jul 19	Original
	968 Dec 22	Not safe to use
Cavenses	764 Jun 4	Probably copied from German source
	774 ?	Cannot identify

TABLE XII.1 (Continued)

Source	Date	Conclusions
Cavenses (con'd)	787 Sep 16	Inaccurate copy of German record
	968 Dec 22	Probably independent
	1033 Jun 29	Original
	1178 Sep 13	Original
Dandulus	693 Oct 5	Copied from Sigebertus
	807 Feb 11	Copied from Laurissenses family from Germany
	828 ?	Copied an unidentifiable record from Sigebertus
	840 May 5	May be independent
	1009 Mar 29	Copied from Sigebertus
	1153 Jan 26	Used what may be source from Bologna; put about 1186
	1191 Jun 23	Probably independent
	1239 Jun 3	Probably independent
Farfenses	807 Feb 11	Probably independent
	939 Jul 19	Not safe to use
	968 Dec 22	Not safe to use
	1033 Jun 29	Not safe to use
Fasti Vindobonenses ("Annals of Ravenna")	118 Sep 3	Perhaps independent; only known record
	393 Nov 20	Perhaps independent
	418 Jul 19	Probably independent

TABLE XII.1 (Continued)

Source	Date	Conclusions
Fasti Vindobon- enses (con'd)	534 Apr 29	Perhaps independent
"Lupus Barensis"	939 Jul 19	A careless copy
	968 Dec 22	Not safe to use
	990 Oct 21 ?	Cannot be identified with certainty
Paschale Campanum	512 Jun 29	Probably independent
Paulus Diaconus	664 May 1	Copied from Bede; editor dated as 679
Procopius	536 Sep 1	Probably not an eclipse
Rampona	59 Apr 30	Not independent
	592 Mar 19 ?	Not safe to use
	664 May 1	Probably not independent
	939 Jul 19	Not safe to use
	1153 Jan 26	Probably from an inde- pendent Bologna source
	1178 Sep 13	Source cannot be traced
	1191 Jun 23	Probably from an inde- pendent Bologna source
	1239 Jun 3	Probably from an inde- pendent Bologna source
Romani	1044 Nov 22	Original

TABLE XII.1 (Concluded)

Source	Date	Conclusions
Romualdus	-----------	Many early records from French and German sources
	1033 Jun 29	Not safe to use
	1093 Sep 23	Probably copied from German source
	1178 Sep 13	Original
Varignana	-----------	Many early records from French and German sources
	1178 Sep 13	Source cannot be traced
	1191 Jun 23	Same source as Rampona
	1239 Jun 3	Same source as Rampona
Villola	-----------	Almost same as Rampona and Varignana for solar eclipses

specific matter of eclipses.

Varignana has one remarkable characteristic. Every date that I noticed was stated clearly.
The dates were not necessarily correct, perhaps
because the sources used may have had errors, but
the chronicler left us in no doubt about the year
that he intended to specify.

Varignana is in Italian throughout.

Villola [ca. 1376]. See the discussion
of Varignana [ca. 1425] above. Villola is in Latin
to about 1300. Thereafter it is in a mixture of
Latin and Italian but mostly Italian. Villola is
the only one of the four related chronicles that
has a tendency to repeat itself.

2. Discussions of the Records of Solar Eclipses

The records of solar eclipses found in
the sources just discussed are listed in Table
XII.1, arranged alphabetically by source. Not all
appearances of often-copied records are listed.

Rampona [ca. 1425, p. 100] has a short
record of the eclipse of 59 Apr 30. There were
many ancient records of this eclipse, and I have
not tried to identify which one was the source
that Rampona used. At least Rampona used a straightforward record and not one of the magical reports
of the eclipse. I used two records of this eclipse
in AAO [pp. 73 and 113].

118 Sep 3 E,I. Reference: Fasti Vindobonenses [ca. 576, p. 285]. This lies outside the
main time interval chosen for this study, but I
include it because it is in a source with records
in the interval. The record says: "Adrian and
Salinator. Under these consuls an eclipse of the
sun took place." The year can be identified as
118 without recourse to the eclipse. I have found
no other record of this eclipse, so this may well

be an independent Italian observation. Since we
know nothing of the history of the record between
the time of the eclipse and the writing of the
source in the form we now have, the record can receive only a low reliability, say 0.2. Place:
Most of Italy. Standard deviation of the magnitude:
0.1.

393 Nov 20 E,I. Reference: Fasti Vindobonenses [ca. 576, p. 298]. This record says: "Theodosius III and Abundantius. Under these consuls
was darkness on the day of the Sun at the 3rd hour[†]
on the 6th calends November (= Oct 27)." From the
names of the consuls, we can determine with little
uncertainty that the year is 393. The eclipse
of 393 Nov 20 was large or total in Italy in the
middle of the morning. The date of the eclipse is
the 12th (xii in Roman numerals) calends December
rather than the 6th (vi) calends November. It is
easy to write November by mistake for December when
giving a date in November, and it is not too hard
to change xii into vi by careless writing and copying. Further, 393 Nov 20 was indeed on the 1st
feria (day of the Sun) as stated while 393 Oct 27
was on the 5th feria. Thus the identification of
the eclipse as 393 Nov 20 seems fairly firm. The
source is probably not contemporaneous at this point.
Reliability: 0.2. Place: Most of Italy. Standard deviation of the magnitude: 0.1.

The Byzantine source Marcellinus [ca. 534]
has a statement about this eclipse that is probably
condensed from Fasti Vindobonenses or, more probably,
from its sources.

418 Jul 19a E,I. Reference: Fasti Vindobonenses [ca. 576, p. 300]. The original text of this
record says: "Honorio XII et Theodosio VIII conss.
Sol eclipsim fecit XIIII. kl. Aug. et a parte Orientis

[†] The S. Gall manuscript has the 3rd hour while the
Vienna copies have the 2nd hour.

apparuit stella ardens per dies XXX."†. The reader
should compare this text with that of the record
418 Jul 19b M,B, from the Byzantine source Marcellinus [ca. 534], in Section XV.4. Mommsen, who
edited both sources, said that Fasti Vindobonenses
was the source used by Marcellinus. This seems unlikely to me; the wording is different and each
record contains statements not in the other. I do
not see how either could have used the other as a
source. There are several ancient records of this
eclipse, and the one in Fasti Vindobonenses seems
independent of all the others, although we have no
reason to suppose that its record is contemporaneous.
Reliability: 0.5. Place: Most of Italy. Standard
deviation of the magnitude: 0.1.

The "fiery star" is probably a comet. It
is interesting that Philostorgius [ca. 425]‡ says
that a "cone-shaped object, which the inexpert called
a comet" could be seen near the sun during the eclipse.

This record appears in the S. Gall manuscript
only.* I suspect that a variant of this manuscript
served as a source for the Belgian source Blandinienses [ca. 1292] and for the Danish source Esromenses [ca. 1307], directly or indirectly. See Sections
VIII.2 and XIII.1.

418 Jul 19b E,I. Reference: Farfenses
[ca. 1099, p. 589]. This record has the largest
dating error that I have found. Farfenses puts
this under 968, which makes an error of 550 years.
To confuse matters still further, there was an

---

†
Consuls Honorius, 12th time, and Theodosius, 8th
time. The sun suffered an eclipse on the 14th
calends August (= Jul 19) and in the eastern region
a fiery star appeared for 30 days.

‡
In connection with the record 418 Jul 19a M,B
in Section XV.4.

*
However, this cannot represent local S. Gall
information. S. Gall was not founded until the
8th century.

eclipse on 968 Dec 22, which <u>Farfenses</u> records
under 969.

Farfenses says: "Sol defecit hora tertia,
14 Kal. Aug."[†] The reader should note that this
is identical with the first part of the record 418
Jul 19 E,BN in Section VIII.2. The latter record
is from the Belgian source <u>Blandinienses</u> [ca. 1292],
which put it under the year <u>951</u>.

I have already shown that the error in
<u>Blandinienses</u> almost surely came from using a source
<u>arranged by years</u> of the "great Easter cycle" of
532 years, or some equivalent arrangement, and that
some annalist made a mistake about the cycle to which
the record belongs. The annalist of <u>Farfenses</u> must
have made a mistake about the great <u>cycle of 532</u>
years and also about the short cycle of 19 years.
To be more explicit, the year 418 was the 419th year[‡]
of the first great cycle and the 1st year of the
current 19-year cycle. 951 was the 420th year of
the second great cycle, so <u>Blandinienses</u> made an
error of 1 in the cycle and <u>1 in the year</u> of the
cycle. 968 is the 19th year of its current 19-
year cycle, but an event in the 19th year just pre-
cedes an event in the 1st year in a table. Thus
the annalist of <u>Farfenses</u> made a mistake of 1 year
about the place <u>in the 19</u>-year cycle, a mistake of
1 cycle of the short 19-year cycle, and a mistake
of 1 cycle of the great cycle.

I gave reasons in the discussion of the
record 418 Jul 19 E,BN, in Section VIII.2, for
suspecting that the record came from one closely
related to the preceding record 418 Jul 19a E,I.
It is probably too much of a coincidence to suspect
that two annalists independently made an error of
1 in the great cycle for this record. It seems
more likely that <u>Farfenses</u> worked from a source in
which the record <u>was alrea</u>dy displaced with respect
to the great cycle and that he made his errors on

[†] The sun was eclipsed at the 3rd hour, on the 14th
calends August (= Jul 19).

[‡] See Section II.8.

top of that one.

512 Jun 29 E,I. Reference: Paschale
Campanum [ca. 585, p. 747]. The source first
gives Easter as the 10th calends May (= Apr 22),
which confirms that the year is 512. It then goes
on to say: "This year on the calends of July (=
Jul 1) the sun suffered an eclipse, and when Vesuvius
erupted on the 8th ides July (= Jul 8), there was
darkness in the vicinity of the mountain." The
printed text gives the date of the eclipse by writing
"in k. Iul." It is unusual to write in in this usage,
and I imagine that someone has mistaken "iii" for
in. If so, the date intended by the annalist is the
3rd calends July, which equals Jun 29 and is correct.
Reliability: 0.5. Place: Naples. Standard deviation of the magnitude: 0.1.

534 Apr 29 E,I. Reference: Fasti Vindobonenses [ca. 576, p. 334], which says: "p.c. Bilisarii
IIII et Stratici IIII tenebrae factae sunt ab hora
diei III usque in horam IIII die Saturnis."† Almost
everyone who has studied this record has tried to
connect it with Bede's records of the eclipses of
538 Feb 15 and 540 Jun 20 (see the discussion of the
records designated 538 Feb 15 E,I and 540 Jun 20
E,I later in this section), requiring that we regard
it as a garbled record of one of those eclipses.

Levison [1936] is the only one I have
seen who goes so far as to give a definite identification. He says that this record "corresponds not
alone in substance but partly in the very words" to
Bede's account of the eclipse of 540 Jun 20 and
concludes that the record from Fasti Vindobonenses

---

† The editor (Mommsen) has a note that six letters
following "Stratici IIII" have been erased. It
was not clear to me whether he meant that they
had been erased and written over, erased and restored by him, or just what. The manuscript apparently says "die Saturnis" and not "7th feria".
This notice occurs only in the S. Gall manuscript.

"no doubt belongs to" the eclipse of the year 540.
After the reference to "Bilisarii" and "Stratici"
the record above can be translated "there was darkness from the 3rd hour of the day to the 4th hour
on Saturn's day." Bede's record can be translated:
"Year 540, an eclipse of the sun on the 12th calends
July (= Jun 20), and stars appeared for about half
an hour from the 3rd hour of the day." Aside from
the coincidence that both eclipses were seen "from
the 3rd hour" of the day, I see no relation in
either substance or words.

In AAO (Section IV.4) I did not try to
identify the eclipse in Fasti Vindobonenses, but I
did use it as an argument that at least one of Bede's
eclipses was observed in Italy and hence that an
Italian source for Bede's records was plausible.

As soon as we free ourselves from an
obsession with Bede's eclipses, we realize that the
record in Fasti Vindobonenses is probably an accurate
but not highly detailed account of the eclipse of
534 Apr 29, rather than a garbled account of one of
Bede's eclipses. We must start by dealing with the
names in the record.

Fasti Vindobonenses is a set of annals
that gives years by means of the consuls. The
eclipse is said to be in the year when Belisarius
and Straticus were each consuls for the fourth time.
It is safe to say that there was no such year.
Paschale [ca. 628], for example, lists the famous
general Belisarius as consul for the years 535, 536,
and 537,[†] and in fact as sole consul for those years.
There is no room for a fourth year in which Belisarius could have been consul. Many other lists
give the consuls for the years around this period.

_____

[†]The Chronicon Paschale uses the Olympiads and the
Indiction in dating, as well as regnal years of
the emperors, but there is no confusion in translating into years of the common era. This should
not be taken to mean that the dates in Paschale
are necessarily correct; that is another matter.
See Section XV.3.

-456-

There are enough differences to establish independence, but all lists except Fasti Vindobonenses are in substantial agreement with Paschale about the consular years of Belisarius.

No other list gives a Straticus as consul at all, much less for four times. In fact, "Straticus" looks suspiciously like a Latinization of the title Στρατηγός , and it is suspicious that the same number IIII is applied to both "Straticus" and Belisarius. It is as if the record said "Belisarius IV and General IV". (The letters "p.c." in the record, standing for "post consulatum", are often used in consular lists to indicate that a person has been consul before.)

It is probable that the scribe who prepared the manuscript of Fasti Vindobonenses (or who compiled the annals) saw a reference to "Belisarius Straticus" as consul and did not realize that this meant one person rather than two. There was some mark that he misinterpreted as IIII.[†] It is unlikely that he made a mistake in reading "p.c.". Since "it. p. c." is commonly used to denote the third year in the lists that use "p.c." for the second year, it is probable that Belisarius was listed as sole consul for the second time in the original record. This year would probably be 536, but there can easily be a random dating error on top of the confusion about consuls.

The record gives two facts about the eclipse. It was from the third hour of the day into the fourth hour, and it was on the day of the week that we call Saturday. Amazingly, there were four eclipses within six years that were large or total in Italy near this time. Looking only at the days of the week, the eclipse of 534 Apr 29 was on Saturday,[‡] the eclipse of 536 Sep 1 was on Monday, the eclipse of 538 Feb 15 was also on Monday, and the eclipse of 540 Jun 20 was on

---

[†] Perhaps he saw "Belisarius ille Straticus". Badly written, "ille" could look like iiii.

[‡] I depart from the custom of numbering the days here because the relevant source named the days.

-457-

Wednesday. The eclipse of 534 Apr 29 agrees with
the weekday, and it is the only one in the possible
time span that does so. Further, it agrees with
the time of day as well as we can judge from the
charts in Oppolzer [1887]. Thus one and only one
eclipse, namely that of 534 Apr 29, agrees with the
record.

It seems safe to accept this identification.
I have given reasons in the preceding section for
taking the place of observation to be most of Italy
for the eclipses in Fasti Vindobonenses. Since the
record has probably been copied more than once, I
shall use the relatively low reliability of 0.2.
The standard deviation of the magnitude receives
the conventional value of 0.1.

536 Sep 1 E,I(?). Reference: Procopius
[ca. 554, De Bello Vandalico, Book II, Chapter 14,
p. 469 of v. 1]. Procopius was a Byzantine writer,
but this observation, if that is what it is, was
made in Italy. Procopius writes: "There was a
grave portent that year. Indeed for the entire
year the sun sent forth his rays without his usual
brilliance, like the moon . . . From this it happened that neither war nor famine nor any manner of
deadly evil ceased to beleaguer mankind. That was
the tenth year of Justinian's reign."

There are parallels to Procopius' description of a prolonged darkening of the sun. There
are Byzantine records (see Section XV.4) that mention a darkness lasting for 17 days in 798. Plutarch
[ca. 100; Life of Caesar] gives a parallel that is
even closer; he says that the sun was dimmed for
the whole year in which Julius Caesar was killed.

This is a magical eclipse; we should
see if it is assimilated, that is, suggested by a
real eclipse but transferred in place or time for
literary or other reasons, before we reject it.
The year stated is roughly equal to our 536. Procopius has put the mention of the darkness at a
place in his narrative where he has not yet mentioned
Belisarius' entry into Rome in December of 536 when

"the city of Rome came back to the power of the
Romans at last after sixty years." [Evagrius, ca.
593, Chapter IV.19]. Thus the time is highly
consistent with the eclipse of 536 Sep 1 and it is
consistent with no other eclipse.

However, as we saw in connection with
the preceding record, there were eclipses that
were large in Italy on 534 Apr 29, 536 Sep 1, 538
Feb 15, and 540 Jun 20. Thus almost any time during these years has a good chance of agreeing
closely with an eclipse by chance, and we cannot
put much stress upon the agreement in dates.

If we had an eclipse that was used
magically but that agreed closely in time and place
with a real eclipse, and if it was unlikely that
the agreement happened by chance, we might be justified in using the record with a low reliability.
As it is, the agreement could well have happened
by chance, so the safe course is not to use this
record.

Procopius was superstitious about celestial events. He was in a position to see the
eclipses of 534 Apr 29, 536 Sep 1, 538 Feb 15, and
540 Jun 20 as total or nearly so. He also wrote
a detailed history of the years from 530 to 550
approximately. It is remarkable that he mentioned
none of the eclipses, unless indeed the passage
under discussion really records an eclipse. It is
unfortunate for the astronomer and for the student
of Bede's sources (see the following two records)
that he did not do so. Perhaps the weather was
bad each time.

538 Feb 15 E,I. Reference: Bede [734?,
Book V, Chapter 24]. "In the year 538, there
happened an eclipse of the sun on the 14th calends
March (= Feb 16), from the first hour until the
third." This record is conveniently discussed
along with the following one.

540 Jun 20 E,I. Reference: Bede [734?,
Book V, Chapter 24]. "In the year 540, there

happened an eclipse of the sun on the 12th calends
July (= Jun 20), and stars appeared for about half
an hour from the third hour of the day."†

I designated these as the records 538 Feb
15 E and 540 Jun 20 E in AAO. There I carelessly
worked from secondary sources and assigned the
place of observation to be Ravenna; the fault is
mine and not that of the sources. Luckily I assigned
a low weight to these records, and the assignment of
place had a negligible effect upon the astronomical
conclusions.

The magnitude of the eclipse of 538 Feb
15 was probably not more than about 0.6 anywhere
in England. It is possible but unlikely that this
eclipse was observed there. The magnitude of the
eclipse of 540 Jun 20 probably reached about 0.8
in southern England, but it is impossible that stars
were seen there. The record of 540 Jun 20 certainly,
and the record of 538 Feb 15 probably, came from outside England.

From a study of Bede's writing, Laistner
[1936] has prepared a tentative list of the documents
used by Bede. We know that the libraries at Jarrow
and Wearmouth had been largely stocked by material
obtained from Rome, and in addition documents from
the papal archives had been copied for Bede's use.
Since Bede's source for the eclipses of 538 and 540
was not English, it most likely was obtained from
Rome. However, materials for the libraries could
have been obtained from other places, or Bede could
have borrowed materials from libraries other than
those at Jarrow and Wearmouth.

---

† It is useful to compare the wording of these records with the record 534 Apr 29 E,I. Bede's record for 538 has: "Anno DXXXVIII, eclypsis solis
facta est XIIII. Kalendas Martias, ab hora prima
usque ad tertiam." His record for 540 has: "Anno
DXL, eclypsis solis facta XII. Kalendas Iulias,
et apparuerunt stellae pene hora dimidia ab hora
diei tertia."

Even if Bede's source was obtained in
Rome, it does not follow that the source was Roman;
Rome had had all the resources of the Empire avail-
able to it. Bede had access to the Chronicle of
Isidore of Seville and to the History of the Franks
of Gregory of Tours,[†] for example, among many non-
Roman works [Laistner, 1936].

However, on the basis of what we know or
believe about the documents available to Bede, an
Italian origin is the most likely. Hence I shall
take "most of Italy" to be the place of observation.
Because this attribution is far from certain, and
because the record that we have is certainly not
original, I shall use the low reliability of 0.05.
I shall use a standard deviation of the magnitude
of 0.1 for the eclipse of 538. Because stars are
mentioned, even though their visibility is clearly
exaggerated, I shall use a standard deviation of
0.01 for the eclipse of 540.

Rampona [ca. 1425, p. 320] says that,
during the time of Pope Gregory (pope 590-604),
the sun arrived like a half moon from early morning
to midday. I suspect that this is an oddly worded
account of the eclipse of 592 Mar 19, but it is not
safe to use it. Even if we could be sure that an
eclipse is meant, Rampona is such a late source
that the reliability of the record would be low.

664 May la E,I. Reference: Rampona
[ca. 1425, p. 349]. This says: "The sun was
eclipsed in Indiction 7." The eclipse is during
the papacy of S. Vitalian (657-672), and the year
which was Indiction 7 in this interval is 664 by
the usual calculation (Section XV.1). I suspect
that the information about the eclipse came from
Bede [734?]; Bede does not give the indiction but
that is easily calculated. Hence I shall not use
this record.

[†] These are the sources cited as Isidorus [ca. 624]
and Gregory of Tours [ca. 592] in this work.

664 May 1b E,I.   Reference: Paulus
Diaconus [ca. 787, p. 166]. In a setting that
seems to be 680, Paulus writes: "In this time,
Indiction 8, the moon suffered an eclipse. The
sun also about that time was eclipsed about the
10th hour of the day, on the 5th nones May." The
day stated is May 3.   Thus we have the same hour
and the same erroneous day for the eclipse that
we find in Bede [734?],[†] and there seems little
question that Paulus derived his information from
Bede's history.

Paulus, however, definitely puts the
eclipse in 680 rather than 664, as we deduce from
the historical setting and the indiction. The
editors of Paulus said that the solar eclipse was
in 679 and the lunar eclipse in 680, but I believe
that they were in error. The clue to the dating
error probably lies in the use of the indiction,
which has a cycle of 15 years. In other words,
Paulus made extensive use of the indiction, and he
accidentally transferred the eclipse from its correct place in one cycle of the indiction to the
corresponding place in the next cycle. I have not
tried to identify the lunar eclipse and do not
know whether Paulus put it in the wrong indiction
cycle or not.

Since Paulus' record is almost surely
derivative, it is not safe to use.

Dandulus [ca. 1340, p. 102] has copied
the report of the eclipse of 693 Oct 5 from the
Belgian source Sigebertus [ca. 1111], with minor
changes in spelling and word order.

Cavenses [ca. 1315] seems to have copied
the record of the eclipse of 764 Jun 4 from the
German source Laurissenses [ca. 829] and to have
added to it a correct calculation of the age of
the moon.

Cavenses also says that there was an

[†] The record 664 May 1 B,E in Chapter VI.

eclipse of the sun on 774 Aug 23, the 3rd feria,
the 9th hour, the 5th of the moon. 774 Aug 23 was
indeed on the 3rd feria but it was on the 11th of
the moon. I can find no plausible eclipse, either
of the sun or the moon; August was not even in the
eclipse seasons in 774.

For 787 Cavenses uses a different German
source. Under 787 it says: "This year there was
an eclipse on the calends October (= Oct 1), day
of the Moon, from the 1st hour to the 5th hour."
787 Oct 1 was indeed on the "day of the Moon" (2nd
feria), but there was no eclipse then, lunar or
solar. It is clear that the annalist at La Cava,
or some intermediate annalist, started by copying
the record 787 Sep 16b E,G, given in Section XI.2,
from Quedlinburgenses [ca. 1025] or from some other
member of the Herolfesfeldenses family.[†] He accidentally omitted the already erroneous numeral 15
from "15th calends October", thus changing the date
from the wrong Sep 17 to the even worse Oct 1, but
he kept the hours of the day given in his source.
He then calculated the day of the week from the
wrong date 787 Oct 1 and inserted the day, which is
now understandably wrong, into the record.

The German source Loiselianos [ca. 814]
has a long passage about astronomical events that
occurred in 807. This passage appears as the report 807 Feb 11 E,G in Section XI.2. Dandulus
[ca. 1340, p. 129] has copied the statement that
the sun was eclipsed once and the moon thrice, but
he omits the provision in the original that the
year meant was from September of 806 to August 807.
Romualdus [ca. 1178, p. 155] has copied the record
of the solar eclipse of 807 Feb 11 but has accidentally changed the date to 807 Feb 13. It is a relief
to turn now to an independent record.

807 Feb 11 E,I. Reference: Farfenses
[ca. 1099]. This has, under 808: "An eclipse of

[†] See the discussion of Hildesheimenses [ca. 1137]
in Section XI.1.

the sun occurred from the 3rd hour to the 6th."
For some reason, all eclipses in this source are
put 1 year too late. Further, all dates that can
be tested in this part of the annals are 1 year
too late, whether they are eclipses or not; I did
not test all dates in the annals, however. Thus I
think there is no question that this refers to the
eclipse of 807 Feb 11. The hours are reasonable
for that eclipse and they occur in no other record
of that eclipse that I have found. This is apparently an independent record. Since it is in the part
of the annals that was compiled in the early 11th
century, it cannot be original. Reliability: 0.5.
Place: Farfa. Standard deviation of the magnitude:
0.1.

We now return to copying. Romualdus
[ca. 1178, p. 155] has copied the records of the
eclipses of 810 Jul 5, 810 Nov 30, and 812 May 14
from some member of the Laurissenses[†] family of
Germany. Dandulus [ca. 1340, p. 148] has taken a
note: "The sun and moon were eclipsed" from the
Belgian source Sigebertus [ca. 1111] (see Section
VIII.2) which Sigebertus put, with slightly different wording, under 833. Dandulus, however, has
put the note under the reign of Michael II (Byzantine emperor from 820 to 829). The eclipses cannot
be identified.

840 May 5a E,I. Reference: Dandulus
[ca. 1340, p. 150]. Dandulus does not specify the
year in this part of his chronicle, but gives only
a reign or some such indication; the editor, however, dated this 838: "Then in the month of May,
at the 6th hour, the sun was obscured and an eclipse
took place." Dandulus puts this shortly before the
death of Louis I, Holy Roman emperor, which occurred
on 840 Jun 20, so there seems no doubt that this is
the eclipse of 840 May 5. Dandulus did not take
this record from Sigebertus, which is his main source
for eclipses. The wording differs from that in any

[†] See the discussion under Laurissenses [ca. 829]
in Section XI.1.

other record that I have found, as does the hour.
Thus there is a reasonable probability that this
record came from a local Venetian source; we know
Dandulus used much local material. Reliability:
0.5. Place: Venice. Standard deviation of the
magnitude: 0.1.

        840 May 5b E,I. Reference: Agnellus
[ca. 841, p. 389]. Agnellus writes: "Also on the
5th day of May, Indiction 3, the sun was made exceedingly dark through the whole earth until the
9th hour." The indiction is correct according to
the rules given for it in Section XV.1. This should
be an original record. Since Agnellus uses a super-
lative in speaking of the darkness, but does not
mention stars or any specific details, I shall take
the standard deviation of the magnitude to be 0.02.
Reliability: 1. Place: Ravenna.

        840 May 5c E,I. Reference: Andreas
Bergomatis [ca. 877, p. 226]. Andreas writes:
"In the 3rd Indiction the sun was obscured in
this world, and stars appeared in the sky, on the
3rd nones May (= May 5), the 9th hour, in the litany
of the Lord, for about half an hour. Great tribu-
lation was caused. Wherever people saw it, many
estimated that so great a one had not come this
century." This is slightly too early for us to
give it full reliability. The eclipse sounds total.
Reliability: 0.5. Place: Bergamo. Standard
deviation of the magnitude: 0.

        Romualdus [ca. 1178, p. 159] reports the
eclipse of 840 May 5 in words almost identical with
those in the French source Senonensis [ca. 1203]
(see the record 840 May 5a E,F in Section X.3).
Since Senonensis became an original source long
before Romualdus, I give it priority.

        939 Jul 19a E,I. Reference: Casinates
[ca. 1042]. This has the entry: "938. Indiction
12. On the 13th stante July (= Jul 19), feria 6,
29th of the moon, the sun was obscured from the 3rd

hour almost to the 9th hour. We looked at the sun, and it had no strength nor splendor nor warmth; we looked at the sky, and its color was changed immoderately; and others said that they saw the sun as if it were halved." I get the Indiction to be 11 for 939. The feria and age of the moon are correct for 939 Jul 19. The statement that the sun was halved seems incompatible with the statement that the sun lacked strength and splendor, but we should perhaps not take the statements too literally. I shall assume that a partial eclipse is meant, and I use the conventional values for one. The record is probably original. Reliability: 1. Place: Monte Cassino. Magnitude: 2/3 with a standard deviation of 1/6; path north of Monte Cassino.

939 Jul 19b E,I. Reference: "Lupus Barensis" [ca. 1102]. This source says: "939. The sun was obscured and stars appeared on the 3rd adstante July (= Jul 29), 3rd feria, 3rd hour, 29th of the moon." This seems to be a case of careless copying. "Lupus" started by omitting "x" out of "xiii. adstante", thus changing the date to Jul 29. "3rd feria" is not correct for either the correct date 939 Jul 19 or for the altered date 939 Jul 29; he probably wrote "3rd" for the feria because he had "3rd adstante" before the feria and "3rd hour" after it. He kept the correct age of the moon. The record is clearly copied and shows no details that I noticed which would cause us to think it independent. Thus I shall not use it.

Rampona [ca. 1425], Farfenses [ca. 1099], and Beneventani [ca. 1130] all have brief notes about a solar eclipse in or near 939. None of these sources can be original for this time, and there is not enough information in the notes to let us judge whether or not they represent independent local records. Thus I shall not use any of them.

968 Dec 22a E,I. Reference: Beneventani [ca. 1130]. This source records: "968. The sun was obscured on the 10th day stante of December

(= Dec 22)." It is conceivable that this record
is derived from some other record of this eclipse
that I have found. However, though short, it
differs enough from other records in its wording
that I shall take it to be independent. Reliability:
0.5. Place: Benevento. Standard deviation of the
magnitude: 0.1.

968 Dec 22b E,I. Reference: Cavenses
[ca. 1315]. Cavenses puts the record under 969:
"This year the sun suffered an eclipse on the 11th
calends January (= Dec 22), between the 3rd and 4th
hours of the day." This is probably a case in which
the nature of the Roman calendar is responsible for
a dating error. The eclipse was in 968 but the
reference calends was that of January of 969; thus
the year intended is ambiguous. Cavenses does not
become an original source until 1033, but this report is independent of others found. Reliability:
0.5. Place: La Cava. Standard deviation of the
magnitude: 0.1.

Casinates [ca. 1042], "Lupus Barensis"
[ca. 1102], and Farfenses [ca. 1099] all have brief
notes under 969 that probably refer to this eclipse.
I cannot identify the specific sources, but it is
not safe to use these records.

"Lupus Barensis" [ca. 1102] says that
the sun was obscured in 987. This is probably a
reference to the eclipse of 990 Oct 21, but it is
a late reference and is not usable. Dandulus [ca.
1340, p. 203] has copied the record of the eclipse
of 1009 Mar 29 from Sigebertus [ca. 1111].

1033 Jun 29a E,I. Reference: Beneventani
[ca. 1130]. Beneventani has three records† of this
eclipse, two under 1033 and one under 1034. The
first one under 1033 says: "An exceedingly dark

---

† But there are only two in any one manuscript.
Thus S. Maxentii [ca. 1140] retains the record
for a single manuscript. See Chapter X.

eclipse of the sun in the month of June." The one
under 1034 says: "On S. Peter's day (= Jun 29)
the sun was obscured." Codex 3 has what seems to
be a composite of these under 1033: "In the month
of June on S. Peter's day, there was an eclipse of
the sun." I believe that the first record is independent of others known. The second one may also
be, but we do not have enough information to judge
with confidence. Hence I shall take the set of
entries in Beneventani to be equivalent to a single
record with a reliability of 0.5. Place: Benevento.
Standard deviation of the magnitude: 0.02, since
the eclipse was "exceedingly dark" but with no
specific indicators of totality.

              1033 Jun 29b E,I. Reference: Cavenses
[ca. 1315]. Cavenses says under 1034: "The
greatest part of the sun was obscured on the feast
of S. Peter." Cavenses has a long gap in its annals
that begins in 976. The eclipse is in the first
entry thereafter; this is the entry that seems to
inaugurate Cavenses as an original source. Saying
"the greatest part" suggests a partial eclipse
with perhaps 0.9 of the sun being eclipsed. Reliability: 1. Place: La Cava. Magnitude: 0.9
with a standard deviation of 0.05; path to the
north of La Cava.

              1033 Jun 29c E,I. Reference: Arnulfus
[ca. 1077, p. 14]. Arnulfus writes: "At that time
in June on the day of S. Peter and Paul in the midday heat an eclipse of the sun appeared, casting
darkness on the earth for 3 hours." It is probable
that the eclipse happened during Arnulfus' lifetime.
However, since he wrote his history about 40 years
later, it is doubtful that this is a strictly contemporaneous record. The eclipse was maximum in
Milan shortly after 12^h, according to the chart in
Oppolzer [1887]. Reliability: 0.5. Place: Milan.
Standard deviation of the magnitude: 0.1.

            Farfenses [ca. 1099] and Romualdus [ca. 1178,
p. 178] have short notices about this eclipse, both
under the year 1034. It is doubtful that either

record is independent.

1044 Nov 22 E,I. Reference: <u>Romani</u>
[ca. 1049, p. 468]. The second entry un<u>der the</u>
year 1046 reads: "And then in the same year on
the feast of S. Cecilia the sun was obscured for
the space of about 3 hours." S. Cecilia's day is
Nov 22. Although the year is stated to be 1046, it
is also stated to be the 12th year of Pope Benedict
VIIII, who was elected in 1032. Probably 1046
(MXLVI) is an accident for MXLIV. Reliability: 1.
Place: Rome. Standard deviation of the magnitude:
0.1.

1067 Feb 16 E,I. Reference: <u>Beneventani</u>
[ca. 1130]. Under 1066 <u>Beneventani</u> has: "On the
16th calends May (= Apr <u>16)</u> appeared the star comet.[†]
On the 3rd day <u>stante</u> of February (= Feb 26) there
was darkness at <u>the 9</u>th hour, and it lasted for 3
hours; . . ." It is interesting that the annalist
mixed the Italian calendrical style with the old
Roman style in adjacent sentences. This suggests,
although it certainly does not prove, that he was
copying. "3rd day <u>stante</u>" is probably an accident[‡]
for "13th day <u>stante</u>" (= Feb 16). Reliability: 0.5.
Place: Bene<u>vento.</u> Standard deviation of the magnitude: 0.1. According to <u>Oppolzer</u>, this is a penumbral eclipse.

<u>Romualdus</u> [ca. 1178, p. 200] has a record
of the ecl<u>ipse of 1</u>093 Sep 23. His record reads
like the record 1093 Sep 23e E,G from <u>Frutolf</u> [ca.
1103] (see Section XI.2) with the feri<u>a added.</u> We
have seen* that Romualdus sometimes calculated the
feria, so its appearance furnishes no presumption
of independence.

[†] This is Halley's comet.

[‡] A similar accident happened with the record 939
Jul 19b E,I above. There is no apparent relation
between the two cases.

* See the discussion of <u>Romualdus</u> in the preceding
section.

It is interesting that I have found no Italian records of the eclipse of 1133 Aug 2, although the magnitude should have reached perhaps 0.85 in parts of Italy.

1153 Jan 26 E,I. Reference: Rampona [ca. 1425, p. 25]. This source has: "In that time the sun was obscured slightly before the 9th hour and it remained in that state long after the 9th hour." A new paragraph that begins immediately says: "At that time 3 moons were seen in the heavens with the sign of the Cross in the center, and after a not-great time 3 suns were seen." Although the year is given as 1155, this must be the eclipse of 1153 Jan 26. The report seems to be independent of others that I have found, so it may represent an independent observation made in Bologna. It is far from being contemporaneous. Reliability: 0.2 only, since this is in a late compilation. Place: Bologna. Standard deviation of the magnitude: 0.1.

The related chronicle Villola [ca. 1376, p. 20] has the item about the 3 moons and the 3 suns and the sign of the Cross[†] under both 1145 and 1155. I did not notice any of the record in Varignana [ca. 1425]. Bolognetti [ca. 1420, p. 26] also had the halos without the eclipse, but only once under 1155. Dandulus [ca. 1340, p. 270], however, puts the entire observation in the papacy of Urban III, who was pope from 1185 Dec 1 to 1187 Oct 20. I have no suggestion about the origin of this error of about 33 years. It does not correspond to any standard calendrical cycle. It could not result from a confusion of popes named Urban. Urban II was pope from 1088 Mar 12 to 1099 Jul 29 and Urban IV from 1261 Sep 4 to 1264 Oct 2.

1178 Sep 13a E,I. Reference: Romualdus [ca. 1178, p. 297]. The last two sentences in Romualdus' Chronicon can be translated thus:

[†] These are halo effects. See Appendix IX.

"Moreover on the 2nd day <u>stante</u> of that month,[†]
the moon in the sign of <u>the Virgin</u> after the middle
of the night suffered an eclipse, and it was darkened almost to the 3rd part of itself.[‡]   And indeed
on the 13th <u>intrante</u> of September (= Sep 13), Indiction XII, <u>the sun</u> in the sign of the Virgin
about the hour".   The chronicle breaks off abrupt-
ly in mid-sentence as the quotation indicates.   It
is not likely that illness prevented Romualdus
from continuing his sentence; he lived until 1181
Apr 1.   Perhaps there once was more to the sentence
and there has been an accident to the manuscript.

Since Romualdus recorded correctly the
partial lunar eclipse of 1178 Aug 30, and since
he has already specified the date 1178 Sep 13, I
believe that there is no reasonable doubt that he
was about to record the solar eclipse of 1178 Sep
13.   Indeed it is likely that he was about to do so
in a contemporaneous fashion.   Reliability: 1.
Place: Salerno.   Standard deviation of the magnitude: 0.1.

The sun is in Virgo on both Aug 30 and
Sep 13.   Therefore the moon on 1178 Aug 30 was in
Pisces and not Virgo.   Romualdus may have simply
made an error.   More likely he was looking at a
calendar which gave the times when the sun enters
the various signs and wrote down Virgo because
that is what appears on the calendar.   In other
words, the sign is calendrical in meaning and is
probably not an astronomical observation.

1178 Sep 13b E,I.   Reference: <u>Cavenses</u>
[ca. 1315].   Under 1178 <u>Cavenses</u> says: "<u>The sun</u>
was obscured for the most part in the middle of

[†]
The context makes this August, so the date is
Aug 30.

[‡]
One cannot infer confidently from the original
whether 1/3 remained bright or whether 1/3 was
darkened.   The astronomical calculations do not
resolve the ambiguity, because the calculated
magnitude [<u>Oppolzer</u>, 1887] is almost exactly 0.5.

the day." This is in the part of the annals that
should be original. Reliability: 1. Place: La
Cava. Magnitude: 2/3 with a standard deviation
of 1/6; path to the west[†] of La Cava.

The four chronicles from Bologna [Bolog-
netti, ca. 1420, Varignana, ca. 1425, Villola, ca.
1376, and Rampona, ca. 1425, all in parallel on p.
43] record an eclipse of the sun in February of 1178.
Villola gives the clue to this dating mystery; it
says the ides of February. The correct date is the
ides of September. It is probable that the record at
some stage of transmission read "ides Sep." which was
misread as "ides Feb.". It would be easy to make
this mistake with some handwritings. It is unlikely
that all four annalists individually made the same
error; therefore it is probable that they had a source
with the error already present. It is unlikely
that the original recorder made such an error.
Therefore the records in the Bologna chronicles
are probably at least third-hand. Since there is
no evidence to indicate local information, it is
best to ignore all four records.

1191 Jun 23a E,I. Reference: Dandulus
[ca. 1340, p. 271]. Dandulus first mentioned the
accession of Pope Celestine III, which was on 1191
Apr 14, and then said: "Then, on the 23rd day of
June, was an eclipse of the sun in the 7th degree
of Cancer, and it lasted for 4 hours." The sun
enters Cancer on Jun 17 according to the medieval
church calendar, so the position of the sun is
"ecclesiastical" rather than astronomical. This
record is independent of any other that I have
found. Reliability: 0.5. Place: Venice. Standard deviation of the magnitude: 0.1.

[†] The chart in Oppolzer shows the path going through
southern Corsica and the Strait of Messina. A
more detailed calculation may show the path actually
to be east of La Cava. If so, I shall change this
assumption.

1191 Jun 23b E,I. Reference: <u>Rampona</u>
[ca. 1425, p. 54]. This gives, under 1191: "That
year the sun was obscured on the 9th calends July
(= Jun 23) from the 3rd to the 9th hour." The
duration seems too long, particularly since the
hours are probably hours of the day, which equal
about 78^m in Bologna on Jun 23. The record
seems to be independent, but we cannot assume that
it is contemporaneous. Reliability: 0.5. Place:
Bologna. Standard deviation of the magnitude: 0.1.

<u>Varignana</u> [ca. 1425, p. 56] has the same
record but with "9th calends July" omitted.

1239 Jun 3a E,I. Reference: <u>Dandulus</u>
[ca. 1340, p. 297]. Dandulus has earlier put the
accession of Jacobo Theopolo as doge in 1228, and
he says this is in Jacobo's 11th year: "That year,
in the month of June, was an eclipse of the sun."
This record is expressed differently from any other
record of this eclipse that I have found, so there
is a good chance that it is independent. Reliability:
0.5. Place: Venice. Standard deviation of the
magnitude: 0.1.

1239 Jun 3b E,I. Reference: <u>Varignana</u>
[ca. 1425, p. 111], which has under 1239: "The
sun was obscured on the 2nd of June from the 6th
hour until the 9th. And it came to be so much
like night that one person could not see another.
A multitude of stars appeared in the sky. This
brought many men to contrition." The other chronicles from Bologna have minor variations of this
record, but <u>Villola</u> and <u>Rampona</u> have the correct
day of the month. The record is still obviously
copied, but it is independent of others known.
Reliability: 0.5. Place: Bologna. Standard
deviation of the magnitude: 0; the record sounds
like that of a total eclipse.

CHAPTER XIII

RECORDS OF SOLAR ECLIPSES FROM
SCANDINAVIAN COUNTRIES

1. Eclipse Reports from Denmark and Sweden

For the purposes of this study, Scandinavia consists of the area occupied by the four
modern countries of Denmark, Iceland, Norway, and
Sweden. To this I shall add the location of one
medieval monastery that is just on the German side
of the present Danish-German border but that is
clearly Danish from the viewpoint of its medieval
annals. I have found no Finnish records.

Still from the restricted viewpoint of
their annals, the four areas fall naturally into
two pairs. There are close relations between the
annals of Iceland and Norway, and there are similarly close relations between annals in Denmark
and Sweden. I shall discuss the annals of Denmark
and Sweden in this section and the annals of Iceland and Norway in the next section. There are
not enough independent sources to warrant estab-
lishing different provenances for the different
pairs.

The Danes, like the English, have a mythical history. Several Danish sources carry their
chronicles back to an eponymous king Dan, whom
Hamsfort [ca. 1585] dates at the year 2673 from
the Creation. Still others carry the chronicles
and the ancestry of their kings from Dan back
through Priam of Troy to Japheth, the third son
of Noah. Danish history becomes reliable in the
late 8th century, and King Harald Klag was baptized
along with many of his court in 826 [Petrus Olai,
ca. 1541].

In spite of this early start, the Danish
and Swedish annals do not provide any earlier useful

eclipse information than do the Icelandic and Norse records. Christianity did not really "take" until about the end of the 10th century, about the same time as in Norway and Iceland. Almost all the annals that we have were compiled, doubtless from older sources, only at about the end of the 13th century, and the oldest useful records of eclipses are for about the middle of the 12th century. Only one Swedish source seems to be appreciably older in its present form.

Brief descriptions of the main sources follow.

Eskinbek [ca. 1323]. This is a chronicle prepared in a monastery at the place in Jutland that Ginzel [1882 and 1884] calls Essenbek. It is in a hand of the first part of the 14th century and it is probable that the work is a chronicle prepared by one person rather than a chronicle prepared contemporaneously with events. It is called Chronologia Rerum Memorabilium ab Anno 1020 usque ad An. 1323, but I think that this is a title given by the editor and not by the compiler.

Esromenses [ca. 1307]. Esrom is the name of a lake in the northern part of the island of Sjaelland, the island on which Copenhagen is located, and I find no other occurrence of the name in the standard sources. Since Langebek (the editor) says that these annals were prepared by a monk of Esrom, I presume that there was once a monastery on or near the lake. As I understand him, Langebek believes that these annals were prepared by one person but that the existing copy, though in a hand of the early 14th century, is not the autograph copy of the compiler. Langebek numbers these among the main annals of Denmark.

Hamsfort [ca. 1585]. Cornelius Hamsfort wrote two books about Danish history. The first is called Series Regum Daniae a Dano ad Fridericum. It runs from the mythical king Dan to Frederic II, king from 1559 to 1588. Hamsfort states in two

places that it was written in 1585. The second is
called Chronologia Rerum Danicarum Secunda; I do
not know why he called it Secunda unless he wrote
it after the other. The last event in the Chronologia is in 1448, but it would be definitely misleading to use this date in citation. Since I saw
no date of writing associated with the Chronologia,
I have borrowed the date 1585 from the other work,
and I hope that use of this date will not be too
misleading.

          Lundense [ca. 1171]. This source is from
Lund in Sweden, which is only about 25 kilometers
east of Copenhagen. It consists of two parts. The
first part consists of lists of kings, prebendaries
of the cathedral at Lund, and other matters, inter-
spersed with a few annalistic notices. The second
part lists, for each day in the year, the names of
people who had died on that day. No year was given
in any case that I noticed. We need only the first
part for this work, and the date 1171 used above is
the date of the last annalistic entry in it.

          I believe that this is the earliest contempor-
aneous Scandinavian source that I have found. Langebek says that it is written in many centuries and
in many hands, starting with the latter part of the
11th century and going into the 14th century (for
the second part only).

          Petrus Olai [ca. 1541]. Petrus Olai was
a monk of the Minorite or Franciscan order in the
monastery at Roskilde. Roskilde was the medieval
capital of Denmark for a considerable time; it is
on the island of Sjaelland about 30 kilometers
west of Copenhagen. All we can say of Petrus
Olai's dates is that he died about 1560. This
source, called Annales Rerum Danicarum, is but
one of several historical works that he wrote. It
is unusual among medieval works (if we may use
"medieval" for a work this late) in that it often
gives the sources for each item of information.
Also, Petrus Olai was often uncertain about the
correct year for an event, as well he might be.

In such cases, he supplied alternate dates.

Ryenses [ca. 1288]. A monastery that
is variously called Rye, Rus regium, or Ruhkloster
was located near the castle of Glücksburg just
south of the modern border between Denmark and
Germany. Since the sources for the Annales Ryenses
were Scandinavian, I have put this in the Scandinavian rather than the German subdivision of the
European provenance.

The annals were compiled by a monk at
Rye in the late 13th century. The editor says
that the sources used by the monk include the
following: (1) Chronicon Danicum from 1074 to
1219. This is in v. III of the Scriptores Rerum
Danicarum Medii Aevi; I did not use it because it
did not seem to contain any independent records
of eclipses. (2) Chronicon Sialandiae. This is
the source that will be discussed next in this
section. (3) Annales Lundenses. Unfortunately
the editor of Ryenses forgot to say where the
Annales Lundenses can be found. I doubt that they
are the same as the source Lundense [ca. 1171]
cited above.

Sialandiae [ca. 1282]. The editor of
this work calls it Incerti Auctoris Danorum &
praecipue Sialandiae, ab An. 1028 ad An. 1307. The
editor of Ryenses [ca. 1288] just discussed referred to this source as Chronicon Sialandiae.
For brevity and for other reasons I prefer the
latter usage.

This source is essentially a chronicle
compiled by one author through the year 1282.
Apparently we know nothing of him. I did not
notice any statement about where he wrote, and I
presume that local references furnish the only
basis for assigning the chronicle to the island
of Sjaelland. Various people then made inserts
in the original chronicle about events that interested them, beginning with 1136, and supplied a
continuation through 1307.

The beginning date 1028 is actually incorrect. The chronicle begins with a lengthy
account of the death of S. Olaf[†] at Stiklestad.
It gives the date as the 4th calends August (= Jul
29), 4th feria, in the year 1028. The correct
year is certainly 1030.[‡]

Stralius [ca. 1314]. A man named Laurentius Stralius compiled these annals, called
Annales Danici, about 1314. The historian Hamsfort
(see the discussion of Hamsfort [ca. 1585] above),
in a copy that the editor Langebek has seen, wrote
a note referring to the "illustrious man Laurentius
Stralius". In spite of this encomium, Langebek
says of Stralius that he is not "fated to know who
he was nor of what country." Langebek allows himself only the conclusion that Stralius may have
been a monk, and, if so, that he may have been
associated with the monastery of Rye.

The source Lundense [ca. 1171] definitely
seems to be associated with a particular place,
namely the cathedral at Lund. Otherwise the sources
do not seem to be associated with a particular
place, at least not so far as most reports of
eclipses are concerned. One can only say of most
reports that they came from Denmark; a few are
perhaps local. The part of Denmark involved seems
to be approximately the triangle whose vertices are
Copenhagen, Glücksburg, and Essenbek. The center
of this triangle is close to the point at $56°.0N$,
$10°.0E$, and the vertices are only about 100 kilometers away from it.

[†] But with no eclipse.

[‡] It has been claimed that there was a solar eclipse
at the death of S. Olaf (see the report 1030 Aug
31 E,Sc in the next section), and hence that the
correct date was 1030 Aug 31. I shall discuss the
matter in more detail in the next section. Here I
only remark that 1030 Jul 29 was on the 4th feria,
while 1030 Aug 31 was on the 2nd feria, 1028 Jul
29 was on the 2nd feria, and 1028 Aug 31 was on
the 7th feria.

Hence, for the purposes of this study, Denmark will be taken as equivalent to the area centered at 56°.0N, 10°.0E, with a radius of about 1°.0 of a great circle. We would, in fact, make but little error if we used only the central point.

There are many Danish sources containing reports of eclipses that I have not mentioned. It is often necessary with other regions to discuss derivative sources at some length just in order to establish that they are derivative. Here, since we are assigning all observations to Denmark and not to any particular place in Denmark, it does not matter much which sources are original and which are derivative. Therefore I shall not discuss the Danish sources that I took to be derivative with regard to their eclipses. One Swedish source needs discussion, however.

Wisbyenses [ca. 1340]. These annals are from Visby on the Swedish island of Gotland, which lies in the Baltic between Sweden and Latvia. Visby is so far geographically from the other places involved that the possible independence of Wisbyenses is important. It contains notices of the eclipses of 1140 Mar 20 and 1263 Aug 5, in almost the identical words used for these eclipses in Ryenses [ca. 1288] and other Danish sources. Wisbyenses was compiled about 1340 in its original form, and then was continued, apparently contemporaneously, until 1525. Ryenses shows that the reports were already in existence before 1288 and therefore that the reports in Wisbyenses cannot be original.

It is of course possible that there were reports originating on Gotland at an early date and that were later used by the Danish annalists as well as the compiler of Wisbyenses. However, there were many places in Denmark where the observations could have been made and apparently only one place on Gotland. Hence the odds are great

TABLE XIII.1

REFERENCES TO SOLAR ECLIPSES FOUND IN
MEDIEVAL DANISH AND SWEDISH SOURCES

Source	Date	Conclusions
Eskinbek	1131 ?	Cannot be identified
	1140 Mar 20	Uses a report from Ryenses
	1178 Sep 13	Same as Ryenses
	1187 Sep 4	Same as Sialandiae
	1230 May 14	Same as Ryenses
	1241 Oct 6	Independent
	1263 Aug 5	Probably independent
Esromenses	418 Jul 19	Perhaps from S. Gall in Switzerland
	733 Aug 14	From England with an odd copying error
	1140 Mar 20	Uses a report from Ryenses
	1187 Sep 4	Same as Ryenses
	1230 May 14	Probably not independent
	1241 Oct 6	Probably not independent
Hamsfort	1187 Sep 4	Independent but cannot tell place of observation
Lundense	1140 Mar 20	Original

TABLE XIII.1 (Concluded)

Source	Date	Conclusions
Petrus Olai	1140 Mar 20	Probably had the original used by Ryenses but now lost
	1230 May 14	Same as Ryenses
Ryenses	1140 Mar 20	Two independent reports
	1178 Sep 13	Independent of other known sources
	1187 Sep 4	Same as Sialandiae
	1230 May 14	Independent of other known sources
	1241 Oct 6	Probably not independent
	1263 Aug 5	Probably independent
	1270 Mar 23	Probably independent
Sialandiae	1187 Sep 4	Earliest of known sources
	1230 May 14	Probably not independent
	1241 Oct 6	Probably not independent
Stralius	1140 Mar 20	Has both reports from Ryenses
	1178 Sep 13	Same as Ryenses
	1187 Sep 4	Same as Ryenses
	1270 Mar 23	Probably not independent

that the reports came from Denmark rather than
Gotland, apart from the evidence of the dates of
compilation. Thus we should not use the reports
in <u>Wisbyenses</u>.

Table XIII.1 contains a listing of the
eclipse reports in the Danish and Swedish sources
that have been discussed, with the exception of
<u>Wisbyenses</u>. The table also contains a brief state-
ment about the assessment made of each report. If
the same report is found in several different sources,
I have not always listed each appearance. In the
rest of this section, I shall present the reasons
for the assessments listed in the table.

418 Jul 19 E,Sc. Reference: <u>Esromenses</u>
[ca. 1307]. <u>Esromenses</u> has the following under
the year 416: "This Easter under Pope & Ozo[†] the
sun was eclipsed at the 3rd hour on the 14th calends
August (= Jul 19), and a star appeared glowing in
the east until the calends of September." Two
errors that are apparently typographical need comment.

The source says "Easter" but it cannot
mean that because a date in July is specified exactly and correctly. I imagine that the annalist
wrote "this Easter" by mistake for "this year", or
as a condensation of "at this year in the Easter
cycle". There is also the mysterious "Papa &
Ozonio". Zozimus, a Greek, was pope from 417 Mar
18 to 418 Dec 26 and he was therefore pope at the
time of the eclipse. I imagine that someone has
mistaken a cursive "Z" for the ampersand.

I have already mentioned this report in
connection with the Belgian report 418 Jul 19 E,BN
in Chapter VIII. The wording is almost identical
in that report and this one, but neither is likely
to be derived from the other for reasons discussed
there. In Section VIII.2 I speculated that the
Belgian and Danish annalists used a common source

[†] The printed edition has "sub Papa & Ozonio" at
this point.

and that that source may have come from S. Gall
in Switzerland. However it almost surely was not
original at S. Gall either. Since I cannot assign
a country of origin to this report, I shall not
use it, just as I did not use the Belgian report.
However this report does come from a source that
seems independent of any usable report of this
eclipse.

          733 Aug 14. Reference: Esromenses [ca.
1307]. Esromenses has an account of this eclipse
that is surely taken from the record 733 Aug 14b
B,E discussed in Section VI.3, and in fact probably
from what has been called a C type text. The annalist at Esrom has made an odd error, however.
Instead of saying that the sun was covered with a
"horrible black shield" he has said that it was
covered in the "horrible center".[†]

          1133 Aug 2 E,Sc(?). Reference: Eskinbek
[ca. 1323]. The reference has, under 1131: "There
was an eclipse of the sun in the autumn." The
chronology of the source is generally good. For
examples, it has the death of Henry (Henry V, Holy
Roman emperor 1106-1125) in 1125 and the election
of Pope Innocent II in 1130; both dates are correct.
However the only eclipse in 1131 visible in Scandinavia was that of 1131 Mar 30. This eclipse was
recorded in Icelandic/Norse sources (see the next
section) but it was not in the autumn. The next
closest eclipse was 1130 Oct 4, which might possibly have been visible at sunrise. 1133 Aug 2 should
have been large in Denmark, but Aug 2 seems early
to call autumn.[‡]

          Since there is no satisfactory identifica-
tion of this eclipse, I give this report a reliability of 0.

[†] " . . in gremio horrendo . . ."

[‡] Autumn was sometimes considered to begin on Aug
7, the day midway between the summer solstice and
the autumn equinox.

1140 Mar 20a E,Sc. Reference: Lundense
[ca. 1171]. Lundense has, under the year 1140:
"On the 12th calends April (= Mar 21), on the 4th
feria, there was an eclipse after nones." The
correct date is the 13th calends, which would have
been written "XIII. Kal. Aprilis"; it is easy to
drop one stroke from the numeral. There are other
sources, none of them Scandinavian, that contain
this much information about the eclipse. However,
Lundense is apparently independent and contemporaneous here, and I give this report a reliability of
1. Place: Lund, Sweden. Standard deviation of the
magnitude: 0.1.

1140 Mar 20b E,Sc and 1140 Mar 20c E,Sc.
Reference: Ryenses [ca. 1288]. There are clearly
two independent Danish reports of this eclipse that
occur in several places. I have chosen Ryenses for
citation because it is the oldest source that contains either one, and in fact it contains both.

Under the year 1140 Ryenses has: "And
on the 4th calends April (= Mar 29) there was darkness over the whole earth." I imagine that "XIII.
Kal." was misread as "IIII. Kal.", thus accounting
for the error in date. Although this report is
rather brief, the wording does not suggest to me
that it was taken from any other known source,
Danish or not. However, the date of compilation
of Ryenses means that the report cannot be original,
so I shall give this report, which I designate as
1140 Mar 20b E,Sc, a reliability of only 0.5.
Place: Denmark. Standard deviation of the magnitude: 0.02. The wording suggests an eclipse that
was larger than usual, but no star or any evidence
of totality is mentioned.

Stralius [ca. 1314] has a similar entry,
but with "III. Kal." in place of "IIII. Kal." This
suggests that he used Ryenses.

Under the year 1137 Ryenses has another
entry: "During Lent there was darkness in all the
earth." Lent is denoted in medieval Latin by
"Quadragesima", with reference to the 40 days of

Lent, not counting Sundays. This is sometimes abbreviated "XLma". Accordingly, the editor of Ryenses suggests the following explanation of the error in date: Perhaps the original source wrote the year as, for example, "MC et quadragesimo" or some such form in which the word or sign for "40th" (in the ablative case) appeared separately from the rest of the number of the year. The annalist might then have thought that "40" referred to Lent.

I do not follow this argument. So far as I can see, it does not explain at all the fact that the annalist used the year as 1137. If he took "40" away to use as the symbol for Lent, he would have 1100 left. If he became confused and used "40" twice, he would have left the year as 1140.

I can find no identification for this record other than 1140 Mar 20. There was no possible eclipse in 1137, whether during Lent or not. Further, in 1140, Easter was on Apr 7 and Ash Wednesday was on Feb 21. Thus 1140 Mar 20, the date of the eclipse, was during Lent, and we need no ingenious explanations. All we need is an accidental transfer of the year, and this has happened with many records.

I shall take this as an independent report, with the designation 1140 Mar 20c E, Sc, and I shall assign it the same characteristics as the record 1140 Mar 20b E, Sc.

Stralius [ca. 1314], like Ryenses, has both reports, but with an amusing difference for the one put under 1137. He says that the eclipse was "near Carnival" instead of "during Lent". I hope that Stralius will forgive me if I wonder whether Carnival[†] was the most important feature that he associated with Lent? Esromenses and Eskinbek have the "c" report but not the "b" one; Esromenses puts it under 1136.

[†] Carnival was on Feb 20 that year and not close to the eclipse.

Petrus Olai [ca. 1541] also has only the
"c" report, but with an interesting difference.
First, he says that the year may be either 1137 or
1144. This suggests that he still had the original
record and that something about its arrangement left
the year uncertain; the fact that his dates span
1140 also suggests that 1140 is a possible date and
confirms the identification. Petrus Olai also has
a longer record that reads: "During Lent there
was darkness in all the earth, and no one moved
from where he was, but was much anguished[†] until
they fainted.[‡] This passage, as well as the multiple dating, suggests that Petrus Olai had a now-
lost source before him.

The Welsh source Brut [ca. 1282], in Section VII.3, also has this eclipse in 1137. There
does not seem to be any relation between the Welsh
and Danish sources.

1178 Sep 13 E,Sc. Reference: Ryenses
[ca. 1288]. Stralius and Eskinbek also report this
eclipse in nearly the same words as Ryenses, which
has under 1178: "In the same year the sun was obscured on the eve of the Holy Cross,* on the 4th
feria." 1178 Sep 13 was indeed on the 4th feria.
I count the three passages as a single record, to
which I give a reliability of 0.5; I do not use 1
because the record cannot be original. Place:
Denmark. Standard deviation of the magnitude: 0.1.

[†] The text has angustiabatur, which looks like the
3rd person singular imperfect passive of a verb
of the 1st conjugation. My dictionary has no
such verb, so I have postulated a deponent verb
angustior.

[‡] The source has evanuerunt and thus does pass from
the singular to the plural as I have translated.
It is possible that evanuerunt has some meaning
other than "fainted".

[*] This refers to the feast of the Exaltation of the
Holy Cross, which is on Sep 14.

1187 Sep 4a E,Sc. Reference: Sialandiae
[ca. 1282]. Ryenses [ca. 1288], Eskinbek [ca. 1323],
Esromenses [ca. 1307], and Stralius [ca. 1314] all
have practically the same report. I cite Sialandiae
particularly because it was the earliest of these
to be compiled. Under the year 1187 it has: "Jerusalem was captured by the pagans. That same year
was an eclipse of the sun." The forces of Saladin
captured Jerusalem on 1187 Oct 2 after a siege of
about 12 days, and I think that there is little
doubt about the identification of the eclipse. Esromenses says the eclipse was on the same day, but I
think that this was an accident. No other source
makes this error. I count all these passages as a
single report with a reliability of 0.5. Place:
Denmark. Standard deviation of the magnitude: 0.1.

1187 Sep 4b E,Sc. Reference: Hamsfort
[ca. 1585]. Hamsfort has under 1187: "With the
moon passing under its disk, the sun on the day
before the nones of September[†] was so much eclipsed
that stars in the sky shone in the daytime just as
at night." Hamsfort adds that the annals giving
this eclipse were supplied by Langius, and that
they had been mentioned by Gerardus Mercator and
Vincentius. Langius is Joannis Langius, a friend
to whom Hamsfort dedicated his work. I do not know
who Vincentius is. Mercator is the Latin form of
the name of Gerhard Kremer, who is best known for
his map projection. He also published a "Chronicle
from the beginning of the world to the year 1568",
in which he made considerable use of eclipses for
chronological purposes, judging from the subtitle.

Since we have no idea whether this report
came from Denmark or Germany or elsewhere, I give
it a reliability of 0.

1230 May 14 E,Sc. Reference: Ryenses
[ca. 1288]. This has, under the year 1230: "There

[†]The date specified for the eclipse is thus 1187
Sep 4.

was an eclipse of the sun and a great pestilence of men and cattle." I do not think that the pestilence was considered to be a consequence of the eclipse. Eskinbek [ca. 1323] and Petrus Olai [ca. 1541] say the same thing in almost identical words. Esromenses [ca. 1307] and Sialandiae [ca. 1282] do not mention the pestilence, and Esromenses puts the eclipse on the 1st feria. This is wrong; the eclipse was really on the 3rd feria. The chronology of the sources seems reliable and the identification of the eclipse seems safe. I count all the records as a single record with a reliability of 0.5. Place: Denmark. Standard deviation of the magnitude: 0.1.

1241 Oct 6 E,Sc. Reference: Eskinbek [ca. 1323], which has: "Year MCCXLI. WALDEMAR II died. An eclipse of the sun on the 2nd nones October." Esromenses, Ryenses, and Sialandiae merely note the eclipse without giving the date, which Eskinbek gives correctly. I take all this as one report with a reliability of 0.5. Place: Denmark. Standard deviation of the magnitude: 0.1.

1263 Aug 5b E,Sc. Reference: Ryenses [ca. 1288]. I find it odd that this is the only eclipse reported in both Danish and Norwegian sources that I have found. The record with "a" following the date appears in the next section. Under 1262 Ryenses has: "There was an eclipse of the sun on the nones of August."[†] Although the year is wrong, the identification is safe. We are probably now at a time when Ryenses is contemporaneous or nearly so, so I give this a reliability of 1. Further, the report is no longer copied into several sources, so this can be taken as an observation at Rye. Standard deviation of the magnitude: 0.1.

[†] August 5.

1263 Aug 5c E,Sc. Reference: Eskinbek
[ca. 1323]. This has under 1263: "There was an
eclipse of the sun on the 2nd nones August (= Aug
4)." Since Eskinbek has the year right but the day
wrong, while Ryenses is the other way around, the
reports are probably independent. However this
report cannot be considered contemporaneous, although it may be local. Reliability: 0.5. Place:
Essenbek. Standard deviation of the magnitude: 0.1.

1270 Mar 23 E,Sc. Reference: Ryenses
[ca. 1288], which has under 1270: "There was an
eclipse of the sun in the early morning of 'Laetare
Jerusalem'." "Laetare Jerusalem" means the fourth
Sunday in Lent and it came on Mar 23 in 1270. The
chart in Oppolzer [1887] shows that the eclipse
should have been in early morning in Denmark. This
seems to be a reliable, contemporaneous, and local
report. Reliability: 1. Place: Rye. Standard
deviation of the magnitude: 0.1. Stralius gives
this eclipse, but only says that it came on a
Sunday in Lent. I shall not use his report.

2. Eclipse Reports from Iceland and Norway

Most of the reports in this section are
drawn from the collections of annals that are
usually called Icelandic annals. Although these
annals were presumably compiled in Iceland in their
present forms, they clearly contain information and
records of local events from both Iceland and Norway.
For this reason I shall use the term "Icelandic/
Norse" in referring to the relevant records and
literature.

The best discussion of the old Icelandic/
Norse literature that I have read is by Vigfusson
[1878]. Unfortunately, most of his discussion and
most of the other relevant discussions deal with
the sagas, whereas we are mostly concerned here
with the annals.

According to Vigfusson, there was apparently
no written Icelandic/Norse literature or history until

the first part of the 12th century; the runes were used only for other purposes. Earlier transmission of history was oral. It seems to me that oral transmission of information can be as accurate as written transmission. Whether transmission of information is accurate or not seems to depend more upon the intent of the transmitter than upon whether the transmission was written or oral.

For examples, the amount of romantic nonsense that Geoffrey of Monmouth and others wrote about the life and career of King Arthur shows that written "history" is not necessarily reliable. In AAO I showed that written transmission of scientific data in the 20th century is frequently in error. By contrast, Schliemann discovered Troy because he decided to accept the geographical information in the Iliad as correct,[†] and thereby showed that such information can be transmitted orally over centuries with considerable fidelity.

There is a problem in using most records that were transmitted orally which is not a direct consequence of the method of transmission: Conditions in which records are transmitted orally are not conditions in which an accurate chronological system is likely to exist. Therefore the inference of dates from oral records may be uncertain.

The example of oral transmission that I find most interesting is related by Neugebauer [1957, p. 165], who attributes it to J. Warren in a source that I have not consulted. This example furnishes a case in which oral transmission is connected with accurate chronology. A certain group

---

[†] Hogarth [1911] says: "No site in the Troad can be brought into complete accordance with all the topographical data to be ingeniously derived from the text of Homer." I am not qualified to judge this point, but I think that Hogarth is prejudiced against Schliemann. He ignores the fact that Schliemann, for all his "want of experience and method", succeeded where all his predecessors had failed.

of people in southern India in the early 1800's
could calculate the beginning, ending, and magnitude of eclipses, using calculations requiring
more than 10 significant figures. Their calculations were carried out by the use of shells or
other markers placed on the ground, apparently by
methods like those used with an abacus. They did
not understand the basis for their calculations
but simply followed a set of rules. These rules
and the astronomical data needed in their execution
were transmitted orally and had apparently been so
transmitted for several centuries.

Ari (1067-1148), called the Historian,
was apparently the first person who began to convert
the historical content of the sagas into written
form. Vigfusson [1878, v. 1, pp. cxxvii et seq.]
wrote: "There are no annals dating before 1150.
The earliest notices, bald and short, are merely
drawn from the works of Ari and Saemend."[†]

Beckman [1912], on the other hand, concluded that the Icelandic/Norse annals became independent Icelandic sources at least as early as
1104; it was not clear to me whether or not he
meant that they had assumed written form that early.
Beckman based his conclusion upon two records. The
first is a record of the solar eclipse of 1131 Mar
30, which was total in Iceland according to the
charts of Oppolzer [1887]. The record says nothing
that implies totality. Beckman said that this record
must have come from Iceland because the calculated
magnitude did not exceed 0.63 in Ireland, Belgium, or
Schleswig. For some reason he did not calculate the
magnitude anywhere in Norway,[‡] the most likely place

[†] Vigfusson did not comment on the fact that many
early notices are drawn from foreign sources such
as Sigebertus [ca. 1111] (see the discussion under
878 Oct 29 below, as well as Chapter VIII). He
was probably thinking about local records only.

[‡] According to the chart in Oppolzer, the magnitude
reached about 0.9 in Trondheim. Evidence shows
that an eclipse of this magnitude is about as
likely to be recorded as is a total eclipse. See
Section XVII.2 for further discussion.

outside of Iceland. The second record used in
Beckman's argument is of an eruption of an Icelandic volcano in 1104.

Beckman is certainly right that the original observation of an eruption in Iceland had to
be made in Iceland. In view of the close relations
between Iceland and Norway, however, this does not
mean that the annal notice had to be written in
Iceland; the information could have been carried
orally from Iceland to Norway. There is nothing
to indicate that the eclipse observation was made
in Iceland. The magnitude was large enough anywhere in Norway to make recording of the eclipse
likely there, weather permitting.

It seems to me that Vigfusson and others
go beyond concluding that the annals in their present form were not written down before the 12th
century. They seem to imply that no written records
existed in Iceland and Norway until then. I find
it somewhat hard to accept this conclusion, for two
reasons.

First, the Norse were in contact with
writing long before the 12th century. The first
Norse raid in England, according to the standard
histories, occurred about 789, and there were raids
and invasions in other parts of the British Isles and
on the continent (including Russia) within a few
years. "Wintering over" began a few decades later,
and large areas remained under Norse occupation until the 12th century or later. It seems unlikely
to me that it took the Norse three centuries to
observe the use of writing by the occupied peoples
and to realize its advantages.

Second, it seems to me that the Church
would have played some role in the introduction of
writing, but I have seen no comment on this possibility. Royal attempts to Christianize Norway
began with Haakon I (died ca. 960), although there
may have been earlier Christians in Norway. S.
Olaf (king from ca. 1015 to 1028) is credited
[Shetelig, 1958] with completing the Christianizing
and with beginning the organization of the Church

in Norway. After S. Olaf there was sporadic
resistance, but Norway remained officially Christian.
In Iceland all inhabitants [Jóhannesson, 1958]
were required to be Christian by legislation passed
in 1000.

　　　　The establishment of the Church in both
Iceland and Norway implies people who could read
and write Latin and it implies the existence of
Easter tables. Thus the necessary conditions existed for the writing of annals in the 11th century,
and many existing annals show strong connections
with Easter tables. Under these circumstances, it
would be surprising if there had been no written
annals in Iceland and Norway in the 11th century,[†]
although none of the Icelandic/Norse annals are
that old in their present form.

　　　　Existence of written annals in the 11th
century would not conflict with the conclusion that
the content of the sagas was not reduced to writing
until the 12th century, and that many of the events
in the annals were taken from them. In fact, entries in the annals for years in the 11th century
are of two sorts. Some are brief notices, consistent with having been written on Easter tables or
similar tables. Others are much longer, as if they
had been taken from chronicles or other extended
sources such as the sagas.

　　　　Storm [1888] has edited the following ten
sets of annals: Annales Reseniani, Annales Vetus-
tissimi, Henrik Høyers Annaler, Annales Regii,
Skálholts Annaler, Annalbrudstykke fra Skálholt,
Lögmanns Annáll, Gottskalks Annaler, Flatøbogens
Annaler, and Oddveria Annall. These will be cited
individually as they are used; Annalbrudstykke fra
Skálholt and Flatøbogens Annaler apparently contain

[†] The annals show the typical kind of development
that was described in Section I.2. They note
early events taken from older chronicles or annals,
going as far back as Julius Caesar in some cases,
and then gradually begin to pick up contemporaneous
notices.

TABLE XIII.2

REFERENCES TO SOLAR ECLIPSES FOUND IN
MEDIEVAL ICELANDIC/NORSE SOURCES

Source	Date	Conclusions
Annales Vetus- tissimi	878 Oct 29	Regii is the assumed source
	1276 ?	Cannot be identified
Gottskalks	878 Oct 29	Regii is the assumed source
	1131 Mar 30	Same entry as in Reseniani and Høyers; cannot tell if this is from Iceland or Norway
	1185 May 1	Regii is the assumed source
	1194 Apr 22	Regii is the assumed source; cannot tell if this is from Iceland or Norway
	1236 Aug 3	Same as Lögmanns; cannot tell if this is from Iceland or Norway
Heming	1064 Apr 19	Only known medieval American eclipse record; may be genuine; did not use because of magical association
Høyers	1131 Mar 30	Same entry as in Reseniani and Gott-skalks; cannot tell if it is from Iceland or Norway

TABLE XIII.2 (Continued)

Source	Date	Conclusions
Høyers (Con'd)	1276 ?	Cannot identify
Lögmanns	1131 Mar 30	Cannot tell if it is from Iceland or Norway
	1157 ?	Probably not an umbral eclipse; cannot be identified
	1236 Aug 3	Same as Gottskalks; cannot tell if it is from Iceland or Norway
	1263 Aug 5	Probably independent
Oddveria	878 Oct 29	Regii is the assumed source
	1185 May 1	Regii is the assumed source
	1226 ?	Probably not an umbral eclipse; cannot be identified
	1263 Aug 5	Not independent
Regii	878 Oct 29	Identical with Sigebertus
	1157 ?	Probably not an umbral eclipse; cannot be identified
	1185 May 1	Assumed independent
	1194 Apr 22	Assumed independent
	1226 ?	Probably not an umbral eclipse; cannot be identified

TABLE XIII.2 (Concluded)

Source	Date	Conclusions
Reseniani	878 Oct 29	Regii is the assumed source
	1131 Mar 30	Same entry as in Gott-skalks and Høyers; cannot tell if it is from Iceland or Norway
Skálholts	878 Oct 29	Regii is the assumed source
	1185 May 1	Regii is the assumed source
	1194 Apr 22	Regii is the assumed source; cannot tell if it is from Iceland or Norway
	1236 Aug 3	Probably independent
Snorri	1030 Aug 31	Alleged "eclipse of Stiklestad"; almost surely did not happen at that battle

no independent records and will not be used.
Vigfusson [1878, v. 2] has also edited the Annales
Regii, which he designates as "ISLENZKIR ANNALAR,
called Annales Regii"; I shall use Annales Regii
as the designation. Storm and Vigfusson do not
always agree about the contents of Annales Regii,
presumably because there is variance among manuscripts. Storm's version ends in 1341, while
Vigfusson says that they end "in 1306, though continued by a later hand."

    Vigfusson and others apparently agree
that Annales Regii are the most important annals
for the period that they cover. According to
them, the other annals were started later; only
occasionally do they give independent information
for the period covered by Regii. For astronomical
purposes, however, it will appear that the other
annals are often independent of Annales Regii.

    Because of the remarks made about Annales
Regii, I shall treat it as the primary source when
it and other annals give the same record. When a
record is not from Annales Regii, I shall attribute
the record to all the annals where it does occur,
treating each on the same basis for lack of information about priority. However, I shall treat a
record that occurs in several annals as a single
record, not a multiple one.

    These annals share a common characteristic
with most other annals: They rarely state where an
eclipse was observed. When we know that an eclipse
was recorded contemporaneously, and when we know
where the annalist worked, we usually infer that
the observation was local. With the Icelandic/
Norse annals, we cannot usually make such an inference. It will be necessary to consider each record
individually in order to decide whether the observation was made in Iceland or Norway.

    The eclipse records in the Icelandic/
Norse sources are listed in Table XIII.2, along
with a brief statement about the conclusions con-
cerning each record. In the rest of this section,
I shall discuss the reasons for the conclusions.
If an eclipse is repeated in many sets of annals,

I have not always noted every appearance.

It is interesting that the reports of eclipses in Icelandic/Norse sources are completely independent of those in Danish and Swedish sources, and conversely, so far as I have been able to discover. In fact, the eclipse of 1263 Aug 5 is the only eclipse that I have found reported in both Norse and Danish sources.

878 Oct 29. The entry in Regii [ca. 1306] for the year 880 states, in Latin, that the sun was darkened at the ninth hour of the day, and that stars appeared in the heaven, according to the edition of Storm. In contrast, the edition of Vigfusson has no entries between 871 and 898. Reseniani [ca. 1295], Ann. Vetust. [ca. 1314], Skálholts [ca. 1356], Gottskalks [ca. 1578], and Oddveria [ca. 1427] give what appears to be a translation of the passage into Icelandic/Norse.

Ginzel [1882 and 1884] took these to be an observation of the eclipse of 878 Oct 29 made in Reykjavik. The eclipse was nearly total there according to Oppolzer [1887], but at about the fifth hour. Hence the observation did not come from there if it is correct. In fact, the entry in Regii is a verbatim copy, including the error in the year, of the record 878 Oct 29 E,BN from the Belgian source Sigebertus [ca. 1111] (see Section VIII.2); many entries in Regii are copies of entries from the same source. That is, the compiler of Regii used either Sigebertus or the sources of Sigebertus. I conclude that none of the Icelandic/Norse records of the eclipse of 878 Oct 29 is independent.

It is interesting that Regii contains none of the other eclipses found in Sigebertus. This suggests that the compiler of Regii used the sources of Sigebertus rather than Sigebertus itself.

1030 Aug 31 E,Sc. Reference: Snorri [ca. 1230]. This is the famous alleged "eclipse of

Stiklestad" that I discussed at length in AAO,
Section IV.4. I decided that the eclipse of 1030
Aug 31 had nothing to do with either the battle
of Stiklestad or the death of S. Olaf "beyond a
reasonable doubt but not beyond all unlikely possibilities." As a result of additional study, I am
now willing to say that I find it almost inconceivable that the eclipse occurred at S. Olaf's
death during the battle of Stiklestad.

At the time of Olaf's death, Norway and
large areas of England formed a common kingdom.
All the Icelandic/Norse annals that I have seen
which cover the year 1030 record the battle of
Stiklestad, the death of S. Olaf, or both, and
many of the English and Danish annals do the same.
None of these annals notes the eclipse of 1030 Aug
31.

In order to appreciate the significance
of this evidence, we must realize that S. Olaf
played an important role in the Christianization
of Norway and that many miracles became connected
with him. Although he was not canonized until
1164, his cult was well established within a few
years of his death, and July 29 was observed as the
anniversary of his death in at least one church
calendar (see AAO, also Section II.3 of this study)
drawn up about 1050. The Swedish source Lundense
[ca. 1171], in which the entries may be contemporaneous, also gives July 29. If a total eclipse of
the sun had occurred at the time and place of his
death, this fact would have made a deep and irre-
vocable impression upon the minds of his followers.

If some annals were already being written
in 1030, as I speculated above, it is almost inconceivable that no contemporary set of annals would
have noted an eclipse happening under such dramatic
circumstances. If the annals were not written down
until 1100 at the earliest, the conclusion is even
stronger. This would mean that the tradition con-
necting the eclipse with Stiklestad and S. Olaf
did not yet exist even 70 years later.

The only known explicit observation of

the eclipse of 1030 Aug 31 comes from one Irish source (see Section VII.1). This source makes no mention of S. Olaf.

The translator of Snorri mentioned in the references stated that S. Olaf must have died on 1030 Aug 31, because of the date of the eclipse. I saw no indication that he allowed room for a different opinion. I did not notice his basis for the statement.

I concluded in AAO that Snorri's record should really be called the "eclipse of Sigvat" if it is accepted at all.[†] I continue to give the "eclipse of Sigvat" the quantitative characteristics that it received in AAO. The reliability is 0.5. Place: Southern Scandinavia or the Baltic Sea. Standard deviation of the magnitude: 0.1.

1064 Apr 19 E,Sc(?). Reference: Heming [ca. 1067]. I include this as a curiosity; it is the only medieval astronomical record I have found that is associated with America. The Saga of Heming, as it says itself, was made during the reign of Harold Hardrada as King of Norway (1047-1066), and the event to be described occurred near the end of the reign, judging by the context.

A man named Lodin left Greenland. When he was about 3 or 3½ days out, ". . a pack of clouds came over our ship; there followed such great darkness that men could not see their hands." Subsequently the clouds opened up and blood fell from them. The connection of this event with an eclipse is fanciful, but no more so than the connection of other events that have been seriously used as total solar eclipses in sober astronomical research. The path of the umbral eclipse of 1064 Apr 19 passed just south of Greenland according to Oppolzer's Canon. Thus there was an umbral eclipse close to

---

[†] Sigvat was a scald, that is, a Teutonic bardic poet, who was contemporary with S. Olaf. A poetic fragment attributed to him apparently contains a reference to the eclipse.

the time and place implied by the saga. Nonetheless, I shall give this report a reliability of zero, because there is doubt that the passage describes an eclipse, and because there is a magical, or at least a fantastic, association.

1131 Mar 30 E,Sc. References: Reseniani [ca. 1295], Høyers [ca. 1310], and Gottskalks [ca. 1578] state that there was an eclipse of the sun on 1131 Mar 30, while Lögmanns [ca. 1430] states that there was an eclipse in 1132. Beckman [1912] and Landmark [1931] state that this report comes from Iceland. According to Oppolzer [1887], the path of the total eclipse of 1131 Mar 30 passed over the northwestern corner of Iceland. It missed the most populated parts of Norway by perhaps 10° of a great circle, but the eclipse would still have been fairly prominent in many parts of Norway.

I think that the annals contain a genuine record of the eclipse of 1131 Mar 30. I did not notice any evidence that would allow locating the observation on the basis of the record. To accept Iceland as the place of observation on the basis of calculation only would be to commit the logical fallacy that I have called the "identification game" (AAO, Section III.2; also Newton [1969]). Therefore I shall not use this record, but I shall calculate it in order to judge the circumstances in Norway more accurately. The calculations will be discussed in Section XVII.2.

1157 (?) E,Sc. References: Regii [ca. 1306] and Lögmanns [ca. 1430] note a myrkr in 1157. There is no umbral eclipse that can be plausibly identified with these records. There were four penumbral eclipses of the sun in 1157 [Oppolzer, 1887], and it is quite plausible that the records refer to one of these. Since there are no clues to the exact date, the time of day, or the place, the record is not useful and I shall ignore it.

1185 May 1 E,Sc. Reference: Regii [ca.

1306] is the record that I take as primary. Skálholts [ca. 1356], Gottskalks [ca. 1578], and
Oddveria [ca. 1427] also note the eclipse. The
record was used in AAO (Section IV.4), where it
was designated as 1185 May 1 E, and there is nothing new to add. Reliability: 0.5. Place: Oslo
to Frankfurt. Standard deviation of the magnitude:
0.1.

1194 Apr 22 E,Sc. Reference: Regii [ca.
1306] is the assumed primary record. Skálholts and
Gottskalks also refer to this eclipse. All note
that there was an eclipse on Apr 22 of the year that
they give as 1193; one record says that it was a
large eclipse. I think that there is little doubt
that the records refer to the eclipse of 1194 Apr
22. The circumstances are remarkably like those
of 1130 Mar 30 E,Sc. Landmark [1931] states that
the report comes from Iceland, and the paths of
the eclipses are almost identical. I know of no
way to determine the place where the observation
was made[†] and shall not use the record. I shall
calculate the eclipse in order to judge the circumstances in Norway, and I shall discuss the calculations in Section XVII.2.

1226 (?) E,Sc. References: Regii [ca.
1306] and Oddveria [ca. 1427] note a darkness at
midday. There is no plausible umbral eclipse.
There were four penumbral eclipses in 1226 [Oppol-
zer, 1887], and the records may refer to one of
these. Because of the detail about midday, it
might be possible to identify the eclipse if the
place were known. As it is, the information that
would be gained from a successful identification
hardly seems worth the bother.

[†]The preceding record for 1185 May 1 proves that
the records in the annals do not necessarily
come from Iceland. That record says explicitly
that the observation was made in Europe.

1236 Aug 3 E,Sc. References: Skálholts [ca. 1356] notes an eclipse in Iceland in the summer of 1236. It is safe to assume that this note is a record of the eclipse of 1236 Aug 3 seen in Iceland. Reliability: 1; the record is perhaps contemporaneous. Standard deviation of the magnitude: 0.1.

Lögmanns [ca. 1430] and Gottskalks [ca. 1578] both note an eclipse on 1236 Aug 3. These probably represent a single primary record that is independent of the record in Skálholts. If so, the observation cannot be located safely and I shall ignore it.

1263 Aug 5a E,Sc. References: Lögmanns [ca. 1430] and Oddveria [ca. 1427]. Lögmanns notes an eclipse seen at "Biorguin" (the original spelling of the modern Bergen) on the nones of August (= Aug 5) of 1263. Part of the entry for 1263 in Oddveria is lost, but what remains refers to "Biorguin" and "August". Since we are now at a time when the records can be contemporary, I count these as a single record of the eclipse of 1263 Aug 5 seen at Bergen, with a joint reliability of 1. Standard deviation of the magnitude: 0.1.

1276 (?) E,Sc. References: Ann. Vetust. [ca. 1314] and Høyers [ca. 1310]. Both references simply note an eclipse of the sun in 1276. The path of the eclipse of 1276 Jun 13 passed close to the North Pole and then went well to the east of Iceland and Norway. The eclipse might have been visible but small in either place. The eclipse of 1279 Apr 12 would have been fairly large in either place, but an error of 3 years in dating seems large at this stage of history. It does not seem possible to identify this eclipse safely. Since the place is not given, the record would be useless even if the eclipse could be identified.

CHAPTER XIV

RECORDS OF SOLAR ECLIPSES FROM SPAIN

The chance turnings of the paths that I
followed in the search for medieval European
sources brought me to the Iberian peninsula last
among the major areas of western Europe. By the
time I reached the Pyrenees I had found so many
records of solar eclipses that additional ones
would have had only a marginal value, at least
for the immediate purposes of this study. If any
reader has persisted to this point, he probably
agrees with this feeling.

Therefore I have made no attempt to
search for records from Portugal and Spain. In
this chapter I shall discuss some early Spanish
records that are included in the German collection
Monumenta Germaniae Historica. In spite of the
small number of records that occur here, I have
reserved a region for Iberian records because it
is probable that a search would reveal a large
number of them.

Hydatius [ca. 468]. The name of the
writer is sometimes spelled Idatius but Hydatius
seems to be more common. Hydatius was a native of
Lemicus in Galicia. The 19th-century editor says
that this is the modern Jinzo de Lima; it is
probably the same as Ginzo on the Lima River. He
traveled extensively in the eastern parts of the
Roman Empire in the early part of the 5th century.
He was made a bishop in 427 and apparently lived
the rest of his life in Galicia.

Hydatius' chronicle is called a continuation of the chronicle of Jerome (Hieronymus),
which was itself a translation of Eusebius [ca.
325] with a continuation to Jerome's own time in
about 375. Hydatius also made an extension of

the Consularia Constantinopolitana discussed in
Section XV.3. He naturally shows a particular
interest in Galician affairs, although he incorporates information from a wider region. The chronicle shows signs that it was made up from notes
even in Hydatius' own time: Some dates are not correct, and for some wrong dates such details as the
feria or the age of the moon correspond to the
date given and not to the correct date. However
in some cases the feria given for an eclipse, for
example, is correct for the actual eclipse date
and not for the date stated. I did not notice
any dating error greater than 2 years, and only
one error that large.

Hydatius used the Olympiads for dating.
The year that he gives for the last entry corresponds to 469, but the listed events actually took
place in 468 and I have used the latter date in
citation. The presumption is that Hydatius died
in 468 or 469.

I used several records from Hydatius in
AAO, Section IV.4. Then I took the place of the
observations to be anywhere in the Iberian peninsula.
Here, in accordance with the convention I have
adopted, I shall take the place to be Galicia.

Isidorus [ca. 624]. Isidorus is often
referred to in English as Isidore of Seville. He
is also known as Isidorus Hispalensis in Latin. He
was bishop of Seville from about 600 until 636, and
he wrote extensively on history, theology, and
science. His best known work is his Historia
Gothorum, Wandalorum, Sueborum, which is the main
source for Spanish history for about 550 to 624.

His own writing is not what concerns us
for the present study. The only data in his history
that I noticed was a copy of the solar eclipse and
other events of 451 from Hydatius. However the editor has followed Isidorus' history with an anonymous continuation to 754; there is no convenient
way to cite the continuation separately. The continuation has several items of interest. It uses

-506-

TABLE XIV.1

REFERENCES TO SOLAR ECLIPSES FOUND IN MEDIEVAL SPANISH SOURCES

Source	Date	Conclusions
Hydatius	402 Nov 11	Independent of other known sources
	418 Jul 19	Probably a local record
	447 Dec 23	Original
	458 May 28	Original
	464 Jul 20	Original
	467 or 468	Fragmentary record; may be the lunar eclipse of 467 Jun 3
Isidorus	655 Apr 12	Independent of other known sources
	718 Jun 3	Independent; may be the eclipse of 720 Oct 6.

what is called the Spanish era; the number of a
year referred to the Spanish era exceeds by 38
the number referred to the Christian era.[†] Even
after we make this correction, the chronology of
the continuation is often wrong by several years.
The continuation also uses a chronology based upon
Byzantine and Islamic rulers, and upon the "years
of the Arabs".

        The records of solar eclipses in these
two sources are tabulated in Table XIV.1 Discussions of the individual records follow.

        402 Nov 11 E,Sp. Reference: Hydatius
[ca. 468]. "Ol. 295.2[‡] (= 402). An eclipse of
the sun occurred on the 3rd ides November (= Nov
11)." Ginzel [1899] must have used a different
edition; according to him the words "on the 2nd
feria" follow the statement of the date. 402 Nov
11 was actually the 3rd feria. This record is
also in Hydatius' continuation of the Byzantine
source Consularia Constantinopolitana [ca. 468;
see Section XV.3]. It is not original but it is
independent of any other record that I have found.
In particular there is no record of this eclipse
in any other of the Byzantine sources discussed in
Section XV.3, so it is probably local to Galicia
and not a record that Hydatius picked up in his
travels. Reliability: 0.5. Place: Galicia.
Standard deviation of the magnitude: 0.1. I
used this record in AAO.

        418 Jul 19 E,Sp. Reference: Hydatius
[ca. 468]. "Ol. 299.2 (= 418). An eclipse of the
sun occurred on the 14th calends August (= Jul 19)
which was the 5th feria." 418 Jul 19 was actually
the 6th feria. I suspect that Hydatius calculated

[†] The Spanish era commemorated the Roman conquest
of Spain under Augustus.
[‡] That is, the 2nd year of the 295th Olympiad.

the weekday using the date 417 Jul 19 by accident.
Thus the only genuine information in Hydatius'
account is the date, and there are several other
records of this eclipse that give the correct date.
Since Hydatius had earlier records that apparently
come from Galicia, I shall assume that this is
also a local record, but I shall lower the reliability to 0.2. Place: Galicia. Standard deviation
of the magnitude: 0.1. I also used this record in
AAO.

447 Dec 23 E,Sp. Reference: Hydatius
[ca. 468]. "01. 306.3 (= 447). An eclipse of the
sun occurred on the 10th calends January (= Dec 23)
which was the 3rd feria." Here all the details are
correct. The record is essentially contemporaneous,
even though Hydatius probably worked from notes.
Reliability: 1. Place: Galicia. Standard deviation of the magnitude: 0.1. I also used this in
AAO.

Hydatius says that there was a sign in
the sun in the year that equals 454. There was a
solar eclipse that should have been large in
Galicia on 453 Feb 24. A sign put in the year 454
could be an eclipse in 453, but it could also be
many other things. It is safest to ignore this
report.

458 May 28 E,Sp. Reference: Hydatius
[ca. 468]. "01. 309.3 (= 459). On the day 5th
calends June (= May 28), the 4th feria, the sun
appeared diminished in the brightness of its orb
in the form of a moon of the 5th or 6th[†] from the
4th hour to the 6th." The weekday is correct for
the eclipse date 458 May 28 rather than for the
day stated, thus the feria was not calculated at a
later time. A moon 5.5 days old has about $\frac{1}{4}$ of its
surface illuminated. This corresponds to a magnitude
of about 0.8.[‡] Reliability: 1. Place: Galicia.

[†] That is, it looked like a moon 5 or 6 days old.
[‡] AAO, Figure V.2.

Magnitude: 0.8 with a standard deviation of 0.1; path north of Spain. I noted the existence of this record in <u>AAO</u> but did not use it.

464 Jul 20 E,Sp. Reference: <u>Hydatius</u> [ca. 468]. "Ol. 311.1 (= 465). On the 13th calends August (= Jul 20), the 2nd feria, the sun was decreased in its light in the form of a 5th moon from the 3rd hour to the 6th." 464 Jul 20, rather than 465 Jul 20, was on the 2nd feria. A moon 5 days has about 1/5 of its surface illuminated, so it will be sufficiently accurate to use 0.85 for the magnitude. Reliability: 1. Place: Galicia. Magnitude: 0.85 with a standard deviation of 0.07; path north of Spain.

For the last two reports, I have taken the standard deviation of the magnitude to be about half the difference between totality and the estimated magnitude.

Hydatius' record for "Ol. 311.4" (= 468) exists only in a fragmentary form. The isolated words "setting of the sun" occur in it. There were no lunar eclipses in 468. The solar eclipses of 468 May 8 and 468 Nov 1 should not have been visible in Spain. Since Hydatius' dates are often 1 year too late, we should also try 467. The solar eclipse of 467 Mar 19 would have been a small eclipse visible only in the morning if at all. The lunar eclipse of 467 Jun 3 would have begun at about 19^h 30^m local time in Galicia, and this is also about sunset there on Jun 3. Thus there is a reasonable chance that Hydatius recorded the lunar eclipse of 467 Jun 3, but there is not enough certainty to let us use the record.

655 Apr 12 E,Sp. Reference: <u>Isidorus</u> [ca. 624, p. 343]. It should be pointed out again that the solar eclipses are actually in a continuation of the history and not in the part written by Isidorus himself. At one point the continuation says: "In his reign, the sun being obscured in the

middle of the day, stars came out in the sky." The reign is that of the Byzantine emperor Constans II, emperor 641-668. There are only three possibilities for a reasonably large eclipse during this time, unless the errors in <u>Oppolzer</u> [1887] are larger than we expect. The eclipses of 644 Nov 5 and 646 Apr 21 are possible, but only if the observation was made in the eastern empire. There are two Byzantine records[†] of the eclipse of 644 Nov 5; neither makes it a large eclipse, but that might be a simple failure to record the indications of one. I have found no Byzantine record of the eclipse of 646 Apr 21. The eclipse of 655 Apr 12 is the only other possible eclipse during the reign of Constans II, and it is the only possibility for an observation in Spain.

A few lines farther on we find the following: "In his times an eclipse of the sun, such that stars appeared in the middle of the day, terrified all Spain and foreshadowed a Gascon invasion with a not-small army." "His" could still refer to Constans II, but I do not believe that one can safely infer this from the text alone. However, the text definitely puts this eclipse a few years after 647 (685 of the Spanish Era), and the only reasonable possibility is the eclipse of 655 Apr 12.

The words used in the two references to an eclipse are quite different but the basic facts are the same. Thus it seems likely that the writer has copied two records of the same eclipse without realizing that fact, although we cannot conclude with surety that the first record refers to 655 Apr 12. I shall assume that we have a record of 655 Apr 12. Because of the likelihood that the record is copied, I shall lower the reliability to 0.5. The writer says "all Spain" and it is safer to use all Spain as the place than to use Seville. Standard deviation of the magnitude: 0.01.

[†] See Section XV.4.

718 Jun 3 E,Sp. Reference: <u>Isidorus</u>
[ca. 624, p. 356]. Here the continuation says:
"At the same time, at the beginning of era 758,
the 100th year of the Arabs, the sun is recognized
by many to be eclipsed from the 7th hour of the
day to the 9th, with stars being seen." This is
also put in the reign of Anastasius II.

The chronology is highly confused.
Anastasius came to the Byzantine throne in 713 and
was deposed in 716, although he was not killed
until 721. The year 758 of the Spanish Era is our
720, so the "beginning" is near 720 Jan 1; the
Spanish year began on Jan 1. Possibly the writer
was still counting Anastasius as the legitimate
ruler.

If we assume that the time is around 720,
the only plausible eclipses are those of 718 Jun 3
and 720 Oct 6. The latter eclipse is possible only
if the errors in <u>Oppolzer</u> are rather large. His
chart shows that the path missed Spain entirely and
that the eclipse would probably have been seen
later in the day than the record says. Further,
it does not seem reasonable to describe either
eclipse as being near the beginning of the year.

The reference to the year of the Arabs
furnishes a resolution of the problem. If the
writer thought that 758 of the Spanish era and
the "100th year of the Arabs" were equivalent, we
can read this as saying that the eclipse was near
the beginning of the 100th year of the Hegira.
The epoch of the era of the Hegira is taken as
either 622 Jul 15 or 622 Jul 16, depending upon
the chronologist. The Moslem year is a lunar calendar with 12 lunar months always; no attempt is
made to keep it approximately adjusted to the solar
year. The rules governing the lengths of the
months make the year equal to 354 11/30 days exactly. An elementary calculation that should be
accurate within a few days puts the beginning of
the year 100 near 718 Aug 2. If the writer did
not know that the Moslem year was lunar, he would
have thought that their year 100 was equal to our
year 721 or, rather, its equivalent 759 of the

-512-

Spanish era; an accident could have put him 1
year in error in his attempted conversion.

        Thus the eclipse of 718 Jun 3, which
should have been large or total in Spain in the
first part of the afternoon, fits the conditions
of the record as well as we can reconstruct them.
No other eclipse fits the conditions well, unless
Oppolzer's approximations happened to cause appreciable error. I shall calculate the circumstances
for 720 Oct 6 as a test. If its path crosses any
part of Spain, I shall ignore this record. If it
does not, I shall take the record as applying to
718 Jun 3, with a reliability of 0.5. Place:
Seville. Standard deviation of the magnitude:
0.01.

        The calculations relating to the eclipse
of 720 Oct 6 will be discussed in Section XVII.2.

CHAPTER XV

RECORDS OF SOLAR ECLIPSES FROM THE
BYZANTINE EMPIRE AND THE HOLY LAND

1. The Byzantine Calendar

        The sources that appeared earlier in this
work used the Roman calendar, with a few exceptions
that are mostly Italian. Almost all sources except
the earliest ones reckoned years by the Christian Era
that was presumably introduced (Section III.4) by
Dionysius Exiguus and made popular by Bede. There
was some variety in the date used for the beginning
of the year; December 25, January 1, March 1, and
March 25 have all been used.

        Although what we customarily call the
Byzantine Empire regarded itself as the Roman Empire,
its calendar differed in some respects from the
Roman calendar. Over the long span of the Byzantine
Empire there was some variety of practice. The
statements that are about to be made represent what
I believe was the most common practice, but they
certainly do not apply universally.

        The Byzantines used the Roman months, with
the same lengths that were already in use under the
first Caesars. They often used the Roman names for
the months, but they often used Greek equivalents,
and sometimes they used both names for clarity. They
often numbered the days of the month from the beginning of the month, just as we do. They rarely used
the Roman method of counting backwards from the
calends, nones, or ides.

        They used both names and numbers for the
days of the week. They rarely used Die Dominica
(or its Greek equivalent 'Ημερα κυριακη) for the first
day. In numbering the days of the week, at least
in most places that I have noted, they simply wrote

"day 3", for example,[†] and did not use an equivalent
of "feria" or of "of the week".

The year began on September 1.

The Byzantines used a variety of ways
for identifying the year. One way was to use the
regnal years of the emperors. Two other ways were
common. One was by the "cycle of the Indiction".
The origin and use of the Indiction are obscure.
The Indiction was a cycle of 15 years; a year of
the Indiction, like the ordinary year,[‡] began on
September 1.

The Byzantines also frequently reckoned
the years by means of an era, and they apparently
began the practice of using an era before western
Europeans adopted the Christian era. The Byzantines
sometimes gave the number of years since the pre-
sumed date of Jesus' birth, but they seemed to do
so in the spirit of giving additional information
rather than in the spirit of using an era. The era
that they used most is called the "Mundane Era of
Constantinople" in modern writing. The year designated by the use of this era is often called "year
of the world". I shall designate a year of this
era by A.M. (Anno Mundi), but it must not be supposed that the Byzantines used an equivalent expression. They commonly wrote ετους (from ετος,
year) followed or preceded by the numeral for the
year.

The Mundane Era began on September 1 of
some year. According to Explanatory Supplement
[1961, p. 430], the year was 5509 B.C. (-5508).
According to all other sources that I have consulted,

[†] They used Greek equivalents, of course.

[‡] Explanatory Supplement [1961, p. 431] says that
a new year of the Indiction began on December 25.
Most sources say that it began on January 1. One
of these days may be correct for usage in the
western empire. It is clear from the Byzantine
records that September 1 was often used in the
eastern empire, as I shall show in a moment.

the year was -5507. We must decide which year is correct.

The last event in Cedrenus [ca. 1100, v. II, p. 637] is the coronation of the emperor Isaac I Comnenus, which took place on September 1, Indiction 11, 6566 A.M., according to Cedrenus. His proclamation as emperor [Cedrenus, v. II, p. 623] took place on June 8, Indiction 10, 6565 A.M.; note that both the Indiction and the year A.M. have changed between June and September. The standard sources place these events in the year 1057 of the common era. The chronology in Cedrenus has been self-consistent for some time before these events, and the dates given are probably correct in the Byzantine system.

If Cedrenus and the standard sources are correct, the date September 1, 6566 A.M. = 1057 Sep 1. The Mundane Era began on September 1, 1 A.M. by definition of the era; thus the era began 6565 years before the date given. 6565 years before 1057 Sep 1 is -5508 Sep 1. Explanatory Supplement is apparently correct.

In order to convert a year A.M. to a year of the common era, do the following:

a. Subtract 5508 from the year A.M. if the month is January to August inclusive.

b. Subtract 5509 from the year A.M. if the month is September to December inclusive.

The Indiction is the remainder when the year A.M. is divided by 15, except that the Indiction is called 15 when the remainder is 0. I do not know if this fact had anything to do with the choice of the Mundane Era or not. Conversely, to find the year from the Indiction, first find the year A.M. by adding the Indiction to some multiple of 15. The trouble is that, of course, one must know the year closely in order to know which multiple to use. The year of the common era can then be found by using the rules above.

The preceding discussion deals with the most common reckoning of the year of the world. Theophanes [ca. 813], for example, gives a different reckoning. On page 211 of the edition cited, he tells us that the emperor Anastasius was crowned on April 14 in the year that was 5999 from Adam by the "calculation of the Romans" but 5983 by the "careful calculation" of the Alexandrians; the Indiction was 14. By the standard histories, the year was 491 of the common era; hence the year "by the calculation of the Romans" is the "year of the world" that was discussed earlier. Theophanes, however, preferred the calculation of the Alexandrians and rarely used the Roman one.†

When the Byzantine writers gave the number of years since the Incarnation they used the equation that the year since the Incarnation is equal to the year of the world (or since Adam) less 5500. The choice of 5500 probably had a numerological origin. This equation was used with either choice for the year A.M. Thus Theophanes considered the year just discussed to be the year 483 of Christ. For some reason the modern editor followed Theophanes, and the years of Christ given in the margins of the edition cited are systematically too small by 7 or 8 years, depending upon the month involved. The reader should be careful about accepting dates given in editions of Byzantine works.

---

† Theophanes said that the day was also the fifth day of the great Easter ('ημερα πεμπτη τη μεγαλη του Πασχα). I do not know what this means unless it means the 5th feria following Easter. This meaning is unlikely since 491 Apr 14 was in fact an Easter Sunday. Perhaps Theophanes meant to write "fifth hour" rather than "fifth day". This would put the coronation during a solemn mass on Easter; such an occasion seems fitting for the coronation of a Christian theocrat. The day was not the fifth day of the Paschal full moon.

2. The Places of Observation for the Byzantine Records

As I said in Section IV.4, few records of a solar eclipse state where the observation was made; the Byzantine records are no exception. On the other hand, the places of earthquakes and of most other events are almost always given. This peculiar practice has a high antiquity; it is already found in Eusebius [ca. 325].

Also in Section IV.4 I decided to use the place where the original form of a record was written as the place of observation. This procedure is plausible, and the errors caused by following it should be random, at least when we are dealing with local sources such as monastic annals. The Byzantine records were mostly written in Constantinople, and Constantinople[†] was the capital of a great empire. It is not obvious that a practice that was safe with monastic annals is also safe with imperial histories.

Certainly the Byzantine histories record a large number of earthquakes, pestilences, and plagues that occurred in places other than Constantinople. However, these events caused large property damage and large loss of life. Events of this sort occurring anywhere in the Empire are matters of serious imperial concern, and it is not surprising that we find them in imperial histories.

We also find a number of events of local interest, such as severe thunderstorms. We would not expect to find histories recording events of this sort outside of the places where they were written. I have not searched the Byzantine sources systematically for thunderstorms, but those that I noted all took place in Constantinople, as we would

---

[†] During the entire period of this study, the city in question was known as Constantinople, the city of Constantine. For this reason I shall call the city by this name instead of using its correct present name when dealing with medieval records.

expect. We must now ask whether a large solar
eclipse is a matter of local or empire-wide interest.

Eclipses were of empire-wide interest in
ancient and medieval China (see Section IV.3 of
AAO or the extensive discussion of Chinese eclipses
by Dubs [1938 through 1955]). The reason is that
an eclipse was interpreted as an omen of immediate
concern to the Chinese emperor, whether it was
large or small and wherever in the empire it was
seen. Watching for eclipses and recording them
was therefore an official government act. It is
probably to this superstitious concern and not to
a spirit of scientific inquiry that we owe the
large number of Chinese eclipse records.

Many people under the Byzantine empire
considered eclipses to be omens, but there is no
evidence of a systematic official concern with
them. Thus we may expect observations of solar
eclipses in the Byzantine sources to be largely
local. Most of the Byzantine sources were written
by people close to the central government and the
locality is thus Constantinople in most cases.
Therefore I shall take Constantinople to be the
place of observation of the solar eclipses in the
absence of specific information about the place.

We may certainly expect that this assumption will be wrong in some cases. Constantinople
was approximately in the center of its empire at
all times during the history of the later Roman
empire. Thus the errors produced by the assumption
should be approximately random.

3. Byzantine Sources

I have found sixteen Byzantine sources
that contain independent records of solar eclipses,
in addition to one source whose single record of
an eclipse is copied. I shall describe these
sixteen sources briefly in this section.

Ginzel [1882 and 1884] lists eclipse
records in the works of Michael Glycas, Nicephorus

Gregoras, and Joannis Zonaras. It seems almost
certain from what Ginzel says either that these
records are not independent or that they are outside of the time period set for this study. Therefore I have not consulted these works.

The Byzantine historian Procopius has a
passage that may be a record of the eclipse of 536
Sep 1. If so, the observation was made in Italy
and I have put the discussion of the passage into
Section XII.2. Several other Byzantine writers
have passages that are clearly derived from this
one. I have omitted these derivative passages
from the discussions in both this chapter and
Section XII.2.

Ammianus Marcellinus [ca. 391]. This
source is outside the time interval of this study,
and I include it only in order to correct an error
that I made in AAO. Ammianus served in the imperial
forces on the Persian front from about 350 until
about 365. He was almost captured [Ammianus, ca.
391, Chapter XIX.8] in 359, but he succeeded in
eluding the enemy and reaching Antioch. From then
until 363 it is not clear whether he was in the
service or not. His work is interesting for its
geographical descriptions and scientific discussions.

Ammianus was a native of Antioch and had
Greek as his mother language. He moved to Rome
about 378 and wrote there in Latin. It is somewhat
arbitrary whether he is considered to be a Greek or
a Roman writer. I include him in Byzantine sources
because his eclipse observation belongs in that
geographical region.

Anna Comnena [ca. 1120]. Anna Comnena
was the daughter of Alexius I Comnenus (1048-1118,
emperor 1081-1118), the first emperor of the
Comnenian dynasty. She married Nicephorus Bryennius,
who had the title of Caesar (the emperor had the
title of Augustus) and who ranked number 3 in the
protocol of the empire. Upon the death of Alexius,
Anna plotted and failed to make her husband emperor

instead of John II Comnenus, the rightful heir.
She failed largely because her husband refused
to take part in the plot; this did not make for
peaceful relations between husband and wife. For
some reason she was spared the usual fate of
would-be usurpers and retired to a convent where
she wrote her history of the reign of Alexius.

Because of her position she was in a
favorable position to know the political history
of the time. As other well-placed politicians
have been known to do, she wrote her political
history from a somewhat biased political viewpoint.
Her description of a solar eclipse is complete
fiction in my opinion. Unfortunately it has been
taken seriously and has been used in determining
some of the chronology of Byzantine history. For
purposes of designation I assign the date 1091 May
21 to her description; it is discussed in the next
section.

Cedrenus [ca. 1100]. This is a relatively
late compendium of history from "the founding of
the world" until the accession of Isaac I Comnenus,
a forerunner of the Comnenian dynasty, in 1057.
Little if any of it is independent. Cedrenus did
something that many medieval historians failed to
do: he listed his sources.

Consularia Constantinopolitana [ca. 468].
This is a composite of a number of annals based
upon consular lists; it has close affiliations with
Paschale [ca. 628] for some periods. At least
until the dedication of New Rome, as Constantinople
was once called, on 330 May 11, the notices could
have come from anywhere in the Empire, eastern or
western. The editor Mommsen calls the part after
395 "Additions by Hydatius"; since Hydatius is a
Spanish chronicler (see Chapter XIV), the records
after 395 are probably but not certainly from
Spain. Only between 330 and 395 are we entitled
to even a presumption that the notices are Byzantine. Unfortunately none of the solar eclipses
occur within this time period. The uncertainty
about the place of the records before 330 is

particularly unfortunate since Consularia Constantinopolitana contains observations of two early
eclipses that are mentioned in no other source.
The eclipses in this source will be listed as a
matter of record even though they cannot be used.

Eusebius [ca. 325]. Eusebius was a
famous Church figure and advisor to Constantine
the Great; he took a leading part in the Council
of Nicaea in 325 as well as in other Church affairs.
He wrote a chronicle in two books. The entire work
is usually cited as Chronicon. The first book is
the chronicle proper, and the second book is a
set of chronological tables, giving the identification of each year in a variety of systems and listing important events in the manner of a set of
annals (see Sections III.1 and III.3). Around 380
S. Jerome translated the Chronicon into Latin and
continued it down to his own time.

Eusebius contains records of several
eclipses including the famous "eclipse of Phlegon"
(see AAO, Section IV.5). All these are outside
the time period adopted for this study. However,
the record of an eclipse near 346 Jun 6 is useful
in the textual study of a writer who is within
the time period.

Georgios Hamartolos [ca. 842]. This
writer is more commonly known as Georgius Monachus
or George the Monk, but I call him Georgios Hamartolos
in order to distinguish him from a later writer also
known as Georgius Monachus; further, the editor of
the cited edition uses this designation for him.
Little is known of Georgios. Since he calls himself
Hamartolos (sinner), we may suspect him of religi-
osity. He is the only independent authority for
much of Byzantine history for much of the first
half of the ninth century.

It is probable that Georgios' own writing
stops with the end of the reign of Theophilus in
842. At least at this point we find: "Up to here
the chronicle of George. Who follow are the logo-

thetes." For this reason I use the date 842 in citation. There are anonymous continuations that will not be distinguished by separate citation.

Georgius Monachus [ca. 948]. We know nothing of this Georgius except that he is not the same person as the preceding Georgios, from whom he copies extensively. He has no independent records of eclipses.

Leo Diaconus [ca. 990]. This is a highly detailed history of the period from 959 to 976 in ten books. We know nothing of the writer except the little that is learned incidentally from his history. He mentions that he was a boy at the time of certain events that we can put in 967, so he was born around 950. References to other events show that he did not write, or at least did not finish, his history until about 990. Hence his work is not strictly contemporary in the sense that it was not written down in diary or annal form at the time. However he wrote about events that he had lived through, and he could call upon partici-pants in those events for guidance in interpreting them.[†]

Leonis [ca. 912]. This denotes an anony-mous work called Leonis Imperatoris Imperium. The

---

[†]
Bury [1898, v. IV, p. 504] says that Leo's history is "a contemporary work in a good sense; depending upon personal knowledge and information derived from living peoples, not on previous writers." This is the first suggestion that I have seen that memories of distant events are more reliable than contemporaneous written records. The legal pro-fession has worked out what is probably the best rule of evidence: Accept evidence only from first-hand sources who can be cross-examined. The "living peoples" could perhaps be cross-examined if Leo were so inclined; this may be what Bury had in mind when he said the work was contemporary "in a good sense".

Leo in question is the emperor Leo VI (emperor 886-912), who is often called Leo the Wise or Leo the Philosopher. This work is called one of the continuations of Theophanes [ca. 813].

Liudprandus [969]. Liudprandus (ca. 922-972) was one of the outstanding Italians of the 10th century, who served under the King of Italy and the Holy Roman emperor. He was made bishop of Cremona in 962. He served as ambassador to the Byzantine court from 948 to 950. In 968 he was sent back to conduct the negotiations that led to the marriage of the future Otto II of the Holy Roman Empire with the daughter of the Byzantine emperor Romanus II. His report of his mission covers the period 968 Jun 4 to 969 Jan 7, and it includes an account of the eclipse of 968 Dec 22 observed in Constantinople.

Marcellinus [ca. 534]. I hope this Marcellinus will not be confused with Ammianus Marcellinus. This one is often called Count Marcellinus of Illyrica; this means that he was a count and that he came from Illyrica, not that he was Count of that province. He was attached in some capacity to Justinian the Great (emperor 527-565) before Justinian became emperor, and afterward he continued to serve in Justinian's court.

His chronicle is one of the many that were intended as continuations of Eusebius (or of Jerome if one prefers). The first version went to 518 and Marcellinus subsequently extended it to 534. An anonymous and presumably contemporary person continued it to 548. Marcellinus' sources include Consularia Constantinopolitana and Consularia Italica (see the reference Fasti Vindobonenses [ca. 576]). Marcellinus is an independent authority for about its last forty years. He is one of the few Byzantine sources who wrote in Latin.

The chronicle has many local references to affairs in Illyrica (roughly the modern Yugoslavia). Marcellinus presumably used these because of his

-525-

interest in his native land. However, I shall take
the place of observation for his independent
eclipses to be Constantinople, in accordance with
the standard practice of this work.

Marinus [ca. 486]. Marinus Neapolitanus
was from Neapolis (the present Nablus) in Samaria,
not Naples in Italy. He succeeded Proclus, often
called the last major pagan philosopher, in the
chair of philosophy in Athens, and he wrote a com-
memorative life of Proclus. He tells us (Chapter
XXXVI) that Proclus died on April 17, 124 years
after the accession of Julian. Marinus was himself
a pagan and it is interesting that he used Julian
the Apostate for dating. Julian acceded in late
361, and the date for Proclus is ambiguous; it
could be either 485 Apr 17 or 486 Apr 17.

I count Marinus as a Byzantine source
since he lived in and under the Byzantine empire.

Odon [1148]. Odon de Deuil was a monk
of the Abbey of S. Denis under the famous Abbé
Suger. When Louis undertook the disastrous Second
Crusade, Odon, by recommendation of Suger, accompanied Louis as his chaplain. Later Odon succeeded
Suger as Abbé. Odon's Histoire de la Croisade de
Louis VII[†] refers to one solar eclipse, which was
observed in the Byzantine dominions. Hence I
count this as a Byzantine source.

Pachymeres [ca. 1308]. Georgius Pachymeres
(ca. 1242-1310) was a native of Nicaea who moved to
Constantinople in 1261. He wrote a contemporary
history of the reign of Michael VIII Palaeologus
(emperor from 1258 or 1259 to 1282; the standard
sources disagree about his year of accession). Since
Pachymeres is outside the period set for this study,
I have not spent much effort on his work. I include

[†] Odon presumably wrote in Latin but I have found
only a French translation of his work.

it only because I believe that the standard edition cited makes a serious error in dating the single solar eclipse that Pachymeres described.

Paschale [ca. 628]. The Chronicon Paschale is a famous set of annals based upon consular lists. It resembles Eusebius [ca. 325] in that it gives the year in each of several chronological systems, including the Olympiads. In spite of the title given it, the Chronicon Paschale does not list the dates of Easter; however, it does start with a discussion of the method of calculating Easter. It is generally believed to be a compilation made by a single unknown person, and it is an independent contemporary authority for the part of the seventh century that it covers. In some of its earlier portions, it apparently used the same sources as Consularia Constantinopolitana [ca. 468].

Philostorgius [ca. 425]. Philostorgius [Philostorgius, ca. 425, Chapter IX.9] was born in Borissus, a town in Cappadocia. At the age of 20 (Chapter X.6) he went to Constantinople, but it is not safe to infer that he moved there; we do not know where he lived and wrote. We can date events near this trip to 388 or 389, so that he was born near 368; thus the latter part of his Ecclesiasticae Historiae is contemporary. His father was apparently an adherent of Eunomius, who was on the "extreme left" of the Arian movement and who had considerable trouble because of his views. Philostorgius' history is mainly a history of the Arian controversy from 300 to 425 approximately.

Philostorgius' original work has been lost. Fortunately many excerpts from his work have been preserved by other writers. In particular, one Photius, a Patriarch of Constantinople in the late ninth century, prepared an extensive epitome of Philostorgius' history, along with extracts from nearly 300 other volumes of historical writings. Much of the work thus extracted by Photius is otherwise lost. The work in the cited edition is actually labelled "Epitome Confecta a Photio Patriarcha". It

seems fairly safe to conclude that Photius was
faithful to his original, and I have taken the
liberty of citing Photius' extract as if it were
the work of Philostorgius. I hope the reader is
not misled thereby.

Simocatta [ca. 625]. Theophylactus
Simocatta was a native of Egypt. He wrote a history of the reign of Mauricius (emperor 582-602),
who was killed by Phocas[†] (emperor 602-610), who
in turn was killed by Heraclius (emperor 610-641);
being Byzantine emperor was sometimes a hazardous
occupation. Simocatta wrote during the reign of
Heraclius, and his history of Mauricius is thus
nearly contemporary. I have invented the date
used in the citation for convenience; it is within
the correct period but has no explicit sanction.

Syncellus [ca. 810]. This writer is
more properly called George the Syncellus, but I
cite him as Syncellus in order to distinguish him
from the other Georges. "Syncellus" means "one
who shares a cell" and the syncellus was a person
assigned to be an assistant or chaplain to the
Patriarch of Constantinople; he was often in fact
a spy in the service of the court. This Syncellus
retired from his duties, whatever they were, about
806. After this, he prepared a chronicle of events
from the Creation to the reign of Diocletian. It
is probable that he would have brought it down to
his time if he had been spared. Certainly he en-
treated his friend Theophanes to continue his
chronicle, and Theophanes [ca. 813] did so.

Syncellus has no events within the time
frame of this study, and he is not an independent
authority for any period. His work is of great
value for the fragments of earlier writers that he
preserved, particularly Eusebius [ca. 325]. The

[†] A column in honor of Phocas is usually considered
the last structure to be erected in the Roman
Forum; it still stands [Romanelli, 1959].

TABLE XV.1

REFERENCES TO SOLAR ECLIPSES FOUND IN
MEDIEVAL BYZANTINE SOURCES

Source	Date	Conclusions
Ammianus Marcellinus	360 Aug 28	Independent; outside of time period but included in order to correct error in AAO
Anna Comnena	1091 May 21 ?	Literary eclipse; standard identification is wrong
Cedrenus	346 Jun 6	Probably from Theophanes
	644 Nov 5	Probably not independent
	693 Oct 5	Probably from Theophanes
	787 Sep 16	From Theophanes
	891 Aug 8	Probably independent
	968 Dec 22	Independent
	990 Oct 21 ?	Probably copied from 891 Aug 8 by accident
Consularia Constantino-politana	291 May 15	Unknown location; independent
	319 May 6	Unknown location; independent
	402 Nov 11	Probably copied from Spanish source Hydatius
Eusebius	346 Jun 6 ?	Outside of time period but needed in textual study; independent of other known sources

TABLE XV.1 (Continued)

Source	Date	Conclusions
Georgios Hamartolos	306 Jul 27	Independent of known sources
	617 Nov 4 ?	Valid observation; cannot date
	693 Oct 5	Independent
	891 Aug 8	Probably independent
Georgius Monachus	891 Aug 8	Copied from Georgios Hamartolos
Leo Diaconus	968 Dec 22	Independent
Leonis	891 Aug 8	Probably independent
Liudprandus	968 Dec 22	Original; Italian writer but a Byzantine observation
Marcellinus	393 Nov 20	May be from an Italian source
	418 Jul 19	Probably independent but cannot locate
	497 Apr 18 ?	Cannot be identified safely
	512 Jun 29	Independent
Marinus	484 Jan 14	Independent
Odon	1147 Oct 26	Independent; French writer but a Byzantine observation

TABLE XV.1 (Concluded)

Source	Date	Conclusions
Pachymeres	1261 Apr 1 ?	Magical eclipse; outside time period; included to correct the standard identification in the literature
Paschale	418 Jul 19	Independent
Philostorgius	418 Jul 19	Independent
Simocatta	590 Oct 4	Independent
Theophanes	346 Jun 6	Independent
	348 Oct 9	Independent
	590 Oct 4	Probably from Simocatta
	644 Nov 5	Independent
	693 Oct 5	Independent
	760 Aug 15	Independent
	787 Sep 16	Independent
	798 Feb 20 ?	Magical 17-day eclipse
	812 May 14	Independent
	813 May 4	Independent

main interest of his chronicle for this study is
his treatment of the birth of Jesus. He makes it
explicit that the birth of Jesus was intended to
be in the year before the year 1; we would call it
the year 0 but Syncellus had no such symbol. In
modern terminology, we would say that early chrono-
logists intended to place the birth of Christ in
the year 1 before Christ, not in the year 1 of
Christ. I did not include him in the count of
sources stated at the beginning of this section.

Theophanes [ca. 813]. Theophanes (ca.
758-817) was a prominent Church figure and the
good friend of Syncellus (see above) who continued
his chronicle. Theophanes did his historical
writing after he had retired from active Church
service. He took a prominent part in the controversy over icons. In 815 he refused to follow
the dictates of the iconoclast emperor Leo the
Armenian and was banished to Samothrace, where he
died.

I mentioned his use of the year A.M. in
Section 1 of this chapter. He also used the regnal
years of the emperors, the regnal years of the
Persian and Moslem rulers, and the years of the
main patriarchates. His chronology seemed to me
to be quite careful, once a person understands
his systems. Although his work is necessarily not
original in the main, he has preserved much material
that is otherwise unknown; at least this is true of
many of his records of eclipses.

4.   Discussions of the Byzantine Records of Solar
     Eclipses

In this section I shall discuss the records
that seem to be independent. Table XV.1 lists most
of the Byzantine records that I have found, whether
they seem to be independent or not. The last column
of the table summarizes the conclusions that I have
reached about each record. I have not listed every
appearance of a record that has been copied many
times, nor have I necessarily listed records of
observations that belong outside the Byzantine

-532-

provenance but that are found in the Byzantine
literature.

291 May 15 M,B. Reference: Consularia
Constantinopolitana [ca. 468]. "Tiberianus and
Dione consuls. Under these consuls there was a
darkness in the middle of the day, and this year
Constantius and Maximinus were elevated to Caesars
on the calends of March." The identification of
the eclipse is certain, but there is unfortunately
no way to know the place of the observation. Therefore I shall not use this eclipse record.

306 Jul 27 M,B. Reference: Georgios
Hamartolos [ca. 842, Chapter IV.180]. A sentence
near the end of this chapter says: "An earthquake
happened in Campania, and 13 cities were struck
down; and an eclipse of the sun occurred at the
3rd hour of the day, so that stars appeared in the
sky." First we must decide upon the place of the
observation.

Since Campania, the district of Italy
that includes Naples, is mentioned, the first im-
pulse is to put the observation there. I believe
that that action is not justified by the record.
It was the custom to give the places of earthquakes
but not the places of eclipses. Thus Campania is
connected to the eclipse only by the accident that
the chronicler put both events into the same sentence,
and we are not entitled to conclude that he meant
to put the eclipse in Campania.

Unfortunately we are not entitled to
assign any place more restrictive than the Roman
Empire. We have a situation here like that which
we have with the early records in Consularia
Constantinopolitana. We cannot assume that the
record was made and preserved in Constantinople,
because the city was not founded until 330 May 11.
While the earlier town of Byzantium was already
there, it would not have been the repository of
official records before that date. Since the Roman
Empire covered such a large area, it is pointless
to use this record.

It is still worthwhile to identify the
eclipse. All that Georgios tells us is that it
was in the reign of Constantine (306-337). The
editor identified it as 324 Aug 6, but that date
is clearly incompatible with the record. According
to Oppolzer [1887], the path on 324 Aug 6 crossed
the Strait of Gibraltar at about the 8th hour of
the day. Even if we allow for the accelerations
that Oppolzer did not include in his calculations,
we must still conclude that the eclipse of 324 Aug
6 would not have been seen anywhere in the Empire
at the 3rd hour. The only eclipse during Constantine's reign that would have been large anywhere
in the Empire at the 3rd hour is the eclipse of
306 Jul 27. The reader who consults Georgios'
work may conclude that the account is late in the
story of Constantine's reign and hence that the
eclipse could not have been in the first year of
it. However, the part of the section on Constantine where the eclipse is found is a collection of
various small notes about events, and the text of
this part is not in chronological order.

In summary, the probability that this
is a record of the eclipse of 306 Jul 27 is fairly
large, say 0.5. However, the place is so uncertain
that the record is not useful.

319 May 6 M,B. Reference: Consularia
Constantinopolitana [ca. 468]. "Licinio V and
Crispo Caes. Under these consuls there was a
darkness at the 9th hour of the day." The editor
(Mommsen) identified the year as 318, but there
is no serious question about the identification of
the eclipse. Again there is no way to tell the
place.

346 Jun 6a M,B. Reference: Theophanes
[ca. 813, p. 57]. "That same year an eclipse of
the sun happened, so that stars appeared in the
sky, at the 3rd hour of the day on the 6th day
of Daisios." Daisios is June. It was not clear
to me whether Theophanes listed this under the
10th or the 11th year of Constantius (this is

Constantius II, emperor 337-361). Cedrenus [ca.
1100, v. 1, p. 523], in a record that seems to be
derived from this one, took it to be the 10th
year. The eclipse was therefore the one of 346
Jun 6. The hour of the day looks reasonable for
this eclipse at Constantinople, judging from the
chart in Oppolzer [1887].

Theophanes cannot be the original source
for this record because of his late date. Both
for this eclipse and for many others, Theophanes
seems to have accurate and fairly detailed information. Hence I shall give most of his reports
a reliability of 0.5, which is the maximum I allow
for a derived record, until we come to his own
time when I shall increase the reliability to 1.
Here,however, we have a situation that does not
apply to most of his records. At this time the
Empire was fairly well united, and the sources
for this period that Theophanes used could themselves have applied to a wider area than Constantinople or even the Eastern Empire.

Instead of using a region rather than a
specific city, I shall take care of the uncertainty
connected with this record by lowering the reliability. I shall use a reliability of 0.2. Place:
Constantinople. Standard deviation of the magnitude:
0.01.

346 Jun 6b M,B(?). Reference: Eusebius
[ca. 325]. This record is actually in Jerome's
continuation. "Year 2362 of Abraham. An eclipse
of the sun happened." In Eusebius' system the
year of Abraham is 2016 years greater than the
year of the common era, if we ignore problems caused
by various conventions about the beginning of the
year. Hence the year indicated is 346 and the
eclipse indicated is that of 346 Jun 6.

Although this is a much older record than
the one in Theophanes, it cannot be the source of
the latter since Theophanes has information not in
this record.

Unfortunately we cannot be sure about
the identification. It is quite possible that
the year differs from 346 by 1 or 2 years. There
was an eclipse on 348 Oct 9 that attained a magnitude near 0.9 in Constantinople and there was
another on 349 Apr 4 that attained a magnitude
near 0.7. Since there are no details in Eusebius
to aid in the identification, we cannot rule out
the two latter eclipses as possibilities. Hence
we must conclude that the eclipse in Eusebius cannot be identified with enough assurance to be
used, although 346 Jun 6 is the most likely date.

348 Oct 9 M,B. Reference: Theophanes
[ca. 813, p. 58]. "That same year the sun again
became obscured at the 2nd hour of the Lord's Day
(κυριακης ημερας)." This is in the year after the
record designated 346 Jun 6a M,B discussed above.
There were no umbral eclipses visible in Constantinople during the year after 346 Jun 6. The
penumbral eclipses of 347 May 26 and 347 Oct 20
were perhaps visible there at about sunset rather
than at the 2nd hour. Further, they were both on
the 3rd feria; hence they do not correspond to
the record. The eclipse of 346 Jun 6 was on the
6th feria, so it does not match. The umbral
eclipse of 348 Oct 9 was at about the 2nd hour on
the 1st feria, while the umbral eclipse of 349
Apr 4 was at about the 7th hour on the 3rd feria.
No other eclipses lie within a reasonable time of
347. Hence the eclipse is almost surely that of
348 Oct 9, and there seems to be an error of 1
year in the dating.

Reliability: 0.2.[†] Place: Constantinople. Standard deviation of the magnitude:
0.1.

[†]This is still during a time when the Empire was
reasonably united and receives the same reliability as the earlier record by Theophanes.

360 Aug 28 M,B. Reference: Ammianus
Marcellinus [ca. 391, Chapter XX.2]. The following
happened during a year that is clearly 360: "That
year throughout the East a darkness was seen, and
stars shone out together from morning until midday."
I used this eclipse in AAO (Section IV.5), where
I wrote that it happened during a battle. I do
not know what caused me to make this blunder; I
can only suppose that I mixed up some notes while
I was writing the text. Although the duration is
clearly exaggerated, it seems safe to accept this
as a genuine record, but I shall lower the reliability to 0.5 because of the exaggeration. Because
of the stars, I shall use 0.01 for the standard
deviation of the magnitude.

At this time Ammianus was either living
in Antioch (see the discussion of Ammianus in the
preceding section) or serving on the Syrian frontier. It is plausible that Ammianus meant "throughout the East" to refer to the region of Syria,
although he may have heard a report from beyond
the frontier. I shall take the place to be Syria,
which is roughly the area from 34° to 37°N and
from 36° to 42°E.

The chart in Oppolzer [1887] shows the
sunrise point for the eclipse of 360 Aug 28 well
to the east of any point in the Empire. For this
reason Bury [1898] suspects that Ammianus made a
mistake in the date of the eclipse, which he may
have done. Bury suggests that the eclipse is
actually that of 364 Jun 16, which was observed by
professional astronomers in Alexandria (see AAO,
Section V.5). I doubt this identification. The
path of totality on 364 Jun 16 passed well to the
north of the Empire, and stars could not have been
seen even in England. If it turns out that the
eclipse of 360 Aug 28 could not have been large
within the Empire, I suspect that Ammianus heard
reports of it from outside the Empire, as I suggested above.

393 Nov 20 M,B. Reference: Marcellinus
[ca. 534]. "At that time indeed there was darkness
at the third hour of the day." The words in this

report are close to but not identical with those used in the report 393 Nov 20 E,I (see Section XII.2). Further, even if this report were independent, there would be no way to locate the observation. Hence I shall not use this record.

402 Nov 11 M,B. Reference: Consularia Constantinopolitana [ca. 468]. "Arcadius V and Honorius V. Under these consuls there was an eclipse of the sun on the 3rd ides November (= Nov 11)." This is in the "additions by Hydatius", hence the report is probably from Spain (see Chapter XIV). I shall assume that this record is not independent and shall not use it.

418 Jul 19a M,B. Reference: Philostorgius [ca. 425, Chapter XII.8]. "When Theodosius was in his boyhood (μειρακιων ηλικιας), and the month of July had advanced to the nineteenth day, about the eighth hour of the day, the sun was so deeply eclipsed that stars could be seen." A few sentences later Philostorgius says that a "cone-shaped" object could be seen during the eclipse, that the "inexpert" called a comet. Theodosius is Theodosius II (401-450, emperor 408-450). Since Theodosius was about 17 at the time, it seems odd to use μειρακιων , which is diminutive for μειραξ , a boy. However, the day and month are correctly stated and the hour of the day is reasonable, so there is no good reason to question the identification, and the usage of μειρακιων may have been standard. This is a contemporary report. Reliability: 1. Place: Constantinople. Standard deviation of the magnitude: 0.01. I used this report in AAO.

418 Jul 19b M,B. Reference: Marcellinus [ca. 534]. The entry for the year that can be equated to 418 contains: "Solis defectio facta est. Stella ab oriente per septem menses surgens

argensque apparuit."[†] Mommsen, the cited editor
of Marcellinus, says that Consularia Italica (see
the discussion of Fasti Vindobonenses [ca. 576]
and the record 418 Jul 19a E,I in Chapter XII) is
the source of this record. This attribution seems
unlikely to me. The Consularia have "sol eclipsim
fecit" followed by the exact and correct date where
Marcellinus has "solis defectio facta est" without
the date. Further, the following references to the
comet are significantly different.

    I conclude that the record in Marcellinus
is probably independent of any other known source,
although it is certainly not original. Unfortunately there is no way to locate the observation.
Hence I shall not use this record.

    418 Jul 19c M,B. Reference: Paschale
[ca. 628]. "Indiction 1. 10. Honorius XII and
Theodosius Augustus VIII. Under these consuls an
eclipse of the sun happened in the month of Panemos,[‡]
on the 14th calends August (= Jul 19), feria 6,[*]
hour 8." The numeral 10 after the Indiction denotes the tenth year of Theodosius; this is 417
or 418 according to the details of the counting
convention. In this historical context, Indiction
1 corresponds to the year 395 × 15 + 1 = 5926
Anno Mundi. Hence the date given is 418 Jul 19,

[†] "An eclipse of the sun occurred. A star appeared
rising out of the east and glowing for seven
months." The star is probably a comet and is
probably the "cone-shaped object" that, according
to Philostorgius (also see Appendix II), could be
seen during the eclipse. "For seven months" is
probably a mistake that Marcellinus made for "in
the seventh month" when transcribing from his
source; other sources say that this comet was
visible for 30 days.

[‡] Panemos equals the Roman July.

[*] The text actually has Παρασκευη , which is the
Greek equivalent for the Jewish "day of preparation" for the Sabbath.

which was the 6th feria.

This report has the week day, which
Philostorgius does not give. It is possible but
unlikely that the compiler of the Chronicon
Paschale calculated it. Further, the entire
wording of the two entries is dissimilar. It
seems safe to conclude that the entry in Paschale
is independent of the entry in Philostorgius, although it is certainly not an original source.
Reliability: 0.5, since the record is secondary.
Place: Constantinople. Standard deviation of the
magnitude: 0.1.

484 Jan 14 M,B. Reference: Marinus
[ca. 486, Chapter XXXVII]. "There were also omens
for a year before his death, for example an eclipse
of the sun which was so great that it made night
out of day. A deep darkness began and stars could
be seen. This happened in the sign of Capricorn
toward the rising point."[†] Since the sun was in
Capricorn, the eclipse happened within a month
after the winter solstice. Proclus died on Apr
17 of either 485 or 486 (see the discussion under
Marinus in the preceding section).

If we do not take Marinus literally that
the eclipse preceded Proclus' death, and merely
search for eclipses visible in the right part of
the world near the time, we find that only the

[†] Αυτη μεν ουν εν αιγοκερωτι εγενετο κατα το ανατολικον
κεντρον. In AAO I translated this as: "This
happened in the sign of Capricorn in the eastern
sky." Rosan, in the translation cited in the
references, gives: "It occurred in the eastern
horn of the sign of Capricorn." Both these are
more specific than the text warrants; the translation given above is closer to the original.
The phrase "toward the rising point" is ambiguous
both textually and astronomically. If the date
484 Jan 14 for the eclipse is correct, the sun
was both in the eastern sky and in the eastern
part (toward the rising) of Capricorn.

eclipses of 484 Jan 14, 486 May 19, 487 Nov 1, and
489 Mar 18 can be considered. The only one of
these that occurred in Capricorn was on 484 Jan 14.
It is also the only one that was nearly total and
the only one that preceded Proclus' death. Thus
it seems safe to accept Marinus' statement in spite
of its magical aspect.

Marinus mentions that another eclipse
"will happen" in the first year after Proclus' death,
but he does not imply that the eclipse will be large
and he gives no helpful details. The predicted
eclipse cannot be identified with assurance, but
it is probably the eclipse of 486 May 19. The
eclipses cannot settle between the dates 485 Apr
17 and 486 Apr 17 for the death of Proclus. The
eclipse of 484 Jan 14 cannot accurately be described
as being within the year before 486 Apr 17, but
there is no reason to assume that Marinus meant to
be rigorously accurate.

I used this eclipse in AAO and I shall
give it the same characteristics that I did there.
Reliability: 0.5, because of the magical aspect.
Place: Athens. Standard deviation of the magnitude: 0.01.

497 Apr 18 M,B(?). Reference: Marcellinus
[ca. 534]. The entry for the year that can be
identified as 497 contains: "An eclipse of the sun
appeared." Unfortunately this brief record cannot
be identified safely because there were two significant eclipses in both Constantinople and Illyrica
within six months of each other. The eclipse of
496 Oct 22, we estimate roughly from the chart in
Oppolzer, had a magnitude of about 0.7 in Constantinople or Illyrica. The eclipse of 497 Apr 18
had a magnitude of perhaps 0.8 in Constantinople
and somewhat less in Illyrica. Hence there is a
good chance that either eclipse would be recorded.

The chronology in Marcellinus is usually
fairly good, but we cannot rely upon it to resolve
an interval as short as six months. Since 497 is
apparently the year intended, and since the eclipse

-541-

of 497 Apr 18 was somewhat larger in Constantinople
than the other one, that identification is more
probable. However the probability is not high
enough to let us use this record.

This record is almost certainly the
source of the Irish record 497 Apr 18 B,I(?); see
Section VII.1.

512 Jun 29 M,B. Reference: Marcellinus
[ca. 534]. "Near these times an eclipse of the
sun happened." This eclipse is probably the eclipse
of 512 Jun 29. However, the eclipse of 511 Jan 15
may have been visible in Constantinople at sunset.
I shall calculate both eclipses in Section XVII.2,
and if it turns out that the eclipse of 511 Jan 15
was significant I shall ignore this record. If I
use it, I shall give it a reliability of 1, since
it is contemporaneous. Place: Constantinople.
Standard deviation of the magnitude: 0.1.

This record is almost surely the source
of the record designated as 512 Jun 29 B,I in
Section VII.1.

590 Oct 4 M,B. Reference: Simocatta
[ca. 625, Chapter V.16]. "Not moved by their
entreaties, the emperor Mauricius went out of the
palace and proceeded one and one-half parasangs
to the Hebdomon, as it is called in Byzantium.
That day there happened a very great eclipse of
the sun. Then there arose a violent roaring
south wind, that almost tore the pebbles from the
depths of the sea." This is an essentially contemporary record that I used in AAO, and there is
nothing to add. Reliability: 1. Place: Constantinople. Standard deviation of the magnitude:
0.02. I use this value in spite of the absence of
specific details because the eclipse is described
as "very great" (μεγιστη).

Theophanes [ca. 813, p. 413] has a
record that seems to be derived from this one.

617 Nov 4 M,B(?). Reference: Georgios
Hamartolos [ca. 842, Chapter IV.227]. "There
happened moreover a severe famine and great mortality; the sun also was darkened and ashes rained."
This is during the reign of Heraclius (emperor 610-
641). It occurs in the early part of the text
that deals with his reign, but it is not safe to
infer the date from this. There were eclipses on
616 May 21, 617 Nov 4, and 634 Jun 1 during Heraclius' reign. 617 Nov 4 probably had the largest
magnitude in Constantinople, according to the
charts in Oppolzer [1887], but it is not legitimate
to infer the date from this. I have found no other
reference to a famine nor to a rain of ashes[†] near
this time. There is no way to date this record
and I shall not attempt to use it.

644 Nov 5 M,B. Reference: Theophanes
[ca. 813, pp. 524-525]. "And an eclipse of the
sun occurred, month Dios fifth, day sixth, hour
ninth." This is listed under the 3rd year of Constans II, emperor 642-668. Dios is the Greek name
for the month equivalent to the Roman November.
Hence the date given is 644 Nov 5, which was the
6th feria. The 9th hour of the day looks reasonable for maximum eclipse at Constantinople. Reliability: 0.5. Place: Constantinople. Standard
deviation of the magnitude: 0.1.

Cedrenus [ca. 1100, v. 1, p. 754] has
the following: "Third year (of Constans) was an
eclipse of the sun." There is not enough information to let us decide whether this is an independent record or not, so I shall ignore it.

693 Oct 5a M,B. Reference: Theophanes
[ca. 813, p. 562]. "That year an eclipse of the
sun happened, month Hyperberetaeus fifth, day
first, hour third, such that some stars shone out."
This was put under the 8th or 9th year of Justinian

[†] Perhaps from a volcanic eruption.

II, emperor 685-695,[†] and Hyperberetaeus is the
same as October. Hence the eclipse was surely the
one on 693 Oct 5, which was on the first feria;
the hour of the day is reasonable. Reliability:
0.5. Place: Constantinople. Standard deviation
of the magnitude: 0.01.

Cedrenus [ca. 1100, v. 1, p. 773] again
has an account that seems to be an abridgement of
this one.

693 Oct 5b M,B. Reference: Georgios
Hamartolos [ca. 842, Chapter IV.240]. "After
the emperor reentered the city, there was an
eclipse of the sun so that stars appeared." This
is sometime during the first reign of Justinian
II, and the eclipse of 693 Oct 5 is the only possibility. Reliability: 0.5. Place: Constantinople.
Standard deviation of the magnitude: 0.01.

760 Aug 15 M,B. Reference: Theophanes
[ca. 813, p. 665]. "A rebellion broke out in
Africa, and a solar eclipse happened, month August
15th, day sixth, hour tenth." This is in the 21st
year of Constantine V Copronymus, emperor 740-775.
760 Aug 15 was on the sixth feria, and the tenth
hour is reasonable on the basis of Oppolzer. We
are now close to the time of Theophanes, but I
shall still give this report a reliability of 0.5.
I shall give his remaining records the reliability
of 1, however. Place: Constantinople. Standard
deviation of the magnitude: 0.1.

787 Sep 16 M,B. Reference: Theophanes
[ca. 813, p. 716]. "Ninth of September month, Indiction 11, day of the Lord,[‡] a very great eclipse

---

[†] He was deposed in 695 but restored in 705, reigning from then until 711. This is in his first
reign.

[‡] 'Ημερα κυριακη.

of the sun happened at the 5th hour of the day,
during mass." This was during the 7th year of
Constantine VI, emperor 780-797. Calculation from
the Indiction according to the rules given in
Section XV.1 yields the year 787. The eclipse
date 787 Sep 16 was on the first feria, so it looks
as if this is a reliable account of that eclipse,
with an accidental error of exactly one week in
reading the date from the calendar. I would guess
that the eclipse was maximum at about the 3rd hour
from Oppolzer, but the difference is not enough to
affect the identification.

It is notable that Theophanes changed
his style of reporting eclipses at this point.
Since we are now well into his adult life, it is
plausible that he has begun to use his own notes.
Reliability: 1. Place: Constantinople. Standard
deviation of the magnitude: 0.02, since this was
a "very great" eclipse.

Cedrenus [ca. 1100, v. 2, p. 23] has
obviously copied from Theophanes (or from Theophanes'
source if Theophanes is not original). He has the
same mistake of giving Sep 9 rather than Sep 16.

798 Feb 20 M,B(??). Reference: Theophanes
[ca. 813, p. 732]. ". . the sun for 17 days gave
off no rays, . ." This is just after the empress
Irene (empress 797-802) seized power. One may
not totally dismiss a story such as this without
study. One must consider the possibility that the
account has been garbled, or that an actual event
has been exaggerated. It is doubtful that this is
an exaggerated account of an eclipse. The eclipse
of 798 Feb 20 may have been a partial eclipse visible
just at sunset in western Europe, but it should not
have been visible at all in Constantinople. There
is no large eclipse at Constantinople within seven
years of this date.

It is possible that a volcanic eruption
or some other event put a large amount of dust into
the air, and the account seems to suggest this kind
of darkening more than an eclipse. It seems more

likely to me that the account was suggested by a
political "sick joke". Irene was cruel and unpopular, and she often blinded people as a punish-
ment. Someone could have made a remark like: "She
is so fond of blinding that she would even blind
the sun." Theophanes uses similar phraseology
later in his account.

It is interesting that Theophanes trans-
mits this story even though he was a responsible
adult at the time of the alleged event. The reader
who is so inclined may take this as evidence that
some physical event underlay the account. Unfortunately adults are not always above the transmission of unsubstantiated rumor. This account
was popular in medieval times and is copied into
annals as far away as England, where it is found
in Diceto [ca. 1202] among other places.[†]

I shall not take this to be a record of
an eclipse.

812 May 14 M,B. Reference: Theophanes
[ca. 813, p. 772]. "14 of May month, day sixth,
a great solar eclipse happened for about three and
a half hours, from the eighth hour until the eleventh
hour." This is in the first year of Michael I (emperor 811-813). Part of our year 812 is in that
year. The eclipse of 812 May 14 was on the 6th
feria, as stated, and the hours of the day look
reasonable. Reliability: 1. Place: Constantinople.
Standard deviation of the magnitude: 0.05, since it
is described as "great" (μεγαλη) but not "very great"
(μεγιστη).

[†] The French chronicler Adonis [ca. 869], after
mentioning Irene in 798, says that Mars could
not be seen from July of 798 until the following
July. Adonis seems to have taken the behavior
of Mars from the source used by S. Florentii
[ca. 1236], who, however, puts the disappearance
from July of 712 to July of 713. See Appendix
III.

813 May 4 M,B. Reference: Theophanes
[ca. 813, p. 780]. In the year following the preceding account we find: "On the 4th of May month
an eclipse of the sun happened near the 12th degree
of Taurus." The position given for the sun was
probably taken from tables or from a calendar rather
than from observation. Reliability: 1. Place:
Constantinople.† Standard deviation of the magnitude: 0.1.

891 Aug 8a M,B. Reference: Georgios
Hamartolos [ca. 842, Chapter V.6]. "Γεγονε δε εκλειψις
'ηλιακη, ωστε νυκτα γενεσθαι 'ωρα ζ' και τους αστερας φαινεσθαι,
. . ."‡ After the eclipse there was a violent
thunderstorm and 7 people were killed by lightning
on the steps of the forum. This report is in an
anonymous continuation of the chronicle of Georgios
Hamartolos. The time is during the reign of "Leo
son of Basil" but there is no direct clue to the
exact year. This Leo is Leo VI (the Wise or the
Philosopher), who reigned from 886 to 912. The
eclipse of 891 Aug 8 was the only large eclipse during his reign, and it was maximum in Constantinople
near midday. It is plausible but not certain that
this anonymous continuation is contemporary; I shall
assume that it is. Place: Constantinople. Standard deviation of the magnitude: 0.01. We cannot
assign a reliability to this record until we can
judge its independence by comparison with the following record.

Georgius Monachus [ca. 948, Section 9 under
"Leo, Son of Basil"] has copied this passage verbatim,

---

† In 812 and 813 Theophanes was probably living in
a monastery a short distance from Constantinople,
and these eclipses may have been observed there.
The error caused by ignoring this possibility
should be small.

‡ "A solar eclipse occurred, so that it seemed
like night at the 6th hour and stars appeared,..."
The exact wording will be needed for the reports
of this eclipse in order to decide about independence. English is poorly suited to render differences among certain Greek tenses.

as well as many paragraphs on either side of it.

891 Aug 8b M,B. Reference: <u>Leonis</u>
[ca. 912, Section 6]. " Γεγονε δε τηνικαυτα και ηλιακη
εκλειψις, 'ωρα της ημερας εκτη, ωστε και τους αστερας
φαινεσθαι, ..."† After the eclipse there was
a thunderstorm and 7 people were killed by lightning on the steps of the forum of S. Constantine.
The eclipse is certainly that of 891 Aug 8. The
report is contemporary or nearly so. Place: Constantinople. Standard deviation of the magnitude:
0.01.

It is hard to decide whether this record
and the preceding record by Georgios Hamartolos
are independent or not. There are many resemblances
in wording, and the Greek tenses used are the same.
On the other hand there are only a certain number
of ways to say that an eclipse happened, if one is
using a "chronicler's" style, and the resemblances
may be the consequence of the form rather than of
copying. Georgios says that it "seemed like night",
which <u>Leonis</u> does not say, but that could be the
result of imagination. <u>Leonis</u> says the "steps of
the forum of S. Constantine"; he could not get this
detail‡ from Georgios.

On the whole I believe that the records
are independent, but I shall back this opinion only
to the extent of giving each record a reliability
of 0.75.

891 Aug 8c M,B. Reference: <u>Cedrenus</u>
[ca. 1100, v. 2, p. 253]. "Κατα τουτον τον καιρον εγενετο
και 'ηλιου εκλειψις περι ωραν εκτην της ημερας, ωστε και αστερας
φανηναι, . . ."* After the eclipse there was a thunder-

---

†"There happened at this time a solar eclipse at
the sixth hour of the day, so that stars appeared..."
‡
There were many great fora in Constantinople, so
that "forum of S. Constantine" is not synonymous
with "forum".
*
"About that time there happened an eclipse of the
sun near the sixth hour of the day, so that stars
appeared, . . ."

storm and 7 people were killed on the steps of
the forum of S. Constantine.

        This report is not contemporary and thus
it cannot receive full reliability. It is hard to
decide whether it represents a source of information
that is independent of the preceding reports. It
does not contain any information that is not in
Leonis; however the wording is distinctly different
and it is expanded in nonessential ways. It is
unusual for a copier to paraphrase his original so
freely. Again I shall compromise. If I felt certain that the report were independent I should give
it a reliability of 0.5 (not 1 because it is late).
As it is I shall use 0.2 for the reliability. Place:
Constantinople. Standard deviation of the magnitude:
0.01.

        968 Dec 22a M,B. Reference: Leo Diaconus
[ca. 990, p. 72]. "While the emperor was doing
these things in Syria, there was an eclipse of the
sun at the winter solstice of a kind that never
happened before, except that which was drawn over
the earth at the Passion of our Lord on account of
the folly of the Jews, as they blindly fastened the
maker of the world to the cross. The eclipse was
of a remarkable kind. It was on the twenty-second
of December, at the fourth hour of the day, in a
clear sky, as darkness spread over the earth and
stars were all seen brightly." The year is unques-
tionably 968 and the exact date of the eclipse is
given. The hour of the day seems correct.

        Leo was about 18 at the time of the eclipse
(see the discussion of him in the preceding section).
Thus he probably saw the eclipse himself and this
account sounds like that of an eyewitness or of an
expert writer. Since Leo probably wrote this more
than 20 years later, it is plausible that some
exaggeration crept into his description. Since this
report is almost surely genuine I shall give it a
reliability of 1. I shall express my doubts about
its accuracy by using 0.2 for the standard deviation
of the magnitude; if there were no doubts, the value
would be 0 because of the definite statements about

the extent of the eclipse.

One could argue that the place is Syria
because of the reference to the emperor in Syria.
The passage did not strike me that way. It seemed
to me rather that Leo was telling what happened at
home while the emperor was away. Since there is
no explicit indication of a place other than Constantinople, I shall follow the convention adopted
in Section 2 of this chapter and take Constantinople
to be the place.

968 Dec 22b M,B. Reference: Cedrenus [ca.
1100, v. 2, p. 375]. "There happened also an eclipse
of the sun on the 22nd of the December month, near
the third hour, so that stars appeared." This is
under the year that Cedrenus identified as 6478 A.M.
and Indiction 13. These data correspond to the year
969. This fact, the total difference in the style
of the reports, and the discrepancy in the hour of
the day all indicate that Cedrenus is independent of
Leo Diaconus here. Cedrenus is still not contemporary.

Reliability: 0.5. Place: Constantinople.
Standard deviation of the magnitude: 0.01.

968 Dec 22c M,B. Reference: Liudprandus
[969, p. 362]. Part of Liudprandus' diplomatic
report reads: "After 4 days, however, namely on
the 11th calends January (= Dec 22), as I was in
place at table eating bread, . . .,† the sun as
though ashamed of such villainy concealed the rays
of his light, and the eclipse that occurred terrified
Michael but did not change him." Apparently Liudprandus was not pleased by the conduct of Michael,
whoever he was. Reliability: 1. Place: Constantinople. Standard deviation of the magnitude: 0.1.

---

† The clause denoted by dots can be translated as
follows: " . . . which spread its heel over me,
 . . ." "Bread" is the only plausible antecedent
for "which" that I can find. The meaning escapes
me. Most of the passage is in rather unusual
Latin, but there is no question that Liudprandus
records that he saw the eclipse personally.

990 Oct 21 M,B(?). Reference: Cedrenus
[ca. 1100, v. 2, p. 434]. "Εγενετο δε και εκλειψι.s
'ηλιου περι μεσην ημεραν, ωs και αστεραs φανηναι."† Ginzel
[1882 and 1884] identified this as the eclipse of
990 Oct 21. I believe that his identification is
open to serious question, for several reasons.

First, according to the charts in Oppolzer
[1887], the path of the eclipse of 990 Oct 21 passed
about through Moscow. It does not seem possible
that stars could have been seen during this eclipse
in any part of the Byzantine dominions. One must
admit that the report could conceivably have originated outside of the Byzantine empire, although
there is no known precedent.

Second, the account of the eclipse def-
intely occurs in the chronicle of Cedrenus long
before events that he dates in years as late as
the year that we call 989. It is possible but unlikely that he displaced the eclipse this much in
time.

Third, the account of this eclipse has
definite similarities to Cedrenus' account designated as 891 Aug 8c M,B. If the reader will compare the two accounts, he will see that they are
identical from the word " εγενετο" on, with two minor
exceptions: (a) Cedrenus uses "sixth hour of the
day" for 891 and "middle of the day" for 990; the
change from one form to the other is almost automatic. (b) Cedrenus uses 'ωστε" for 891 where he
uses "ωs " for 990; the words are similar in spelling
and almost identical in meaning. These changes are
ones that a person copying a record could make without thought.

For these reasons I conclude that Cedrenus
accidentally copied the record of 891 Aug 8 in a
temporal position where it did not belong. I shall
not use this record.

---

† "An eclipse of the sun happened about the middle
of the day, so that stars appeared."

1091 May 21 M,B(?). Reference: <u>Anna</u>
<u>Comnena</u> [ca. 1120]. One of the main problems faced
by <u>Alexius</u> I Comnenus (emperor from 1081 - 1118)
was a war with the people whose name is spelled in
English sometimes as Petchenegs and sometimes as
Patzinaks. At one point in the war, after they had
suffered some reverses (Book VII, Chapter 2), the
Petchenegs sent a peace delegation to ask for an
audience with Alexius. Alexius was apparently not
anxious to negotiate. While he was discussing the
matter with his council, an "under-clerk" named
Nicholas told Alexius that a solar eclipse was predicted for that day. Alexius sent word to the
delegation that he would look for a sign in the
heavens; if a sign appeared, he would know that
they were not to be trusted. "Before two hours
had gone by, the light of the sun failed and all
of its disk became dark as the moon intercepted
it." The Patzinaks were stupefied. The war continued for some more time until Alexius finally
dealt them a severe defeat. Within a few decades
thereafter, the Petchenegs disappeared from history.

I hope that admirers of Anna will forgive
me if I doubt the truthfulness of this story. Anna
wrote it about 30 years after the event. Portrayal
of an uneducated people who are stupefied by an
eclipse is a literary device that was already
ancient in Anna's time; <u>Herodotus</u> [ca. -446] uses
it. The device of taking advantage of an uneducated people by predicting an eclipse has been
used by <u>Clemens</u> [1889] among others; Anna's use of
this device is, however, the oldest that I have
seen.

The obvious procedure is to start by
using independent information about the war in
order to find an approximate date for the alleged
eclipse. If an eclipse actually happened near
this time, we could accept the reality of the
eclipse observation, while staying free to doubt
the use made of the eclipse if we choose to remain
sceptical. Unfortunately we cannot follow this
procedure. Anna is our only authority for the
Patzinak war [<u>Buckler</u>, 1929, pp. 211 and 231], and
we can find the chronology of the war only by

dating the eclipse.

If this is so, I believe that we have
little chance of finding an accurate chronology
for the war. Anna's work is not self-consistent.
If the rest of her account is correct, the eclipse
did not happen. If the eclipse did happen, other
statements must be wrong and it will turn out that
we cannot identify the eclipse confidently in order
to date the event.

We can bracket the war by independent
evidence. Anna places the war between the death
of Robert Guiscard, which we know happened in
1085, and the opening of the Crusades, which we
know happened in 1096. Within this interval,
there were three eclipses that were total or nearly
so in Constantinople. Their dates are 1086 Feb 16,
1091 May 21, and 1093 Sep 23. In Oppolzer [1887],
the paths of all three eclipses go almost exactly
through Constantinople.

Since there had already been some cam-
paigning that spring before the alleged eclipse,
we know that the eclipse cannot be early in the
year. Therefore 1086 Feb 16 cannot be correct.[†]
Since there was considerable action on the day of
the eclipse, not to mention almost two hours of
waiting, the eclipse cannot have been early in the
morning. Thus 1091 May 21 cannot be correct, because that eclipse was maximum in Constantinople
shortly after sunrise.[‡] This leaves only 1093 Sep
23, which was late enough in the year and which
was maximum in Constantinople at about mid-afternoon.

However, Anna gives one more clue about
the chronology of the war. The severe defeat (Book
VIII, Chapter 5), which came after the "eclipse",
was on the 29th day of April and on the 3rd day of
the week; Anna failed to tell us the year. Now

[†] There is an additional reason for discarding
this date that will appear shortly.

[‡] There is also an additional reason for discarding
this date.

Apr 29 was on the 3rd feria in 1085, 1091, and
1096. Only 1091 is allowed by the dates of Robert
Guiscard and the First Crusade, and the date given
for the defeat must be 1091 Apr 29.[†] Since the
eclipse was before the defeat, the eclipse cannot
have been on 1093 Sep 23 (nor on 1091 May 21, for
that matter). Thus all possible eclipses have
been eliminated.

Buckler [1929, p. 435] writes: "Chalandon
has worked out her (meaning Anna's) chronology so
carefully that we will not attempt to do it again;
we will merely accept his pronouncement that the
eclipse mentioned in VII.2 is that of August 1,
1087." If Anna's account is correct, this identification is not correct. The path of the eclipse
of 1087 Aug 1 passed through northern Africa. It
is doubtful that the magnitude in Constantinople
exceeded about 3/4. An eclipse this small has a
fair chance of passing unnoticed (unless it was
being watched for because it had been predicted),
and it would not be a very impressive sign. However, Anna is quite specific in her statement that
the eclipse was total. I have not tried to locate
the work of Chalandon that Buckler cites.

There is another point about the date
1087. Anna's chronology is indeed confusing. It
is hard to be certain about the time that elapsed
between the eclipse and the major defeat on 1091
Apr 29. I did not study the intervening events
carefully, but it did not seem likely that they
could occupy most of four years. This tends by
itself to eliminate the date 1087 Aug 1, and the
argument applies with even more force to the date
1086 Feb 16.

Thus on the one hand we have a romantic
literary eclipse that is doubtful on its very
face. On the other hand we have a simply stated
date (April 29 on the 3rd feria). The date has no
apparent importance or emotional significance, and

[†] Standard histories give 1091 for the year of this
defeat. I imagine that the calculation just outlined is the basis of this. We can eliminate 1096
Apr 29 because it is too close to the Crusade.

there is no visible reason why Anna should shift
the defeat from some other date to this one; however we must admit the possibility that she simply
made up a date for purposes of verisimilitude.
The date and the eclipse are inconsistent. Since
a choice must be made, I unhesitatingly choose the
date.†

It is possible that there was a typographical error in connection with the date. In
the printed edition cited, the numbers involved in
the date are written out as words, and an error in
reading Anna's manuscript is thus unlikely. We do
not know how Anna's source wrote the date, however.
If we discard the date because there may have been
an error in it, we are left with 1093 Sep 23 as
the only possible identification of the eclipse.
The defeat that we earlier dated as 1091 Apr 29
would then be changed to the spring of either 1094
or 1095. This choice is uncomfortably close to
the opening of the First Crusade and we cannot make
it with confidence; events between the defeat and
the Crusade took some time.

In summary, adoption of the date 1091 Apr
29 for the defeat is the best choice open to us.
We can then reject the eclipse totally, or we can
accept it as a literary eclipse arising from assimilation. If we do the latter, we cannot date the
eclipse, because there are three large eclipses
within the time interval allowable for assimilation.

I shall not use this eclipse report.

---

† If Anna invented the date, she could certainly
invent the eclipse. Buckler [1929, p. 211] seems
impressed by the quality of Anna's description of
the eclipse. The description is clear and explicit,
but it implies neither observation nor expert
knowledge. It could have been written by anyone
with an elementary knowledge of astronomy or astrology.   Clemens [1889] is far more detailed and
realistic.

Perhaps the most interesting feature about Anna's report is what it implies about the ability to predict solar eclipses. In Anna's account, the emperor Alexius is willing to base his conduct of diplomatic negotiations upon the prediction of an eclipse, and Anna expects her readers to find this conduct completely credible, with no explanation or justification required. Whether Anna's account is true or false, it could be written only at a time when the ability to predict eclipses accurately is taken for granted by people who are not themselves astronomers.

1147 Oct 26 M,B. Reference: Odon [1148, p. 336]. "On that day the sun saw a crime that it could not endure; but in order not to put that crime at the level of the treason done to our Savior, it shone with half its face and veiled the other half." The crime was the "betrayal" (that is, a betrayal from the Crusaders' viewpoint) and consequent defeat of the German Crusader army near the town of Dorylaeum, which is the modern Eskisehir about 200 kilometers west of Ankara. This action took place in October of 1147, hence the eclipse is certainly that of 1147 Oct 26.

Odon was with the French army and it would be possible to determine just where the French army was on that day by the expenditure of enough effort, but the effort is not justified for a report of such low precision. After the French learned of the German defeat in central Asia minor, they turned to follow the western and southern coasts of Asia Minor until they reached Syria. I shall take the place to be at latitude 41°N, longitude 30°E; at this point they would still have a choice of routes.

I do not believe that the word "half" is intended to be an estimate of magnitude. I confess to being influenced by the fact that the magnitude, on the basis of the chart in Oppolzer, must have been about 0.9. I shall use the values that I have adopted as conventional for a partial eclipse, namely a magnitude of 2/3 with a standard deviation of 1/6; the path was south of the point of observation

Reliability: 1. I use this reliability even
though I suspect Odon of a small amount of assimilation. He almost surely saw the eclipse. It
could have occurred exactly as he says, but Odon
perhaps "adapted" it slightly for dramatic purposes.

The German annals Herbiopolenses [ca.
1215] and several other German sources have accounts
of the eclipse of 1147 Oct 26 that are discussed in
Section XI.2 under the designation 1147 Oct 26a
E,G. There are considerable grounds for believing
that they record an observation made in Mediterranean
regions. Because of the great uncertainty in the
place of observation, I do not use these German
records.

1261 Apr 1 M,B(?). Reference: Pachymeres
[ca. 1308, Book I, Chapter 13]. The chapter opens
with the death of the emperor Theodore II Lascaris,
and then says: "This event was indeed foretold.
For the sun was eclipsed at the third hour of the
sixth day and deep darkness seized everything, so
that stars appeared." The standard sources are
about evenly divided between the years 1258 and
1259 for the death of Theodore. In his chronological summary in the cited edition of Pachymeres,
the editor Possini assigns the event to August of
1259.

This is clearly a magical eclipse that
should be viewed with suspicion. It should not
be accepted at all unless there is an eclipse
that clearly and unambiguously matches the record;
if this happens, we can assume that Pachymeres
merely reported a genuine observation in magical
terms.

Possini identified the eclipse as that
of 1255 Dec 30. He took the place of observation
to be Nicaea; I presume he did this because
Pachymeres is believed to have lived in Nicaea until 1261, when he moved to Constantinople. For
purposes of identification, it does not matter
whether the place was Nicaea or Constantinople.
The eclipse of 1255 Dec 30 was maximum at either
place at about the 11th hour of the 5th day, in

clear contradiction to the record.

Possini gets around this little problem by putting the eclipse at the 3rd hour "post meridiem" on the 6th day (inclusive) after the coronation of Theodore. In order to do this he has to put Theodore's coronation on Christmas Day of 1255.

I suppose it is possible that an eclipse occurring 6 days after the coronation of an emperor could be taken as an omen of his death at an early age. However, I think that Possini is stretching the possibilities. Every standard reference that I have consulted puts the accession of Theodore in 1254, not 1255. If the references are correct, the eclipse could not have been on the 6th day after his coronation. Further, Pachymeres does not say "on the 6th day after the coronation" or on the 6th day after anything.[†] He simply says "the 6th day". In every Byzantine record that I have seen, this usage means the 6th day of the week. Similarly, his usage about the hour does not mean the 3rd hour of the afternoon, it means the 3rd (unequal) hour of the day and hence it means slightly before midmorning.

I have no hesitation in rejecting Possini's identification.

The eclipse of 1255 Dec 30 is the only eclipse that could have been large in Constantinople during the reign of Theodore, and there was probably no other that was visible as even a small partial eclipse. Hence there was no eclipse that could have served as Pachymeres' omen. The only remaining possibility is that the eclipse was assimilated. There are three eclipses close enough to the event to make assimilation conceivable. The eclipse of 1261 Apr 1 was nearly total at the 5th hour of the 6th day, the eclipse of 1263 Aug 5 was nearly total at the 10th hour of the 1st day, and the eclipse of 1267 May 25 was nearly total at the 5th hour of the 4th day.

[†] Further, Pachymeres does not mention the coronation anywhere nearby in the text.

The eclipse of 1261 Apr 1 is the only possibility that can be supported seriously, and it is plausible that this eclipse was assimilated to the death of Theodore. However, the conservative thing is to ignore Pachymeres' record entirely, and I shall do so.

5. Records from the Holy Land

I have found records of 3 solar eclipses made by writers from western Europe who were in the Holy Land for a Crusade or who were serving under the Christian kings of Jerusalem. Since the records are not part of the Islamic culture it seems preferable to set up a separate category for these records, in spite of their small number.

De Expugnatione [ca. 1188]. This is an anonymous work whose full title is De Expugnatione Terrae Sanctae per Saladinum Libellus.† The writer was an Englishman who became a member of the garrison of Jerusalem. The final siege of Jerusalem, in which the author was wounded, began on 1187 Sep 21 and the city surrendered on 1187 Oct 2. Before he described the siege of Jerusalem, the writer described the siege of Ascalon, but not nearly in the same detail. Ascalon surrendered on 1187 Sep 4 after a brief action.

Fulcher [ca. 1127]. The writer is known in Latin as Fulcher Carnotensis and his name is often rendered in English as Foucher of Chartres.‡ Fulcher was a French priest who was present at the council at Clermont where Urban II spoke on 1095 Nov 26 in a call for a crusade. Fulcher accompanied

---

† A Small Book About the Capture of the Holy Land by Saladin. This title may have been furnished by a modern editor.

‡ I thank Professor Harold Fink of the University of Tennessee for calling my attention to the eclipse records in Fulcher.

TABLE XV.2

REFERENCES TO SOLAR ECLIPSES IN REPORTS
FROM THE HOLY LAND

Source	Date	Conclusions
De Expugnatione	1187 Sep 4	Original
Fulcher	1113 Mar 19	Original
	1124 Aug 11	Original

Robert of Normandy in the (First) Crusade. After
a short time he became chaplain to Baldwin, who
subsequently became the first king of Jerusalem.[†]
Thus Fulcher began to live in Jerusalem in 1100,
and it is believed that he lived there until his
death sometime after 1127. The work used here is
his History of Jerusalem. Fulcher's book is considered to be one of the most reliable histories
that we have of the First Crusade and of the subsequent history of the kingdom of Jerusalem down
to the close of the book in 1127.

The records of solar eclipses found in
these sources are listed in Table XV.2.

1113 Mar 19 M,HL. Reference: Fulcher
[ca. 1127, Book II, Chapter 48]. The passage says:
"In the year 1113 from the Incarnation of the Lord,
while we had the 28th day of the moon in the month
of March, we saw the sun, from early morning to the
1st hour[‡] or more, diminished by a lack of one part,
and that part that began to waste away at the top
at last reached in its roundness the lower part.
But still the sun lost not all its brightness, and
what was not lacking was by estimate about the 4th
part of its considerably horned form." One manuscript says "5th part". The numbers would be
written IV^a and V^a respectively. I believe that
a copyist is more likely to drop a stroke before
the "V" than he is to add one, although this is
far from certain. It should be adequate to assume
that about 3/4 of the area of the sun was eclipsed.
This corresponds to a magnitude of about 0.8.[*]
The 28th day of the moon in March of 1113 was on
1113 Mar 19, so the dating is accurate. Reliability:
1. Place: Jerusalem. Magnitude: 0.8 with a
standard deviation of 0.1; path assumed southeast
of Jerusalem.

[†]This of course refers only to the medieval Christian
kingdom.

[‡]One manuscript says "3rd hour".

[*]Figure V.2 of AAO.

I am basing the assumption about the
path upon the chart of the eclipse in Oppolzer
[1887]. This seems to contradict Fulcher's statement; if the lower part of the sun remained bright
the path should be north of the observer. However,
it is possible to read Fulcher's statement in a
way that makes the bottom be the eclipsed part;
the passage is somewhat ambiguous on this point.

1124 Aug 11 M,HL. Reference: Fulcher
[ca. 1127, Book III, Chapter 37]. The relevant
passage says: "Thereupon the sun appeared to us
for almost an hour with a colorful light, changed
into a new hyacinth form or into a kind of horned
eclipsed moon. This was on the 3rd ides August
(= Aug 11), as the 9th hour was passing." Since
he says "horned" rather than "considerably horned",
the magnitude is probably not as great as in the
eclipse of 1113 Mar 19. Reliability: 1. Place:
Jerusalem. Magnitude: 2/3 with a standard deviation of 1/6; path assumed north of Jerusalem.

1187 Sep 4 M,HL. Reference: De Expugnatione [ca. 1188, p. 238]. The source says: "In
the year 1187, in the month of September, on the
4th day of the month, the 6th feria, at the 9th
hour, the sun was darkened, and in that darkness
the greater number of those native to Ascalon came
to the camp of the Egyptians." The weekday is
correct and the hour looks reasonable. I think
the sense of the passage is that the natives of
Ascalon, who would have been Moslem by birth and
sympathy, took advantage of the darkness to join
the besieging "Egyptian" (Saladin's) force. If so,
the eclipse is somewhat literary; the darkness
should not have been nearly great enough to allow
such an action.

While the author puts the darkness at
Ascalon, perhaps for literary effect as I have
just said, he probably observed the eclipse at
Jerusalem. The writer was definitely in Jerusalem
a short time after the eclipse. It is possible
that he fought at Ascalon and was then able to

join the Jerusalem garrison, but this seems unlikely. Ascalon fell after a short siege. Few
soldiers escape from a victorious siege, which
implies the surrounding of the captured place.
Luckily Ascalon and Jerusalem are only about 60
kilometers apart and it matters little which place
we use. Reliability: 1, in spite of the literary
aspect. Place: Jerusalem. Standard deviation
of the magnitude: 0.1.

CHAPTER XVI

GENERAL REMARKS ABOUT THE
ASTRONOMICAL ANALYSIS

1. Coordinates of the Places of Observation

      The rest of this work will deal with
the astronomical, as opposed to the textual, analysis of the records. It is not feasible to avoid
astronomical jargon in the remaining chapters.

      In Chapters VI through XV, the places
of observation have been identified by name. In
order to use the places in the astronomical analysis,
it is necessary to have their latitudes and longitudes.

      Some observations have been assigned to
specific points, some to anywhere on a line, and
some to anywhere in an area. The method of analysis
used in AAO involved distinguishing between lines
and areas. The method of analysis to be used here
does not require the distinction, and we can use the
term "region" to apply either to a line or an area.

      We need the coordinates (latitude and
longitude) of the center point of a region. We
also need a quantity to be called the "radius" of
a region. The radius will enter only into determining the statistical weight to be given to an
observation, and hence it is not needed to high
accuracy. The radius is not to be interpreted in
a strictly geometric sense. It is an estimate of
the standard error made in the position of any point
in the region if the coordinates of the center are
used in place of the coordinates of the point.

TABLE XVI.1

REGIONS USED AS PLACES OF OBSERVATION

Name Used	Center Latitude (deg)	Longitude (deg)	Radius (deg)
Anjou	47.5N	0.5W	1.0
Belgium	50.8N	4.4E	0.5
Burgundy	46.5N	5.5E	1.0
Denmark	56.0N	10.0E	1.0
England	53.0N	1.0W	1.5
Galicia	42.5N	7.5W	1.0
Iceland	65.0N	20.0W	1.5
Ireland	53.4N	7.5W	1.5
Italy ("most of")	43.4N	12.0E	3.0
Man, Isle of	54.2N	4.5W	0.1
Orléans to Tours	47.6N	1.3E	0.2
Oslo to Frankfurt	55.0N	10.0E	3.0
Oxfordshire	51.8N	1.2W	0.2
Peterborough or Canterbury	52.0N	0.5E	0.6
Scandinavia and Baltic	59.0N	13.0E	3.0
Spain	40.0N	5.0W	3.0
Syria	35.5N	39.0E	2.0
Western Wales	52.5N	4.5W	0.6

TABLE XVI.2

COORDINATES OF SPECIFIC PLACES
OF OBSERVATION

Place	Latitude (deg)	Longitude (deg)
Aachen, Germany	50.8N	6.1E
Admont, Austria	47.6N	14.5E
Angers, France	47.5N	0.5W
Angoulême, France	45.7N	0.2E
Arras, France	50.3N	2.8E
Athens	38.0N	23.7E
Augsburg, Germany	48.4N	10.9E
Autun, France	47.0N	4.3E
Bamberg, Germany	49.9N	10.9E
Beauvais, France	49.4N	2.1E
Benevento, Italy	41.1N	14.8E
Bergamo, Italy	45.7N	9.7E
Bergen, Norway	60.4N	5.3E
Beze, France	47.5N	5.3E
Blandain, Belgium	50.6N	3.3E
Bologna, Italy	44.5N	11.3E
Bourbourg, France	51.0N	2.2E
Brauweiler, Germany	49.8N	7.5E
Brechin, Scotland	56.7N	2.7W
Burton-upon-Trent, England	52.8N	1.6W

TABLE XVI.2 (Continued)

Place	Latitude (deg)	Longitude (deg)
Bury S. Edmunds, England	52.2N	0.7E
Cambrai, France	50.2N	3.2E
Canterbury, England	51.3N	1.1E
Cluny, France	46.4N	4.6E
Coggeshall, England	51.9N	0.7E
Cologne, Germany	50.9N	7.0E
Constantinople	41.0N	29.0E
Corvei, Germany	51.8N	9.4E
Dijon, France	47.3N	5.0E
Disibodenberg, Germany	49.8N	7.7E
Douai, France	50.4N	3.1E
Durham, England	54.8N	1.6W
Egmond aan Zee, The Netherlands	52.6N	4.6E
Eichstet, Germany	48.9N	11.2E
Engelberg, Switzerland	46.8N	8.4E
Erfurt, Germany	51.0N	11.0E
Essenbek, Denmark	56.4N	10.2E
Farfa, Italy	42.2N	12.7E
Farnham, Surrey, England	51.2N	0.8W
Flavigny, France	47.5N	4.5E

TABLE XVI.2 (Continued)

Place	Latitude (deg)	Longitude (deg)
Fosse, Belgium	50.4N	4.7E
Freising, Germany	48.4N	11.8E
Fulda, Germany	50.6N	9.7E
Gembloux, Belgium	50.6N	4.7E
Gottweih, Austria	48.4N	15.6E
Harsefeld, Germany	53.4N	9.5E
Heilsbronn, Germany	49.4N	10.8E
Herzogenrath, Germany	50.8N	6.1E
Hildesheim, Germany	52.2N	10.0E
Hradisch, Czechoslovakia	49.1N	17.5E
Ingolstadt, Germany	48.8N	11.4E
Jarrow, England	55.0N	1.5W
Jerusalem	31.8N	35.2E
Kremsmünster, Austria	48.1N	14.1E
La Cava, Italy	40.7N	14.7E
Lambach, Austria	48.1N	13.9E
Laon, France	49.6N	3.6E
Liège, Belgium	50.6N	5.6E
Limoges, France	45.8N	1.2E
London, England	51.5N	0.2W

TABLE XVI.2 (Continued)

Place	Latitude (deg)	Longitude (deg)
Lorsch, Germany	49.6N	8.6E
Lund, Sweden	55.7N	13.2E
Lyon, France	45.8N	4.8E
Magdeburg, Germany	52.1N	11.6E
Malmesbury, England	51.6N	2.1W
Margan, Wales	51.6N	3.7W
Melk, Austria	48.2N	15.4E
Melrose, Scotland	55.6N	2.7W
Merseburg, Germany	51.4N	12.0E
Metz, France	49.1N	6.2E
Milan, Italy	45.5N	9.2E
Monte Cassino, Italy	41.5N	13.8E
Mont S. Michel, France	48.6N	1.5W
Munich, Germany	48.1N	11.6E
Naples, Italy	40.8N	14.2E
Neresheim, Germany	48.8N	10.4E
Nevers, France	47.0N	3.2E
Osney, England	51.8N	1.2W
Oxnead, England	51.8N	1.2W
Paderborn, Germany	51.7N	8.7E

TABLE XVI.2 (Continued)

Place	Latitude (deg)	Longitude (deg)
Pegau, Germany	51.2N	12.2E
Pohlde, Germany	51.7N	10.4E
Pont-à-Mousson, France	48.9N	6.0E
Prague, Czechoslovakia	50.1N	14.4E
Prum, Germany	50.2N	6.4E
Quedlinburg, Germany	51.8N	11.2E
Ravenna, Italy	44.4N	12.2E
Regensburg, Germany	49.0N	12.1E
Reichenau, Germany	47.7N	9.1E
Reichersberg, Austria	48.3N	13.4E
Rome	41.9N	12.5E
Rouen, France	49.4N	1.1E
Rye[a], Germany	54.8N	9.6E
S. Albans, England	51.8N	0.4W
S. Amand-les-Eaux, France	50.4N	3.4E
S. Benoit-sur-Loire, France	47.8N	2.3E
S. Blasien, Germany	47.8N	8.1E
S. David's, Wales	51.9N	5.3W
S. Gall, Switzerland	47.4N	9.4E
S. Omer, France	50.8N	2.2E

[a] In Germany but included with Danish sources.

TABLE XVI.2 (Continued)

Place	Latitude (deg)	Longitude (deg)
S. Trudperti, Germany	47.9N	7.8E
Salerno, Italy	40.7N	14.8E
Salzburg, Austria	47.8N	13.0E
Saumur, France	47.3N	0.1W
Sazava, Czechoslovakia	49.9N	14.9E
Schaffhausen, Switzerland	47.7N	8.6E
Scheftlarn, Germany	47.9N	11.5E
Sens, France	48.2N	3.3E
Seville, Spain	37.4N	6.0W
Stade, Germany	53.6N	9.5E
Stuttgart, Germany	48.8N	9.2E
Tewkesbury, England	52.0N	2.2W
Tours, France	47.4N	0.7E
Trier, Germany	49.8N	6.6E
Vendôme, France	47.8N	1.1E
Venice, Italy	45.4N	12.3E
Vigeois, France	45.4N	1.5E
Vormezeele, Belgium	50.8N	2.9E
Weingarten, Germany	49.0N	8.5E
Weltenburg, Germany	48.9N	11.9E

TABLE XVI.2 (Concluded)

Place	Latitude (deg)	Longitude (deg)
Winchester, England	51.1N	1.3W
Wissembourg, France	49.0N	8.0E
Worcester, England	52.2N	2.2W
Worms, Germany	49.6N	8.4E
Würzburg, Germany	49.8N	10.0E
Xanten, Germany	51.7N	6.4E
Zwettl, Austria	48.6N	15.2E
Zwiefalten, Germany	48.2N	9.4E

The coordinates of the center points, and the radii, are listed in Table XVI.1 for the various regions that have been assigned to one or more observations. Since great accuracy is not needed, all the information in Table XVI.1 is taken by eye from maps. The radii are given in degrees of a great circle.

The coordinates of specific points are listed in Table XVI.2. The coordinates are taken from the Times Atlas [1955] whenever the places can be found there. Times Atlas lists latitudes and longitudes in degrees and minutes; they are converted to decimal fractions and rounded to the tenth of a degree in Table XVI.2. In order to guard against the possibility of typographical errors in the Atlas lists, I tested values against the largest-scale maps that I could find, usually road maps. It is probable that the Atlas contains an error in the latitude of Salzburg; all road maps agree that the latitude is about 47° 48′ rather than the Atlas value. With this exception, I used the values listed in the Atlas.

Coordinates not listed in the Atlas were taken from road maps whenever the places could be found there; most of the road maps used were on a scale of 1:1000000. Coordinates for places that could not be found in either the Atlas or on a road map were taken from Ginzel [1882 and 1884]. For the very few places that could not be found in any of these sources, I used the coordinates of a place that I judged to be nearby on the basis of the annals themselves. These places have been noted specifically in the discussions of the sources in the preceding chapters.

2. Computation of the Circumstances of Solar Eclipses

The program used for calculating the circumstances of solar eclipses in this work is basically the same program as that used in AAO. The only changes in the program since it was used for AAO have been made to increase the ease of running it. This program calculates both general circumstances and local circumstances. The general circumstances calculated are: (a) the Besselian elements [Explanatory Supplement, 1961, pp. 216-219] and certain useful auxiliary quantities as functions of time, and (b) the central line and the limits of the umbral zone. Local circumstances include: (a) local times of sunrise, apparent noon, and sunset, (b) the times of beginning and end of the local eclipse, if the sun does not rise or set eclipsed, and (c) the magnitude as a function of time during the local eclipse. Local circumstances can be calculated for any number of points.

The calculations can be performed for any value of the lunar acceleration. They can also be performed in effect for any value of the earth's spin acceleration, although the spin acceleration does not enter directly into the calculations. The longitude that is entered for a point is the ephemeris longitude, and the difference between the geographical longitude and the ephemeris longitude is a function of the spin acceleration assumed.

The program is based directly upon the discussion of Chapter 9 of Explanatory Supplement [1961]. There are two main differences from the calculations outlined there. First, only about the largest third of the perturbation terms used in the standard ephemerides were retained. Second, certain derivatives needed in calculating the limits of the umbral zone, which are calculated numerically in Explanatory Supplement, are calculated here by analytic approximations.

The testing of the program was described in AAO (Chapter VII). The most delicate test is the position of the central line of the umbral zone. The average error in this position was about 0°.1 of a great circle, and the maximum error observed in about 50 cases was 0°.4. In order to guard against accidental alteration of the program, I maintain two copies of the program, with which I occasionally run certain standard cases and examine the output digit by digit.

Penumbral eclipses of course have no

umbral path on the surface of the earth. For these the umbral path on the fundamental plane is calculated instead. The fundamental plane is the plane through the center of the earth and normal to the symmetry axis of the shadow.

3. Variables Used in the Analysis; Some Notation

The two parameters that we wish to estimate from the analysis are the accelerations $\dot{n}_M$ of the moon's mean motion and $\dot{\omega}_e$ of the earth's spin. $\dot{n}_M$ will be the acceleration with respect to ephemeris time (ET) as opposed to solar time. It will always be expressed in units of seconds of arc per century per century ($''/cy^2$). Instead of using $\dot{\omega}_e$ directly, I shall use a quantity y defined by:

$$y = 10^9 (\dot{\omega}_e/\omega_e) + 25 . \qquad (XVI.1)$$

In this, ω_e is the earth's spin. y will be expressed in units of cy^{-1} (reciprocal centuries), so that $y - 25$ means the spin acceleration in units of parts in 10^9 per century. In other words, -25 parts in 10^9 per century has been chosen as a convenient reference value, and y measures the derivation from the reference. The units of y will usually be omitted in the rest of this work. The reference value is close to the estimate that was made in AAO for the medieval period.

A convenient reference value for the acceleration of the moon will also be chosen. Spencer Jones' value, which is used in the standard national ephemerides, is $-22.44''/cy^2$ [Spencer Jones, 1939]. Van Flandern [1970] has estimated that the present value is $-52 \pm 16''/cy^2$. In AAO (Section XIV.3) I estimated $-42.3 \pm 6.1''/cy^2$ for an epoch fairly near to the mean epoch of the observations that will be used here. The value $-32.44''/cy^2$ appears to be both convenient and probably close to the correct value, and I shall use it as the reference.

Thus I shall define a quantity x by:

$$x = \dot{n}_M + 32.44 \quad ''/cy^2 . \qquad (XVI.2)$$

x is clearly the deviation from the reference
value -32.44 $''/cy^2$. The units of x and of $\dot{n}_M$
will usually be omitted in the rest of this work.

        In AAO, Chapter XIII, I implicitly re-
garded y as a linear function of x and used a
method of finding this function that involved the
geometry of the umbral path of an eclipse. The
method used there necessarily fails with a penumbral eclipse, which has no umbral path. In addition, it fails with about one umbral eclipse out
of ten.[†] Thus it is desirable to find a new method
of procedure.

        We need to decide on a quantity, Q say,
that satisfies two requirements: (a) each observation must furnish an estimate of Q, and (b) it must
be possible to calculate Q theoretically as a function of x and y for the circumstances of each
observation.

        The first possibility that comes to mind
for Q is the magnitude of an eclipse. There are
two difficulties involved in using the magnitude.
First, according to the definition given in Section
IV.5, the magnitude is 1 everywhere within the
umbral zone and thus it does not vary with x and
y there. This difficulty can be overcome by using
the more complex definition given on p. 245 of
Explanatory Supplement [1961], which allows the
magnitude to vary within the umbral zone while
agreeing with the simpler definition outside it.
Second, the magnitude as modified is a maximum on
the center line of the umbral path and thus it is
near a maximum for most of the observations. It is
not desirable to fit to a variable near its maximum;

[†] Failing means here that the method does not provide
any information, not that it provides wrong information.

we want to use a variable that varies strongly
with changes in conditions.

A better candidate is the minimum distance r_m from the observer to the symmetry axis
of the shadow; this is the distance at the time
when the eclipse is greatest. Every observation
that has been selected for use implicitly yields
an estimate of r_m, and r_m is clearly a function
of x and y. Thus r_m satisfies the requirements for Q, but it is nonetheless a poor choice.
The reason is that r_m is one of the polar coordinates of the observer in a particular coordinate
system, and if we use it we will be using it near
the origin where there is a singular point. That
is, it is not an analytic function of x and y
in the region of interest. This consideration
suggests replacing r_m by a rectangular coordinate,
which would not be singular near the origin.

Accordingly, consider the plane normal
to the symmetry axis of the moon's shadow and passing through the position of the observer at maximum
eclipse; this plane is defined even if the observer
is on the axis. We shall define a coordinate system
$\xi\eta\zeta$ whose origin is the intersection of the shadow
axis with the plane. The positive ζ-axis points
toward the moon and sun. To find the $\xi\eta$ axes,
first calculate the velocity vector of the moon
relative to the observer. The ξ-axis is parallel
to the projection of this vector onto the plane,
and the η-axis completes a right-handed set. The
value of η used for an observation will be the
value of the coordinate η of the observer at the
instant of maximum eclipse; η is positive if the
observer is northward of the shadow axis.

η meets the requirements for the quantity
needed in the analysis: An observation furnishes
an estimate of η, η is a function of x and y,
and in fact it is an analytic function in the region
of interest. Further, the correct values of x
and y are probably close enough to 0 that we can
take η to be a linear function of x and y with
sufficient accuracy. We have defined η only for
the instant of maximum eclipse, but it would be

-578-

possible to generalize the definition for any time if we wished to do so.

Each record of a solar eclipse furnishes an observed estimate of η. Let $\eta_{e,i}$ denote the estimate furnished by the ith record. We need procedures for finding the value of $\eta_{e,i}$ from the information furnished in the record. We also need to estimate $\sigma_{e,i}$, the a priori estimate of the standard error in $\eta_{e,i}$. The procedures for finding $\eta_{e,i}$ and $\sigma_{e,i}$ will be developed in the next section.

$\eta_{e,i}$ and $\sigma_{e,i}$ will be expressed in units of the earth's equatorial radius.

4. Deducing Coordinates and Standard Deviations From the Observations

We must deduce $\eta_{e,i}$ and $\sigma_{e,i}$, the values of the coordinate η and its expected standard deviation for the ith record, from the information contained in the ith record. We start from the estimates of the magnitude and the auxiliary circumstances given in the discussions of the records in Chapters VI through XV.

The definition of the magnitude on p. 245 of Explanatory Supplement [1961], when put into the notation used here, reads:

$$\mu = (L_1 - r_m)/(L_1 + L_2) . \qquad (XVI.3)$$

In this, μ is the magnitude, r_m is the distance defined in the preceding section, L_1 is the radius of the penumbra cast on the plane defined in the preceding section, and L_2 is the radius of the umbra cast on the same plane. L_1 is a positive quantity while L_2, which is much smaller in size than L_1, is positive for annular eclipses and negative for total eclipses. Equation XVI.3 is the generalization of Equation IV.1 in Section IV.5 that was mentioned in the preceding section. μ is less than 1 for an observer outside the umbral zone, and it is less than 1 for an annular eclipse even within the umbral zone. For a total eclipse, for which $L_2 < 0$, $\mu = 1$ when $r_m = |L_2|$, which happens on the edge of the umbral zone. Within the umbral zone, $r_m < |L_2|$ and $\mu > 1$.

The solution of Equation XVI.3 for r_m is:

$$r_m = (1 - \mu)L_1 - \mu L_2 . \qquad \text{(XVI.4)}$$

From the definitions that have been given, it follows that $r_m = |\eta|$.

Most of the records merely indicate that the eclipse was large but furnish no specific estimate of the magnitude. For these records, $\eta_{e,i}$ is as likely to be positive as negative and hence the best estimate of $\eta_{e,i}$ is 0. When a record furnishes a specific estimate of the magnitude, we should calculate r_m from Equation XVI.4 using the estimated magnitude. We then attach a sign to $\eta_{e,i}$ in accordance with the position of the observer with respect to the shadow axis. The sign has already been deduced tentatively from the charts in Oppolzer [1887] but, as I said in Section IV.5, the final assignment of sign should be made from more precise calculations.

For convenience, an approximation will be made in using Equation XVI.4. It should be remembered that the plane used in defining L_1 and L_2 is the plane normal to the shadow axis through the position of the observer. The values of L_1 and L_2 can be made available from the calculations, but they are awkward to use because they depend upon the position of the observer and hence they vary with time during an eclipse. L_1 and L_2 do not differ much from the radii ℓ_1 and ℓ_2 of the penumbra and umbra on the fundamental plane. ℓ_1 and ℓ_2 are independent of the observer and are characteristic only of the eclipse. They are in fact included in the list of Besselian elements and are readily available. Use of ℓ_1 and ℓ_2 in place of L_1 and L_2 in Equation XVI.4 should introduce only a small random error into the calculations.

We must also estimate the standard
error $\sigma_{e,i}$ to be associated with the value of
$\eta_{e,i}$ for each eclipse. In making this estimate
we may use approximations freely since the error
estimate is not needed with high accuracy. Two
components enter into the estimate of $\sigma_{e,i}$.
These arise from the uncertainty in the magnitude
and from the uncertainty in the observer's position.
The uncertainty in magnitude gives rise to two
distinct cases that must be discussed in turn.

In the first case, the standard deviation
of the magnitude assigned to a record in Chapters
VI through XV is 0. In the terms used in this
chapter, this means simply that the observer was
within the umbral zone at the time of maximum eclipse,
with a probability of about $2/3$. Hence his distance
from the shadow axis was probably less than or equal
to $|L_2|$. Instead of using an individual value of
L_2 for each eclipse, it will suffice in this application to use a standard value for all eclipses,
after replacing L_2 by ℓ_2.

The largest value of $|\ell_2|$ found by in-
specting several pages of Oppolzer [1887] was 0.0287.
ℓ_2 depends mostly upon the mean anomaly of the moon
at the time of the eclipse and hence the values of
$|\ell_2|$ should have approximately a sinusoidal distri-
bution. Thus the standard deviation of $|\ell_2|$ should
be about $0.0287/\sqrt{2} = 0.02$ approximately. This is
the value of $|L_2|$ that will be used in estimating
$\sigma_{e,i}$ in the first case.

The possible values of r_m for an observer
in this case are distributed with uniform probability
from 0 to $|L_2|$. Hence the standard deviation is
$|L_2|/\sqrt{3} = 0.0287/\sqrt{6} = 0.012$ approximately. That is,
0.012 will be used as the contribution of the uncertainty in magnitude to $\sigma_{e,i}$ in those cases described
by "standard deviation of the magnitude: 0" in
Chapters VI through XV.

In the second case, the standard deviation
assigned to the magnitude is some number σ_μ greater

than 0. For these records we use Equation XVI.4, in principle, with L_1 and L_2 replaced by ℓ_1 and ℓ_2. The desired contribution to $\sigma_{e,i}$ is then the value of r_m that we get if we replace μ by $(1 - \sigma_\mu)$. That is,

$$\sigma_{e,i} = \sigma_\mu \ell_1 - (1 - \sigma_\mu)\ell_2 . \qquad \text{(XVI.5)}$$

We can use convenient approximations in evaluating $\sigma_{e,i}$.

ℓ_1 is nearly the same for all eclipses and is about 0.55. $|\ell_2|$, on the other hand, has a maximum value of about 0.0287 and ℓ_2 may have either sign. Hence the average value of the second term in Equation XVI.5 is much smaller than the first term unless σ_μ is small. The most common value of σ_μ is 0.1. For this and larger values I shall neglect the second term and use:

$$\sigma_{e,i} = 0.55 \, \sigma_\mu \quad \text{if} \quad \sigma_\mu \geq 0.1. \quad \text{(XVI.6)}$$

For smaller values, we should recognize that we were thinking implicitly of total eclipses in the assignment of σ. For total eclipses, as opposed to annular ones, ℓ_2 is negative and we can obtain reasonable results if we use $\ell_2 = -0.02$ in Equation XVI.5. For $\sigma_\mu = 0.01$, which is the most common small value, this choice leads to

$$\sigma_{e,i} = 0.025 \quad \text{if} \quad \sigma_\mu = 0.01. \quad \text{(XVI.7)}$$

If σ_μ lies between 0.01 and 0.1, I shall interpolate linearly between the value from Equation XVI.7 and the value 0.055 that we obtain from Equation XVI.6 with $\sigma_\mu = 0.1$.

It should be remembered that $\sigma_{e,i}$ in Equations XVI.5 through XVI.7 does not mean the total value of $\sigma_{e,i}$ but only the contribution from the uncertainty in the magnitude.

Finally we must estimate the contribution
to $\sigma_{e,i}$ that results when an observation is assigned
to a region rather than to a specific point. This
contribution is just the radius listed for each region
in Table XVI.1, after it is converted from degrees of
a great circle to units of the earth's radius. In
other words, each degree of the quantity called
"radius" in Table XVI.1 contributes 0.01745 to $\sigma_{e,i}$.
In principle, we should use only the projection of
the radius onto the fundamental plane, but this refinement seems unnecessary for a quantity that is
not needed accurately.

The reader should remember that the total
$\sigma_{e,i}$ is not the sum of the two contributions. Instead it is the square root of the sum of the squares.

5. Weights Assigned to the Eclipse Reports

"Best estimates" of the accelerations $\dot{n}_M$
and $\dot{\omega}_e$ are to be inferred from the records of
solar eclipses that were discussed in Chapters VI
through XV. In the inference process we shall use
each record that was assigned a reliability greater
than 0 and we shall assign a weight to each record
that is to be used. The inference process will be
described in the next section. It is convenient
first to explain the method of assigning the weights.

The weight assigned to each record depends
upon two factors, the reliability and the estimate
of $\sigma_{e,i}$ discussed in the preceding section.

Records with low expected accuracy should
receive low weight, and vice versa. That is, the
weight should be low if $\sigma_{e,i}$ is large. In the
method of least squares, which is the inference
method that will be used, the relative weight is
$1/\sigma_{e,i}^2$.

This assignment of weight reflects the
limitations on accuracy imposed by the nature of
the data. In dealing with the eclipse records we
must also consider the fidelity with which the data
available to us preserve the data originally recorded.

The number that I have called the reliability
represents my attempt to assess this fidelity.
When the reliability has been taken as 1, it is
almost certain that the data are those originally
recorded. When the reliability has been taken as
0, it is almost certain either that the data do
not correspond to an observation of any identifiable eclipse or that they do correspond but are
copied from a known record and hence do not come
from an independent observation. Intermediate
values of the reliability are used in intermediate
cases. The weight assigned to a record should be
proportional to the reliability in addition to re-
flecting the value of $\sigma_{e,i}$.

Let us use R_i to denote the reliability
assigned to the ith record. Then the weight assigned
to the ith observation in the inference of the accelerations will be W_i, defined by:

$$W_i = R_i/\sigma_{e,i}^2 . \qquad \text{(XVI.8)}$$

6. The Inference Method

The inference method to be used is the
method of least squares. This is a standard method
and I shall outline it here only to the extent
needed to explain the notation and the terms that
will be used in later discussion.

Let $\eta_{th,i}$ denote a value of the coordinate
η calculated from the eclipse program described in
Section XVI.2. I shall calculate a value of $\eta_{th,i}$
for each record with a weight greater than 0 for
each of four pairs of values of x and y, as
follows:

$$
\begin{aligned}
&(1) \quad x = -10 \;, \quad y = -5, \\
&(2) \quad x = +10 \;, \quad y = -5, \\
&(3) \quad x = -10 \;, \quad y = +5, \\
&(4) \quad x = +10 \;, \quad y = +5.
\end{aligned}
\qquad \text{(XVI.9)}
$$

If we assume that $\eta_{th,i}$ can be represented by a linear function of x and y with sufficient accuracy in the ranges of interest, we can write it in the form:

$$\eta_{th,i} = \eta_{0,i} + A_i x + B_i y . \qquad \text{(XVI.10)}$$

For $\eta_{0,i}$, I shall use the mean of the four values of η calculated with the four pairs of accelerations listed in Equation XVI.9. In order to find A_i, we note that A_i means $\partial\eta/\partial x$ and that we can estimate two values of the partial derivative numerically from the four values of η that have been found. I shall use the mean of these two values for A_i, and I shall find the value of B_i in a similar fashion.

If the observations were perfect, we could find a pair of values for x and y that would make $\eta_{th,i}$ equal to $\eta_{e,i}$ for each observation. Since the observations are not perfect, we cannot make them equal for all observations. Instead, we choose x and y in the way that minimizes the sum of the squares of the discrepancies. The "best estimate" of x and y means the pair of values that does this.

In order to find the "best" x and y, we form the following function F of x and y:

$$F = \tfrac{1}{2} \sum_i W_i (A_i x + B_i y - Z_i)^2 . \qquad \text{(XVI.11)}$$

In Equation XVI.11, W_i is the weight defined by Equation XVI.8 and Z_i is the quantity defined by

$$Z_i = \eta_{e,i} - \eta_{0,i}. \qquad \text{(XVI.12)}$$

That is, Z_i is the discrepancy between the observation and the circumstances that correspond to the reference values of the accelerations.

The goal is now to minimize F. Before doing this, we find it convenient to introduce some other quantities. Define:

$$a_i = A_i \sqrt{W_i} \ , \quad b_i = B_i \sqrt{W_i} \ , \quad z_i = Z_i \sqrt{W_i} \ .$$

$$(XVI.13)$$

In terms of the variables in Equations XVI.13, F is:

$$F = \tfrac{1}{2} \sum_i (a_i x + b_i y - z_i)^2 .$$

Thus z_i can be regarded as an "observed" quantity; it has no dimensions, each observation now has the same weight, and the estimated standard deviation of each z_i is 1. The pair of values of x and y that minimizes F can be written in terms of the matrix

$$\begin{pmatrix} \sum_i a_i^2 & \sum_i a_i b_i \\[2ex] \sum_i a_i b_i & \sum_i b_i^2 \end{pmatrix} .$$

$$(XVI.14)$$

The pair xy is given by:

$$\begin{pmatrix} x \\ y \end{pmatrix} = \begin{pmatrix} M_{xx} & M_{xy} \\ M_{yx} & M_{yy} \end{pmatrix} \begin{pmatrix} \sum_i a_i z_i \\ \sum_i b_i z_i \end{pmatrix} .$$

$$(XVI.15)$$

The matrix $\underline{\underline{M}}$ in Equation XVI.15 is the inverse of the matrix XVI.14, and M_{xx} , and so on, are its coefficients. $\underline{\underline{M}}$ is clearly symmetric.

In the remaining discussion, x and y will denote the values from Equation XVI.15.

Define the $\underline{i}$th residual r_i by

$$r_i = a_i x + b_i y - z_i \ , \qquad (XVI.16)$$

and define the standard deviation $\sigma(r_i)$ of the residuals by

$$\sigma(r_i) = (\Sigma r_i^2)^{\frac{1}{2}} . \qquad\qquad (XVI.17)$$

If the original error estimates $\sigma_{e,i}$ were well chosen, we should find $\sigma(r_i) = 1$. We get a useful test of the validity of the error estimates by comparing $\sigma(r_i)$ with unity.

We must expect that the values of x and y will contain some error. The best estimates σ_x and σ_y of the respective standard errors are

$$\sigma_x = \sigma(r_i)\sqrt{M_{xx}}, \quad \sigma_y = \sigma(r_i)\sqrt{M_{yy}} . \quad (XVI.18)$$

If $\sigma(r_i)$ is unity, as it should be, the errors reduce simply to $\sqrt{M_{xx}}$ and $\sqrt{M_{yy}}$.

CHAPTER XVII

RECORDS THAT REQUIRE SPECIAL HANDLING

1. The Lunar Eclipse of 755 Nov 23

The record designated 755 Nov 23 B,E requires special handling because it is the only record of a lunar eclipse that will be used in the astronomical analysis. The record was discussed in Section V.1.

The center of the eclipse occurred at $18^h 44^m$, according to the calculations of Oppolzer [1887]; this time is subject to some error because Oppolzer neglected the accelerations. The longitude of the sun at $18^h 44^m$ on 755 Nov 23 was 245°.03 according to Naval Observatory [1953]. Since the sun moves slowly, this value should be rather accurate, and 65°.03 should thus be a good estimate of the longitude of the moon at the center of the eclipse. Both longitudes are taken with respect to the mean equinox of date. The latitude of the moon with respect to the ecliptic of date [Naval Observatory, 1953] was -0°.23.

The coordinates just given are geocentric coordinates, while the observation was made from Durham, England. When we correct for parallax at Durham, we find that the apparent latitude and longitude of the moon become -0°.97 and 65°.42 respectively.

A drawing of that portion of the sky is given in Figure XVII.1. The calculated appearance of the moon is shown by the large circle. The four stars shown are β Taurus, 109 Taurus, 114 Taurus, and 123 Taurus. The star positions are calculated in the ecliptic coordinates of the ancient date by the method described in Section VIII.1 of AAO; they have been corrected for proper motions and abberation.

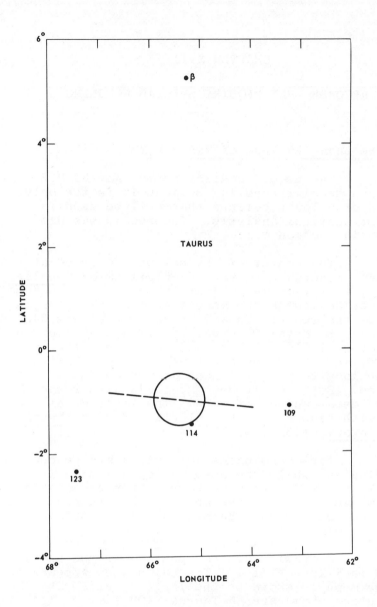

Figure XVII.1. Approximate configuration of the
moon and several stars near the center of the lunar
eclipse of 755 Nov 23. The stars are all in Taurus.
The dashed line shows the path of the center of
the moon. The coordinates are the mean ecliptic
coordinates of date.

It appears from Figure XVII.1 that the star occulted by the moon during the eclipse was 114 Taurus. It was almost surely not Aldebaran (α Taurus), contrary to the statement of Johnson [1889]. Aldebaran is about 15° to the west and several degrees to the south of the position shown for the moon.

The conclusion that the occulted star was 114 Taurus is not a happy one. The apparent visual magnitude of 114 Taurus [Smithsonian Astrophysical Observatory, 1966; this is also the source used for current stellar positions and proper motions] is only 4.8. A star of nearly the 5th magnitude seems too faint to merit the adjective "bright" that the record used. There should have been no difficulty in seeing a star this faint during the eclipse, but the record implies that the star could be seen within a degree of the full moon both before and after the eclipse. However, there is no reasonable alternative to 114 Taurus. The only "bright" star nearby is β Taurus, which is certainly too far to the north. 109 Taurus was also occulted by the moon, but probably some time before the eclipse; further, it is even fainter (magnitude 5.1) than 114 Taurus.

If we proceed on the assumption that the correct star is 114 Taurus, we see from Figure XVII.1 that we must decrease the longitudes assigned to the sun and moon by about 0°.25, to 244°.78 and 64°.78, respectively. Thus the value of the ephemeris time that we use must be decreased by about 6 hours. In view of the low precision of the observation, we may round times to the nearest hour and take the ephemeris time to be 14^h on 755 Nov 23. The a priori estimate of the standard deviation of the time from Section V.1 is 4^h. At 14^h, the longitude of the moon from Naval Observatory [1953] was 62°.32, and the uncertainty in this value arising from the uncertainty in time is 2°.0. Since the longitude inferred from the record is 64°.78, we need an acceleration of -135 ± 110 in addition to the value -22.44 used in the standard tables.

In sum, the record 755 Nov 23 B,E yields

the rounded estimate:

$$\dot{n}_M = -157 \pm 110 \quad ''/\text{cy}^2. \qquad (\text{XVII.1})$$

2. Identifications Held in Abeyance

The Byzantine chronicler Marcellinus [ca. 534] recorded an eclipse under the year that we call 512, but gave no details to help in the identification. Oppolzer [1887] shows that the path of the eclipse of 512 Jun 29 went almost exactly through Constantinople, and no other eclipse path near this time did so. For this reason, I used the identification 512 Jun 29 M,B for the record in Section XV.4.

However, we cannot rely upon the chronology being exact to the year. We also find from Oppolzer that the eclipse of 511 Jan 15 should have been fairly large in Constantinople at sunset, when a partial eclipse would be more prominent than usual. Thus, in Section XV.4, I decided to calculate the circumstances of the eclipse of 511 Jan 15 at Constantinople more carefully than Oppolzer did. If it should turn out to be large, I would conclude that Marcellinus' eclipse was not identifiable.

The circumstances were calculated for the four combinations of accelerations specified in Equation XVI.9 in Section XVI.6. The values found for the magnitude at sunset ranged from 0.3 to 0.75. Thus the eclipse of 511 Jan 15 was probably obvious in Constantinople, weather permitting, and Marcellinus' eclipse cannot be identified.

We had a similar problem in connection with the record 718 Jun 3 E,Sp in Chapter XIV. Isidorus [ca. 624] said that there was an eclipse from the 7th to the 9th hour of the day, and that stars could be seen, in a year that is near 720. In Oppolzer, the only eclipse path that crossed any part of Spain near this time was on 718 Jun 3, although the path of the eclipse of 720 Oct 6 came fairly close to the southern tip. Thus, the identification of the eclipse was made contingent upon careful calculation of the circumstances of 720 Oct 6.

For three of the four combinations of accelerations used, the path of totality of the eclipse of 720 Oct 6 crossed some part of Spain. The hour of maximum eclipse was slightly outside the hours stated by Isidorus, but not enough so to let us exclude the alternate date. Thus we cannot identify the eclipse in Isidorus.

These are the only records in the preceding chapters that I accepted on a contingent basis. In three other cases, I decided to calculate the circumstances even though the records clearly do not contain enough information to permit their use. The purpose of the calculations was to see if we could reach any further conclusions, with the understanding that the records would not be used even if we did.

The Irish sources Scotorum [ca. 1650] and Ulster [ca. 1498] contain references to an eclipse in the morning sometime around the year 592. We cannot even be sure whether the records refer to one or to two eclipses. I discussed these records under the designation 592 Mar 19 B,I in Section VII.1. The eclipse of 594 Jul 23 was the greatest eclipse in Ireland near 592, and it came about 1^h after sunrise there. However, we cannot exclude 590 Oct 4, 591 Sep 23, and 592 Mar 19 without careful calculation, and the eclipse cannot be identified safely.

The magnitude of each of the eclipses depends upon the accelerations used, and the values about to be given are the greatest found for some combination specified in Equation XVI.9. We find 0.75 for 590 Oct 4, 0.34 for 591 Sep 23, and 0.63 for 592 Mar 19. It is not too likely that the eclipse of 591 Sep 23 was observed in Ireland, but there is a reasonable probability that the others were. Thus we must continue to allow at least three possible dates for the eclipse or eclipses recorded in Scotorum and Ulster.

Icelandic/Norse sources record the eclipses of 1131 Mar 30 and 1194 Apr 22. All authorities that I have consulted say that the observations were made in Iceland (see Section XIII.2). Since the records say nothing about the place of observation, it is probable that the writers base their conclusions upon calculation, and Beckman [1912] explicitly does so for the eclipse of 1131. Since the use of calculation to find a place of observation is merely a variant of the "identification game", we cannot accept the conclusion that the observations were made in Iceland. However, it is instructive to make careful calculations of the circumstances of the two eclipses. In Section XIII.2 I decided to calculate the circumstances for Trondheim and for the center of Iceland.

It happens that the magnitudes are substantially independent of the accelerations used for both eclipses at both places. Both eclipses were total or nearly so in Iceland. The magnitude at Trondheim reached about 0.8 for 1131 Mar 30 and about 0.84 for 1194 Apr 22. Since Beckman [1912, p. 17] explicitly assumed that an eclipse is likely to be recorded if the magnitude exceeds 0.75, he had no basis for excluding Norway as a possible place of observation, and we have no basis either. The magnitude at Trondheim exceeded 0.75 for both eclipses for all combinations of accelerations tested.

3. Assignments of Time and Place for Which There Is No Local Eclipse

In Chapters VI through XV, I concluded that there were 379 records of solar eclipses with enough information to be useful, and I assigned dates and places of observation to these records. Two of the dates were tentative and were abandoned on the basis of the considerations of the preceding section; this leaves 377 records. Calculations of the circumstances yielded eclipses for all 377 assignments, for all combinations of accelerations specified in Equation XVI.9, with only six exceptions.

Only two of the exceptions, both for the same eclipse, pose a fundamental problem, and they will be discussed first.

I concluded that there were two independent records of the penumbral eclipse of 810 Jul 5. These were designated as the records 810 Jul 5 E,CE in Section IX.1, assigned to Gottweih, Austria and 810 Jul 5 E,G in Section XI.2, assigned to Lorsch, Germany. However, I was puzzled by the fact that all records that I found that seemed to refer to this eclipse, whether they were independent or not, had the day of the year wrong.

The calculations fail to yield even the smallest eclipse on 810 Jul 5 at either place. In fact, the eclipse was probably not visible in any place closer than Iceland. We must explain this puzzling circumstance.

It is unlikely that there was a mistake in the year. Two independent sources (Laurissenses [ca. 829] and Gottwicensis [ca. 1054]) give the year as 810. Further, Laurissenses accurately remarks that there were two lunar eclipses that same year, on June 21 and December 15.†

There is a piece of evidence that I did not appreciate until I realized that the eclipse of 810 Jul 5 was invisible in Europe. This is a letter to Charlemagne from a monk named Dungal that appears on pages 635-636 of vol. V of the Bouquet series (see Section X.1). Apparently Dungal is answering a letter from Charlemagne in which Charlemagne said that many people had told him about two eclipses of the sun in 810; Charlemagne did not remember reading that such a thing had ever happened before and he was writing to enquire about it. Charlemagne's letter does not seem to be extant.

Bouquet omits much of the text of the letter and the part that he gives has little astronomical information. However, Dungal twice refers to 810 as the

---

†The dates from Oppolzer are 810 Jun 20 and 810 Dec 14. Laurissenses used the dates of the following days in both instances.

year of the eclipses and he twice refers to it as
the year preceding the one in which he is writing.
He gives the dates of the eclipses as June 7 and
November 30 and he twice says that the second eclipse
was in the seventh month following the first. That
is, six lunations elapsed between the two eclipses.†

This information did not come from annals
and it is unlikely that there is confusion about the
dates. The year is given twice and is probably not a
scribal error. June might possibly have been written
in error for July except that Dungal twice says that
six months elapsed between the eclipses. Since the
second date is certainly correct, June must be the
month intended. Thus it must have been widely believed at the time that there had been an eclipse on 810
Jun 7. The following is perhaps the explanation:

The day 810 Jun 7 was an overcast "dark day"
of a kind that happens occasionally. Because it happened to be on "Luna 1", the onlookers concluded that
a solar eclipse was responsible; they knew in principle that a solar eclipse should come on the last day
of a moon (and 810 Nov 30 was in fact on "Luna 29"),
but, as Dungal pointed out, the ecclesiastical moon is
sometimes inexact. The confusion about the date that
we find in secondary sources came from repeated copying of the record into various annals.

A record that I designated at 1067 Feb 16
E,I appears in Section XII.2. The record actually
gives the date as 1066 Feb 26. There was an eclipse
on 1066 Mar 28, but it was not visible anywhere north
of the Equator. Two errors, both easy ones to make,
are needed in order to let us assume that the record
refers to 1067 Feb 16, when there was a penumbral
eclipse.

I assigned the place of observation to
be Benevento, near Naples. The calculations do not
yield a local eclipse there for any combination of

---

† Apparently Dungal thought that 810 Jun 7 was a true
astronomical new moon and hence that 810 Nov 30 began the seventh true lunation thereafter.

accelerations. However, the eclipse was almost
visible there and we need only to assume that the
record in fact originated from somewhere to the
north. In view of the political situation of the
time, it is plausible that the record came from
England or Normandy, although other points of origin are certainly possible. There is no safe way
to use this record, although it poses no fundamental
problem.

There was no local eclipse for the place
(Syria) assumed with the record 360 Aug 28 M,B
with 3 of the 4 combinations of accelerations used.
Here also there is no fundamental problem. There
was a fairly large eclipse with one pair of accelerations, and it may well be that the date and
place assigned are correct.[†] Further, the circumstance that prevents a local eclipse in some of
the calculated cases is the occurrence of sunrise.
We need only assume that the writer heard an account of the eclipse from places a short distance
to the east in order to explain the record for any
combination of accelerations.

I used this record in AAO, and a minor
modification of procedure would allow me to use
it here. Since there is a wealth of observational
material available, no serious consequences should
result from ignoring the record and, for simplicity,
I shall do so.

The combination of accelerations $x = +10$
and $y = -5$ yields no local eclipse for the records
402 Nov 11 E,Sp and 484 Jan 14 M,B, but all other
combinations used do yield a local eclipse. Thus
the calculations yield enough information to let
us evaluate the quantities $\eta_{0,i}$, A_i, and B_i (see
Equation XVI.10 in Section XVI.6) and we could
readily use the records so far as this problem is

[†] Since the considerations of this section and the
next one affect the estimation of the accelerations,
I am writing them in ignorance of the "best estimates" of the accelerations that will be found
later in this work.

concerned. However, I shall not use them for a separate reason that will appear in the next section.

The eclipses of 810 Jul 5 and 1067 Feb 16 were penumbral. No difficulty was found with any other penumbral eclipse.

4. Problems Caused by Sunrise or Sunset

There are a few records in which sunrise gives rise to a non-analytic type of constraint. When this happens, we find that maximum eclipse occurs between sunrise and sunset for some combinations of the accelerations. For other combinations, however, we calculate that the magnitude of the eclipse either falls steadily from sunrise on or that it is still rising when sunset occurs, depending upon the record. When this happens, the definition of η that was given in Section XVI.3 fails.

It would be possible to generalize the definition of η in a way that would include the occurrence of maximum eclipse at sunrise or sunset. If we did so, the laws that govern the dependence of η upon the accelerations would depend upon whether the time of maximum eclipse occurred during the day or at one end of the day. To put it another way, η would depend upon the accelerations in a non-analytic fashion. It would be possible to modify the procedures in a way that would allow us to overcome this problem,[†] just as we could have overcome some of the problems raised in the preceding sections. Since the problem records are few in comparison with the total number of records available, overcoming the special problems does not seem to be worth the trouble.

[†] Note that there would be no problem if maximum eclipse occurred at sunrise (or sunset) for all accelerations used. It is only the mixture of cases that causes a problem.

Problems with sunrise or sunset occur for the following records: 348 Oct 9 M,B, 402 Nov 11 E,Sp, 484 Jan 14 M,B, 841 Oct 18 E,F, 1186 Apr 21a B,E, 1186 Apr 21b B,E, 1230 May 14 B,E, and 1230 May 14 E,G. Two of these were also discussed in the preceding section. Thus we have dropped 14 records altogether out of the 379 records kept in Chapters VI through XV, leaving 365 records for use in the final inference of the accelerations. All but 4 of the 14 records could be used if we wished to take the trouble.

The record 1018 Apr 18 E,G, discussed in Section XI.2, describes the eclipse as occurring at sunset. For that reason I stated the intention of using the magnitude at sunset in the estimation of the accelerations. It develops that the maximum eclipse occurred slightly before sunset, for all combinations of the accelerations that were used, but that the sun set still eclipsed in accordance with the record. It should make only a trifling difference whether I use the maximum eclipse or the eclipse at sunset. Since it is simpler to use the maximum eclipse, I shall use it, in spite of the intention expressed earlier.

5. Records That Will Be Changed from "Partial" to "Unspecified"

In order to use a record in which the eclipse was definitely partial, it is necessary to know whether the observer was to the north or the south of the path. In some cases we can infer the direction from a description of the shape of the visible part of the sun. It is risky to do so, however, because the description may be ambiguous, and it is always easy for the recorder to make an error that cannot be detected.

On the other hand, reasonable or even unreasonable values of the accelerations cannot change the direction from the path to the observer if the distance is fairly large. Thus the safest way to assign the direction is on the basis of calculation, provided that all calculations give

the same answer. The direction shows up in the
calculations as the sign of η .

With three records in which the eclipse
was stated to be partial, η is positive for some
combinations of the accelerations and negative for
others. With these, the only safe course is to
take $\eta = 0$ as the best estimate of η that the
record affords. This is equivalent to taking the
nature of the eclipse to be unspecified. The three
records are 458 May 28 E,Sp, 1133 Aug 2h E,CE, and
1147 Oct 26b E,G.

The magnitudes originally estimated from
the last two records are 0.97 and 0.9, respectively.
Thus it is not surprising that we should be unable
to resolve the direction for them. The magnitude
estimated for 458 May 28 E,Sp was only 0.8. This
is such an old record that the accelerations can
change the geometry appreciably, and we cannot be
sure of the direction even though the eclipse was
far from total.

6. Was the Corona Recorded in Ancient or Medieval
Times?

In Section V.1 I called attention to the
fact that an annalist at Gembloux [Gemblacensis,
ca. 1148] used the phrase circumfusus caligine
with respect to the eclipses of 1140 Mar 20 and
1147 Oct 26. I suggested "surrounded by a mist"
as a likely translation of the phrase, and I suggested that this might be a reference to the corona.
I also pointed out that the eclipse of 1147 Oct 26
was annular rather than total according to Oppolzer
[1887], and that I would return to this point after
calculating the circumstances of the eclipses.

The calculations show that Oppolzer was
correct and that the eclipse of 1147 Oct 26 was
annular by a safe margin. It is doubtful that
its magnitude exceeded 0.97 anywhere, and this
should make observation of the corona impossible.

Thus we must conclude that the annalist

did not see the corona on 1147 Oct 26 and that the
phrase had some other meaning for that eclipse.
Since he used the same phrase for 1140 Mar 20, we
cannot infer that he recorded the corona for this
eclipse either, even though it was total somewhere.

Therefore there is no clear reference to
the corona in any ancient or medieval record that
I have found. The most likely reference is perhaps
the remark by Plutarch that was mentioned in Section
V.1, but the meaning of Plutarch's remark is far
from certain. He may merely have meant the small
rim of the sun that remains visible during an annular
eclipse.

CHAPTER XVIII

CONCLUSIONS AND SUMMARY OF THE
ASTRONOMICAL ANALYSIS

1. The Coefficients of Condition

It is desirable to provide some information about the calculations performed for each individual eclipse record. This will allow an interested reader to test the accuracy of the calculations that I have performed. It will also facilitate any independent use that he may wish to make of the records. There are several ways of providing useful information about the calculations.

Each observation that is to be used contributes one term to the summation in the right member of Equation XVI.11. The coefficients W_i, A_i, B_i, and Z_i provide a highly condensed summary of the calculations for each eclipse as well as of the properties of the record itself. Accordingly, I shall give the coefficients in Equation XVI.11 for each record. These coefficients, which I call the coefficients of condition, are listed in Tables XVIII.1 through XVIII.9. The observations are listed chronologically. Table XVIII.1 contains all observations made before 500. After 500, a table is devoted to each century through the 13th.

I have listed $W_i^{\frac{1}{2}}$ rather than W_i itself in the tables. There are two reasons for this. First, $W_i^{\frac{1}{2}}$ has a smaller range than W_i and it is thus easier to tabulate. Second, if one wishes to use an observation in estimating the accelerations, he must start by calculating a_i, b_i, and z_i as defined in Equations XVI.13, and for this he needs $W_i^{\frac{1}{2}}$. The reader who wishes to use the observations with his own assignment of weights rather than mine can readily do so with the aid of Tables XVIII.1 through XVIII.9.

TABLE XVIII.1

COEFFICIENTS OF CONDITION FOR ECLIPSES
BEFORE 500

Designation			$100A_i$	$100B_i$	$100Z_i$	$W_i^{\frac{1}{2}}$
118	Sep 3	E, I	-0.148	+0.295	+ 8.92	5.9
346	Jun 6a	M, B	+0.495	-0.760	-10.90	17.9
393	Nov 20	E, I	-0.390	+0.660	+ 5.05	5.9
418	Jul 19a	E, I	+0.025	0.000	- 3.00	9.3
418	Jul 19	E, Sp	+0.280	-0.410	- 4.50	7.7
418	Jul 19a	M, B	-0.242	+0.415	- 1.68	40.0
418	Jul 19c	M, B	-0.242	+0.415	- 1.68	12.9
447	Dec 23	B, W	+0.258	-0.385	- 7.22	4.1
447	Dec 23	E, Sp	+0.298	-0.435	- 1.88	17.2
458	May 28	E, Sp	+0.598	-0.925	+ 6.72	17.2
464	Jul 20	E, Sp	+0.310	-0.520	- 1.31	20.8

TABLE XVIII.2

COEFFICIENTS OF CONDITION FOR ECLIPSES
IN THE 6TH CENTURY

Designation			$100A_i$	$100B_i$	$100Z_i$	$W_i^{\frac{1}{2}}$
512 Jun 29	E,I		+0.440	-0.690	-13.65	12.9
534 Apr 29	E,I		+0.315	-0.470	-18.10	5.9
538 Feb 15	E,I		+0.148	-0.205	-13.18	2.9
540 Jun 20	E,I		+0.292	-0.495	- 5.92	3.9
563 Oct 3	E,F		-0.118	+0.235	+13.50	12.9
590 Oct 4	E,F		-0.555	+0.850	+12.21	18.2
590 Oct 4	M,B		-0.578	+0.895	+ 4.22	35.7
592 Mar 19	E,F		+0.358	-0.545	+ 0.79	8.6

TABLE XVIII.3

COEFFICIENTS OF CONDITION FOR ECLIPSES
IN THE 7TH CENTURY

Designation			$100A_i$	$100B_i$	$100Z_i$	$W_i^{\frac{1}{2}}$
603 Aug	12	E,F	−0.300	+0.500	− 3.85	12.2
644 Nov	5	M,B	−0.440	+0.660	+ 2.90	12.9
655 Apr	12	E,Sp	+0.128	−0.225	− 0.42	12.2
664 May	1	B,E	−0.090	+0.170	+ 2.75	11.6
664 May	1a	B,I	−0.048	+0.105	+ 3.02	11.6
664 May	1b	B,I	−0.048	+0.105	+ 3.02	3.7
688 Jul	3	B,I	+0.140	−0.240	− 1.46	4.7
693 Oct	5a	M,B	−0.235	+0.400	0.00	28.3
693 Oct	5b	M,B	−0.235	+0.400	0.00	28.3

TABLE XVIII.4

COEFFICIENTS OF CONDITION FOR ECLIPSES
IN THE 8TH CENTURY

Designation	$100A_i$	$100B_i$	$100Z_i$	$W_i^{\frac{1}{2}}$
733 Aug 14a B,E	+0.050	-0.050	- 1.85	11.6
733 Aug 14b B,E	+0.045	-0.050	- 4.70	18.6
753 Jan 9 B,E	-0.055	+0.060	- 5.95	10.0
753 Jan 9 B,I	-0.098	+0.125	- 4.58	7.3
755 Nov 23 B,E	+0.909	0*	- 1.14	1
760 Aug 15 E,CE	-0.280	+0.430	+ 9.40	5.8
760 Aug 15 M,B	-0.220	+0.330	+ 7.00	12.9
764 Jun 4 B,I	+0.130	-0.220	- 1.70	11.6
764 Jun 4 E,CE	-0.028	+0.025	+ 9.02	5.8
764 Jun 4 E,F	+0.078	-0.145	+ 9.58	12.9
764 Jun 4 E,G	+0.035	-0.080	+ 6.90	12.9
787 Sep 16 E,CE	-0.065	+0.130	-13.40	5.8
787 Sep 16a E,G	-0.035	+0.080	-14.60	18.2
787 Sep 16b E,G	-0.045	+0.100	-18.10	12.9
787 Sep 16 M,B	-0.155	+0.270	- 5.65	35.7

---

*This is a simultaneous lunar eclipse and occultation.

TABLE XVIII.5

COEFFICIENTS OF CONDITION FOR ECLIPSES
IN THE 9TH CENTURY

Designation	$100A_i$	$100B_i$	$100Z_i$	$W_i^{\frac{1}{2}}$
807 Feb 11 B,W	+0.048	−0.095	+ 3.42	8.1
807 Feb 11a E,CE	+0.168	−0.285	+10.48	12.9
807 Feb 11b E,CE	+0.178	−0.305	+11.08	5.8
807 Feb 11 E,F	+0.095	−0.170	+ 5.35	12.9
807 Feb 11 E,G	+0.135	−0.240	+ 7.75	18.2
807 Feb 11 E,I	+0.185	−0.330	+14.85	12.9
809 Jul 16 B,E	+0.125	−0.180	+12.55	11.6
810 Nov 30a E,CE	−0.238	+0.365	+ 5.48	12.9
810 Nov 30b E,CE	−0.220	+0.340	+ 3.95	5.8
810 Nov 30 E,G	−0.248	+0.385	+ 6.42	18.2
812 May 14a E,G	+0.178	−0.265	−22.72	12.9
812 May 14b E,G	+0.162	−0.245	−23.88	12.9
812 May 14 M,B	+0.022	−0.015	− 4.82	26.3
813 May 4 M,B	+0.118	−0.175	+ 2.18	18.2
818 Jul 7a E,G	+0.198	−0.335	+23.08	12.9
818 Jul 7b E,G	+0.192	−0.325	+22.38	12.9
840 May 5a E,F	+0.095	−0.170	− 4.55	12.9
840 May 5b E,F	+0.092	−0.165	− 0.58	40.0
840 May 5c E,F	+0.065	−0.120	− 4.80	18.2
840 May 5d E,F	+0.125	−0.210	− 1.60	18.2

TABLE XVIII.5 (Continued)

Designation	$100A_i$	$100B_i$	$100Z_i$	$W_i^{\frac{1}{2}}$
840 May 5a E,G	+0.060	-0.120	- 2.65	8.1
840 May 5b E,G	+0.060	-0.120	- 4.75	8.1
840 May 5c E,G	+0.052	-0.105	- 6.92	40.0
840 May 5d E,G	+0.068	-0.125	- 9.08	83.3
840 May 5e E,G	+0.042	-0.085	- 4.02	58.9
840 May 5a E,I	+0.040	-0.090	+ 1.30	12.9
840 May 5b E,I	+0.042	-0.095	+ 2.88	35.7
840 May 5c E,I	+0.060	-0.110	+ 0.50	58.9
865 Jan 1 B,I	+0.058	-0.115	- 0.22	11.6
878 Oct 29 B,E	-0.070	+0.130	+ 0.70	25.0
878 Oct 29 B,I	-0.098	+0.175	+ 1.92	11.6
878 Oct 29 B,W	-0.092	+0.165	+ 2.22	8.1
878 Oct 29a E,F	-0.075	+0.140	+ 3.95	7.7
878 Oct 29b E,F	-0.058	+0.115	+ 1.58	18.2
878 Oct 29c E,F	-0.045	+0.090	+ 3.00	40.0
878 Oct 29d E,F	-0.062	+0.115	+ 3.22	18.2
878 Oct 29e E,F	-0.052	+0.105	+ 1.38	28.3
878 Oct 29f E,F	-0.058	+0.115	+ 1.38	12.9
878 Oct 29a E,G	-0.022	+0.055	+ 2.42	12.9
878 Oct 29b E,G	-0.025	+0.060	+ 1.65	12.9

TABLE XVIII.5 (Concluded)

Designation	$100A_i$	$100B_i$	$100Z_i$	$W_i^{\frac{1}{2}}$
878 Oct 29c E,G	-0.020	+0.050	+ 0.60	83.3
878 Oct 29d E,G	-0.038	+0.085	+ 1.12	18.2
885 Jun 16 B,I	+0.200	-0.310	+ 1.80	15.2
885 Jun 16 B,SM	+0.175	-0.270	+ 0.25	12.9
891 Aug 8 E,CE	-0.122	+0.175	- 0.42	12.9
891 Aug 8 E,F	-0.062	+0.085	- 4.52	12.9
891 Aug 8a M,B	-0.310	+0.470	- 0.05	34.6
891 Aug 8b M,B	-0.310	+0.470	- 0.05	34.6
891 Aug 8c M,B	-0.310	+0.470	- 0.05	17.9

TABLE XVIII.6

COEFFICIENTS OF CONDITION FOR ECLIPSES
IN THE 10TH CENTURY

Designation	$100A_i$	$100B_i$	$100Z_i$	$W_i^{\frac{1}{2}}$
939 Jul 19 E,CE	+0.135	−0.200	− 7.90	18.2
939 Jul 19 E,G	+0.118	−0.175	+ 4.93	10.9
939 Jul 19a E,I	+0.140	−0.210	−16.76	10.9
961 May 17a E,G	+0.205	−0.320	− 3.00	18.2
961 May 17b E,G	+0.205	−0.320	− 3.00	18.2
968 Dec 22a E,BN	−0.082	+0.145	− 2.68	18.6
968 Dec 22b E,BN	−0.082	+0.145	− 2.68	4.1
968 Dec 22 E,CE	−0.080	+0.140	− 1.70	18.2
968 Dec 22 E,F	−0.105	+0.180	+ 0.05	40.0
968 Dec 22 E,G	−0.080	+0.140	− 1.80	12.9
968 Dec 22a E,I	−0.068	+0.125	+ 1.52	12.9
968 Dec 22b E,I	−0.070	+0.120	+ 1.85	12.9
968 Dec 22a M,B	+0.025	−0.030	+ 0.30	35.7
968 Dec 22b M,B	+0.025	−0.030	+ 0.30	28.3
968 Dec 22c M,B	+0.025	−0.030	+ 0.30	18.2
990 Oct 21a E,G	−0.145	+0.250	+ 7.30	18.2
990 Oct 21b E,G	−0.145	+0.250	+ 7.50	18.2

TABLE XVIII.7

COEFFICIENTS OF CONDITION FOR ECLIPSES
IN THE 11TH CENTURY

Designation	$100A_i$	$100B_i$	$100Z_i$	$W_i^{\frac{1}{2}}$
1009 Mar 29a E,BN	+0.058	−0.105	+28.02	18.2
1009 Mar 29b E,BN	+0.055	−0.100	+27.60	12.9
1018 Apr 18 E,BN	−0.068	+0.125	−27.78	5.8
1018 Apr 18 E,G	−0.082	+0.145	− 3.50	18.2
1023 Jan 24 B,E	+0.158	−0.245	− 0.48	18.2
1023 Jan 24 B,I	+0.138	−0.215	− 3.32	11.6
1023 Jan 24 E,BN	+0.182	−0.275	+ 3.22	12.9
1023 Jan 24 E,F	+0.195	−0.310	+ 5.00	12.9
1030 Aug 31 B,I	−0.128	+0.215	+14.28	11.6
1030 Aug 31 E,Sc	−0.090	+0.150	+ 2.40	9.3
1033 Jun 29 E,BN	+0.052	−0.065	− 7.72	12.9
1033 Jun 29a E,F	+0.048	−0.065	+ 4.00	23.8
1033 Jun 29b E,F	+0.028	−0.035	− 4.72	18.2
1033 Jun 29c E,F	+0.048	−0.065	− 2.38	17.9
1033 Jun 29d E,F	+0.052	−0.075	− 7.38	83.3
1033 Jun 29e E,F	+0.058	−0.085	− 2.18	18.2
1033 Jun 29a E,G	+0.018	−0.015	− 9.52	18.2
1033 Jun 29b E,G	+0.025	−0.030	− 2.65	18.2
1033 Jun 29c E,G	+0.022	−0.025	− 4.38	18.2
1033 Jun 29a E,I	−0.010	+0.030	+ 8.00	25.2

TABLE XVIII.7 (Continued)

Designation	$100A_i$	$100B_i$	$100Z_i$	$W_i^{\frac{1}{2}}$
1033 Jun 29b E,I	-0.010	+0.030	+ 3.67	26.3
1033 Jun 29c E,I	+0.025	-0.030	+ 0.85	12.9
1037 Apr 18 E,BN	+0.190	-0.290	+ 8.10	12.9
1037 Apr 18a E,F	+0.198	-0.305	+ 8.89	28.3
1037 Apr 18b E,F	+0.198	-0.305	+ 7.59	22.1
1039 Aug 22 E,BN	-0.200	+0.310	- 4.60	12.9
1039 Aug 22a E,F	-0.220	+0.340	- 4.45	18.2
1039 Aug 22b E,F	-0.230	+0.350	- 0.60	18.2
1039 Aug 22a E,G	-0.225	+0.340	- 6.60	8.1
1039 Aug 22b E,G	-0.212	+0.325	- 8.82	18.2
1044 Nov 22a E,F	-0.115	+0.200	- 2.60	18.2
1044 Nov 22b E,F	-0.112	+0.195	- 2.88	12.9
1044 Nov 22c E,F	-0.112	+0.195	- 2.68	18.2
1044 Nov 22a E,G	-0.082	+0.145	- 7.52	18.2
1044 Nov 22b E,G	-0.080	+0.140	- 7.20	18.2
1044 Nov 22 E,I	-0.098	+0.165	- 2.02	18.2
1087 Aug 1 E,CE	+0.062	-0.095	-33.28	12.9
1093 Sep 23a B,E	-0.152	+0.235	+11.32	5.8
1093 Sep 23b B,E	-0.142	+0.215	+ 7.62	5.8
1093 Sep 23 E,BN	-0.172	+0.265	+10.18	18.2

TABLE XVIII.7 (Concluded)

Designation	$100A_i$	$100B_i$	$100Z_i$	$W_i^{\frac{1}{2}}$
1093 Sep 23a E,CE	-0.208	+0.325	+ 5.92	8.1
1093 Sep 23b E,CE	-0.198	+0.305	+ 3.32	18.2
1093 Sep 23c E,CE	-0.200	+0.310	+10.20	40.0
1093 Sep 23a E,G	-0.180	+0.280	+ 4.05	18.2
1093 Sep 23b E,G	-0.200	+0.310	+ 7.70	18.2
1093 Sep 23c E,G	-0.198	+0.305	+ 9.02	12.9
1093 Sep 23d E,G	-0.198	+0.305	+ 9.02	18.2
1093 Sep 23e E,G	-0.192	+0.295	+ 6.02	18.2
1098 Dec 25a E,G	+0.080	-0.110	-11.65	18.2

TABLE XVIII.8

COEFFICIENTS OF CONDITION FOR ECLIPSES
IN THE 12TH CENTURY

Designation	$100A_i$	$100B_i$	$100Z_i$	$W_i^{\frac{1}{2}}$
1109 May 31 E,BN	+0.045	−0.070	+12.80	18.2
1113 Mar 19 M,HL	+0.070	−0.110	+ 1.42	18.2
1118 May 22 E,BN	+0.110	−0.180	+25.70	18.2
1124 Aug 11a B,E	−0.070	+0.130	+19.40	5.8
1124 Aug 11b B,E	−0.070	+0.120	+19.05	5.8
1124 Aug 11c B,E	−0.068	+0.125	+19.28	5.8
1124 Aug 11 E,BN	−0.082	+0.145	+ 0.23	10.9
1124 Aug 11 E,CE	−0.102	+0.175	+15.62	18.2
1124 Aug 11a E,G	−0.088	+0.155	+15.98	18.2
1124 Aug 11b E,G	−0.082	+0.145	+13.58	18.2
1124 Aug 11 M,HL	−0.148	+0.255	+ 6.83	10.9
1133 Aug 2a B,E	−0.120	+0.190	+ 5.10	20.8
1133 Aug 2b B,E	−0.115	+0.170	+ 6.85	28.3
1133 Aug 2c B,E	−0.112	+0.175	+ 4.42	24.4
1133 Aug 2d B,E	−0.120	+0.190	+ 5.10	28.3
1133 Aug 2 B,I	−0.095	+0.140	+ 6.50	11.6
1133 Aug 2 B,SM	−0.100	+0.150	+ 1.60	18.6
1133 Aug 2a E,BN	−0.138	+0.215	+ 4.78	40.0
1133 Aug 2b E,BN	−0.135	+0.210	+ 5.05	40.0
1133 Aug 2c E,BN	−0.132	+0.205	+ 5.52	18.2

TABLE XVIII.8 (Continued)

Designation	$100A_i$	$100B_i$	$100Z_i$	$W_i^{\frac{1}{2}}$
1133 Aug 2d E,BN	-0.138	+0.215	+ 4.28	40.0
1133 Aug 2e E,BN	-0.130	+0.200	+ 2.25	83.3
1133 Aug 2a E,CE	-0.165	+0.260	+ 1.15	31.2
1133 Aug 2b E,CE	-0.165	+0.260	+ 3.20	83.3
1133 Aug 2c E,CE	-0.168	+0.265	+ 2.42	83.3
1133 Aug 2f E,CE	-0.162	+0.255	+ 2.32	18.6
1133 Aug 2g E,CE	-0.158	+0.255	- 0.42	12.9
1133 Aug 2h E,CE	-0.155	+0.250	- 0.35	31.2
1133 Aug 2i E,CE	-0.165	+0.260	- 1.25	18.2
1133 Aug 2a E,F	-0.130	+0.200	+ 5.55	40.0
1133 Aug 2b E,F	-0.138	+0.215	+ 6.68	18.2
1133 Aug 2c E,F	-0.135	+0.210	+ 6.15	40.0
1133 Aug 2d E,F	-0.150	+0.240	+ 9.75	83.3
1133 Aug 2e E,F	-0.132	+0.205	+ 8.32	18.2
1133 Aug 2f E,F	-0.132	+0.205	+ 8.32	8.1
1133 Aug 2b E,F	-0.138	+0.215	+ 1.18	40.0
1133 Aug 2c E,G	-0.155	+0.250	+ 3.05	12.9
1133 Aug 2d E,G	-0.155	+0.250	+ 5.05	12.9
1133 Aug 2e E,G	-0.140	+0.220	+ 3.75	83.3
1133 Aug 2f E,G	-0.150	+0.240	+ 2.05	12.9

TABLE XVIII.8 (Continued)

Designation	$100A_i$	$100B_i$	$100Z_i$	$W_i^{\frac{1}{2}}$
1133 Aug 2g E,G	-0.142	+0.225	+ 0.68	40.0
1133 Aug 2h E,G	-0.148	+0.225	+ 4.12	28.3
1133 Aug 2i E,G	-0.155	+0.240	+ 2.70	12.9
1133 Aug 2j E,G	-0.155	+0.240	+ 2.70	83.3
1133 Aug 2k E,G	-0.152	+0.235	+ 2.72	83.3
1133 Aug 2ℓ E,G	-0.148	+0.235	- 0.22	12.9
1133 Aug 2m E,G	-0.148	+0.225	+ 4.22	12.9
1140 Mar 20a B,E	+0.065	-0.110	+ 0.65	26.3
1140 Mar 20b B,E	+0.068	-0.115	+ 0.38	28.3
1140 Mar 20c B,E	+0.062	-0.105	+ 1.78	28.3
1140 Mar 20d B,E	+0.068	-0.115	- 0.42	18.2
1140 Mar 20e B,E	+0.065	-0.110	+ 0.65	18.6
1140 Mar 20f B,E	+0.062	-0.115	+ 0.82	18.2
1140 Mar 20g B,E	+0.068	-0.115	- 0.42	12.9
1140 Mar 20 B,SM	+0.062	-0.105	- 5.02	12.9
1140 Mar 20 B,W	+0.072	-0.125	- 1.02	12.9
1140 Mar 20a E,BN	+0.058	-0.105	+ 3.72	83.3
1140 Mar 20b E,BN	+0.060	-0.100	+ 3.35	35.7
1140 Mar 20 E,F	+0.060	-0.110	+ 2.50	12.9
1140 Mar 20 E,G	+0.055	-0.090	+ 8.65	18.2

TABLE XVIII.8 (Continued)

Designation	$100A_i$	$100B_i$	$100Z_i$	$W_i^{\frac{1}{2}}$
1140 Mar 20a E,Sc	+0.032	−0.065	− 1.52	18.2
1140 Mar 20b E,Sc	+0.040	−0.070	− 2.50	21.4
1140 Mar 20c E,Sc	+0.040	−0.070	− 2.50	21.4
1147 Oct 26a E,BN	−0.165	+0.250	+ 4.80	83.3
1147 Oct 26b E,BN	−0.165	+0.250	+ 4.20	26.3
1147 Oct 26c E,BN	−0.155	+0.240	+ 2.85	18.2
1147 Oct 26 E,CE	−0.182	+0.275	+ 6.02	12.9
1147 Oct 26a E,F	−0.165	+0.250	+ 7.80	18.2
1147 Oct 26b E,F	−0.168	+0.255	+ 6.58	18.2
1147 Oct 26c E,F	−0.178	+0.275	+ 9.68	18.2
1147 Oct 26b E,G	−0.160	+0.240	− 1.20	26.3
1147 Oct 26c E,G	−0.162	+0.245	+ 3.68	18.2
1147 Oct 26d E,G	−0.168	+0.255	+ 3.68	18.2
1147 Oct 26e E,G	−0.162	+0.245	+ 2.98	18.2
1147 Oct 26f E,G	−0.178	+0.275	+ 5.28	12.9
1147 Oct 26 M,B	−0.182	+0.285	+11.91	10.9
1153 Jan 26 E,CE	+0.170	−0.250	+ 1.25	12.9
1153 Jan 26a E,F	+0.138	−0.205	− 7.92	18.2
1153 Jan 26b E,F	+0.132	−0.195	− 9.58	18.2
1153 Jan 26a E,G	+0.160	−0.240	− 2.70	18.2

TABLE XVIII.8 (Continued)

Designation	$100A_i$	$100B_i$	$100Z_i$	$W_i^{\frac{1}{2}}$
1153 Jan 26b E,G	+0.162	−0.245	− 1.18	18.2
1153 Jan 26c E,G	+0.148	−0.225	+ 0.26	26.3
1153 Jan 26d E,G	+0.142	−0.215	− 5.92	18.2
1153 Jan 26 E,I	+0.178	−0.265	+ 1.38	8.1
1178 Sep 13a B,E	−0.085	+0.140	− 6.55	18.2
1178 Sep 13b B,E	−0.085	+0.150	− 3.30	26.3
1178 Sep 13c B,E	−0.082	+0.145	− 6.52	12.9
1178 Sep 13 B,SM	−0.075	+0.120	−10.15	26.3
1178 Sep 13a B,W	−0.078	+0.135	− 5.92	8.0
1178 Sep 13b B,W	−0.082	+0.135	− 5.18	8.1
1178 Sep 13a E,F	−0.100	+0.170	− 1.65	12.9
1178 Sep 13b E,F	−0.092	+0.155	− 6.78	18.2
1178 Sep 13c E,F	−0.098	+0.165	+ 1.92	26.3
1178 Sep 13d E,F	−0.100	+0.170	− 4.15	18.2
1178 Sep 13e E,F	−0.105	+0.170	− 0.10	35.7
1178 Sep 13a E,I	−0.122	+0.205	− 2.08	18.2
1178 Sep 13b E,I	−0.120	+0.200	+16.53	10.9
1178 Sep 13 E,Sc	−0.080	+0.130	−15.45	12.2
1185 May 1a B,E	+0.062	−0.095	+ 9.72	18.2
1185 May 1b B,E	+0.060	−0.090	− 8.31	10.9

TABLE XVIII.8 (Continued)

Designation	$100A_i$	$100B_i$	$100Z_i$	$W_i^{\frac{1}{2}}$
1185 May 1c B,E	+0.060	-0.090	+ 9.40	18.2
1185 May 1d B,E	+0.065	-0.100	+10.00	8.1
1185 May 1a B,SM	+0.062	-0.095	+ 2.88	40.0
1185 May 1b B,SM	+0.070	-0.110	+ 4.40	12.9
1185 May 1a B,W	+0.078	-0.115	+ 7.52	8.1
1185 May 1b B,W	+0.072	-0.115	+ 8.52	8.1
1185 May 1a E,F	+0.070	-0.110	+16.30	12.9
1185 May 1 E,Sc	+0.030	-0.040	+ 6.35	9.3
1187 Sep 4 B,E	-0.128	+0.195	- 1.23	10.9
1187 Sep 4 E,BN	-0.135	+0.210	+16.85	18.2
1187 Sep 4a E,CE	-0.158	+0.245	+11.12	18.2
1187 Sep 4b E,CE	-0.158	+0.245	+13.32	18.2
1187 Sep 4c E,CE	-0.160	+0.240	+12.20	28.3
1187 Sep 4a E,G	-0.152	+0.235	+14.22	18.2
1187 Sep 4b E,G	-0.158	+0.245	+14.02	18.2
1187 Sep 4c E,G	-0.140	+0.210	+11.05	18.2
1187 Sep 4d E,G	-0.142	+0.225	+10.38	18.2
1187 Sep 4e E,G	-0.140	+0.210	+14.15	18.2
1187 Sep 4f E,G	-0.158	+0.245	+14.32	18.2
1187 Sep 4a E,Sc	-0.122	+0.185	+ 6.88	12.2

TABLE XVIII.8 (Continued)

Designation	$100A_i$	$100B_i$	$100Z_i$	$W_i^{\frac{1}{2}}$
1187 Sep 4 M,HL	-0.190	+0.300	+ 8.45	18.2
1191 Jun 23a B,E	-0.015	+0.020	-14.25	7.7
1191 Jun 23b B,E	-0.008	+0.005	-14.83	7.7
1191 Jun 23c B,E	-0.015	+0.010	+ 4.35	8.1
1191 Jun 23d B,E	-0.020	+0.020	+ 2.75	28.3
1191 Jun 23e B,E	-0.010	+0.010	+ 4.50	18.2
1191 Jun 23 B,SM	-0.010	0.000	- 2.35	18.2
1191 Jun 23 B,W	-0.002	-0.005	+ 3.78	8.1
1191 Jun 23a E,BN	-0.030	+0.040	+ 5.15	18.2
1191 Jun 23b E,BN	-0.035	+0.050	+ 4.75	18.2
1191 Jun 23a E,CE	-0.072	+0.115	+ 6.12	18.2
1191 Jun 23b E,CE	-0.068	+0.105	+ 7.48	18.2
1191 Jun 23c E,CE	-0.070	+0.100	+ 6.65	12.9
1191 Jun 23d E,CE	-0.065	+0.100	+ 6.60	18.2
1191 Jun 23a E,F	-0.025	+0.030	+ 3.57	31.2
1191 Jun 23b E,F	-0.010	+0.010	+10.20	18.2
1191 Jun 23c E,F	-0.020	+0.020	+ 9.65	18.2
1191 Jun 23d E,F	-0.028	+0.035	+ 5.32	23.8
1191 Jun 23e E,F	-0.040	+0.050	+ 7.00	18.2
1191 Jun 23a E,G	-0.052	+0.075	+ 7.78	12.9

TABLE XVIII.8 (Concluded)

Designation	$100A_i$	$100B_i$	$100Z_i$	$W_i^{\frac{1}{2}}$
1191 Jun 23b E,G	-0.060	+0.090	+ 7.40	26.3
1191 Jun 23c E,G	-0.060	+0.090	+ 6.90	12.9
1191 Jun 23d E,G	-0.040	+0.060	+ 4.05	18.2
1191 Jun 23e E,G	-0.048	+0.065	- 9.07	7.7
1191 Jun 23f E,G	-0.045	+0.060	+ 5.80	58.9
1191 Jun 23a E,I	-0.068	+0.105	+11.38	12.9
1191 Jun 23b E,I	-0.068	+0.105	+13.08	12.9
1194 Apr 22 B,E	+0.010	-0.030	+ 6.14	10.9

TABLE XVIII.9

COEFFICIENTS OF CONDITION FOR ECLIPSES
IN THE 13TH CENTURY

Designation	$100A_i$	$100B_i$	$100Z_i$	$W_i^{\frac{1}{2}}$
1207 Feb 28a B,E	+0.132	−0.205	−11.08	12.9
1207 Feb 28b B,E	+0.132	−0.205	−10.58	18.2
1207 Feb 28c B,E	+0.132	−0.205	−10.78	12.9
1207 Feb 28 E,BN	+0.142	−0.225	− 7.98	18.2
1207 Feb 28 E,F	+0.145	−0.230	− 8.85	12.9
1207 Feb 28a E,G	+0.158	−0.245	− 1.92	18.2
1207 Feb 28b E,G	+0.160	−0.250	− 0.25	18.2
1207 Feb 28c E,G	+0.142	−0.225	− 5.58	18.2
1207 Feb 28d E,G	+0.160	−0.250	− 0.05	18.2
1230 May 14 E,Sc	+0.040	−0.070	+ 0.55	12.2
1232 Oct 15 E,G	−0.045	+0.080	− 1.11	10.9
1236 Aug 3 E,Sc	+0.012	−0.015	+ 3.32	16.4
1239 Jun 3a B,E	+0.058	−0.085	−15.98	18.2
1239 Jun 3b B,E	+0.058	−0.085	−16.78	18.2
1239 Jun 3a E,CE	+0.022	−0.035	− 6.98	18.2
1239 Jun 3b E,CE	+0.025	−0.030	− 7.35	18.2
1239 Jun 3a E,G	+0.032	−0.045	+ 9.80	10.9
1239 Jun 3b E,G	+0.020	−0.040	− 8.10	18.2
1239 Jun 3a E,I	+0.028	−0.035	− 3.22	12.9
1239 Jun 3b E,I	+0.032	−0.045	− 1.88	58.9

TABLE XVIII.9 (Continued)

Designation	$100A_i$	$100B_i$	$100Z_i$	$W_i^{\frac{1}{2}}$
1241 Oct 6a B,E	−0.122	+0.195	+ 8.48	18.2
1241 Oct 6b B,E	−0.122	+0.195	− 9.04	10.9
1241 Oct 6c B,E	−0.120	+0.190	+ 9.35	18.2
1241 Oct 6d B,E	−0.122	+0.195	+ 8.98	12.9
1241 Oct 6 E,BN	−0.125	+0.200	+ 6.05	18.2
1241 Oct 6a E,CE	−0.125	+0.200	+ 2.10	40.0
1241 Oct 6b E,CE	−0.122	+0.195	+ 1.18	40.0
1241 Oct 6c E,CE	−0.122	+0.195	+ 1.28	83.3
1241 Oct 6d E,CE	−0.125	+0.190	+ 1.50	83.3
1241 Oct 6e E,CE	−0.115	+0.180	− 0.15	35.7
1241 Oct 6a E,G	−0.125	+0.200	+ 3.25	35.7
1241 Oct 6b E,G	−0.125	+0.200	+ 3.15	18.2
1241 Oct 6c E,G	−0.118	+0.185	+ 2.02	83.3
1241 Oct 6d E,G	−0.125	+0.200	+ 2.55	40.0
1241 Oct 6e E,G	−0.122	+0.195	+ 2.58	18.2
1241 Oct 6f E,G	−0.112	+0.175	+ 0.88	83.3
1241 Oct 6g E,G	−0.122	+0.195	+ 2.18	83.3
1241 Oct 6h E,G	−0.125	+0.200	+ 3.05	58.9
1241 Oct 6i E,G	−0.125	+0.200	+ 3.05	83.3
1241 Oct 6k E,G	−0.125	+0.200	+ 4.05	58.9

TABLE XVIII.9 (Concluded)

Designation	$100A_i$	$100B_i$	$100Z_i$	$W_i^{\frac{1}{2}}$
1241 Oct 6ℓ E,G	-0.118	+0.185	+ 1.48	18.2
1241 Oct 6 E,Sc	-0.105	+0.160	- 0.75	12.2
1245 Jul 25 E,G	+0.050	-0.090	+23.15	18.2
1255 Dec 30 B,E	+0.008	-0.015	-17.88	18.2
1255 Dec 30 B,CE	+0.045	-0.080	-14.25	18.2
1261 Apr 1 B,E	+0.098	-0.145	-27.92	18.2
1261 Apr 1 E,G	+0.120	-0.190	-19.15	12.9
1263 Aug 5a B,E	-0.118	+0.185	+13.58	18.2
1263 Aug 5b B,E	-0.112	+0.175	+13.38	18.2
1263 Aug 5 E,BN	-0.115	+0.180	+11.30	18.2
1263 Aug 5 E,CE	-0.112	+0.175	-10.99	10.9
1263 Aug 5 E,G	-0.112	+0.175	+ 7.08	18.2
1263 Aug 5a E,Sc	-0.088	+0.135	+ 3.88	18.2
1263 Aug 5b E,Sc	-0.102	+0.155	+ 4.78	18.2
1263 Aug 5c E,Sc	-0.098	+0.145	+ 3.58	12.9
1267 May 25 E,CE	+0.055	-0.100	-16.55	18.2
1267 May 25 E,G	+0.060	-0.100	-17.30	18.2
1270 Mar 23 E,Sc	+0.012	-0.025	- 7.18	18.2
1288 Apr 2 B,E	+0.022	-0.045	+35.98	18.2
1288 Apr 2 B,W	+0.022	-0.045	+16.37	10.8

For convenience, I have tabulated $100A_i$, $100B_i$, and $100Z_i$. Thus, for the record 118 Sep 3 E,I, for example, $A_i = -0.00148$, $B_i = +0.00295$, and $Z_i = +0.0892$. Z_i is in units of the earth's radius. A_i and B_i are in the units appropriate to use with the units specified for $\dot{n}_M$ and $\dot{\omega}_e$ in Section XVI.3.

The record 755 Nov 23 B,E in Table XVIII.4 provides the only exception to the preceding remarks. That record is the record of the simultaneous lunar eclipse and occultation that was analyzed in Section XVII.1. The entries for this record in Table XVIII.4 represent the normalized form of Equation XVII.1.

I selected 379 records of solar eclipses for analysis in Chapters VI through XV. Before embarking upon the estimation of the accelerations, I dropped 14 of these for reasons described in Chapter XVII. Tables XVIII.1 through XVIII.9 contain only the remaining 365 records of solar eclipses, plus the single record of a lunar eclipse.

It should be noticed that the number of eclipse records tends to increase strongly through the 12th century and then decreases in the 13th century. It is doubtful that there was an actual decline in the number of eclipse records made in the 13th century. The apparent decline suggested by Table XVIII.9 is probably a consequence of the method that I followed. It should be remembered that 1200 was the principal cut-off point that I adopted. I kept records after 1200 (see the preface) only if they occurred in works that also contained records before 1200. Thus Table XVIII.9 contains only a small fraction of the number of records from the 13th century that have been preserved in sources that I did not use.

2. The Distribution of the Values of z_i; Exceptionally Large Values of z_i

The sign of z depends upon whether the

TABLE XVIII.10

DISTRIBUTION OF THE VALUES OF z_i

Interval From	To	Number of Values in the Interval	Number Expected if $\sigma = 1.25$
-8.999	-8	0	< 1
-7.999	-7	1	< 1
-6.999	-6	1	< 1
-5.999	-5	1	< 1
-4.999	-4	1	< 1
-3.999	-3	5	3
-2.999	-2	10	17
-1.999	-1	32	57
-0.999	-0.5	25	48
-0.499	0	63	57
0	0.499	57	57
0.5	0.999	52	48
1	1.999	74	57
2	2.999	30	17
3	3.999	6	3
4	4.999	4	< 1
5	5.999	1	< 1
6	6.999	1	< 1
7	7.999	0	< 1
8	8.999	1	< 1

observer is north or south of the path of the
eclipse.[†] Since he is just as likely to be north
as to be south, the expected mean value of the
normalized coordinate z is 0, even before we
adjust the computations according to the best
estimates of the accelerations.

The process of estimating the accelerations minimizes the sum of the squares of the values
of z_i. The values of z_i were calculated for the
accelerations $\dot{n}_M$ = -32.44 and $10^9(\dot{\omega}_e/\omega_e)$ = -25 in
the standard units. If these were the correct
accelerations, we should expect 0 to be the most
probable value of $|z_i|$. Since these are almost
surely not the correct accelerations, we expect
the most probable $|z_i|$ to be different from 0.

Since the mean of z_i should be 0 while
the most probable value of $|z_i|$ is different from
0, we expect z_i to have a bimodal distribution.
However, the accelerations used to calculate z_i
are believed to be close to the correct values,
and the bimodal nature of z_i may not be apparent.

Table XVIII.10 shows the distribution of
the values of z_i derived from Tables XVIII.1 through
XVIII.9, excluding the value of z_i for the lunar
eclipse of 755 Nov 23, which is not on a basis
comparable to the values from solar eclipses. Two
values of z_i were 0.000; that is, they were 0 to
the number of significant figures used. These
values were divided between the intervals adjacent
to 0.

The bimodal character of z_i is not apparent from Table XVIII.10. This suggests that the
use of the correct accelerations would not greatly
alter the values of z_i.

Rather surprisingly, Table XVIII.10 suggests that the mean value of z_i is observably

---

[†]
Aside from the few observations that show the
eclipse to be definitely partial. For these the
sign of z should be unbiased, just as it should
be for the other observations.

different from 0, and calculation of the mean supports this conclusion. The sum of all the values of z_i is 136.9 and the mean is 0.38. Now we expect the standard deviation of z_i to be near unity.[†] If we take a sample of 365 values from a population with zero mean and a standard deviation of unity, the standard deviation of the mean of the sample should be $1/\sqrt{365} \approx 0.052$. The difference from the observed mean of 0.38 is striking.

The discrepancy between the expected and observed means is probably a consequence of correlations between observations. Europe is a small part of the earth's surface and few paths of eclipses cut across it. Most European observers lie on the same side of the umbral path of a particular eclipse. Thus there is strong correlation between the values of z_i that relate to the same eclipse. The sum of the values of z_i for reports of the eclipse of 1133 Aug 2 alone is 48.1, more than 1/3 of the total for all records. In view of the correlations that exist, the mean value of z_i is probably not statistically significant, but I have made no explicit calculations on the point.

The last column in Table XVIII.10 gives the number of values that we would expect in each interval if the values of z_i followed the Gaussian or "normal" law of error with a standard deviation of $1\frac{1}{4}$. The distribution actually found agrees rather well for $|z_i| < 4$; thus the standard deviation of z_i is fairly close to $1\frac{1}{4}$.

The residual r_i is the difference between the "observed" value of $\eta_{e,i}$ and the value that would be calculated using the "correct" accelerations, while z_i is the difference obtained with the nominal accelerations of -32.44 and -25. The process of finding the "correct" accelerations necessarily makes the sum of r_i^2 less than the sum of z_i^2 and hence it makes the standard deviation of r_i less

---

[†] See the discussion following Equations XVI.13 in Section XVI.6.

than the standard deviation of z_i, which is about $1\frac{1}{4}$. If the weights W_i are well chosen, the standard deviation of r_i is 1. It is clear from the preceding considerations that the standard deviation of r_i is close to 1 and hence that the weights are generally well chosen. It does not seem necessary to make an accurate calculation of the standard deviation, and I shall assume in the remaining discussion that the standard deviation of r_i is unity.

Table XVIII.10 shows that there are many more values of $|z_i|$ above 4 than we expect from Gaussian statistics. The expected number of $|z_i| > 4$ is less than 1, while the table shows 11 such values. Further, we found in Section XVII.3 that there are three reports for which there was no eclipse at the time and place assumed. These represent values of z_i outside the range of Table XVIII.10. Thus there are altogether 14 large values of z_i, and this matter requires some investigation.

Four of the large values of z_i occur with records that I took to be records of totality; these are the records 840 May 5d E,G, 1033 Jun 29d E,F, 1133 Aug 2d E,F, and 1147 Oct 26a E,BN. Another large value occurs with the record 1093 Sep 23c E,CE, for which I took the magnitude to be within 0.01 of totality. Each of these records will now receive a brief discussion.

840 May 5d E,G from Xantenses [ca. 873]. This record says that stars could be clearly seen, just as at night. This eclipse must have been total or very close to totality. I assigned the observation to Xanten because the annalistic entries changed character about 830 and seemed to become local and contemporaneous. I must have been in error about the place of the observation; the report probably originated elsewhere in the Frankish dominions.

1033 Jun 29d E,F from Elnonenses Minores [ca. 1061]. The report does not say that the eclipse was total nor does it mention the appearance of stars. I took the eclipse to be total because of the description of the sun during recovery. It is

-630-

probable that I was wrong and that the description
is merely more vivid than the situation warranted.

1093 Sep 23c E,CE from Bernoldus [ca. 1100].
I assigned a magnitude within 0.01 of totality because Bernoldus says that a circle appeared in the
sun and that it shone very darkly. I feel now that
I was not warranted in taking the eclipse to be so
near totality on the basis of the description.

1133 Aug 2d E,F from Honorius [ca. 1137].
The details given by Honorius make it almost certain
that this is a report of a total eclipse. However,
the calculations make it almost certain that the
eclipse was not quite total at Autun, which was the
writer's home. Thus it is probable that the writer
Honorius saw the eclipse when he was away from home.

1147 Oct 26a E,BN from Gemblacensis [ca.
1148]. I expressed some reservation about the
interpretation that I adopted in Section VIII.2.
It appears that my reservation was justified.

In all the other cases, I made no assumption about the magnitude of the eclipse. In assigning the weight, I used the value of 0.1 for the
standard deviation of the magnitude. The number
of large deviations merely means that the probability
of observing an eclipse does not follow the Gaussian
law and there is no reason to expect that it should.

The number of large deviations does not
constitute a basis for altering any quantitative
values assigned in Chapters VI through XV and I
have not done so. The only change that I have
made has been to drop certain records for reasons
discussed in Chapter XVII. The reasons had mainly
to do with convenience, and it is doubtful that
dropping the records in question has any significant
impact upon the values of the accelerations that
will be estimated.

In sum, the standard deviation of the
normalized values z_i is close to unity. The values
do not follow Gaussian statistics for two main
reasons. First, the values of z_i that correspond

to different records of the same eclipse are highly correlated. Second, we expect more large deviations than we would have with Gaussian statistics.

The record 1191 Jun 23a E,F is a record of a partial eclipse in which the writer describes the appearance of the sun at maximum eclipse. I pointed out in the discussion of this record in Section X.3 that the description was incompatible with the position of the path given by Oppolzer [1887], and I reserved judgment until detailed calculations of the eclipse path were made. The detailed calculations support the path given by Oppolzer. It is probable that the chronicler accidentally wrote "northern" when he meant "southern"; this change would reconcile the description and the calculations. It is possible that the chronicler observed the eclipse with some visual aid that inverted the image and that he forgot to allow for this in his record.

3. <u>Estimates of the Accelerations from the Medieval Records</u>

If we make a small change in the value used for the acceleration of the moon in calculations, we change slightly the calculated time of a conjunction of the moon with the sun. This in turn means that we change the orientation of the earth with respect to the moon-sun axis at the time of conjunction even if we make no change in the spin acceleration of the earth. The main effect of this in eclipse calculations is to shift the eclipse path in longitude. A shift in longitude can be compensated exactly by changing the spin acceleration $\dot{\omega}_e$ assumed.

Thus to first order a change in $\dot{n}_M$ has the same effect upon the circumstances of a solar eclipse as a change in $\dot{\omega}_e$. If the effects of $\dot{n}_M$ and $\dot{\omega}_e$ were exactly the same, we would not be able to estimate both of them; we could estimate only some function of them. Luckily the effects are not exactly the same. Because of the obliquity of the ecliptic and the inclination of the moon's orbit, a change in the lunar acceleration $\dot{n}_M$ makes small changes in latitude, and these changes cannot be compensated by changing $\dot{\omega}_e$.

The fact that we must rely upon small effects in order to estimate both $\dot{n}_M$ and $\dot{\omega}_e$ means that we have low sensitivity and hence that we need a large body of data. The seriousness of the problem shows up in the determinant of the matrix (XVI.14) in Section XVI.6. The determinant is typically less than 0.01 times the product of the diagonal elements. A small determinant means that the elements M_{xx} and M_{yy} in the inverse matrix are large. We see from Equations XVI.18 that the errors in the estimated values of x and y are correspondingly large.

In order to have usefully small errors in the estimated accelerations, we must use all of the eclipses in Tables XVIII.1 through XVIII.9, including the lunar eclipse of 755 Nov 23. The estimates are:

$$\dot{n}_M = -78.9 \pm 15.9 \quad ''/cy^2,$$

$$10^9(\dot{\omega}_e/\omega_e) = -42.8 \pm 10.1 \quad cy^{-1}. \tag{XVIII.1}$$

We must now attach an epoch to these estimates.

The epoch that we attach should be a mean epoch of all the observations, but we cannot assign equal weight to all observations in calculating the mean epoch. Neither can we use the weights W_i listed in Tables XVIII.1 through XVIII.9; the effect of an observation upon the estimates depends upon factors other than the weight. It is plausible that the importance of an observation is proportional to a_i^2 for that observation, since a_i^2 tells us the contribution of the observation to the diagonal matrix elements. If we use a_i^2 to weight the epochs, we find

$$\text{Weighted mean epoch} = 976. \tag{XVIII.2}$$

This is the epoch that I shall associate with the accelerations in Equations XVIII.1 Any plausible set of weights gives about the same epoch.

For the approximate epoch 1000 in AAO I found the following accelerations:

$$\dot{n}_M = -42.3 \pm 6.1 \quad ''/cy^2,$$

$$10^9(\dot{\omega}_e/\omega_e) = -22.5 \pm 3.6 \quad cy^{-1}. \qquad \text{(XVIII.3)}$$

The epochs are sufficiently close together that we can directly compare the values in Equations XVIII.1 and XVIII.3. The difference in the values of $\dot{n}_M$ is 36.6 and the sum of the standard deviations is 22.0. The difference in the values of $10^9(\dot{\omega}_e/\omega_e)$ is 20.3 and the sum of the standard deviations is 13.7. In both cases the difference is roughly $1\frac{1}{2}$ times the sum of the standard deviations. Such a difference is somewhat large but not disturbingly so.

4. A Sudden Change in the Behavior of the Accelerations

Finding the accelerations $\dot{n}_M$ and $\dot{\omega}_e$ is interesting, but that alone does not meet the main goal of this work. The main goal is to test a tentative conclusion reached in AAO. This conclusion is connected with a parameter that is well determined by the eclipse observations. It is desirable to explain the physical significance of the parameter in question.

The accelerations $\dot{n}_M$ and $\dot{\omega}_e$ are taken using ephemeris time as the time base. Ephemeris time is the kind of time kept by the orbital motions of the planets. However, the path of an eclipse upon the surface of the earth depends upon the orientation of the earth with respect to the sun. This orientation is a measure of solar time. Thus, with respect to the specialized problem of interpreting ancient or medieval eclipses, solar time is more fundamental than ephemeris time.

If we use solar time as the base, there
is no acceleration of the earth's spin, by definition. Instead, we find that there is an acceleration of the earth in its orbit, which is the same
as an acceleration in the apparent yearly motion
of the sun through the stars. Thus it is customary to speak of accelerations of the moon and the
sun when one is using solar time as opposed to
ephemeris time.

The relations needed to use solar time as
the time base are given in many places, including
Equations I.1 on page 5 of AAO. There I used a
cumbersome notation adopted for historical reasons.
Here, let us use ν_M for the angular velocity of
the moon and ν_S for the apparent angular velocity
of the sun in its yearly motion, and let us use
a prime to denote a derivative with respect to
solar time. Then the accelerations $\nu_M{}'$ and $\nu_S{}'$
are related to $\dot{n}_M$ and $\ddot{\omega}_e$ by

$$\nu_M{}' = \dot{n}_M - 1.7373 \times 10^9 \ (\dot{\omega}_e/\omega_e) \quad ''/cy^2,$$

(XVIII.4)

$$\nu_S{}' = \quad - 0.1300 \times 10^9 \ (\dot{\omega}_e/\omega_e) \quad ''/cy^2.$$

Let us use D to denote the mean elongation
of the moon from the sun, that is, the angle from
the position of the sun to the position of the moon.
If we neglect small effects, we can say that a conjunction of the moon with the sun occurs whenever
D is a multiple of 360°. A solar eclipse can occur
only when D is such a multiple, although an eclipse
does not occur at every multiple. An eclipse thus
depends primarily upon D and not upon the solar and
lunar positions individually.

In sum, we expect eclipses to give us a
sensitive estimate of D'', the acceleration term in
the elongation with respect to solar time, but we
do not expect them to give us any parameter independent of D'' with much sensitivity. We easily find
from Equations XVIII.4 that

-635-

TABLE XVIII.11

ESTIMATES OF THE ACCELERATION PARAMETER D″

Time Interval	D'' $''/\text{cy}^2$	$\sigma(D'')$ $''/\text{cy}^2$
118 - 464	7.0	5.2
512 - 592	-13.6	8.9
603 - 693	6.4	9.6
733 - 787	13.8	15.6
807 - 840	-12.1	9.9
865 - 891	2.9	6.4
939 - 990	-13.0	11.5
1009 - 1044	39.0	8.5
1087 - 1098	2.0	25.3
1109 - 1147	-16.4	2.4
1153 - 1194	- 9.9	7.9
1207 - 1288	-10.9	4.3

$$D'' = \dot{n}_M - 1.6073 \times 10^9 (\dot{\omega}_e / \omega_e). \qquad \text{(XVIII.5)}$$

This is -1.6073 times the quantity that I tabulated in Table XIV.4 and plotted in Figure XIV.2 of AAO. For simplicity, I shall word the remaining discussion as if D'' were the quantity used in AAO.

In AAO I formed 14 independent estimates of D'' for epochs ranging from -700 to $+1050$. These estimates are plotted in Figure XIV.2 on page 283 of AAO. They show with high statistical significance that D'' was far from constant during the time range involved, and they suggest strongly that there was a sudden change in the behavior of D'' about the epoch 700. My main goal in the present work is to test this result with additional data.

Therefore it is desirable to make a change of parameter from x and y as defined in Section XVI.3. We should take D'' as a new parameter. It does not matter whether we keep x or y as the second parameter. We must now repeat the analysis using D'' and y, say, as parameters instead of x and y. The steps are simple and it should not be necessary even to sketch them.

In terms of D'' and y, the matrix corresponding to (XVI.14) becomes almost diagonal. This means that the estimate of D'' has almost no interaction with the second parameter and hence that the accuracy of estimating D'' is about an order of magnitude greater than the accuracy of estimating $\dot{n}_M$ and $\dot{\omega}_e$ simultaneously, for a given sample of observations.

We can take advantage of the increased accuracy by dividing the total set of records into several samples. I have divided the records into 12 samples, with the results summarized in Table XVIII.11. The division has been made on a chronological basis. The first column in Table XVIII.11 gives the years of the first and last records used in each sample. The second column gives the estimate of D'' formed from the records lying between the years in the first column. The third column

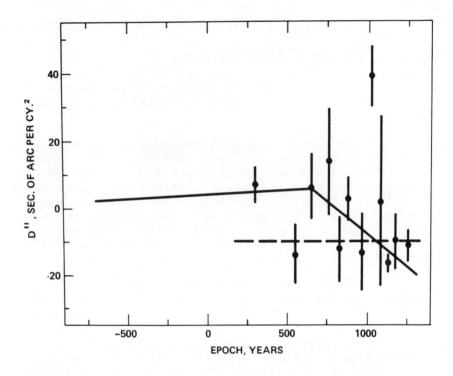

Figure XVIII.1.   Time behavior of the acceleration
term D''.   The ordinate D'' is the acceleration
term in the mean lunar elongation, taken with respect to solar time, and the abscissa is time in
years.   The length of a bar is proportional to the
standard deviation associated with each individual
point.   The points plotted are taken from the data
gathered in this study.   The pair of straight lines
shows the behavior of D'' as inferred from the results in AAO.   The agreement is good, and it seems
to be well confirmed that there was a sizeable
change in the behavior of D'' near 700.   The dashed
line is the best-fitting constant value of D''.

gives the statistical estimate of the standard
deviation of D''.

In Figure XVIII.1, to which the reader
should now refer, I started by drawing the two
straight lines. These are drawn to fit the values
of D'' obtained from Table XIV.4 of AAO (the reader
should remember the factor of $-1.60\overline{73}$). I merely
drew the lines by eye, with no attempt to make a
best fit. The solid circles in the figure are the
values of D'' from Table XVIII.11, plotted against
the mean of the years in the first column. The
distance from each circle to the end of the corresponding bar is $\sigma(D'')$, the estimate of the standard
deviation of the value of D''.

It seems clear to the eye that the values
of D'' fit the pair of straight lines better than
they fit a constant. In order to test this, I
used the "non-central χ^2 test". This test says that
the pair of straight lines fits the data better than
a constant does, with a confidence level of 50 to
1. The χ^2 test is based upon the assumption of
Gaussian statistics, which the population does not
necessarily obey. While the confidence level quoted
should not be taken literally, the test nonetheless
confirms strongly that the eye is not deceived.
The pair of straight lines is a much better fit to
the data than any constant.

It should be emphasized that there is
almost no overlap between the data used in AAO and
those used in this study. The agreement between
AAO and this study is not a consequence of shared
data.

The main goal of this study has now been
met. The observations discovered in the course of
this study confirm the conclusions reached in AAO.
There was a large change in the behavior of an
important astronomical parameter, and the change
occurred near the year 700. If I have made no
substantial systematic error in the astronomical
calculations, the change was in the acceleration
of the moon or in the acceleration of the earth or,

more probably, in both.

Ideally, I should use the data of this study and the data from AAO simultaneously in finding $\dot{n}_M$ and $\dot{\omega}_e$ and D''. To do so would take us far beyond the scope of this study, and I shall defer the simultaneous analysis of all data to another work. Here I shall only remark that the use of all data would increase the statistical significance of the pair of straight lines in Figure XVIII.1 to a very large value, perhaps to 10^4 to 1 or even more.

However, a point that received only brief mention in AAO is worth more attention now that the change near 700 has been confirmed. This point will be the subject of the next section.

5. The Variation of the Instantaneous Acceleration

Over the course of the centuries the acceleration of the earth's spin causes the earth to be in an angular position different from that calculated by means of the standard ephemerides. The acceleration $\dot{\omega}_e$ has been calculated by setting the difference in position equal to $\frac{1}{2}\dot{\omega}_e T^2$, in which T is time in centuries from some reference epoch. The reference epoch chosen has been 1900 Jan 0.5 in ephemeris time. Thus $\dot{\omega}_e$, when evaluated for some epoch, does not mean the acceleration of the earth's spin that existed at that epoch. It means instead the mean acceleration between that epoch and the reference epoch. The mean is not the mean with respect to time; it is a mean defined in the particular way just stated.

In the remaining discussion, a bar will be drawn over the symbol for an acceleration if the symbol denotes the mean in the sense just defined. Thus the displacement of the earth from the position given by the standard ephemeris is $\frac{1}{2}\bar{\omega}_e T^2$. A symbol without a bar will denote the instantaneous value of the acceleration. Since $\frac{1}{2}\bar{\omega}_e T^2$ is the displacement, the instantaneous acceleration is the second derivative of this

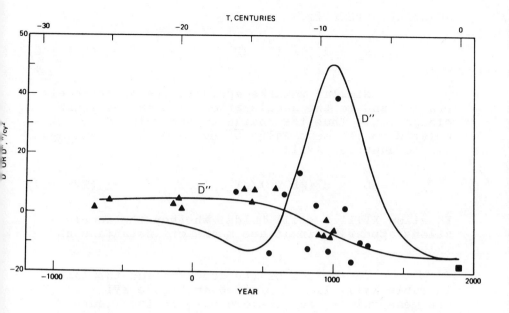

Figure XVIII.2. Time behavior of the mean and the
osculating values of D″. $\overline{D}″$ denotes the mean value
of the acceleration term in the lunar elongation.
The mean is taken between the epoch plotted and the
reference epoch 1900 Jan 0.5. T denotes time in
centuries from the reference epoch. D″ denotes
the instantaneous value of the acceleration term
in the elongation. $\overline{D}″$ was denoted by D″, without
the bar, in Figure XVIII.1. The triangles are
estimates of $\overline{D}″$ from AAO. The circles are estimates
of $\overline{D}″$ found in this work. The square is a speculation about the value of $\overline{D}″$ at the epoch 1900. The
curve labelled D″ shows the time variation of D″ as
derived from the curve labelled $\overline{D}″$.

-641-

quantity. That is,

$$\dot{\omega}_e = d^2(\tfrac{1}{2}\overline{\dot{\omega}}_e T^2)/dT^2 .$$ (XVIII.6)

Similar remarks apply to the lunar acceleration and to the acceleration term in the lunar
elongation. Thus the instantaneous value of D'' is
related to the mean value $\overline{D}''$ by an equation analogous to Equation XVIII.6:

$$D'' = d^2(\tfrac{1}{2}\overline{D}'' T^2)/dT^2 .$$ (XVIII.7)

Equation XVIII.5 still holds, whether the accelerations appearing in it are means or instantaneous
values.

The values of the acceleration tabulated
in Table XVIII.11 and plotted in Figure XVIII.1
are mean values, and the ordinate in the figure
should now be understood to be $\overline{D}''$. We should now
like to consider D'' rather than the mean.

In deriving D'' from the mean acceleration
between a past epoch and the present epoch, we are
hampered somewhat by not knowing the acceleration
at the present epoch. Luckily, the inference of
D'' at a distant epoch is not critically dependent
upon the value at the present epoch. Thus any
plausible value for the present value of D'' or of
$\overline{D}''$ will serve for the purposes of this section.

Tidal friction accounts for essentially
all of the lunar acceleration, so far as we know.
If Spencer Jones [1939] is correct, the current
lunar acceleration is -22.44. On page 286 of AAO
I showed that the contribution of tidal friction to
$10^9(\dot{\omega}_e/\omega_e)$ is about $1.147\dot{n}_M$, or about -25.7. Also
I speculated on page 286 that the part of $10^9(\dot{\omega}_e/\omega_e)$
that does not arise from tidal friction is about
+23. If these speculations are correct, the present
value of $10^9(\dot{\omega}_e/\omega_e)$ is about -2.7, and the present
value of D'' is about -18. For the present epoch,
D'' and $\overline{D}''$ are of course equal.

The continuation of the trend shown in Figure XVIII.1 would yield the value of -44 at present. It is unlikely that D'' is this large in magnitude. The value just obtained is more plausible.

Now consider Figure XVIII.2. The triangles plotted there are the values of $\overline{D}''$ from AAO and the circles are the values obtained in this study. The square shown at 1900 is the value just reached by speculation. The curve labelled $\overline{D}''$ is the function

$$\overline{D}'' = -4 + (T/8) - 11\tanh \tfrac{1}{4}(T + 9.75). \qquad (XVIII.8)$$

I have not found the parameters in Equation XVIII.8 by any formal process, nor do I suggest that the form used has any phenomenological basis. I have merely chosen Equation XVIII.8 by inspection as being about the smoothest possible function that is reasonably consistent with the values found for $\overline{D}''$.

In order to estimate the time behavior of D'' we substitute from Equation XVIII.8 into Equation XVIII.7 and perform the operations indicated. The result is the curve labelled D'' in Figure XVIII.2.

We should not attach too much importance to the time variation shown by the curve labelled D''. Nonetheless, it seems fairly safe to conclude that there have been 3 main regimes of D'' in the past 2000 years. Until about 700, D'' was reasonably constant and about 0 or slightly negative. Between about 700 and about 1300, D'' was perhaps about +30. Around 1300, D'' changed again, to a level of about -15. So far as we can judge from the meager evidence available, D'' has stayed at roughly this value since about 1300.

These considerations would not be changed greatly if we used the value of Van Flandern [1970] for the present lunar acceleration rather than the value of Spencer Jones [1939]. Van Flandern's value is -52. If we use this, the value obtained for D'' at the present epoch would be about +6. This value might require still another change in the level of D'' since 1300.

6. Summary

In this work I examined a large number of medieval European documents with the main purpose of discovering records of solar eclipses and with the secondary purpose of discovering other useful scientific data. Because medieval European historical documents were transmitted freely over most of Europe, and because they were frequently copied without indication of source, it is necessary to study simultaneously documents from most of Europe, including the Byzantine Empire, in order to decide which records are independent and which are derived. In order to interpret many of the records, it is necessary to study the evolution of medieval historical writing, and it is necessary to study the Easter problem and its relation to the annalistic form.

I selected 379 records of solar eclipses that seem to be independent and for which the date and place of observation can be determined with useful confidence.

I used these records in forming estimates of the accelerations of the moon and of the earth for the mean epoch 976. The results, given in Section XVIII.3, will be repeated for convenience:

$$\dot{n}_M = -78.9 \pm 15.9 \ ''/\text{cy}^2,$$
$$10^9(\dot{\omega}_e/\omega_e) = -42.8 \pm 10.1 \ \text{cy}/^{-1}.$$

These are somewhat larger in magnitude than other recent estimates, but not disturbingly so.

Finally, I used the records in order to trace the behavior of the parameter D'' through the medieval period; D'' is the acceleration term in the lunar elongation, taken with respect to solar time. The results are plotted in Figure XVIII.2. They confirm the conclusion of AAO that there was a sudden change near the year 700. It is possible but less certain that there has been another large change since 1300.

APPENDIXES

APPENDIX I

A CATALOGUE OF OBSERVATIONS OF LUNAR ECLIPSES

APPENDIX I

A CATALOGUE OF OBSERVATIONS
OF LUNAR ECLIPSES

In AAO I studied ancient measurements
of the magnitudes of lunar eclipses that had been
made by professional astronomers. The conclusion
was that the magnitude of a lunar eclipse is quite
insensitive to the acceleration of the moon and
that it is independent of the acceleration of the
earth. The measurements of magnitude made by professional astronomers were not accurate enough for
them to be useful, in view of the low sensitivity.
Therefore we can be quite sure that the few estimates made by casual observers in the period of
this study would not be useful.

Measurements of the times associated
with lunar eclipses, in contrast, turned out to
be highly useful in AAO. Many of the observations
found in the course of this study have times associated with them. Nonetheless I am afraid to use
them.

Most of the observations give time only
to the hour, and hence they have low precision.
That is not the reason for not using them. A statistical treatment would give results of high accuracy, because of the large number of observations,
if we could be sure of making no systematic error
in treating the observations. I am afraid to use
the times stated in the observations, because I am
not sufficiently sure of what was meant and thus
there is a large risk of introducing systematic
error. An example should suffice.

At the equinox, to choose a simple example, the "4th hour of the night" ran from 3
hours after sunset to 4 hours after sunset. If an

-649-

observer said that the moon was eclipsed "from the 4th hour of the night", what did he mean? If he meant literally that the eclipse began during the 4th hour, we should assign 3 1/2 hours as the average of such observations. However, he might have meant that "4" was the nearest number on whatever scheme he was using for estimating time; in this case, we should use 4 hours rather than 3 1/2. Finally, even if he knew better, he would have a strong tendency to write "4th hour" if the time point called "4 hours" had passed and the time called "5 hours" had not passed. In other words, there is a strong tendency to errors of a certain sort rather than to random errors.

Even if we could be certain of what was meant by the time, we would usually be uncertain of the phase of the eclipse to which the time is meant to apply. If a person is merely looking at the moon from time to time, as opposed to systematically watching for the beginning of a predicted eclipse, it seems inevitable that he will not notice an eclipse until some time after it has begun. On the other hand, there is no way, I think, by which he would put down a time before the beginning except by typographical error. In other words, the times recorded, even if we were sure what they meant, would almost surely apply to a time that is systematically late by an unknown amount.

We could neglect the times and simply use the fact that the eclipse was observed, just as we do for most of the solar eclipses. This would be equivalent to saying that the zone of eclipse for a lunar eclipse is approximately half the earth; the large number of observations would still give a useful result in spite of the low precision if they are unbiased in this regard. Put another way, this would be equivalent to saying that the center time of the eclipse was at local midnight.

This procedure would give a valid result provided that an eclipse occurring after midnight is as likely to be observed as one before midnight.

-650-

This is probably not true for casual observers, as opposed to dedicated ones, in our culture. There are surely more people awake between sunset and midnight in our society than between midnight and dawn. I simply do not know the extent to which this was true in medieval Europe, and I am afraid of a bias in the observations toward the first half of the night.

Thus I shall not use the lunar eclipses in this study, with the exception of one that is discussed in Section V.1. It is possible that the lunar eclipses will turn out to be useful for other purposes. Therefore I shall catalogue them.

Astronomical observations besides those of solar and lunar eclipses are tabulated in the next two appendices. Some of them, like meteors, are not related to time in an accurately known way and therefore they cannot be used in a study of the astronomical accelerations. Others, like conjunctions of the moon with a star, cannot be used unless the times are measured to an accuracy of a fraction of an hour. Since they were not measured to this accuracy, they cannot be used either. Thus, with the one exception mentioned, the only observations that can be used in the astronomical analysis are those of the solar eclipses.

The place where a solar eclipse was observed is important to the use of the observation, and that is the reason for the extensive study directed at finding the origins of the records in the main text of this work. The place is relatively unimportant for lunar eclipses, and I have made no particular effort to find which sources are original when there are several records of the same lunar eclipse. I have omitted records that are obviously copied.

The records are tabulated in Tables A.I-1 through A.I-8. In the tables I have largely but not entirely preserved the geographical regions used in arranging the solar eclipses; for example, I have put all records from any part of the British Isles in the same table.

The first column of each table gives the date of the eclipse. If the first column contains a question mark or several dates, it means that the identification of the eclipse is uncertain at least to the extent indicated. The second column gives the sources in which the records are found, along with comments. The date given in the source forms part of the comment. If no date appears, it means that the date given in the source is correct to the extent that it was given. For examples, the first line in Table A.I-1 says "Scotorum in 670", meaning that the source Scotorum gave only the year and that the year cannot be correct. The second line has no date after "Ulster" although it has dates after the other sources named. This means that Ulster gave at least the correct year. It may or may not have given more of the date; if it did, whatever part of the date it gave was correct.

In some cases, more than one date is given after a source in the second column. In most of these cases, there are variant readings in different manuscripts. In a few of these cases, it is not sure which date the source intended.

When the difference between the date in column 1 and that given by the source is 1 day, it may not be an error in dating. There is an automatic ambiguity in giving the dates of lunar eclipses, which necessarily come between sunset and sunrise. That is: Did the writer assign the eclipse to the day just finished or to the coming day? Further, the dates in column 1 are on the basis of Greenwich Mean Time, while the observers used local time.

In identifying the sources in column 2, I have generally given only the name needed in the citation and not the accompanying date. The date associated with the citation is still enclosed within square brackets when it appears, but it appears only when it is needed in order to distinguish between sources with the same name.

Many of the eclipses involved were partial. The smallest eclipse that I noticed is that of 1155 Jun 16, which appears in the French records in Table

A.I-4. Only about 0.07 of the moon's diameter was covered by the earth's shadow, according to the calculations of <u>Oppolzer</u> [1887]. This is so small that one wonders whether the record may have been inspired by a prediction rather than an observation.

TABLE A.I-1

LUNAR ECLIPSES IN SOURCES FROM THE BRITISH ISLES

Date	Sources and Comments
672 May 17, 672 Nov 10, or 673 May 6	Scotorum in 670
691 Nov 11	Ulster; Brut in 692; Cambriae in 690
?	Ulster in 717
726 Dec 13 ?	Ulster on 725 Dec 15
734 Jan 24	Ulster on 734 Jan 22; Bede on 734 Jan 31
752 Jul 31	Melrose
753 Jan 24	Bede
755 Nov 23	Simeon on 755 Nov 24, simultaneous eclipse and occultation; Melrose on 756 Nov 24
762	Ulster, 2 possible eclipses
773 Dec 4	Ulster
788 Feb 26 or 789 Feb 14	Ulster on 788 Feb 19
796 Mar 28	Anglo-Saxon Chronicle on 795 or 796 Mar 28; Domitiani Latini on 795 May 27
800 Jan 15	Anglo-Saxon Chronicle on 800 Jan 16
?	Anglo-Saxon Chronicle on 802 Dec 20
806 Sep 1	Anglo-Saxon Chronicle; Scotorum and Ulster in 807
809 Dec 25	Brut on 810 Dec 25
828 Dec 25	Anglo-Saxon Chronicle on 827 or 829 Dec 25
831 Oct 24	Brut on 831 Oct 25 or 831 Dec 8

Date	Sources and Comments
865 Jan 15	Ulster and Scotorum
878 Oct 15	Ulster
?	Anglo-Saxon Chronicle in 904; Melrose in 905
921 Dec 17	Ulster on 922 Dec 18
1023 Jan 9	Ulster on 1023 Jan 10
1078 Jan 30	Anglo-Saxon Chronicle
1110 May 5	Anglo-Saxon Chronicle on 1110 May 4 or 5
1117 Dec 11	Anglo-Saxon Chronicle, Melrose
1121 Apr 4	Anglo-Saxon Chronicle on 1121 Apr 4 or 5
?	Matthew Paris on 1135 Jul 29
1135 Jan 1	Ralph of Coggeshall on 1136 Jan 1
1179 Aug 19	Gervase
1185 Apr 16 ?	Diceto on 1186 Apr 16; eclipse not visible in England.
1186 Apr 5	Gervase
1189 Feb 3	Gervase
1200 Jan 3	Diceto on 1200 Jan 4
1208 Feb 3	Wendover on 1208 Feb 2
1218 Jul 9	Wendover; observation may have come from Holy Land.
1230 Nov 22	Wendover
?	Worcester in 1243
?	Oxenedes in 1247
1255 Jul 21 ?	Oxenedes
1258 May 18	Florence on 1258 May 19; Oxenedes on 1258 May 29

TABLE A.I-1 (Concluded)

Date	Sources and Comments
1265 Dec 24	Florence
1270 Sep 30	Florence; Oxenedes on 1271 Sep 30
1272 Aug 10	Oxenedes on 1271 Aug 10
1276 Nov 23	Florence
1280 Mar 18	Florence. Says S. Edmund's Day; probably means S. Edward's Day, which is Mar 18.
1281 Mar 7	Florence
1281 Aug 31	Florence
?	Cambriae on 1287 Jun 19
1287 Oct 22	Florence
1288 Oct 11	Florence, says "about total" but Oppolzer gives as total.
1291 Feb 14	Florence on 1291 Feb 15; Wykes on 1290 Feb 14

LUNAR ECLIPSES IN SOURCES FROM BELGIUM AND
THE NETHERLANDS

Date	Sources and Comments
806 Sep 1	Blandinienses on 806 Sep 2
807 Feb 26	Blandinienses
807 Aug 21	Blandinienses on 807 Aug 22
?	Blandinienses on 810 Apr 25
?	Laubacenses has undated lunar eclipses in 893 and 894.
1096 Feb 11	Sigebertus
1096 Aug 6	Sigebertus on 1096 Aug 7
1110 May 5	Formosolenses
1117 Dec 11	Anselmus [ca. 1135]; Leodienses
?	Formosolenses in 1119
1129 Oct 29	Egmundani

LUNAR ECLIPSES IN SOURCES FROM CENTRAL EUROPE

Date	Sources and Comments
?	Sangallensis lists lunar eclipses in 882, 893, and 894 with no supporting details.
1122 Mar 24	Cosmas
1128 Nov 8	Wissegradensis on 1128 Nov 9
?	Wissegradensis in 1131, probably 1131 Sep 8
1132 Mar 3	Wissegradensis on 1132 Mar 4
1133 Feb 21	Wissegradensis on 1133 Feb 22
1142 Feb 12 ?	Gradicenses in 1142
1154 Jan 1	Mellicenses
1161 Feb 12	Admuntenses; Cremifanenses, Engelbergenses
1204 Apr 16	S. Rudberti [ca. 1286], wrote 16th calends April, meant 16th calends May.
1204 Oct 10	S. Rudberti [ca. 1286] on 1204 Oct 11
?	Reicherspergenses in 1223
1238 Jan 2 ?	S. Rudberti [ca. 1286] wrote 1238, 3rd nones June, probably meant 3rd nones Jan., correct date is 4th nones Jan.
1258 May 18	Pragensium on 1258 May 19

LUNAR ECLIPSES IN FRENCH SOURCES

Date	Sources and Comments
580 and 582	Gregory of Tours. No eclipse in 580, two in 582
590 Oct 18 ?	"Fredegarius Scholasticus" in 590
788 Feb 26	Flaviniacenses et Lausonenses on 788 Feb 25
795 ?	Flaviniacenses et Lausonenses in 795, two eclipses
803 Nov 2 ?	Flaviniacenses et Lausonenses in 803
806 Sep 1	Sithienses on 806 Sep 2
807 Feb 26	Sithienses
807 Aug 21	Sithienses on 807 Aug 22
809 Dec 25	Sithienses on 809 Dec 26
810 Jun 20	Sithienses on 810 Jun 21
810 Dec 14	Sithienses on 810 Dec 15
817 Feb 5	Sithienses
820 Nov 23	Sithienses on 820 Nov 24
843 Mar 19	Nithardus on 843 Mar 20
878 Oct 15	Mettenses
955 Sep 4	Floriacenses on 956 Sep 2
1009 Oct 6	Lemovicenses on 1010 Oct 7
1044 Nov 8	S. Benedicti on 1044 Dec 8
1056 Apr 2	Godellus on 1055 Apr 3, Nivernenses on 1056 Apr 4
?	S. Maxentii on 1062 Aug 1
?	S. Maxentii in 1071 Oct

TABLE A.I-4 (Concluded)

Date	Sources and Comments
1075 Sep 27	Vindocinense wrote 5th calends November instead of 5th calends October.
1077 Feb 10	S. Albini
1078 Jan 30	Besuenses
?	S. Maxentii on 1082 Nov 11
1098 Dec 11	S. Maxentii
1107 Jul 6	S. Maxentii on 1107 Jul 4
1110 May 5 ?	S. Maxentii in 1110
1117 Jun 16	S. Maxentii on 1117 Jun 15
1129 Oct 29	Burburgensis
1149 Mar 26	Praemonstratensis on 7th feria, 1149 Mar 22. 1149 Mar 26 was 7th feria.
1154 Dec 21	Praemonstratensis on 1154 Dec 22; Robertus de Monte on 1st feria before Christmas, eclipse was night between 3rd and 4th feria.
1155 Jun 16	Robertus de Monte on 1155 Jun 17. Eclipse was so small (magnitude 0.07) that one wonders if eclipse was calculated.
1160 Aug 18	Bellovacenses on 1160 Aug 19
1161 Feb 12	Vindocinense
1178 Mar 5	Aquicinctina
1189 Feb 3	Elnonenses Maiores
1192 Nov 21	Aquicinctina estimated magnitude at 2/3.
1193 May 18	Elnonenses Maiores
1193 Nov 10	Aquicinctina on 1193 Nov 11

TABLE A.I-5

LUNAR ECLIPSES IN GERMAN SOURCES

Date	Sources and Comments
806 Sep 1	Loiselianos on 806 Sep 2
807 Feb 26	Loiselianos
807 Aug 21	Loiselianos on 807 Aug 22
809 Dec 25	Laurissenses on 809 Dec 26
810 Jun 20	Laurissenses on 810 Jun 21
810 Dec 14	Laurissenses on 810 Dec 15
817 Feb 5	Laurissenses
820 Nov 23	Laurissenses on 820 Nov 24
824 Mar 18	Laurissenses on 824 Mar 5. Calculation?
828 Jul 1	Laurissenses
828 Dec 25	Laurissenses
831 Oct 24	Xantenses
832 Apr 18	Xantenses
835 Feb 17	Xantenses
842 Mar 30	Fuldenses
865 Jan 15 ?	Xantenses in 866 Jan; 864 Jan 27 also possible
878 Oct 15	Fuldenses
?	Augustani in 976
995 Jul 14 ?	Augustani in 995
1074 Oct 7	Naufragia Ratisbonensia on 1074 Oct 8; Augustani
1096 Aug 6	Augustani on 1096 Aug 7, taken for a sign because moon's age not correct; Frutolf on 1096 Aug 8
1098 Dec 11	S. Blasii

TABLE A.I-5 (Continued)

Date	Sources and Comments
1110 May 5	Hildesheimenses on 1110 May 6
1117 Jun 16	S. Disibodi
1117 Dec 11	S. Disibodi
1124 Feb 1	Hildesheimenses; S. Disibodi on 1124 Feb 3
?	Hildesheimenses on 1124 Mar 3. Calculated?
1128 Nov 8	S. Blasii on 1128 Nov 9
1132 Mar 3	S. Blasii on 1132 Mar 2
1138 Oct 20	S. Blasii on 1138, 13th calends of an omitted month. 1138 Oct 20 is 13th calends November.
1142 Feb 12	S. Blasii
1154 Jan 1	S. Stephani
1160 Aug 18 ?	Magdeburgenses in 1160
?	Magdeburgenses in 1162
?	Theodorus in 1171
1172 Jan 13	Magdeburgenses
1179 Aug 19	S. Trudperti
1193 May 18	S. Trudperti
1204 Apr 16	Scheftlarienses Maiores
1208 Feb 3	Scheftlarienses Maiores and S. Stephani, both on 1208 Feb 4
1215 Mar 17	Colonienses on 1214 Mar 17
1216 Aug 28	S. Trudperti on 1216 Aug 29
1241 Apr 27	Scheftlarienses Maiores said 1240 on S. Vitus' Day, meant 1241 on day of S. Vitalis, martyr.

TABLE A.I-5 (Concluded)

Date	Sources and Comments
1291 Feb 14	S. Stephani on 1291, 16th calends Feb.; meant 16th calends March (= Feb 14).

LUNAR ECLIPSES IN ITALIAN SOURCES

Date	Sources and Comments
?	Paulus Diaconus; see record of solar eclipse 664 May 1b E,I in Section XII.2.
1033 Dec 8	Arnulfus after solar eclipse of 1033 Jun 29
1042 Jan 9	Casinates; Cavenses on 1037 Jan 9
1089 Jun 25 ?	Romualdus says in 1089 at moonrise, 1088 Dec 30 is possible.
1095 Feb 22	Farfenses on 1096 Feb 22
1096 Aug 6	Cavenses; Romualdus on 1096 Aug 7
1099 Nov 30	Romualdus on 1099 Dec 1
1107 Jan 11	Cavenses
1117 Jun 16	Romualdus and Varignana have eclipse in July.
1117 Dec 11	Romualdus has 1118 Dec 11 and gives weekday for the wrong date.
1118 Nov 30	Romualdus; lists eclipses on 1118 Nov 30 and 1118 Dec 11.
1178 Mar 5	Romualdus; says 3/4 eclipsed while Oppolzer gives 2/3.
1178 Aug 30	Romualdus; says 1/3 eclipsed while Oppolzer gives 1/2.
1208 Feb 3	Rampona and Varignana; moon became red and black and blue.

TABLE A.I-7

LUNAR ECLIPSES IN SCANDINAVIAN AND SPANISH SOURCES

Date	Sources and Comments
451 Sep 26	Hydatius
462 Mar 2	Hydatius
?	Esromenses in 695; is surely copied.
734 Jan 24	Esromenses on 736 Jan 31; is badly copied from Bede.
1078 Jan 30	Esromenses on 1078 Jan 31
1146 Nov 20	Lundenses on 1146 Nov 26, the 4th feria; feria is correct for 1146 Nov 20 so day of month was accidentally written wrong.

TABLE A.I-8

LUNAR ECLIPSES IN SOURCES FROM THE BYZANTINE EMPIRE AND THE HOLY LAND

Date	Sources and Comments
622 Feb 1 ?	Theophanes in 622. Eclipse of 622 Jul 28 was probably not visible in Constantinople.
1117 Jun 16	Fulcher; thought it was a sign and not an eclipse because it was on the 13th of the moon.
1117 Dec 11	Fulcher

APPENDIX II

A CATALOGUE OF OBSERVATIONS OF COMETS

APPENDIX II

A CATALOGUE OF OBSERVATIONS OF COMETS

It is often difficult to decide whether
a bright new heavenly object mentioned in an old
source is a nova, a comet, or in some cases even
a meteor. Part of the trouble arises from the
fact that a comet used to be called a star; "the
star called comet" is often used in medieval
sources and one may suspect that all but "the star"
was omitted sometimes. In some cases, "new star"
was used to denote a comet. Philostorgius [ca. 425]
used it for the object listed under "388 or 389"
in Table A.II-8, for example.

A description of motion affords the most
definite means of discrimination when one is given.
A rapid motion indicates a meteor while a slow motion indicates a comet. Unfortunately the time
scale of the motion is not always given even when
motion is mentioned. Further, one may not safely
infer that an object was stationary and hence a
nova merely because motion is not mentioned.

Observations of objects that were probably comets are listed in Tables A.II-1 through
A.II-8, but we must expect that further study
might show that some of the objects were something
else. The tables contain two columns. The first
column gives a date while the second column gives
the source and comments. I have made no concerted
effort to determine which records are original but
I have omitted obvious copies.

The dates given for the comets do not
have the firmness of those given for eclipses.
The dates are no longer calculable in general and
are merely those to be inferred from the source.
When I give a date with no indication of uncertainty, it merely means that I believe that I know how
to interpret the dating indications given by the
writer. It does not at all mean that the date
listed is necessarily correct. It should be noted

that some of the dates are fragmentary; in these cases no more detail can be inferred with confidence.

Comets were widely regarded as omens. I suspect some of the comets of being "magical" comets, that is, comets invented to add drama to a situation because of their monitory aspect. The comet reported in Theophanes [ca. 813] in 632 is highly suspicious. It does not follow that Theophanes invented the comet himself if it is invented; he probably merely followed an old account.

I am not presenting the tabulations in this appendix or in the remaining appendices as research results. They are merely lists of information that I have come across in the course of research with other purposes and that I preserve in case they may have research value for others. Hence I have made no systematic effort to discover what may have been done earlier. Needham [1959, p. 430] says that Williams[†] has made what is still the most extensive list of Chinese observations of comets. I have not seen any reference to an extensive catalogue of European observations. However, I have seen references to a catalogue of cometary orbits by J.G. Porter in the Memoirs of the British Astronomical Association, v. 39, 1961.

<hr />

[†] Williams, J., Observations of Comets from -611 to +1640, Extracted from the Chinese Annals, Strangeways and Walden, London, 1871.

TABLE A.II-1

OBSERVATIONS OF COMETS IN BRITISH SOURCES

Date	Source and Comments
678, from August	Bede, for 3 months
729	Bede, 2 reported but probably the same one before and after perihelion
868	Florence, Melrose
891 May 10	Anglo-Saxon Chronicle, Domitiani Latini, year established by Easter data
905 Oct 20	Anglo-Saxon Chronicle
975	Anglo-Saxon Chronicle, Melrose
995	Anglo-Saxon Chronicle
1066	Anglo-Saxon Chronicle
1097 Oct	Anglo-Saxon Chronicle, Florence, Margan, seen for about 15 days
1106 Feb 16	Anglo-Saxon Chronicle, Florence, Ralph of Coggeshall, Worcester, visible for a varying number of days, about 25
1109 Dec	Eadmer, near the Milky Way
1110 Jun 8	Anglo-Saxon Chronicle, Florence, for most of June
1113 or 1114 May	Anglo-Saxon Chronicle, Matthew Paris
1119 Sep 28	Florence
1132 Oct 8	Florence, for 5 days
1145 or 1146	Bermondsey, Diceto, near Ascension Day
1165	Melrose, two recorded but probably same one

TABLE A.II-1 (Concluded)

Date	Source and Comments
1180 Jul	<u>Melrose</u>
1187 Dec 20	<u>Margan</u>
1198 Nov	<u>Ralph of Coggeshall</u>, for 15 days
1222 Jun	<u>Florence</u>
1264 Aug	<u>Florence</u>, all month

TABLE A.II-2

OBSERVATIONS OF COMETS IN SOURCES FROM BELGIUM AND THE NETHERLANDS

Date	Source and Comments
842	Sigebertus, in Aquarius
941	Leodienses
944	Sigebertus
1005	Sigebertus; in the south. This may be the nova of 1006; see Section V.2.
1017 Mar	Sigebertus, for 4 months
1066 near Easter	Sigebertus
1097 Oct	Sigebertus, in west, all first week of October
1106 Feb	Sigebertus says all of February. Egmundani says a meteor detached itself from a comet on 1106 Feb 16 and fell to earth; it lists another comet in all of February, 1107.
1110 Jun	Leodienses, Sigebertus

TABLE A.II-3

OBSERVATIONS OF COMETS IN SOURCES FROM CENTRAL EUROPE

Date	Source and Comments
868	Sangallenses
891	Sangallensis
911 or 912	911 in Sangallenses, 912 in Sangallensis
942	Cosmas, for 14 nights
975	Sangallenses, in the autumn
989 Aug 10	Sangallenses says on S. Lawrence's Day.
1003 Feb	Sangallenses, apparently first seen near sunset and then near sunrise
1155 May 5	Admuntenses
1222	Pragensium, called a "star of unusual brightness" in the west. Could be a nova; see Section V.2.
1265	Mellicenses. Seen all autumn. Described as a "star of unusual brilliance" that "poured out smoke like a furnace"; see Section V.2.

TABLE A.II-4

OBSERVATIONS OF COMETS IN FRENCH SOURCES

Date	Source and Comments
563	Gregory of Tours, all year
580	Gregory of Tours
582	Gregory of Tours, had an extra-ordinarily long tail
595	"Fredegarius Scholasticus"
817 Feb 5	Sithienses, tail was like a sword, date confirmed by a lunar eclipse.
841 Dec to 842 Feb	Nithardus. First seen in Pisces, then moved into Lyra and Andromeda, then into Arcturus and vanished.
844	S. Maxentii, for 20 days
864 May	Floriacenses, about first 20 days of May
868 Jan 29	Senonensis, for 25 days there-after. First seen in Ursa Minor, then it moved almost "to the triangle".
874 or 875 Jul	Rainaldus, S. Florentii
892	Rainaldus, in Scorpio for 80 days
905 May	Floriacenses, for 23 days. First seen in the north, crossed the Zodiac between Leo and Gemini (sic).
912	Besuenses, says "comets were seen"
918 Mar	Hugo Floriacensis. Hugo's chronology is questionable here.

TABLE A.II-4 (Concluded)

Date	Source and Comments
942 Oct	Rainaldus, in the west for 21 days
995 Aug to Oct	Besuenses, S. Florentii, for 80 days
1004 or 1006	1004 in Besuenses, 1006 in Mosomagenses, may be same or different
1046	Godellus
1066 Apr 24	Senonensis has most extensive description. Appeared in Taurus, moved to "extreme" part of Gemini. Some sources give the year as 1067, and some careless sources have comets in both years.
1097 Oct 6	S. Maxentii, for 7 nights
1106 Feb and Mar	S. Maxentii and many others
1110 Jun	S. Albini
1135	Burburgensis
1145 May	Senonensis

TABLE A.II-5

OBSERVATIONS OF COMETS IN GERMAN SOURCES

Date	Source and Comments
729	Quedlinburgenses
817 Feb 5	Fuldenses, Vita Hludowici Imperatoris, in Sagittarius
837 or 838 Apr	There may be 2 comets. Fuldenses says it appeared in Libra on 837 Apr 11 and was seen for 3 nights. Vita Hludowici Imperatoris says it appeared in the middle of the Easter "festivity" in Virgo, and was seen for 25 days, crossing Leo, Cancer, and Gemini. Easter was Apr 1 in 837 and Apr 14 in 838. Thus it is probably the same comet, in 838, seen by the Fuldenses writer before the Vita writer.
839 ?	Fuldenses says there was a comet in Aries in 839, and one in Aquarius on 841, on the 8th calends January (= Dec 25). Vita says a comet appeared in Scorpio on Jan 1 of a year that is probably meant to be 839. This may be 2 comets all together, one in 839 perhaps and one about 840 Dec 25 perhaps.
842	Xantenses, during Lent, in the west with its tail to the east
869 Feb 17	Xantenses, in the northwest. This may be the same as that under 868 in Augienses and Fuldenses.

TABLE A.II-5 (Concluded)

Date	Source and Comments
875 Jun 6	<u>Fuldenses</u>, in the north at 1st <u>hour of</u> the night
882 Jan 18	<u>Fuldenses</u>, 1st hour of the <u>night</u>
905	<u>Corbeienses</u>, at Pentecost
907	<u>Corbeienses</u>, near Easter
912 or 914	<u>Quedlinburgenses</u> in 912, <u>Wirziburgenses</u> in 914. May be the same or different.
937 Oct 18 to Nov 1	<u>Widukindus</u>
941	<u>Corbeienses</u>
975	<u>Corbeienses</u>, for many nights in the northeast near Gemini
1010	<u>Quedlinburgenses</u>
1018 Aug	<u>Thietmarus</u>
1066	<u>Augustani</u>, in Gemini
1097	<u>Augustani</u>, rather faint
1106	<u>Colonienses</u>, for 3 weeks in <u>Lent</u>
1110 Jun 9 to Jun 30	<u>Corbeienses</u>, <u>Hildesheimenses</u>, <u>tail</u> toward the east
1132	<u>S. Blasii</u>
1145 May	<u>Brunwilarenses</u>, more than 14 <u>nights</u>
1208	<u>S. Stephani</u>
1214 Mar 6	<u>Colonienses</u>
1217	<u>S. Stephani</u>

TABLE A.II-6

OBSERVATIONS OF COMETS IN ITALIAN SOURCES

Date	Source and Comments
157	Varignana
390	Fasti Vindobonenses, for 30 days
418	Fasti Vindobonenses, in the east for 30 days
428 Mar 5	Fasti Vindobonenses
540	Procopius [Persian Wars, II.4], for 40 days, tail to the west
551 or 552	Agnellus, for 30 days, may be a nova
565 ?	Agnellus, August to Oct 1, put just before death of Justinian; Fasti Vindobonenses
595 ?, Jan	Paulus Diaconus, in the east and west all month
606 Apr to May	Paulus Diaconus
606 Nov to Dec	Paulus Diaconus
677 Aug	Romualdus, Paulus Diaconus, for 3 months
684 Dec 25 – 685 Jan 6	Romualdus, Varignana, dated by means of dates of Pope Benedict II (684 Jun 26 to 685 May 8)
ca. 728 Jan	Romualdus, for 10 days
840 May or Jun	Agnellus; this may be a "magical" comet.
875 Jun	Andreas Bergomatis, Novaliciense. Latter has 874 but Emperor Louis II died the following August. For 14 days, appeared in Aries.
877 Mar	Novaliciense, for 15 days, in Libra

Date	Source and Comments
990	Varignana, first in the east and then in the west a few days later
1018	Beneventani
1066	Romualdus, in the east for many days and then in the west for many days
1098 Oct	"Lupus Barensis"
1105 or 1106 Feb	Beneventani, Romualdus
1110 Jun	Beneventani, Romualdus. All month. Romualdus says it was in the north.
1222 Aug	Villola, tail to the west
1264 Jul 25 to Oct 1	Cavenses
1298 or 1299 Jul	Cavenses
1300 Oct	Rampona

TABLE A.II-7

OBSERVATIONS OF COMETS IN SOURCES FROM SCANDINAVIA AND SPAIN

Date	Source and Comments
442 Dec	Hydatius, appeared at the beginning of the month
452 Jun 18	Hydatius, in the east, then in the west in August
1222	Regii
1240	Regii
1273	Regii

OBSERVATIONS OF COMETS IN SOURCES FROM THE
BYZANTINE EMPIRE AND THE HOLY LAND

Date	Source and Comments
388 or 389	Philostorgius, first seen in the Zodiac near Venus; in 40 days it moved into Ursa Major and then vanished. Was very bright. Marcellinus compared it to Venus in brightness.
390	Marcellinus, seen for 30 days
418	Marcellinus, Philostorgius, for many months. Seen during eclipse of 418 Jul 19; see record 418 Jul 19a M,B in Section XV.4.
422 Mar	Paschale, for 10 nights, rose after cockcrow
442	Marcellinus, seen for a long time
466	Theophanes, seen for 40 days. It may not be a comet. It is described as a cloud in the shape of a trumpet.
518 or 519	Cedrenus, Paschale, Theophanes, seen in the east, tail pointed west
530 or 531	Cedrenus, Theophanes, in the west for 20 days
626 Mar	Paschale, in the west after sunset
632	Theophanes, for 30 days, extended from south to north. I think this is a "magical" comet, used to herald the rise of Islam.
734	Theophanes
744	Theophanes, in the east

TABLE A.II-8 (Concluded)

Date	Source and Comments
760	Theophanes, in the east for 10 days and the west for 21 days. Note that it was recognized as a single comet.
762	Theophanes, in the east
between 813 and 820	Georgios Hamartolos, dated only in the reign of Leo V. Had the form of 2 moons joined together.
905	Leonis, seen for 40 days. It may be magical; it was seen at the birth of the imperial heir.
974 or 975	Cedrenus, Leo Diaconus, seen for 80 days from August to October
1042 Oct 6	Cedrenus, seen for the rest of the month
1106 Feb 5	Fulcher, seen for 40 days thereafter, first seen at the "place of winter setting", tail was white like a veil of linen

APPENDIX III

A CATALOGUE OF OBSERVATIONS OF METEORS AND METEORITES AND OF MISCELLANEOUS ASTRONOMICAL MATTERS

APPENDIX III

A CATALOGUE OF OBSERVATIONS OF METEORS AND METEORITES AND OF MISCELLANEOUS ASTRONOMICAL MATTERS

There are a small number of observations of meteors, mostly in showers, and of meteorites. There are so few of these that they can all be put into Table A.III-1. A number of showers recorded in Chinese, Korean, and Japanese sources are listed by Imoto and Hasegawa [1958]. A correlation of Imoto and Hasegawa with Table A.III-1 should prove interesting.

There are also a few observations of miscellaneous astronomical matters, such as occultations and conjunctions. These are listed in Table A.III-2. Several of these observations report light seen in the circuit of the new moon or, on 1096 Aug 6, during a total lunar eclipse. The account of 1178 Jun 18 sounds quite dramatic, but it may be only a meteor that happened to pass one horn of the new moon.

The dates listed in Tables A.III-1 and A.III-2 are those that can be inferred from the sources with reasonable confidence, but they are not necessarily correct.

TABLE A.III-1

OBSERVATIONS OF METEORS AND METEORITES

Date	Source and Comments
452 or 454	Cedrenus, Marcellinus, meteorites fell in Thrace.
532	Theophanes, a shower
587	"Fredegarius Scholasticus", a meteorite
600	"Fredegarius Scholasticus"
744	Henry of Huntingdon, a shower
763 Mar	Theophanes, a shower
839 May 8	Agnellus, a shower; recorded with less detail in Fuldenses
855 Oct 17	Fuldenses, a shower all night
913 Feb 2	Sangallenses, a shower until midnight
975	Corbeienses
998	Quedlinburgenses, a meteorite fell in Magdeburg, another "beyond the river Albiam".
1029 Oct 31	Cedrenus
1032 Jul 28	Cedrenus
1093 Aug 1	Sigebertus
1095 Apr 4 or 5	Anglo-Saxon Chronicle, S. Albini, Lupus; widely reported and probably spectacular. Matthew Paris has showers in 1094 and 1096; he probably found two misdated accounts of this one.
1101	S. Maxentii, on the "16th" but does not say of what
1106 Feb 16	Egmundani, a meteor that was detached from a comet

TABLE A.III-1 (Concluded)

Date	Source and Comments
1106 or 1107 Feb 12	Egmundani, Sigebertus, meteors in Bari (Italy) in the day-time
1123 Apr 5	S. Maxentii, Cosmas, a shower
1143 Jun 15	S. Blasii, a meteorite

TABLE A.III-2

MISCELLANEOUS ASTRONOMICAL OBSERVATIONS

Date	Source, Event, Comments
577 Nov 10	Gregory of Tours, light on the moon, year is doubtful.
712 Jul	S. Florentii, Mars could not be seen from 712 Jul to 713 Jul.
807 Jan 31	Loiselianos, the moon occulted Jupiter
807 Mar 17-25	Loiselianos, a sunspot
1044 Nov	Rodulfus, Venus was agitated up and down. Rodulfus is regarded with suspicion by his editors.
1058	Rampona, a bright light within the circle of the new moon; is in 1086 in Cavenses.
1074 Oct 7	Naufragia Ratisbonensia, the moon "completed its course" with Mars and Jupiter on the same night as a lunar eclipse; double occultation?
1096 Aug 6	Cavenses, a bright light on the moon during a total eclipse
1170 Sep 13	Gervase, close conjunction of two unnamed planets
1178 Jun 18	Gervase, the upper horn of the new moon seemed to split in two and a flame shot from it. The astronomical moon was about 1 day old on this date.
1180 Dec 21	Robertus Autissiodorensis, conjunction of the new moon with a bright star at the 6th hour

TABLE A.III-2 (Concluded)

Date	Source, Event, Comments
1285	Worcester, Jupiter and Saturn in conjunction in Aquarius; it had not happened since the Incarnation. This suggests that the "star in the East" had already been interpreted as a conjunction by 1285.

APPENDIX IV

A CATALOGUE OF RECORDS OF FAMINE, PLAGUE, AND PESTILENCE

APPENDIX IV

A CATALOGUE OF RECORDS OF FAMINE, PLAGUE,
AND PESTILENCE

Famine, plague, and pestilence were ever-present threats in medieval Europe, just as famine at least still is in many parts of the world. In the German records, there is an occurrence on the average of about one each 15 years. I have found more in German sources than in those from other areas, but that is probably a consequence of the fact that I have examined more sources from Germany than from other areas.

Medieval Europe must have lived just barely above the subsistence level, so that any serious interference with agriculture resulted in famine. Thus we often find that floods, severe winters, or drought were followed by famine. There are a few famines reported to be so bad that people were reduced to cannibalism. In Italy in 1102 (Table A.IV-3) we find it reported that fruit sold for 100 denarii during a famine. The denarius probably played the monetary role of our cent, so that 100 denarii was probably the equivalent of one dollar. In purchasing power it was probably many dollars, indeed a high price to pay for a piece of fruit.

The reports are divided into three tables. Table A.IV-1 contains the reports from French sources, Table A.IV-2 those from German sources, and Table A.IV-3 those from all other areas.

Plagues of locusts are included in these tables.

As in the two preceding appendices, the dates are those inferred from the sources and they are not necessarily correct. In at least one instance the date given is probably wrong. It is highly probable that the event listed under 868 in Table A.IV-1 is the same as that under 869 in Table A.IV-2.

FAMINE, PLAGUE, AND PESTILENCE IN FRENCH SOURCES

Date	Source, Event, Comments
547	Gregory of Tours, famine after a severe winter
559	Gregory of Tours, two plagues of locusts
582	Gregory of Tours, great plague, symptoms described briefly
584	Gregory of Tours, severe plague, worst around Narbonne
585	Gregory of Tours, famine in almost all Gaul
609	"Fredegarius Scholasticus", famine throughout Gaul and Germany
808	Mettenses, pestilence after a mild winter
868	Rainaldus and Senonensis, unheard-of-famine, worst in Aquitania and Burgundy; 56 died in Sens alone. See the German reports for 869.
873 Aug 19	Elnonenses, a swarm of locusts. Rainaldus apparently dated it 868 Aug 17.
875	Rainaldus, famine
919	Hugo Floriacensis, famine
1085	S. Maxentii, plagues of locusts
1177	Gaufredus, "intolerable" drought, famine, and death
1197	S. Vincentii, great famine and mortality

APPENDIX A.IV-2

FAMINE, PLAGUE, AND PESTILENCE IN GERMAN SOURCES

Date	Source, Event, Comments
823	Laurissenses, pestilence through all France
862	Quedlinburgenses, great famine and death in Germany and elsewhere
868	Quedlinburgenses, widespread famine and death. May be the same famine as the following.
869	Xantenses, after severe floods, there was a famine so great in Burgundy and Gaul that men were reduced to eating men. See the French reports for 868.
873	Fuldenses says there was famine in Italy and Germany and many died of hunger. Quedlinburgenses says there was famine in Germany and an incredible swarm of locusts; it was not clear if the locusts and famine were connected.
895 and 897	Fuldenses, many died of famine in Bavaria in both years.
991-994	Augustani, Corbeienses, Hildesheimenses, and Quedlinburgenses. There seem to have been several years of almost continuous famine, and not just one year's famine reported under various dates. The situation was brought about by a combination of floods, severe winters, and droughts.
1006	Hildesheimenses, widespread great famine

Date	Source, Event, Comments
1017	Quedlinburgenses, pestilence great enough to stop a military campaign
1025	Corbeienses; many died of famine and pestilence.
1045	Corbeienses, great famine
1054	Hildesheimenses, famine
1056	Hildesheimenses, famine
1060	Augustani, great famine and death
1069	Augustani, great famine
1083	Hildesheimenses, pestilence
1093-1094	Augustani, Hildesheimenses, famine and pestilence over all Germany
1099	Augustani, famine after a long winter
1101	Augustani, famine after a "changeable" winter
1146	Scheftlarienses Maiores, famine
1177	Halesbrunnenses, severe winter and pestilence followed by drought and famine
1196	Scheftlarienses Maiores, great famine
1198	Neresheimenses, great famine for 3 years, may include the preceding
1211	S. Stephani, Scheftlarienses Maiores, famine and pestilence in Bavaria
1217	S. Stephani, famine in Bavaria, Austria, and Hungary

TABLE A.IV-2 (Concluded)

Date	Source, Event, Comments
1225	<u>Scheftlarienses Maiores</u>, widespread famine and pestilence

TABLE A.IV-3

FAMINE, PLAGUE, AND PESTILENCE IN SOURCES OTHER THAN GERMAN OR FRENCH

Date	Source, Event, Comments
ca. 150	Varignana, famine in Rome
ca. 230	Georgios Hamartolos, famine in Rome, cannibalism
ca. 333	Cedrenus, great famine in the East
542	Procopius [Persian Wars, II. 22], great plague in Constantinople that lasted 4 months
ca. 557	Cedrenus, bubonic plague in Constantinople
570	Paulus Diaconus, pestilence in Liguria
591	Paulus Diaconus, drought from January to September followed by famine; plague of locusts in Trento
594	Varignana, pestilence in Rome, stopped by an archangel upon the intercession of Pope Gregory
ca. 676	Paulus Diaconus, a devastating pestilence
ca. 681	Paulus Diaconus, a pestilence for 3 months, worst in Ticino
681	Brut, mortality in all Britain
685	Brut, mortality in Ireland
700	Ulster, famine in Ireland so bad that man ate man

TABLE A.IV-3 (Concluded)

Date	Source, Event, Comments
733	Theophanes, plague in Syria
ca. 745	Cedrenus, pestilence in Sicily and Calabria
793	Domitiani Latini, famine
872 Aug	Andreas Bergomatis, plague of locusts, probably in 875
941	Leodienses, famine
1054	Cedrenus, plague
1085	Cavenses, many deaths from famine
between 1086 and 1093	Ryenses, such a famine that people fought over wild plants
1102	Beneventani, many died from famine, fruit sold for 100 denarii
1114 Apr and May	Fulcher, a plague of locusts
1117	Fulcher, another plague of locusts
1126	Wissegradensis, widespread famine
1145	Reicherspergenses, great famine. He also explains that Saturn is the coldest planet. The season in which the sun passes Saturn will be colder than average. It is not clear if this is related to the famine.
1149	Egmundani, severe winter followed by pestilence
1166	Reicherspergenses, famine. Dandulus, pestilence
1225	Mellicenses, pestilence among the cattle with resulting mortality of men (from hunger?)
1230	Ryenses, pestilence of man and beasts

APPENDIX V

A CATALOGUE OF RECORDS OF UNUSUAL OR
FANCIFUL EVENTS

APPENDIX V

A CATALOGUE OF RECORDS OF UNUSUAL OR
FANCIFUL EVENTS

These are a few events that I have noted in case they might prove useful to someone. There are so few that one table suffices for all areas. Many of the reports are of monstrous births. Several concern dragons. It is hard to tell whether these are fictional or not. In 1093 and 1208 dragons are mentioned in connection with solar eclipses. Now eclipses can occur only when the sun and moon are near nodes of the moon's orbit. The nodes were anciently called the head and tail of the dragon in many astronomical works, for reasons that I do not know. It is possible that someone thought that a reference to the sun being in the head of the dragon, for example, meant that a dragon had been seen trying to swallow the sun. It is also conceivable that auroras were sometimes interpreted or described as the flames from a dragon.

As usual in accounts that do not deal with eclipses, the dates listed are not guaranteed.

TABLE A.V-1

UNUSUAL OR FANCIFUL EVENTS

Date	Source, Event, and Comments
364 Oct	Theophanes, girl born with 2 heads, lived 1 month
496	Marcellinus, India sent to the emperor a gift of an elephant and 2 giraffes
between 518 and 527	Georgios Hamartolos, a woman in Cilicia who was a cubit taller than any man
after 590	Simocatta, a boy without eyes, eyelids, eyebrows, or fore-arms. Later, a boy with 4 legs.
ca. 591	Cedrenus, a very large dog born with 6 feet and a lion's head
ca. 595	Cedrenus, a child born with 4 feet and 2 heads
ca. 602	Dandulus, two animals who were a man and woman down to the waist were seen above water in the Nile; though adjured in the name of God, they stayed visible for many hours.
ca. 690	Brut, the milk and butter turned to blood. This may be a cattle disease.
757	Mettenses, Emperor Constantine sent King Pepin a pipe organ.
807	Fuldenses, Harun al-Rashid gave Charlemagne a clock that struck the hours.

Date	Source, Event, Comments
879	Ulster, body of a woman thrown out of the sea was 195 feet tall, with fingers 7 feet long, whiter than a swan.
ca. 942	Rampona, a woman lived in Gascony who was divided from the waist up.
ca. 956	Floriacenses, a dragon without a head was seen in the sky.
974	Leo Diaconus, twins joined from waist to shoulders
998	Hildesheimenses, male quintuplets in Bavaria
1064	Diceto, female twins joined from waist to shoulders
1093	Hildesheimenses, a dragon was seen; it is in the same sentence as the solar eclipse of 1093 Sep 23.
1099	Scotorum, girls joined from waist to shoulders; one wonders if such accounts are independent.
1126	Mortui-Maris, in Spain a monster was born, upper parts man and lower parts dog.
1130 Oct 8	Wissegradensis, a serpent was seen flying over Bohemia.
1133	Anselmus [ca. 1135], a child with 2 heads born in S. Riquier
1166	Blandinienses, a child with 3 heads
1171	S. Rudberti, [ca. 1286], a calf born in the mountains with 2 heads, 8 feet, and 2 tails

TABLE A.V-1 (Concluded)

Date	Source, Event, Comments
1196	Elnonenses Maiores, an early example of inflation; grain rose to the unheard-of price of 50 solidi per rasera.
1208	Dunstable, monsters fought with the sun and moon, and a horrible eclipse happened. If this is a fanciful description of an eclipse, it cannot be identified.
1233 Jun	Wendover, two dragons fought in the sky.
1277	Rampona, a woman from Modena named Antonia had 42 children, including quintuplets, triplets, and 11 sets of twins. She was exiled from Modena.

APPENDIX VI

A CATALOGUE OF OBSERVATIONS OF AURORAS

APPENDIX VI

A CATALOGUE OF OBSERVATIONS OF AURORAS

There are a moderate number of observations of auroras, but not enough to warant putting
them into different tables for different geographical regions. The place where an aurora is observed
is usually not important, although we expect to find
more in northwestern than in southeastern Europe.

It may be possible to use auroras occasionally for chronological purposes. For example, an
aurora on 1117 Dec 16 is well dated and was seen in
Jerusalem. One seen that far to the southeast must
have been spectacular, and indeed we find it reported
in many western sources.

It is possible that some of the events
listed were not auroras. Lightning displays and
auroras could easily be confused. The event listed
under 1039 or 1040 Apr 6 is particularly suspicious,
since the light is recorded as appearing in the southeast.

The dates, as usual, are those inferred
from the sources and they may be in error in some
cases.

TABLE A.VI-1

OBSERVATIONS OF AURORAS

Date	Source and Comments
450 Apr 4	Hydatius; year inferred from weekday
512	Marcellinus
525	Agnellus, year uncertain
552 Jul 25	Rampona, year uncertain
556	Cedrenus; Theophanes; probably the same event, year uncertain
560 Nov 11	Agnellus
563	Gregory of Tours
580	Gregory of Tours, year uncertain
582	Gregory of Tours, on Easter night
584	Gregory of Tours, many this year
585	Gregory of Tours, two this year
590	Gregory of Tours, the accompanying Easter data indicate 591
600	"Fredegarius Scholasticus"
654	Xantenses
664 summer	Ulster
710 or 714	Brut in 710; Cambriae in 714
762 autumn	Ulster
765	Melrose
793	Domitiani Latini says dragon flames seen
807 Feb 26	Loiselianos, very bright, date fixed by a lunar eclipse the same night
817 Oct	Xantenses

TABLE A.VI-1 (Continued)

Date	Source and Comments
829 Dec 3	Weissemburgenses Minores
836 Feb	Xantenses
839 Mar 25	Xantenses
840	Xantenses, 2 nights probably early in the year
841	Lugdunenses, very bright
846	Weissemburgenses Minores
861	Rainaldus
890 Jan 1	Scotorum; Ulster
923 Feb 1	Senonensis; Hugo Floriacensis
937 Feb 24	Senonensis
965	Rainaldus
966 May 12	S. Florentii, his chronology is often poor; S. Maxentii has 968 and very bad chronology; this may be the same as below.
971	Suevicum
975	Leo Diaconus, apparently in Constantinople
978	Melrose
979 Oct 28	Sigebertus, is often wrong by 5 years
992 Oct 21 and Dec 26	Quedlinburgenses, may be off 2 years
1032	Domitiani Latini, description is ambiguous
1034 Sep	Cedrenus
1039 or 1040 Apr 6	Sigebertus, in southeast and could be lightning
1093 Aug 1	Matthew Paris

TABLE A.VI-1 (Concluded)

Date	Source and Comments
1098 Sep 21, 26, or 27	Sigebertus, Augustani, and others. Sep 27 is the most likely date.
1117 Dec 16	Fulcher, observed in Holy Land; Anselmus [ca. 1135] and other western sources, date confirmed by an eclipse.
1117 Dec, after Christmas	Hildesheimenses, may be different from above
1128 Nov 19	Wissegradensis, date confirmed by an eclipse
1132 Jan 14	Wissegradensis, date confirmed by an eclipse
1137 Oct	Bertholdus [ca. 1138]
1138 Feb 26	Wissegradensis
1138 Oct 14, 15, and 16	Wissegradensis
1139 Mar	Wissegradensis says the 7th nones March, but there is no such day.
1162 Sep 15	Reicherspergenses
1173 Feb 10 or 11	Gervase; Magdeburgenses; Theodorus
1174 Nov 4	Worcester; Diceto
1177 Nov 29	Diceto; Gervase
1188 Oct 12	Gervase
1188 Dec 20	Gervase
1192 Jan 15	Aquicinctina
1195 Sep 10	Gervase
1204 Apr 1	Wendover
1205 Jun 24	Ralph of Coggeshall; this may not be an aurora.
1263	S. Stephani

APPENDIX VII

A CATALOGUE OF RECORDS OF VOLCANIC ACTIVITY

APPENDIX VII

A CATALOGUE OF RECORDS OF VOLCANIC ACTIVITY

The few records of volcanic activity that have been found can all be put into a single table. Six of the eleven records in the table concern Vesuvius. All concern Mediterranean regions except for two that concern Germany. At least one of these seems suspicious: Widukindus [ca. 973] reports that the mountain near Quedlinburg erupted in flames in 937, but the annals from Quedlinburg itself [Quedlinburgenses, ca. 1025] do not mention this remarkable event.

TABLE A.VII-1

RECORDS OF VOLCANIC ACTIVITY

Date	Source, Event and Comments
472	Marcellinus, Vesuvius
505 Nov 9	Paschale Campanum, Vesuvius
512 Jul 8	Paschale Campanum, Vesuvius
585	Gregory of Tours, two islands erupted
685 Mar	Paulus Diaconus, Romualdus, Vesuvius
ca. 726	Cedrenus, Thera
937	Widukindus, in Quedlinburg, Germany. Local records have no such event.
996	Diceto, in Sicily
1037 Jan 27	Cavenses, Vesuvius
1117	Pegavienses, earth boiled up in Swabia and then collapsed
1140	Herbipolenses, Vesuvius

APPENDIX VIII

A CATALOGUE OF RECORDS OF EARTHQUAKES

APPENDIX VIII

A CATALOGUE OF RECORDS OF EARTHQUAKES

Earthquake activity, unlike volcanic
activity, seems to be spread over most of Europe.
The records of earthquakes are listed in Tables
A.VIII-1 through A.VIII-8. The tables are arranged
by geographic region, divided in the same way as
the tables in Appendix I. It should be noticed
that the records are arranged according to the area
where the record was written, not according to the
area where the earthquake occurred. Although earthquakes may occur almost anywhere, they are definitely
concentrated in the Mediterranean regions.

In Table A.VIII-7, the term Dacia is probably a name for Denmark or some part of it, rather
than being the former Roman province of the same
name.

As usual, the dates listed in the tables
are those inferred from the sources, and I have done
little research aimed at determining the correct
dates. One date, however, has caused chroniclers
some trouble. It is intimately connected with reports of the solar eclipse of 1133 Aug 2, and research was needed into the date for the main purposes of this work. The results that concern the
date of the earthquake may be summarized here.

The Anglo-Saxon Chronicle [ca. 1154] put
the eclipse on 1135 Aug 2. William of Malmesbury
[ca. 1142] put it on 1132 Aug 5, and he put the
earthquake two days later, in the early morning.
However, shortly after the sentence in which he
gave the time of the earthquake, he mentioned the
earthquake and the eclipse in the same sentence,
and this caused some later writers to think that
the two were on the same day. We should not be
surprised, in fact, to find that the continuator
of Florence [ca. 1118] seemed to think that there
were two earthquakes, one on the day of the eclipse
and another two days later. The combination of the

errors in the Anglo-Saxon Chronicle and in William, plus some correct records, plus the wrong reading of William, gives rise to an almost infinite variety of possibilities in dating the eclipse and the earthquake.

I do not know why the Chronicle and William misplaced the year of the eclipse, but it is easy to show that the chroniclers worked from authentic records. The Chronicle indeed has the correct day of the year.[†] William says that the eclipse was on the "nones of August" and on the 4th feria. Now 1132 Aug 5 (the nones of August) was on the 6th feria, while 1133 Aug 2, the actual day of the eclipse, was on the 4th feria. Thus William must have had a record that was basically correct, but for which he somehow misread the year. He should have written "the 4th nones August, on the 4th feria"; the need for two "4th's" in quick succession probably caused one of them to be omitted. He then says that the earthquake was in the early morning of the 6th feria following the eclipse.

Hence the earthquake was in the early morning of 1133 Aug 4, the 6th feria, local time.

[†] The chronicler perhaps confused the year of the eclipse with the year in which King Henry I died.

TABLE A.VIII-1

EARTHQUAKES IN BRITISH SOURCES

Date	Source, Place if Given, Comments
ca. 685	Brut, Cambriae, in Brittany
974	Florence
1048 May 1	Anglo-Saxon Chronicle; Melrose puts it in Worcester.
1060 Jul 4	Anglo-Saxon Chronicle
1076 Mar 27	Matthew Paris
1076 Apr 22	Wykes
1081 Mar 27	Matthew Paris, probably the same event as 1076 Mar 27 but recorded twice
1089 Aug 11	Anglo-Saxon Chronicle
1110	Florence
1113	Matthew Paris, near Antioch
1117 Jan 3	Waverley; later he says there was an earthquake in Lombardy, but it is unclear if the two are the same. Florence and others have an earthquake in Lombardy in 1117.
1119 Sep 28 or 29	Anglo-Saxon Chronicle, Margan
1122 Jul 25	Anglo-Saxon Chronicle
1129 Dec 5	Anglo-Saxon Chronicle
1133 Aug 4	William of Malmesbury; see the main text for a discussion of the dating.
1165 Jan 25	Gervase. This is probably the correct date. Both 1164 and 1165 are given in other sources, and both Jan 25 and 26.

Date	Source, Place if Given, Comments
1170 Jun 29	Melrose, in the east, probably means the Orient
1184 Jan 16	Wykes, about midnight, in many parts of England
1185 Apr 15	Waverley, many places in England; Ralph of Coggeshall has it on Apr 17. Matthew Paris puts it at Lincoln.
1201 Jan 8	Florence
1201 May 22	Diceto, in Dorsetshire
1246 Feb 19	Wykes, in all the west, presumably of Great Britain. This is probably the event that Florence has on 1247 Mar 1 and Worcester on 1247 Mar 10.
1275 Sep 15	Brut, about vespers, in Wales. Florence has probably the same event on 1275 Sep 11.
1287 Mar 15	Cambriae

TABLE A.VIII-2

EARTHQUAKES IN SOURCES FROM BELGIUM OR
THE NETHERLANDS

Date	Source, Place if Given, Comments
950	Sigebertus, many earthquakes in Gaul and Germany
1000	Leodienses, described as very great
1013 Nov 16	Leodienses
1081 Mar 27	Leodienses
1095 Sep 10	Sigebertus
1117 Jan 3	Anselmus ⌊ca. 1135⌋
1117 May 3	Anselmus ⌊ca. 1135⌋, at vespers, greater than the one in January
1121 Dec 10	Anselmus [ca. 1135]; is 1122 Dec 10 in French sources
1129 Nov 2	Egmundani

TABLE A.VIII-3

EARTHQUAKES IN SOURCES FROM CENTRAL EUROPE

Date	Source, Place if Given, Comments
944 Apr 16	Sangallenses
1021 May 12	Mellicenses
1117 Jan 2 or 3	Cosmas, Mellicenses, from Bohemia to Lombardy, much damage
1155 Jan 18	Engelbergenses
1158 Jan 21	Engelbergenses
1161 Feb 22	Engelbergenses
1161 Jul 16	Engelbergenses
1162 Feb 25	Engelbergenses, 2 in the same night
1162 Dec 2	Engelbergenses
1163 Sep 27	Admuntenses
1170 Apr 1	Engelbergenses, one in the late afternoon, one near midnight
1175 Apr 30	Engelbergenses
1201 May 4	Admuntenses, Lambacenses, S. Rudberti [ca. 1286], in many places. S. Rudberti says quakes lasted 1 1/2 years in the village of Longou.
1267 May 8	Lambacenses, S. Rudberti [ca. 1286]
1267 Oct 30	Lambacenses

TABLE A.VIII-4

EARTHQUAKES IN FRENCH SOURCES

Date	Source, Place if Given, Comments
580	Gregory of Tours, in Burdegalensis, probably Bordeaux
582	Gregory of Tours, in Anjou
584	Gregory of Tours, in Anjou
ca. 590	Gregory of Tours; data connected with the date are self-contradictory.
801 Apr 30	Sithienses; Mettenses says it was serious in Italy, Gaul, and Germany.
803	Mettenses, probably at Aachen, many killed
842 Nov 5	Nithardus, felt in most of Gaul
ca. 849, Feb 15	Floriacenses
1000 Mar 29	Elnonenses; also in the Belgian source Leodienses
1079 Jul 17	Nivernenses, described as "everywhere"
1083 Mar 21	S. Albini, under 1082 but auxiliary data prove 1083
1083 Oct 18	S. Maxentii
1098 Oct 4	Rainaldus
1102 Jan 30	Rainaldus
1105 Apr	S. Maxentii
1116	Mortui-Maris, in Antioch
1117 Jan 2 or 3	S. Maxentii, in Italy
1122 Dec 10	Mortui-Maris; date is 1121 Dec 10 in Belgian sources.

TABLE A.VIII-4 (Concluded)

Date	Source, Place if Given, Comments
1155 Jan 18	<u>Robertus de Monte</u>, in Burgundy
1163 Aug 1 or 2	S. <u>Albini</u>
1165 Jun 20	S. <u>Albini</u>, S. <u>Florentii</u>
1207 Feb 25	S. <u>Albini</u>

TABLE A.VIII-5

EARTHQUAKES IN GERMAN SOURCES

Date	Source, Place if Given, Comments
801 Apr 30	Laurissenses, in Italy
845 Mar 22 and 29	Xantenses, 2 earthquakes in Worms
858 or 859 Jan 1	Fuldenses and Xantenses, in Worms and Mainz; part of a church collapsed in Mainz.
867 Oct 9	Fuldenses, many places, presumably in Germany
880 or 881 Dec 30	Fuldenses, great in Mainz
944 Apr 16	Herimannus, Suevicum
954	Augienses, many places in Germany and Gaul
994	Augustani, church in Augsburg collapsed
998	Quedlinburgenses, all Saxony
1012	Quedlinburgenses
1013	Quedlinburgenses, a hole opened in the earth in Lüneburg. This need not be an earthquake.
1021 May 12	Hildesheimenses and many other sources, in Bavaria
1035 May 23	Corbeienses
1046 Nov 9	Naufragia Ratisbonensia
1048 Oct 13	Naufragia Ratisbonensia, Augustani
1065 Mar 27	Augustani, in Italy
1092 Jun 26	Augustani, 3 earthquakes in 1 day in Hungary

Date	Source, Place if Given, Comments
1112 Apr 20	Aquenses
1117 Jan 3	Bertholdus [ca. 1138], Corbeienses, and others. In Saxony. S. Blasii says there were 2, and many buildings collapsed.
1117 Jul 1	Hildesheimenses, "unheard-of magnitude", worst in Italy
1127 Mar 25	S. Blasii
1127 Apr 13	Scheftlarienses Maiores, described as "night and day"
1128 Jun 29	S. Blasii
1134 Jan 12	S. Blasii
1141 Apr 24	Aquenses
1167 Jan 20	Colonienses
1179 Aug 1	Colonienses, Brunwilarenses
1183 Apr 30	S. Stephani
1189 Feb 27	S. Disibodi
1189 Aug 15	Scheftlarienses Maiores
1201 or 1202 May 4	Monacensis, Scheftlarienses Maiores
1215 Aug 28	Colonienses
1222 Jan 11	Colonienses, in Cologne
1222 Dec 25	Colonienses, much damage in Lombardy, continued for 2 weeks
1236 or 1237 Sep 16	Neresheimenses, Scheftlarienses Maiores, Augustani Minores

TABLE A.VIII-6

EARTHQUAKES IN ITALIAN SOURCES

Date	Source, Place if Given, Comments
429 Aug 26	Fasti Vindobonenses, on the "day of the Sun", so either the year or the day is wrong by 1.
455 Sep 7	Fasti Vindobonenses, on the "day of Venus", so either the year or the day is wrong.
492 May 26	Fasti Vindobonenses
ca. 520	Romualdus, near Vienne in Gaul
801	Farfenses
990 Oct 25	Beneventani, many killed, many buildings collapsed
1005	Casinates, persisted for 15 days, some damage
1044 Apr 19	Beneventani
1088 Sep 10	Romualdus, in Apulia
1094 Jan 14 and 17	Beneventani, much damage to the city, presumably Benevento
1115 Dec	Romualdus, in Syria
1117 Jan 13	Dandulus, in Venice, sulfur waters emitted, houses and one church burned
1117 Apr	Romualdus, across northern Italy and into Transalpine Gaul
1168 or 1169 Feb 4	Cavenses, Romualdus, in southern Italy and Sicily
1190 Apr 2	Villola
1222 or 1223 Dec 25	Rampona; Villola has earthquakes on both dates, in Bologna.
1279 Apr 24 and 30	Dandulus, in Venice

TABLE A.VIII-7

EARTHQUAKES IN SCANDINAVIAN AND
SPANISH SOURCES

Date	Source, Place if Given, Comments
452	Hydatius, in Galicia
454	Hydatius, in Galicia
1076 Apr 22	Esromenses
1173	Esromenses, in Dacia
1198	Esromenses, in Dacia
1272	Eskinbek, near "John before the Latin gate"

TABLE A.VIII-8

EARTHQUAKES IN BYZANTINE SOURCES OR FROM
THE HOLY LAND

Date	Source, Place if Given, Comments
ca. 333	Cedrenus, on Cyprus
ca. 340 and 341	Cedrenus, more than one at Antioch; Consularia Constantinopolitana says "all year".
ca. 348	Cedrenus, destroyed much of Beirut
358 Aug 24	Ammianus Marcellinus, in Nicomedia (modern Izmit); Consularia Constantinopolitana
368 Oct 11	Consularia Constantinopolitana, in Nicaea; Paschale
394 Sep to Nov	Marcellinus, many widespread earthquakes
396	Marcellinus, for many days
402	Marcellinus, at Constantinople
407 Apr 1	Paschale, apparently at Constantinople
408	Marcellinus, the forum of Peace at Rome rumbled for 7 days.
417 Apr 20	Paschale
419	Marcellinus, in Palestine
422	Paschale
ca. 435	Cedrenus, a series of earthquakes at Constantinople for 3 years. Evagrius says an earthquake made many openings in the earth in the time when Attila was king of the Huns.

Date	Source, Place if Given, Comments
447	Marcellinus, many and wide-spread
467	Marcellinus, in Ravenna
ca. 477, on Sep 25	Cedrenus, many killed in Con-stantinople; Theophanes has it on 478 Sep 25.
480	Marcellinus, persisted for 40 days. I suspect this is an ex-aggerated account of the pre-ceding.
494	Marcellinus, four cities in Asia Minor damaged
499	Marcellinus, in Pontic province, probably northeast Asia Minor
ca. 502	Cedrenus, at Constantinople
518	Marcellinus, 24 buildings des-troyed in Dardania
ca. 524	Cedrenus, in Cilicia
526	Marcellinus, in Antioch, may be the same as the preceding
ca. 528, on Nov 29	Cedrenus, in Antioch, specific-ally mentioned as following another large one. Theophanes has it on 528 Nov 29 as I understand his chronology.
543 Sep 6	Cedrenus, Cyzicus (near Con-stantinople) severely damaged
ca. 550, on Jul 9	Cedrenus, all around the eastern end of the Mediterranean
554 Aug 15	Cedrenus; Theophanes, severe in Constantinople and Nicomedia
555 Jul	Cedrenus, in the hills, does not say which ones

TABLE A.VIII-8 (Continued)

Date	Source, Place if Given, Comments
ca. 557–558	Cedrenus, apparently a series that lasted for most of a year. The walls of Constantinople were damaged.
ca. 560	Cedrenus, through much of Asia Minor
582 or 583 May 10	Cedrenus
ca. 717	Cedrenus, in Syria
740 Oct 26	Cedrenus, very severe in Constantinople, shocks continued for 11 months.
745 or 746 Jan 18	Cedrenus, Theophanes, in Palestine and Syria
790	Cedrenus
967	Leo Diaconus, in the summer, in Constantinople and in Galatia
975	Leo Diaconus, on the evening of "Demetrius' Day"
984 or 985 Oct	Cedrenus
1009 Jan–Mar	Cedrenus, a series for 2 months
1032 Aug 13	Cedrenus
1034	Cedrenus, for 40 days at Jerusalem
1036 Dec 18	Cedrenus
1037 Nov 2	Cedrenus
1040 Feb 2	Cedrenus
1041 Jun 10	Cedrenus
1114 Aug 10	Fulcher; this and the following were at Mamistra; I do not know its modern equivalent.

TABLE A.VIII-8 (Concluded)

Date	Source, Place if Given, Comments
1114 Nov 13	Fulcher

Note: In the 10th century and after, the place is probably Constantinople if it is not given.

APPENDIX IX

A CATALOGUE OF METEOROLOGICAL RECORDS

APPENDIX IX

A CATALOGUE OF METEOROLOGICAL RECORDS

The meteorological records are listed in seven tables, arranged by the area to which the source belongs; this is not necessarily the area where the event occurred. A few observations, or types of observation, need short discussions.

The tables include some instances in which wolves entered cities. These are included here because such incursions perhaps resulted from severe weather and near-starvation conditions in the countryside.

The tables include what are probably tidal waves, or tsunamis if the reader prefers. They are included here mainly as a matter of convenience, although a genuine wave perhaps resulted from tectonic rather than meteorological activity. However, the source of the flooding was not always clear. In some cases the flooding may have been produced by the weather.

There are also a large number of observations of multiple suns or moons, or of rings around the sun or moon, and other observations that were probably caused by refraction or reflection phenomena. Minnaert [1954] gives a good discussion of these phenomena. They can be caused either by the sun or the moon. For simplicity I shall refer only to the sun. Minnaert gives a theoretical and quantitative discussion of the events; here it is sufficient to give only a qualitative description.

First there are the halos. The primary halo has a radius of 22°. It is usually the brightest and hence it is the most often seen. There are other and larger halos. Three halos are recorded on several occasions. Matthew Paris [ca. 1250] records 4 halos on 1104 Jun 7 (in Table A.IX-1), but he is a late derivative source and may be in error. Wissegradensis [ca. 1142] also has 4 halos, on 1135 Mar 7.

-739-

Next there is the "parhelic" circle. This
is a circle having the same elevation above the horizon as the sun at all points, hence its center is
the zenith.

Finally, so far as frequent phenomena are
concerned, there is a vertical pillar of light
through the sun. It rarely extends more than about
15° from the sun and hence it usually does not reach
to the primary halo. It is usually brighter above
the sun than below. There is as yet no quantitative
explanation of the pillar.

Any of these apparitions can occur alone
or in almost any combination with the others. Further, almost any one can occur only partially, and
almost any part of any one can occur in a detached
manner. Thus we can have a variety of configurations.
The configuration that seems to have been most striking to a medieval Christian is the one that is usually called "the sign of the cross in the sun" in
the medieval annals. It results when the parhelic
circle and the vertical pillar are prominent. Many
medieval sources contain drawings of this apparition.

There is often an intensification of light
at or near where the parhelic circle and the vertical pillar cross a halo, although the intensifications
seem to result from independent phenomena and not
just from the combination of the causes for halo,
circle, and pillar. Sometimes conditions are such
that only the intersection points appear bright.
Under these conditions we have "multiple" suns.

All these appearances are caused by finely
divided ice at high altitudes in the atmosphere.
Hence they are the result of meteorological conditions and are included in this appendix. The reader
should remember that "moon" can be substituted for
"sun" throughout the preceding discussion.

Three records, all in Table A.IX-5, need
discussion.

It is recorded that a block of ice mea-

suring 15 feet by 7 feet by 2 feet fell from the
sky near the time of the summer solstice in 824.
This seems to be a record size for a hailstone.
The reader is at liberty to doubt the accuracy of
the measurements.

On 841 Jul 28 it is recorded that there
were three circles in the sky, but only the smallest one surrounded the sun. There was one in the
north, and there was also one in the west that
seemed to touch the sun. The latter may be an
occurrence of the full parhelic circle. I do not
know how to identify the one in the north. It may
be what Minnaert (pp. 199-200) calls the circumzenithal arc. He says that only one half of the
circle can ever be seen according to the accepted
theory, but he cites a claim that the full circle
was seen once in the 19th century. The 841 record
may be another instance.

In 1116, in December, it is recorded that
a second moon rose from the west as the full moon
was rising from the east. If the year is correct,
the date is near Dec 21; we cannot assume that the
moon was literally full. The "moon" rising in the
west may have been the lunar equivalent of what
Minnaert (p. 203) calls the "anthelion" and that
he lists among "very rare and doubtful" phenomena.
Since there is no satisfactory theory of the vertical pillar, we may have to allow the possibility
that it is part of a vertical circle passing through
the zenith. A mock sun (or in this case, a moon)
might then conceivably arise where this hypothetical
extension of the pillar meets the parhelic circle.

The reader is cautioned for the last time
that the dates in the tables have not been subjected
to careful investigation, and they may be in error.

TABLE A.IX-1

METEOROLOGICAL REPORTS FROM THE BRITISH ISLES

Date	Source, Events, Comments
762	Ulster, great snow
776	Anglo-Saxon Chronicle, red cross in sky after sunset
806 Jun 4	Domitiani Latini, cross about the moon, gives a drawing. Near dawn.
806 Aug 30	Domitiani Latini, halo around the sun at the 4th hour
909 May 6	Scotorum, two suns
969	Four Masters, two equal suns
1014 Sep 28	Anglo-Saxon Chronicle, Domitiani Latini, sea destroyed many villages in England
1072	Worcester, severe cold
1089	William of Malmesbury [ca. 1125], bad harvest
1091 Oct 17	William of Malmesbury [ca. 1125], storm blew off the roof of Bow Church in London
1092	William of Malmesbury [ca. 1125], Salisbury Cathedral struck by lightning on S. Osmund's Day
1092 and 1093	Four Masters, great snow and cold both years
1093	William of Malmesbury [ca. 1125], bad floods and ice
1094	William of Malmesbury [ca. 1125], crops failed with ensuing famine
1097 Sep 29	Florence, cross in the sky

Date	Source, Events, Comments
1099 Nov 3	Anglo-Saxon Chronicle and Melrose, probably a tidal wave with many drowned; William [ca. 1125] has a flood in the Thames in 1099.
1104 Jun 7	Ralph of Coggeshall, many circles around sun; Matthew Paris says 4.
1105 or 1106	Wykes and Florence respectively; two full moons seen on Holy Thursday
1114 Oct 10	Eadmer, extremely low tide in Medway and Thames
1156 Oct	Wykes, cross about the moon
1164 Sep 18	Wykes, 3 circles around the sun at the 9th hour; they disappeared and soon 2 suns appeared.
1172 Dec 25	Diceto, Ralph of Coggeshall, tremendous thunderstorm
1177 Dec 1	Diceto, damaging windstorm
1178 Jan 12	Waverley, great flood in Holland
1179 Jan 7	Gervase, great storm
1184 Feb 1	Wykes, thunderstorm
1191	Wykes, sign of the cross with a crucified figure on it
1200	Matthew Paris, dated "before Christmas", 5 moons
1201 Aug	Wykes, flooding
1205	Worcester, I believe in December, two full moons seen in the daytime
1208 May	Wendover, 3 crosses in the sky seen in Holland
1217 Apr 26	Worcester, 2 suns

Date	Source, Event, Comments
1218 Feb 8, Sep 14, and Nov 30	Wendover, thunderstorms
1218 Dec 12	Wendover, windstorm
1228	Wendover, thunderstorms all summer
1230 Mar 23	Wendover, terrible thunderstorm; following summer was rainy.
1233 Apr 8	Wendover, 4 suns around the natural sun
1250, ca. Sep 29	Worcester, tidal wave, men and beasts drowned
1270 Mar 25	Wykes, rains such as had not been seen since Noah
1284 Apr 9	Wykes, great storm and darkness
1285 May 8	Florence, two moons
1286 Jan 1	Wykes, tidal wave from North Sea

TABLE A.IX-2

METEOROLOGICAL REPORTS FROM BELGIUM AND
THE NETHERLANDS

Date	Source, Event, Comments
764	Laubacenses, great freeze
1129 May 13	Anselmus [ca. 1135], great tempest at Cologne. Year is given as 1128 but 1129 is probably correct.
1134 Oct 1	Anselmus [ca. 1135], a wave that exterminated 3 counties with the people and animals therein
1149	Egmundani, a severe winter followed by pestilence
1152 Mar 22	Egmundani, circle around moon and rays "like a cross reached from the moon to the circle"
1248	Egmundani, a winter with no snow but followed by heavy floods

TABLE A.IX-3

METEOROLOGICAL REPORTS FROM CENTRAL EUROPE

Date	Source, Event, Comments
821-822	Mellicenses, snow lasted from 821 Sep 22 to 822 Apr 12
913 Apr 13	Sangallenses, heavy snow
944	Sangallenses, wet summer
1091	Sasavensis, neither snow nor rain all winter
1118 Sep	Cosmas, rain this month as not see since the flood
1125 May 20	Cosmas, great snow and freeze
1135 Mar 7	Wissegradensis, 4 circles around the sun
1139 Jul 19	Wissegradensis, darkness that lasted a week
1162	Reicherspergenses, good crops this year
1166 Jun 24	Admuntenses, Cremifanenses, hailstones as big as eggs
1210 May 30	Mellicenses, great rain followed by floods that killed many
1210 Aug 9	Mellicenses, rain for a week that badly damaged the crops
1222 May 22	S. Rudberti [ca. 1286], devastating hail the size of eggs

TABLE A.IX-4

METEOROLOGICAL REPORTS FROM FRENCH SOURCES

Date	Source, Event, Comments
547	Gregory of Tours, severe winter followed by famine
563	Gregory of Tours, 4 suns
580	Gregory of Tours, a wolf entered Poitiers, Loire flooded heavily, a great wind flattened houses.
582 Jan	Gregory of Tours, severe thunderstorm
582	Gregory of Tours, rain of blood in Paris, wolves entered Bordeaux.
584	Gregory of Tours, roses bloomed in January, vines became diseased.
587	"Fredegarius Scholaticus", excessive floods in Burgundy
589	Gregory of Tours, unusually warm autumn, roses bloomed in November.
800 Jul 6 and 9	Mettenses, freezes on both days
806 Jun 5 ?	Rainaldus, cross, cannot tell if sun or moon. Chronological data are inconsistent.
806 Aug 19 or 30	Rainaldus, halo about the sun; chronology is confused; see Table A.IX-1.
808	Mettenses, very mild winter followed by a pestilence
868	Senonensis, famine that was finally relieved by a good crop

Date	Source, Event, Comments
959	Folcwinus, cross seen
1037	S. Benedicti, the Loire flooded with much damage
1055 Aug 1 ?	Senonensis, Godellus, severe tempest that ruined the crops and killed many people and animals; the chronological data are self-contradictory.
1058 Nov 1	Godellus, a rain of blood
1077 Dec 28	Mosomagenses, severe thunderstorm
1096 Aug 7	S. Maxentii, a cross in the sky
1126	Mortui-Maris, severe winter followed by famine
1128 Dec 9	Mosomagenses, many suns seen
1135 Oct 28	Rotomagensi, a violent wind that blew down a church tower
1137	Gaufredus, a great drought for 6 months
1139 Jul 2	Laudunense, a great tempest
1141 Jan 19	Laudunense, a great wind
1142	Gaufredus, severe winter
1144 Jan 19	Rotomagensi, a great wind. Could this be the same as 1141 Jan 19?
1156	Robertus de Monte, a great flood at Rome
1156	Mortui-Maris, cross about the moon
1157	Mortui-Maris, 3 moons with a cross in the center
1159	Mortui-Maris, 3 suns in the west

TABLE A.IX-4 (Concluded)

Date	Source, Event, Comments
1177	Gaufredus, "intolerable" drought followed by famine
1193	Aquicinctina, rain at every new moon from March to December except October
1195 Oct 10	Elnonenses, wind that destroyed trees and buildings

TABLE A.IX-5

METEOROLOGICAL REPORTS FROM GERMAN SOURCES

Date	Source, Event, Comments
722	Quedlinburgenses, good crops
744	Xantenses, a rain of ashes
764	Quedlinburgenses, a severe winter
806 Jun 4	S. Maximini, cross about the moon near dawn
806 Aug 30	S. Maximini, halo around the sun
824, ca. Jun 21	Laurissenses, near Autun a tempest during which a piece of ice fell that measured 15 feet by 7 feet by 2 feet
838 Dec 26	Xantenses, a windstorm followed by flooding from the sea
839 Nov	Quedlinburgenses, wind that damaged trees and buildings. There is no such date as that given; Nov 8 is probably meant.
841 Jul 28	Xantenses, 3 circles in the sky. Smallest one surrounded the sun but others did not.
860 Feb 5	Xantenses, thunderstorm
869 Feb 15	Xantenses, storm followed by floods and famine
869 Jul 1	Augienses, lightning storm
873	Corbeienses, floods, also locusts
942	Corbeienses, floods
968	Corbeienses, floods

Date	Source, Event, Comments
974	Corbeienses, severe drought all summer
981	Corbeienses, year's drought
991	Augustani, floods, crops ruined
993-994	Augustani, Corbeienses, Hildesheimenses, Quedlinburgenses, severe drought followed by severe winter and famine
1011 Jul 30	Quedlinburgenses, apparently a hailstorm
1012 Aug 10	Quedlinburgenses, great storm
1013 May 15	Quedlinburgenses, a tempest that destroyed buildings
1013 Dec 15	Quedlinburgenses, great flood
1020 Jul 18	Quedlinburgenses, halos around the sun for several hours
1044 Nov 22 - Dec 6	Naufragia Ratisbonensia, sun not seen between these dates
1046	Corbeienses, heavy snow, cold, and storms
1063 Mar 21	Augustani, severe freeze and heavy snow
1076 Nov 1 - 1077 Apr 1	Augustani, hard winter with much snow, bad crops followed
1099	Augustani, long winter followed by bad crops and famine
1101	Augustani, bad winter and famine
1116, ca. Dec 21	Hildesheimenses, at the full moon, a second moon rose from the west.

TABLE A.IX.5 (Concluded)

Date	Source, Event, Comments
1119	Hildesheimenses, people eaten by wolves
1121 or 1122 Dec 25	Hildesheimenses, wind damaged buildings
1132	Hildesheimenses, wind knocked buildings over
1133 Jun 29	Hildesheimenses, solar halo seen at Paderborn
1144 Jan 20	Scheftlarienses Maiores, wind caused much destruction
1144 May 7	Herbipolenses, hailstorm
1164 Feb 17	Colonienses, tidal wave in Hanover
1167 Apr 3	Colonienses, hailstorm
1173 May 30	Theodorus, heavy snow
1177	Halesbrunnenses, severe winter followed by dry summer and famine
1179	Magdeburgenses, very severe winter, trees did not leaf until Jun 19.
1179 Mar 7	Theodorus, multiple suns; editor thinks year should be 1177.
1190 May	Colonienses, hail in Mainz devastated more than 100 villas; much rain and floods all summer.
1197 Jun 20	S. Trudperti, solar halo for about 5 hours
1222	Colonienses, a rain of blood in Rome

METEOROLOGICAL REPORTS FROM ITALIAN SOURCES

Date	Source, Event, Comments
365 Jul 21	Fasti Vindobonenses, sounds like a tidal wave
ca. 509, Oct 17	Paulus Diaconus, floods in Venice and Lombardy as not seen since Noah, also a thunderstorm and a pestilence that year
591	Paulus Diaconus, drought from January to September, and consequent famine, also a plague of locusts
604	Paulus Diaconus, severe winter and many died
708	Rampona, grain, oil, and beans rained from the sky in Campania.
797	Farfenses, floods at Rome
822 Jun	Varignana, the great piece of ice found in German sources for 824
839 May	Agnellus, a rain of blood
864	Andreas Bergomatis, very severe winter, did not thaw for 100 days, vines severely damaged
872 Aug	Andreas Bergomatis, vines and olives badly damaged, a plague of locusts; the wine was "turbulent".
ca. 877	Dandulus, red rain in Brescia for 3 days
ca. 885	Dandulus, heavy floods in Venice damaged homes and churches.

TABLE A.IX-6 (Concluded)

Date	Source, Event, Comments
944 Feb 19	Farfenses, heavy snow
1006	Beneventani, drought for 3 months
1029	Beneventani, floods
1031	Beneventani, more floods
1063 Oct 5	Beneventani, hailstorm
1079	Beneventani, river Calor in southern Italy froze and people could walk across.
1091	Beneventani, drought for 7 months
1105	Beneventani, flood
1112 Jun 20	Beneventani, sea receded several hundred feet at Naples; many fish killed
1115 Jun	Dandulus, rain of blood in Ravenna and Parma
ca. 1155	Rampona, 3 moons, 3 suns, cross about the moon

TABLE A.IX-7

METEOROLOGICAL REPORTS FROM THE BYZANTINE EMPIRE

Date	Source, Event, Comments
351 May 7	Paschale, cross in the sky seen at Jerusalem
365 Jul 21	Consularia Constantinopolitana, the sea left its bounds.
401	Marcellinus, severe winter
443	Marcellinus, severe winter, much snow, many died from cold.
545	Theophanes, a wave that drowned many in Thrace
ca. 544, Jul 19	Cedrenus, severe thunderstorm
762-763	Theophanes, a severe winter; the freeze began in October.
1008-1009	Cedrenus, severe winter, the rivers froze.

REFERENCES

The term "standard sources" when used in the text of this study means one or more of the following reference works:

Encyclopaedia Britannica, Eleventh Edition, Encyclopaedia Britannica, Inc., New York, 1911.

Encyclopaedia Britannica, in continuous revision, Encyclopaedia Britannica, Inc., Chicago, as published in 1958.

New Catholic Encyclopedia, McGraw-Hill Book Co., New York, 1967.

Webster's Biographical Dictionary, 1st Ed., G. & C. Merriam Co., Springfield, Mass., 1953.

Webster's Geographical Dictionary, G. & C. Merriam Co., New York, 1949.

The alphabetical list of the other references follows. The list is limited to works that have been directly consulted in the course of preparing this study. Some works mentioned in the text have been referred to by others but have not been consulted directly by me. These works are identified near where they are mentioned; they are not included in the following list.

Most of the sources in this list have been used as potential sources of astronomical data, as contrasted with sources of needed background information. Such sources are identified by a Roman numeral at the end of the citation. The numeral gives the number of the chapter in which the source is discussed.

The word "Saint" or its equivalent, when it appears as part of a name, is designated by "S." regardless of the language and regardless of whether it is masculine or feminine, singular or plural. All names beginning with "S." are alphabetized separately before the other S's.

AAO = Newton [1970].

Ademarus, Historia Francorum, Book III, ca. 1028.
There is an edition by G. Waitz in Monumenta
Germaniae Historica, Scriptores, v. IV, G.H.
Pertz, ed., Hahn's, Hannover, 1841. X.

Admuntenses, Annales, ca. 1250. There is an edi-
tion by W. Wattenbach in Monumenta Germaniae
Historica, Scriptores, v. IX, G.H. Pertz, ed.,
Hahn's, Hannover, 1851. IX.

Adonis, Chronicon, ca. 869. There is an edition
by M. Bouquet in Recueil des Historiens des
Gaules et de la France (Rerum Gallicarum et
Francicarum Scriptores), with the part pertaining to the reigns of Pepin and Charles
in v. V, Chez Martin, Coignard, Mariette,
Guerin, et Guerin, Paris, 1744. X.

Agnellus, Liber Pontificalis Ecclesiae Ravennatis,
ca. 841. There is an edition by O. Holder-
Egger in Monumenta Germaniae Historica,
Scriptores Rerum Langobardicarum et Italicarum,
Saec. VI-IX, L. Bethmann and G. Waitz, eds.,
Hahn's, Hannover, 1878. XII.

Albertus, Annales a Condito Orbe Usque ad Annum
Iesu Christi 1256, 1256. There is an edition
by I.M. Lappenberg in Monumenta Germaniae
Historica, Scriptores, v. XVI, G.H. Pertz,
ed., Hahn's, Hannover, 1859. XI.

American Ephemeris and Nautical Almanac, U.S.
Government Printing Office, Washington, published annually.

Ammianus Marcellinus, Res Gestae, ca. 391. There
is an edition with parallel English translation
by John C. Rolfe in the Loeb Classical Library,
in 3 vols., William Heinemann, Ltd., London,
1935. XV.

Andreas Bergomatis, Historia, ca. 877. There is
an edition by G. Waitz in Monumenta Germaniae
Historica, Scriptores Rerum Langobardicarum
et Italicarum, Saec. VI-IX, L. Bethmann and
G. Waitz, eds., Hahn's, Hannover, 1878. XII.

Anglo-Saxon Chronicle, ca. 1154. There is an edition of the six main Anglo-Saxon texts by
Benjamin Thorpe in Rerum Britannicarum Medii
Aevi Scriptores, no. 23, v. 1, Longman, Green,
Longman, and Roberts, London, 1861. There is
a collated translation of the same texts,
with occasional notes about the single Latin
text, by D. Whitelock with D.C. Douglas and
S.I. Tucker in The Anglo-Saxon Chronicle, Rut-
gers Univ. Press, New Brunswick, N.J., 1961.
See also Domitiani Latini [ca. 1057]. VI.

Anna Comnena, Syntagma Rerum ab Imperatore Alexio
Comneno Gestarum, commonly called Alexiadis,
ca. 1120. There is an edition by L. Schopen
with Latin translation by P. Possini and notes
by C. DuCange, in 2 vols., in Corpus Scriptorum
Historiae Byzantinae, B.G. Niebuhr, ed., Weber's,
Bonn, 1839. XV.

Annales Vetustissimi, ca. 1314. There is an edition
by Gustav Storm in Islandske Annaler indtil
1578, Grøndahl & Sons, Christiana, 1888. XIII.

Annals of Ravenna: see Fasti Vindobonenses [ca.
576].

Anselmus, Gesta Episcoporum Tungrensium, Traiectensium, et Leodiensium, ca. 1048. There is an
edition by Rudolfus Koepke in Monumenta Germaniae Historica, Scriptores, v. VII, G.H.
Pertz, ed., Hahn's, Hannover, 1846. VIII.

Anselmus (Anselmus Gemblacensis), Continuatio
Sigeberti Chronica, ca. 1135. There is an
edition by L.C. Bethmann in Monumenta Germaniae
Historica, v. VI, G.H. Pertz, ed., Hahn's,
Hannover, 1844. VIII.

Aquenses, Annales, ca. 1196. There is an edition
     by G.H. Pertz in Monumenta Germaniae Historica,
     Scriptores, v. XVI, G.H. Pertz, ed., Hahn's,
     Hannover, 1859. XI.

Aquicense, Anon. Auctarium et Continuatio Chronica
     Sigeberti, ca. 1166. There is an edition by
     L.C. Bethmann in Monumenta Germaniae Historica,
     Scriptores, v. VI, G.H. Pertz, ed., Hahn's,
     Hannover, 1844. X.

Aquicinctina, Continuatio Chronica Sigeberti, ca.
     1237. There is an edition by L.C. Bethmann
     in Monumenta Germaniae Historica, Scriptores,
     v. VI, G.H. Pertz, ed., Hahn's, Hannover,
     1844. X.

Aristotle, On the Heavens, Bk. II, Chapter XIV,
     ca. -325. There is an edition with a parallel
     translation by W.K.C. Guthrie in the Loeb
     Classical Library, Harvard Univ. Press,
     Cambridge, Mass., 1939, reprinted 1960.

Arnulfus, Gesta Archiepiscoporum Mediolanensium,
     ca. 1077. There is an edition by L.C. Beth-
     mann and W. Wattenbach in Monumenta Germaniae
     Historica, Scriptores, v. VIII, G.H. Pertz,
     ed., Hahn's, Hannover, 1848. XII.

Asser, Annales Rerum Gestarum Alfredi Magni, ca.
     893. There is an edition with introduction
     and notes by W.H. Stevenson, Clarendon Press,
     Oxford, 1904. VII.

Augienses, Annales, ca. 954. There is an edition
     by G.H. Pertz in Monumenta Germaniae Historica,
     Scriptores, v. I, G.H. Pertz, ed., Hahn's,
     Hannover, 1826. XI.

Augustani, Annales, ca. 1104. There is an edition
     by G.H. Pertz in Monumenta Germaniae Historica,
     Scriptores, v. III, G.H. Pertz, ed., Hahn's,
     Hannover, 1839. XI.

Augustani Minores, Annales, ca. 1321. There is an
     edition by G.H. Pertz in Monumenta Germaniae
     Historica, Scriptores, v. X, G.H. Pertz, ed.,
     Hahn's, Hannover, 1852. XI.

Baldericus: see Cameracensium [ca. 1042].

Beckman, N., Quellen und Quellenwert der Isländischen
Annalen, Xenia Lideniana, pp. 16-39, Stockholm,
1912.

Bede, De Temporum Ratione (or Liber de Temporibus
Maior), 725. There is an edition in the source
cited as Jones [1943].

___, Historia Ecclesiastica Gentis Anglorum, 734?
I write "734?" rather than the customary 731,
because Bede may have prepared a "2nd edition"
in 734. There is an edition by C. Plummer
in Baedae Opera Historica, 2 vols., Clarendon
Press, Oxford, 1896. VI.

Bellovacense, Auctarium et Continuatio Chronica
Sigeberti, ca. 1163. There is an edition by
L.C. Bethmann in Monumenta Germaniae Historica,
Scriptores, v. VI, G.H. Pertz, ed., Hahn's,
Hannover, 1844. X.

Beneventani, Annales, ca. 1130. There is an edition
by G.H. Pertz in Monumenta Germaniae Historica,
Scriptores, v. III, G.H. Pertz, ed., Hahn's,
Hannover, 1839. XII.

Bermondsey (Bermundeseia), Annales Monasterii de,
ca. 1433. There is an edition with notes by
H.R. Luard, ed., in Rerum Britannicarum Medii
Aevi Scriptores, no. 36, v. 3, Longmans,
Green, Reader, and Dyer, London, 1866. VI.

Bernoldus, Chronicon, 1100. There is an edition
by G.H. Pertz in Monumenta Germaniae Historica,
Scriptores, v. V, G.H. Pertz, ed., Hahn's,
Hannover, 1844. IX.

Bertholdus, Annales, 1080. There is an edition by
G.H. Pertz in Monumenta Germaniae Historica,
Scriptores, v. V, G.H. Pertz, ed., Hahn's,
Hannover, 1844.

Bertholdus, Liber de Constructione Monasterii
Zwivildensis, ca. 1138. There is an edition
by Otto Abel in Monumenta Germaniae Historica,
Scriptores, v. X, G.H. Pertz, ed., Hahn's,
Hannover, 1852. XI.

Besuenses, Annales, ca. 1174. There is an edition
by G.H. Pertz in Monumenta Germaniae Historica,
Scriptores, v. II, G.H. Pertz, ed., Hahn's,
Hannover, 1829. X.

Blandinienses, Annales, ca. 1292. There is an edi-
tion by L.C. Bethmann in Monumenta Germaniae
Historica, Scriptores, v. V, G.H. Pertz, ed.,
Hahn's, Hannover, 1844. VIII.

Bolognetti, Cronaca Detta Dei Bolognetti, ca. 1420.
There is an undated edition by L.A. Muratori,
revised and corrected by Albano Sorbelli, in
Rerum Italicarum Scriptores, v. XVIII, in 3
parts, Stamperia di S. Lapi, Citta di Castello,
1906. XII.

Brunwilarenses, Annales, ca. 1179. There is an edition by G.H. Pertz, based upon information
supplied by L.C. Bethmann, in Monumenta Germaniae
Historica, Scriptores, v. XVI, G.H. Pertz, ed.,
Hahn's, Hannover, 1859. XI.

Brut (Brut y Tywysogion), or The Chronicle of the
Princes, ca. 1282. There is a translation
with preface and notes by J.W. ab Ithel, Rerum
Britannicarum Medii Aevi Scriptores, no. 17,
Longman, Green, Longman, and Roberts, London,
1860. VII.

Buckler, Georgina, Anna Comnena, Oxford University
Press, London, 1929.

Burburgensis, Anon. Continuatio Chronica Sigeberti,
ca. 1164. There is an edition by L.C. Bethmann
in Monumenta Germaniae Historica, Scriptores,
v. VI, G.H. Pertz, ed., Hahn's, Hannover,
1844. X.

Burton, Annales de, ca. 1262. There is an edition
in Annales Monastici, H.R. Luard, ed., in
Rerum Britannicarum Medii Aevi Scriptores, no.
36, v. 1, Longman, Green, Longman, Roberts,
and Green, London, 1864. VI.

Bury, J.B., The History of the Decline and Fall of the Roman Empire by Edward Gibbon, Edited in 7 Volumes, Methuen and Co., London, 1898.

Cambriae, Annales, ca. 1288. There is an edition with introduction and notes by J.W. ab Ithel in Rerum Britannicarum Medii Aevi Scriptores, no. 20, Longman, Green, Longman, and Roberts, London, 1860. VII.

Cameracensium, Gesta Episcoporum Cameracensium, ca. 1042. There is an edition by L.C. Bethmann in Monumenta Germaniae Historica, Scriptores, v. VII, G.H. Pertz, ed., Hahn's, Hannover, 1846. X.

Casinates, Annales, ca. 1042. There is an edition by G.H. Pertz in Monumenta Germaniae Historica, Scriptores, v. III, G.H. Pertz, ed., Hahn's, Hannover, 1839. XII.

Cavenses, Annales, ca. 1315. There is an edition by G.H. Pertz in Monumenta Germaniae Historica, Scriptores, v. III, G.H. Pertz, ed., Hahn's, Hannover, 1839. XII.

Cedrenus, Georgius, Synopsis Istorion, ca. 1100. There is an edition by Immanuel Bekker, with a Latin translation by Guilielmus Xylandrus made in 1566, in 2 vols., in Corpus Scriptorum Historiae Byzantinae, B.G. Niebuhr, ed., Weber's, Bonn, Prussia, 1838. XV.

Celoria, G., Sull 'Eclissi Solare Totale del 3 Giugno 1239, Memorie del Reale Istituto Lombardo di Scienze e Letteri, Classe di Scienze Matematiche e Naturali, 13, pp. 275-300, 1877a.

___, Sugle Eclissi Solari Totali del 3 Giugno 1239 e del 6 Ottobre 1241, Memorie del Reale Istituto Lombardo di Scienze e Letteri, Classe di Scienze Matematiche e Naturali, 13, pp. 367-382, 1877b.

Cicero, M. Tullius, De Oratore, Book II, 12.52,-54. There is an edition by C.F.A. Nobbe in M. Tullii Ciceronis Opera Omnia, Tauchnitz, Leipzig, 1850.

Clarius, Chronico S. Petri Vivi Senonensi, ca.
    1124. There is an edition by G. Waitz in
    Monumenta Germaniae Historica, Scriptores,
    v. XXVI, G. Waitz, ed., Hahn's, Hannover,
    1882. X.

Clemens, S.L., A Connecticut Yankee in King Arthur's
    Court, Chapter VI, Harper and Bros., New York,
    1889.

Colonienses, Annales Colonienses Maximi, ca. 1238.
    There is an edition by K. Pertz in Monumenta
    Germaniae Historica, Scriptores, v. XVII,
    G. H. Pertz, ed., Hahn's, Hannover, 1861. XI.

Consularia Constantinopolitana ad A. CCCXCV cum
    Additamento Hydatii ad A. CCCCLXVIII, ca. 468.
    There is an edition by Theodor Mommsen in
    Chronica Minora, Saec. IV V VI VII, v. 1,
    (of 3 vols.); this volume is listed as vol.
    9 of Monumenta Germaniae Historica, Auctorum
    Antiquissimorum, Weidmann's, Berlin, 1892.
    XV.

Corbeienses, Annales, ca. 1148. There is an edi-
    tion by G.H. Pertz in Monumenta Germaniae
    Historica, Scriptores, v. III, Hahn's,
    Hannover, 1839. XI.

Corbeienses Notae, ca. 1241. There is an edition
    by G. Waitz in Monumenta Germaniae Historica,
    Scriptores, v. XIII, G. Waitz, ed., Hahn's,
    Hannover, 1881. XI.

Cosmas (Pragensis), Chronica Boemorum, ca. 1125.
    There is an edition by Rudolfus Koepke in
    Monumenta Germaniae Historica, Scriptores,
    v. IX, G.H. Pertz, ed., Hahn's, Hannover,
    1851. IX.

Cremifanensis, Continuatio (of Mellicenses [ca.
    1564]), ca. 1216. There is an edition by
    W. Wattenbach in Monumenta Germaniae Historica,
    Scriptores, v. IX, G.H. Pertz, ed., Hahn's,
    Hannover, 1851. IX.

Dandulus, Andrea, Chronica per Extensum Descripta,
ca. 1340. There is an edition by L.A. Muratori
(1728), revised and corrected by Ester Pastorello,
in Rerum Italicarum Scriptores, v. XII, Nicola
Zanichelli, Bologna, 1925. XII.

De Expugnatione Terrae Sanctae per Saladinum Libellus,
ca. 1188. There is an edition by Joseph
Stevenson in Rerum Britannicarum Medii Aevi
Scriptores, no. 66, Her Majesty's Stationery
Office, London, 1875. XV.

De Fundatione Scheftlariensi, ca. 1341. There is
an edition by Philip Jaffé in Monumenta Germaniae
Historica, Scriptores, v. XVII, G.H. Pertz, ed.,
Hahn's, Hannover, 1861.

Diceto, Ralph of, Ymagines Historiarum, ca. 1202.
There is an edition with notes by William Stubbs
in Rerum Britannicarum Medii Aevi Scriptores,
no. 68, in 2 vols., Longman and Co., London,
1876. VI.

Domitiani Latini, Annales, ca. 1057. (This is the
Latin half of the bi-lingual F text of the
Anglo-Saxon Chronicle.) There is an edition
by F.P. Magoun, Jr. in "Annales Domitiani
Latini: An Edition" in Mediaeval Studies,
An Annual Published by the Pontifical Insti-
tute for Mediaeval Studies of Toronto, IX,
pp. 235-295, 1947. VI.

Dreyer, J.L.E., A History of the Planetary Systems
from Thales to Kepler, Cambridge University
Press, Cambridge, 1905, Chapter X. Reprinted
in Theories of the Universe, M.K. Munitz, ed.,
The Free Press, Glencoe, Ill., 1957.

Dubs, H.H., The History of the Former Han Dynasty,
a critical translation with annotations,
Waverly Press, Baltimore, v. 1, 1938, v. 2,
1944, and v. 3, 1955.

Dunstable (Dunstaplia), Annales Prioratus de, ca.
1297. There is an edition in Annales Monastici,
H.R. Luard, ed., in Rerum Britannicarum Medii
Aevi Scriptores, no. 36, v. 3, Longmans, Green,
Reader, and Dyer, London, 1866. VI.

Eadmer, Historia Novorum in Anglia, ca. 1122.
    There is an edition by Martin Rule in Rerum
    Britannicarum Medii Aevi Scriptores, no. 81,
    Her Majesty's Stationery Office, London, 1884.

Egmundani, Annales, ca. 1315. There is an edition
    by G.H. Pertz in Monumenta Germaniae Historica,
    Scriptores, v. XVI, G.H. Pertz, ed., Hahn's,
    Hannover, 1859. VIII.

Einhard, Vita Karoli Magni, ca. 820. There is an
    edition by M. Bouquet in Recueil des Historiens
    des Gaules et de la France, v. V, Chez Martin,
    Coignard, Mariette, Guerin, et Guerin, Paris,
    1744. XI.

Einhard or Annales Einhardi: see the discussion of
    Fuldenses [ca. 901] and of Laurissenses [ca.
    829] in Section XI.1.

Ekkehardus: see Frutolf [1103].

Elnonenses Maiores, Annales, ca. 1224. There is an
    edition by G.H. Pertz in Monumenta Germaniae
    Historica, Scriptores, v. V, G.H. Pertz, ed.,
    Hahn's, Hannover, 1844. X.

Elnonenses Minores, Annales, ca. 1061. There is an
    edition by G.H. Pertz in Monumenta Germaniae
    Historica, Scriptores, v. V, G.H. Pertz, ed.,
    Hahn's, Hannover, 1844. X.

Engelbergenses, Annales, ca. 1175. There is an
    edition by G.H. Pertz in Monumenta Germaniae
    Historica, Scriptores, v. XVII, G.H. Pertz,
    ed., Hahn's, Hannover, 1861. IX.

Engolismenses, Annales, ca. 870. There is an edi-
    tion by G.H. Pertz in Monumenta Germaniae
    Historica, Scriptores, v. XVI, G.H. Pertz,
    ed., Hahn's, Hannover, 1859. X.

Erphesfurdensis, Annales S. Petri Erphesfurdensis,
    ca. 1182. There is an edition by G.H. Pertz
    in Monumenta Germaniae Historica, Scriptores,
    v. XVI, G.H. Pertz, ed., Hahn's, Hannover,
    1859. XI.

Eskinbek, Chronologia Rerum Memorabilium ab Anno
    1020 usque ad An. 1323, ca. 1323. There is
    an edition by Jacob Langebek in Scriptores
    Rerum Danicarum Medii Aevi, v. II, A.H. &
    F.C. Godiche, Copenhagen, 1773. XIII.

Esromenses, Annales Rerum Danicarum, ca. 1307.
    There is an edition by Jacob Langebek in
    Scriptores Rerum Danicarum Medii Aevi, v. I,
    A.H. & F.C. Godiche, Copenhagen, 1772. XIII.

Ethelwerd, Chronicon, ca. 975. There is a transla-
    tion by J.A. Giles in Six Old English Chronicles,
    J.A. Giles, ed., Bell and Daldy, London, 1872.
    VI.

Eusebius Pamphili (sometimes called Eusebius of
    Caesarea), Chronicon, ca. 325. There is a
    collated edition of the main texts edited
    by A. Schoene, Weidmann's, Berlin, in 2 vols.;
    v. 1 is dated 1875 and v. 2, is dated 1866
    (sic). Chronicon is the title generally used
    in citation. Eusebius divided his work into
    two books. In Schoene's edition, Book I is
    called Chronicorum Liber Prior and Book II
    is called Chronicorum Canonum. XV.

___, (also called Eusebius of Caesarea), De Vita
    Constantina, Book III, Chapters 18-19, ca.
    338. There is an edition by H. Valesii, F.
    Vigeri, B. Monfauconii, and A. Maii (these
    names are Latinized forms for most of which
    I do not know the original forms) in Patro-
    logiae Cursus Completus, Series Graeca, v. 20,
    pp. 1075-1079, J.-P. Migne, Paris, 1857.

Evagrius Scholasticus, Ecclesiasticae Historiae,
    ca. 593. There is an edition by J.-P. Migne,
    with a parallel translation into Latin by H.
    Valesius, in Patrologiae Cursus Completus,
    Series Graeca Prior, v. 86, part 2, L. Migne,
    Paris, 1865.

Explanatory Supplement to The Astronomical Ephemeris
    and The American Ephemeris and Nautical Almanac,
    H.M. Stationery Office, London, 1961.

Farfenses, Annales, ca. 1099. There is an edition
by G.H. Pertz in Monumenta Germaniae Historica,
Scriptores, v. XI, G.H. Pertz, ed., Hahn's,
Hannover, 1854. XII.

Fasti Consulares: see Fasti Vindobonenses [ca. 576].

Fasti Vindobonenses, ca. 576. There is an edition
by Theodor Mommsen, in which Fasti Vindobonenses
is collated with several other consular lists
under the general title Consularia Italica, in
Chronica Minora, Saec. IV V VI VII, v. 1 (of
3 vols.). This volume is v. 9 of Monumenta
Germaniae Historica, Auctorum Antiquissimorum,
Weidmann's, Berlin, 1892. The group that
Mommsen calls Consularia Italica has been called
Fasti Consulares and also Annals of Ravenna.
XII.

Flaviniacenses et Lausonenses, Annales, ca. 985.
There is an edition by G.H. Pertz in Monumenta
Germaniae Historica, Scriptores, v. III, G.H.
Pertz, ed., Hahn's, Hannover, 1839. X.

Florence of Worcester (Florentius Wigorniensis),
Chronicon ex Chronica, ca. 1118. There is an
edition with notes by B. Thorpe in Publications
of the English Historical Society, London,
1848. There is a translation with notes by
T. Forester in Bohn's Antiquarian Library,
Henry G. Bohn, London, 1854. VI.

Floriacenses, Annales, ca. 1060. There is an edi-
tion by G.H. Pertz in Monumenta Germaniae
Historica, Scriptores, v. II, G.H. Pertz, ed.,
Hahn's, Hannover, 1829. X.

Folcwinus, Gesta Abbatum Sithiensium, ca. 962.
There is an edition by O. Holder-Egger in
Monumenta Germaniae Historica, Scriptores,
v. XIII, G. Waitz, ed., Hahn's, Hannover,
1881. X.

Formoselenses, Annales, ca. 1136. There is an edi-
tion by G.H. Pertz in Monumenta Germaniae
Historica, Scriptores, v. V, G.H. Pertz, ed.,
Hahn's, Hannover, 1844. VIII.

Fossenses, Annales, ca. 1384. There is an edition
   by G.H. Pertz in Monumenta Germaniae Historica,
   Scriptores, v. IV, G.H. Pertz, ed., Hahn's,
   Hannover, 1841. VIII.

Fotheringham, J.K., A solution of ancient eclipses
   of the sun, Monthly Not. Roy. Astro. Soc.,
   81, pp. 104-126, 1920.

Four Masters, Annals of the Kingdom of Ireland by
   the Four Masters (occasionally known as the
   Annals of Donegal), ca. 1636. There is an
   edition, with an introduction and an English
   translation, edited by John O'Donovan, Hodges,
   Smith, & Co., Dublin, in 7 vols., 2nd Ed.,
   1856. VII.

Frazer, Sir J.G., The Golden Bough, abridged edition, MacMillan Co., New York, 1922; reprinted
   1948.

"Fredegarius Scholasticus", Chronicon, ca. 641.
   There is an edition by M. Bouquet in Recueil
   des Historiens des Gaules et de la France,
   v. II, M. Bouquet, ed., Chez Martin, Coignard,
   Mariette, Guerin, et Guerin, Paris, 1739. X.

Frutolf, Chronicon Universale, 1103. There is an
   edition (in which the work is ascribed to
   Ekkehardus Uraugiensis) by G. Waitz in Monumenta Germaniae Historica, Scriptores, v. VI,
   G.H. Pertz, ed., Hahn's, Hannover, 1844. XI.

Fulcher (Carnotensis), Historia Hierosolymitana,
   ca. 1127. There is an edition by Heinrich
   Hagenmeyer, Carl Winters Univeritätsbuchhandlung,
   Heidelberg, 1913. XV.

Fuldenses, Annales, ca. 901. There is an edition
   by G.H. Pertz in Monumenta Germaniae Historica,
   Scriptores, v. I, G.H. Pertz, ed., Hahn's,
   Hannover, 1826. XI.

Gaimar, Geffrei, Lestorie des Engles, ca. 1100.
   There is an edition with notes by Sir Thomas
   D. Hardy in Rerum Britannicarum Medii Aevi
   Scriptores, no. 91, v. I, H.M. Stationery
   Office, London, 1888. VI.

Galbertus Brugensis, Passio Karoli Comitis Flandriae,
ca. 1128. There is an edition by Rudolfus
Koepke in Monumenta Germaniae Historica, Scriptores, v. XII, G.H. Pertz, ed., Hahn's, Hannover, 1856. VIII.

Garstensis, Continuatio (of Mellicenses [ca. 1564]),
ca. 1257. There is an edition by W. Wattenbach
in Monumenta Germaniae Historica, Scriptores,
v. IX, G.H. Pertz, ed., Hahn's, Hannover, 1851.
IX.

Gaufredus, Chronicon Vosiensis, ca. 1182. There is
an anonymous edition in Recueil des Historiens
des Gaules et de la France, v. XII, edited by
the Benedictine monks of S. Maur, Chez la
veuve Desaint, Paris, 1781. X.

Gemblacensis, Sigeberti Continuatio Gemblacensis,
ca. 1148. There is an edition by L.C. Bethmann
in Monumenta Germaniae Historica, Scriptores,
v. VI, G.H. Pertz, ed., Hahn's, Hannover, 1844.
VIII.

Georgios Hamartolos, Chronicon Breve, ca. 842.
There is an edition by J.-P. Migne, with a
parallel Latin translation by Eduard de Muralto,
in Patrologiae Cursus Completus, Series Graeca
Posterior, v. 110, J.-P. Migne, Paris, 1863.
XV.

Georgius Monachus, Vitae Recentiorum Imperatorum,
ca. 948. There is an edition with parallel
Latin translation by F. Combefisius in Scrip-
tores Post Theophanes, which is v. 109 of
Patrologiae Cursus Completus, Series Graeca
Posterior, J.-P. Migne, ed., J.-P. Migne,
Paris, 1863. XV.

Gervase of Canterbury, Chronica, ca. 1199. There
is an edition with preface and notes by William
Stubbs in Rerum Britannicarum Medii Aevi
Scriptores, no. 73, v. 1, Longman and Co.,
London, 1879. VI.

Ginzel, F.K., Astronomische Untersuchungen über
    Finsternisse, Sitzungberichte der Kaiserlichen
    Akademie der Wissenschaften, Wien, Math. -
    Naturwiss. Classe, in 3 parts; Part I: v. 85,
    663-747, 1882; Part II: v. 88, 629-755, 1884;
    Part III: v. 89, 491-559, 1884.

___, Spezieller Kanon der Sonnen- und Mondfinstern-
    isse, Mayer and Müller, Berlin, 1899.

Giraldus Cambrensis, Expugnatio Hibernica, ca. 1200.
    There is an edition with notes by J.F. Dimock
    in Rerum Britannicarum Medii Aevi Scriptores,
    no. 21, v. 5, Longmans, Green, Reader, and
    Dyer, London, 1867. VII.

Gottskalks Annaler, ca. 1578. There is an edition
    by Gustav Storm in Islandske Annaler indtil
    1578, Grøndahl & Sons, Christiana, 1888. XIII.

Gottwicensis, Codex Gottwicensis Herimanni Chronici,
    ca. 1054. This denotes the parts of Herimannus
    [1054] that are peculiar to the codices at
    Gottweih, as given by the editor of Herimannus.
    I have invented the date for purposes of cita-
    tion only. IX.

Gradicenses, Annales, ca. 1145. There is an edi-
    tion by W. Wattenbach in Monumenta Germaniae
    Historica, Scriptores, v. XVII, G.H. Pertz,
    ed., Hahn's, Hannover, 1861. IX.

Gregory of Tours, De Cursu Stellarum Ratio Qualiter
    ad Officium Implendum, ca. 580. There is an
    edition in Sancti Georgii Florentii Libri
    Miraculorum Aliaque Opera Minora, v. IV, H.L.
    Bordier, ed., La Société de l'Histoire de
    France, Publ. No. 125, Chez Jules Renouard
    et Cie., Paris, 1864. (The year is a guess;
    we know only that this preceded the Historiae
    Ecclesiasticae.)

___, Historiae Ecclesiasticae Francorum Libri Decem,
    ca. 592. There is an edition by M. Bouquet in
    Recueil des Historiens des Gaules et de la
    France, v. II, M. Bouquet, ed., Chez Martin,
    Coignard, Mariette, Guerin, et Guerin, Paris,
    1739. X.

Gundechar, Liber Pontificalis Eichstetensis, ca.
    1074, with continuation to 1162. There is
    an edition by L.C. Bethmann in Monumenta
    Germaniae Historica, Scriptores, v. VII, G.H.
    Pertz, ed., Hahn's, Hannover, 1846. XI.

Halesbrunnenses, Annales, ca. 1241. There is an
    edition by G.H. Pertz in Monumenta Germaniae
    Historica, Scriptores, v. XVI, G.H. Pertz,
    ed., Hahn's, Hannover, 1859. XI.

Halesbrunnenses Notae, ca. 1133. There is an edi-
    tion by G.H. Pertz in Monumenta Germaniae
    Historica, Scriptores, v. XVI, G.H. Pertz,
    ed., Hahn's, Hannover, 1859. XI.

Hamsfort, Cornelius, Chronologia Rerum Danicarum
    Secunda, ca. 1585. There is an edition by
    Jacob Langebek in Scriptores Rerum Danicarum
    Medii Aevi, v. I, A.H. & F.C. Godiche, Copen-
    hagen, 1772. XIII.

Heimo, Chronographia, 1135. Parts of this work
    are edited by G.H. Pertz in Monumenta Germaniae
    Historica, Scriptores, v. X, G.H. Pertz, ed.,
    Hahn's, Hannover, 1852. XI.

Heming, Saga of, ca. 1067. There is a translation
    by Sir G.W. Dasent in Appendix E to The
    Orkneyinger's Saga, in Rerum Britannicarum
    Medii Aevi Scriptores, no. 88, v. 3, Her
    Majesty's Stationery Office, London, 1894.
    XIII.

Henry of Huntingdon, Historia Anglorum, 1154. First
    prepared in 1130 and revised in 1154. There
    is a translation with notes by Thomas Forester
    in Bohn's Antiquarian Library, Henry G. Bohn,
    London, 1853. There is an edition with notes
    by Thomas Arnold in Rerum Britannicarum Medii
    Aevi Scriptores, no. 74, Longman and Co.,
    London, 1879. VI.

Herbipolenses, Annales, ca. 1215. There is an edi-
    tion by G.H. Pertz in Monumenta Germaniae His-
    torica, Scriptores, v. XVI, G.H. Pertz, ed.,
    Hahn's, Hannover, 1859. XI.

Herimannus, Chronicon, 1054. There is an edition
    by G.H. Pertz in Monumenta Germaniae Historica,
    Scriptores, v. V, G.H. Pertz, ed., Hahn's,
    Hannover, 1844. XI.

Herodotus, History, Book I, ca. -446. There is a
    translation by George Rawlinson, first pub-
    lished in 1858, reprinted by Tudor Publishing Co.,
    New York, 1947.

Herolfesfeldenses (Annales Herolfesfeldenses): see
    the discussion of Hildesheimenses [ca. 1137]
    in Section XI.1.

Hildesheimenses, Annales, ca. 1137. There is an
    edition by G.H. Pertz in Monumenta Germaniae
    Historica, Scriptores, v. III, G.H. Pertz,
    ed., Hahn's, Hannover, 1839. XI.

Hogarth, D.G., Troy and Troad: The Site of Troy,
    in Encyclopaedia Britannica, 11th Ed., v. 27,
    p. 316, Encyclopaedia Britannica, Inc., New
    York, 1911. (Much of Hogarth's article is
    reproduced in the 1958 printing of the Britan-
    nica, apparently over the name of J.L. Myres.)

Holland, A.W., Einhard, in Encyclopaedia Britannica,
    11th Ed., v. 9, Encyclopaedia Britannica Co.,
    New York, 1910.

Honorius Augustodunensis, Summa Totius Mundi, ca.
    1137. There is an edition by Roger Wilmans
    in Monumenta Germaniae Historica, Scriptores,
    v. X, G.H. Pertz, ed., Hahn's, Hannover, 1852.
    X.

Høyers, Henrik, Annaler, ca. 1310. There is an
    edition by Gustav Storm in Islandske Annaler
    indtil 1578, Grøndahl & Sons, Christiana,
    1888. XIII.

Hsi Tse-tsung, A New Catalog of Ancient Novae (in
    an anonymous translation), Smithsonian Contri-
    butions to Astrophysics, 2, no. 6, U.S. Govern-
    ment Printing Office, Washington, 1958.

Hugo Flaviniacensis, Chronicon Virdunensis, ca. 1102. Part of this is edited anonymously in Recueil des Historiens des Gaules et de la France, v. XI, Chez L.F. Delatour et Cie., Paris, 1767. X.

Hugo Floriacensis, Opera Historica, ca. 1137. Portions of Hugo's works are edited by G. Waitz in Monumenta Germaniae Historica, Scriptores, v. IX, G.H. Pertz, ed., Hahn's, Hannover, 1851. X.

Humboldt, A. von, Kosmos, Entwurf einer physischen Weltbeschreibung, v. 3, Cotta, Stuttgart, 1850.

Hydatius, Continuatio Chronicorum Hieronymianorum, ca. 468. There is an edition by Theodor Mommsen in Chronica Minora, Saec. IV V VI VII, v. 2; this volume is Monumenta Germaniae Historica, Auctorum Antiquissimorum, v. XI, Weidmann's, Berlin, 1894. XIV.

Imoto, S. and Hasegawa, I., Historical Records of Meteor Showers in China, Korea, and Japan, Smithsonian Contributions to Astrophysics, 2, no. 6, 1958.

Isidorus (Isidore of Seville), Historia Gothorum, Wandalorum, Sueborum (with continuations), ca. 624. There is an edition by Theodor Mommsen in Chronica Minora, Saec. IV V VI VII, v. 2; this volume is Monumenta Germaniae Historica, Auctorum Antiquissimorum, v. XI, Weidmann's, Berlin, 1894. XIV.

Jacobs, Joseph and Adler, Cyrus, History of Calendar, in Jewish Encyclopedia, v. III, pp. 498-501, Funk and Wagnalls, New York, 1906.

Joannis Asserii: See S. Neots [ca. 914].

Jóhannesson, T., History of Iceland, Encyclopaedia Britannica, v. 12, pp. 45-47, Encyclopaedia Britannica, Inc., Chicago, Ill., 1958.

Johnson, S.J., Eclipses, Past and Future, Parker and Co., London, 1889.

Jones, Charles W., Bedae Opera de Temporibus,
    published by The Mediaeval Academy of America,
    Cambridge, Mass., printed by George Banta
    Publishing Co., Menasha, Wisc., 1943.

Juvavenses, Annales Juvavenses Maiores, ca. 975.
    There is an edition by G.H. Pertz in Monumenta
    Germaniae Historica, Scriptores, v. I, G.H.
    Pertz, ed., Hahn's, Hannover, 1826. IX.

Kozai, Y., Determination of Love's number from
    satellite observations, Trans. Roy. Soc.,
    v. A262, pp. 135-136, 1967.

Laistner, M.L.W., The Library of the Venerable
    Bede, which is Chapter IX of Bede, His Life,
    Times, and Writings, A. Hamilton Thompson,
    ed., Russell and Russell, London, 1936. (The
    title page says 1932, but this cannot be cor-
    rect; 1936 is my guess as to the correct year.)

Lambacensis, Auctarium et Continuatio (of Cremi-
    fanensis [ca. 1216]), ca. 1283. There is an
    edition by W. Wattenbach in Monumenta Germaniae
    Historica, Scriptores, v. IX, G.H. Pertz, ed.,
    Hahn's, Hannover, 1851. IX.

Lambertus (Lambertus Aschafnaburgensis), Annales,
    ca. 1077. There is an edition by G.H. Pertz
    in Monumenta Germaniae Historica, Scriptores,
    v. III, G.H. Pertz, ed., Hahn's, Hannover,
    1839. XI.

Lambertus Parvus, Annales, ca. 1193. There is an
    edition by G.H. Pertz in Monumenta Germaniae
    Historica, Scriptores, v. XVI, G.H. Pertz,
    ed., Hahn's, Hannover, 1859. VIII.

Landmark, J.D., Solmørket over Stiklestad, Kong.
    Norske Vidensk. Selsk. Skrifter for 1931, no.
    3, pp. 1-62, 1931.

Laubacenses, Annales, ca. 926. There is an edition
    by G.H. Pertz in Monumenta Germaniae Historica,
    Scriptores, v. I, G.H. Pertz, ed., Hahn's,
    Hannover, 1826. VIII.

Laudunense, Auctarium Chronica Sigeberti, ca. 1145.
There is an edition by L.C. Bethmann in Mon-
umenta Germaniae Historica, Scriptores, v. VI,
G.H. Pertz, ed., Hahn's, Hannover, 1844. X.

Laureshamenses, Annales, ca. 803. There is an edi-
tion by G.H. Pertz in Monumenta Germaniae
Historica, Scriptores, v. I, G.H. Pertz, ed.,
Hahn's, Hannover, 1826. XI.

Laurissenses, Annales, ca. 829. These are edited
under the title Annales Einhardi by G.H.
Pertz in Monumenta Germaniae Historica, Scriptores, v. I, G.H. Pertz, ed., Hahn's, Hannover,
1826. XI.

Lemovicenses, Annales, ca. 1060. There is an edition by G.H. Pertz in Monumenta Germaniae
Historica, Scriptores, v. II, G.H. Pertz,
ed., Hahn's, Hannover, 1829. X.

Leo Diaconus, Historia, ca. 990. There is an edition by C.B. Hasius with a parallel Latin
translation in Corpus Scriptorum Historiae
Byzantinae, B.G. Niebuhr, ed., Weber's, Bonn,
1828. XV.

Leodienses, Annales, ca. 1121. There is an edi-
tion by G.H. Pertz in Monumenta Germaniae
Historica, Scriptores, v. IV, G.H. Pertz, ed.,
Hahn's, Hannover, 1841. VIII.

Leonis Imperatoris Imperium, ca. 912. Anonymous.
There is an edition with a parallel Latin
translation by F. Combefisius in Scriptores
post Theophanem, which is v. 109 of Patrologiae Cursus Completus, Series Graeca Posterior, J.-P. Migne, ed., J.-P. Migne, Paris,
1863. XV.

Levison, Wilhelm, Bede As Historian, which is
Chapter V of Bede, His Life, Times, and
Writings, A. Hamilton Thompson, ed., Russell
and Russell, London, 1936. (The title page
says 1932, but this cannot be correct; 1936
is my guess.)

Liudprandus, Relatio de Legatione Constantinopolitana, 969. There is an edition by G.H. Pertz in Monumenta Germaniae Historica, Scriptores, v. III, G.H. Pertz, ed., Hahn's, Hannover, 1839. XV.

Lögmanns Annáll, ca. 1430. There is an edition by Gustav Storm in Islandske Annaler indtil 1578, Grøndahl & Sons, Christiana, 1888. XIII.

Loiselianos, Annales Loiselianos, ca. 814. There is an edition by M. Bouquet in Recueil des Historiens des Gaules et de la France, v. V, Chez Martin, Coignard, Mariette, Guerin, et Guerin, Paris, 1744.

Lugdunenses, Annales, ca. 841. There is an edition by G.H. Pertz in Monumenta Germaniae Historica, Scriptores, v. I, G.H. Pertz, ed., Hahn's, Hannover, 1826. X.

Lundense, Necrologium, ca. 1171. There is an edition by Jacob Langebek in Scriptores Rerum Danicarum Medii Aevi, v. III, A.H. & F.C. Godiche, Copenhagen, 1774. XIII.

"Lupus Barensis", Chronicus, ca. 1102. There is an edition by G.H. Pertz in Monumenta Germaniae Historica, Scriptores, v. V, G.H. Pertz, ed., Hahn's, Hannover, 1844. XII.

MacCarthy, B., Introduction to the Annals of Ulster, in v. 4 of the Annals of Ulster, in 4 vols., William M. Hennessy, ed., H.M. Stationery Office, Dublin, 1901.

Magdeburgenses, Annales, ca. 1188: There is an edition by G.H. Pertz in Monumenta Germaniae Historica, Scriptores, v. XVI, G.H. Pertz, ed., Hahn's, Hannover, 1859. XI.

Manniae, Chronicon Regum Manniae, ca. 1266. There is an edition by Jacob Langebek in Scriptores Rerum Danicarum Medii Aevi, v. III, A.H. & F.C. Godiche, Copenhagen, 1774. VII.

Marbacenses, Annales, ca. 1375. There is an edition
by Roger Wilmans in Monumenta Germaniae Historica,
Scriptores, v. XVII, G.H. Pertz, ed., Hahn's,
Hannover, 1861. XI.

Marcellinus, Chronicon, ca. 534. There is an edi-
tion with notes by Theodor Mommsen in Chronica
Minora, Saec. IV V VI VII, v. 2 (of 3 vols.);
this volume is vol. 11 of Monumenta Germaniae
Historica, Auctorum Antiquissimorum, Weidmann's,
Berlin, 1894. XV.

Margan, Annales de, ca. 1232. There is an edition
with notes by H.R. Luard in Rerum Britannicarum
Medii Aevi Scriptores, no. 36, v. 1, Longman,
Green, Longman, Roberts, and Green, London,
1864. VII.

Marianus Scotus, Chronicon, ca. 1082. There is
an edition by G. Waitz in Monumenta Germaniae
Historica, Scriptores, v. V, G.H. Pertz, ed.,
Hahn's, Hannover, 1844. XI.

Marinus Neapolitanus, Vita Procli, ca. 486. There
is an edition with Latin translation by J.F.
Boissonade, publisher not given, Leipzig,
1814; reprinted by Adolf M. Hakkert, Amsterdam,
1966. There is also a translation into English
by L.J. Rosan in The Philosophy of Proclus,
Cosmos Greek-American Printing Co., New York,
1949. XV.

Matthew Paris (Matthaei Parisiensis), Historia
Anglorum (sometimes called Historia Minor),
ca. 1250. There is an edition with notes
by Sir Frederic Madden in Rerum Britannicarum
Medii Aevi Scriptores, no. 44, in 3 vols.,
Longmans, Green, Reader, and Dyer, London,
1866. Part of Matthew's work is sometimes
wrongly attributed to a fictitious person
called Matthew of Westminster. VI.

Mattingly, Garrett, The Armada, Houghton Mifflin
Co., Boston, 1959.

Mellicenses, Annales, ca. 1564. There is an edition by W. Wattenbach in Monumenta Germaniae Historica, Scriptores, v. IX, G.H. Pertz, ed., Hahn's, Hannover, 1851. IX.

Melrose (Mailros), Chronica de, ca. 1275. There is an edition with notes by Joseph Stevenson, Edinburgh Printing Co., Edinburgh, 1835. VII.

Mettenses, Annales, ca. 905. There is an edition by M. Bouquet in Recueil des Historiens des Gaules et de la France, Chez Martin, Coignard, Mariette, Guerin, et Guerin, Paris. Years 687-749 are in v. II, 1739, 750-813 in v. V, 1744, 829 and 830 in v. VI, 1749, 841-877 in v. VII, 1749, and 878-905 in v. VIII, 1752. X.

Mercati, A., The new list of the Popes, Medieval Studies, An Annual Published by the Pontifical Institute of Mediaeval Studies of Toronto, IX, pp. 71-80, 1947.

Minnaert, M., The Nature of Light and Colour in the Open Air, translated by H.M. Kremer-Priest, revised by K.E. Brian Jay, Dover Publications, Inc., New York, 1954.

Monacensis, Annalium Salisburgensium Additamentum, Continuatio Codicis Monacensis, ca. 1322. There is an edition by W. Wattenbach in Monumenta Germaniae Historica, Scriptores, v. XIII, G. Waitz, ed., Hahn's, Hannover, 1881. XI.

Mortui-Maris, Chronicon, ca. 1180. There is an anonymous edition in Recueil des Historiens des Gaules et de la France, v. XII, Chez la veuve Desaint, Paris, 1781. X.

Mosomagenses, Annales, ca. 1371. There is an edition by G.H. Pertz in Monumenta Germaniae Historica, Scriptores, v. III, G.H. Pertz, ed., Hahn's, Hannover, 1839. X.

Naufragia Ratisbonensia, ca. 1358. There is an
    edition by Philip Jaffé in Monumenta Germaniae
    Historica, Scriptores, v. XVII, G.H. Pertz,
    ed., Hahn's, Hannover, 1861. XI.

Naval Observatory (U.S.), Ancient Sun and Moon, un-
    published tables prepared by the staff of the
    Naval Observatory, 1953. The preparation of
    the tables is described by Woolard, E.W.,
    Theory of the Rotation of the Earth Around
    Its Center of Mass, Astronomical Papers Pre-
    pared for the Use of the American Ephemeris
    and Nautical Almanac, XV, Part I, U.S. Gov-
    ernment Printing Office, Washington, D.C.,
    1953.

Necrologici Fuldenses, Annales Necrologici Fuldenses,
    ca. 1064. There is an edition by G. Waitz
    in Monumenta Germaniae Historica, Scriptores,
    v. XIII, G. Waitz, ed., Hahn's, Hannover,
    1881. XI.

Needham, J., Science and Civilization in China,
    v. 3, (with the collaboration of Wang Ling),
    Cambridge University Press, Cambridge, 1959.

Neresheimenses, Annales, ca. 1296. There is an
    edition by Otto Abel in Monumenta Germaniae
    Historica, Scriptores, v. X, G.H. Pertz, ed.,
    Hahn's, Hannover, 1852. XI.

Neugebauer, O., The Exact Sciences in Antiquity,
    Brown University Press, Providence, R. I.,
    2nd Ed., 1957.

Newton, R.R., A satellite determination of tidal
    parameters and earth deceleration, Geophys.
    Jour. Roy. Astron. Soc., 14, pp. 505-539,
    1968.

___, Secular accelerations of the earth and moon,
    Science, 166, pp. 825-831, 1969.

___, Ancient Astronomical Observations and the
    Accelerations of the Earth and Moon, The Johns
    Hopkins Press, Baltimore, Md. 21218, 1970.

Nithardus, Historiarum Libri IIII, ca. 843. There is
an edition by G.H. Pertz in Monumenta Germaniae
Historica, Scriptores, v. II, G.H. Pertz, ed.,
Hahn's, Hannover, 1829, X.

Nivernenses, Annales, ca. 1188. There is an edition
by G. Waitz in Monumenta Germaniae Historica,
Scriptores, v. XIII, G. Waitz, ed., Hahn's, Hannover, 1881. X.

Novaliciense, Chronicon, ca. 1048. There is an edition
by L.C. Bethmann in Monumenta Germaniae Historica,
Scriptores, v. VII, G.H. Pertz, ed., Hahn's, Hannover, 1846.

Oddveria Annall, ca. 1427. There is an edition by
Gustav Storm in Islandske Annaler indtil 1578,
Grøndahl & Sons, Christiana, 1888. XIII.

Odon de Deuil, Histoire de la Croisade de Louis VII,
1148. There is a translation into French by M.
Guizot in Collection des Mémoires Relatifs à
l'Histoire de France, v. 24, M. Guizot, ed.,
Imprimerie de A. Belin, Paris, 1825. XV.

O'Donovan, John, Introduction to the Annals of the
Kingdom of Ireland by the Four Masters, John
O'Donovan, ed., Hodges, Smith, & Co., Dublin,
in 7 vols., 2nd Ed., 1856.

Ó Maílle, T., The Language of the Annals of Ulster,
Manchester University Press, Manchester, 1910.

Oppolzer, T.R. von, Canon der Finsternisse, Kaiserlich-
Königlichen Hof- und Staatsdruckerei, Wien, 1887.
Reprinted (with the introduction translated into
English by O. Gingerich) by Dover Publications,
Inc., New York, 1962.

Osney (Oseneia), Annales Monasterii de, ca. 1347;
see Wykes [ca. 1289].

Oxenedes, Johannis de, Chronica, ca. 1292. There is
an edition with notes by Sir Henry Ellis in
Rerum Britannicarum Medii Aevi Scriptores, no.
13, Longman, Brown, Green, Longmans, and Roberts,
London, 1859. VI.

Pachymeres, Georgios, De Michaele et Andronico
    Palaeologis Libri XIII, ca. 1308. There is
    an edition by Immanuel Bekker, with a paral-
    lel Latin translation by Peter Possini, and
    with critical notes and a chronological sum-
    mary by Possini, in Corpus Scriptorum Historiae
    Byzantinae, B.G. Niebuhr, ed., Weber's, Bonn,
    1835. Bekker divides this into two works,
    namely a history of Michael in 6 books and a
    history of Andronicus in 7, but this does
    not seem to have been the intention of
    Pachymeres. XV.

Palidenses: see Theodorus [ca. 1182].

Paschale, Chronicon, ca. 628. There is an edition
    by L. Dindorf, with critical notes and a
    parallel Latin translation by C. DuCange, in
    Corpus Scriptorum Historiae Byzantinae, B.G.
    Niebuhr, ed., Weber's, Bonn, 1832. XV.

Paschale Campanum, ca. 585. There is an edition
    by Theodor Mommsen, in which Paschale Campanum
    is collated with other sources under the gen-
    eral title Consularia Italica, in Chronica
    Minora, Saec. IV V VI VII, v. 1; this volume
    is Monumenta Germaniae Historica, Auctorum
    Antiquissimorum, v. IX, Weidmann's, Berlin,
    1892. XII.

Paulus Diaconus, Historia Gentis Langobardorum,
    ca. 787. There is an edition by L. Bethmann
    and G. Waitz in Monumenta Germaniae Historica,
    Scriptores Rerum Langobardicarum et Italicarum,
    Saec. VI-IX, Hahn's, Hannover, 1878. XII.

Pegavienses, Annales, ca. 1227. There is an edi-
    tion by G.H. Pertz in Monumenta Germaniae
    Historica, Scriptores, v. XVI, G.H. Pertz,
    ed., Hahn's, Hannover, 1859. XI.

Petrus Olai, Annales Rerum Danicarum, ca. 1541.
    There is an edition by Jacob Langebek in
    Scriptores Rerum Danicarum Medii Aevi, v. I,
    A.H. & F.C. Godiche, Copenhagen, 1772. XIII.

Philostorgius, Ecclesiasticae Historiae, ca. 425.
There is an edition by J.-P. Migne, with a
parallel Latin translation by H. Valesius,
in Patrologiae Cursus Completus, Series
Graeca Prior, v. 65, J.-P. Migne, Paris,
1864. More accurately, what appears in this
edition is the "Epitome Confecta a Photio
Patriarcha"; the original has been lost. XV.

Pictish Chronicle, ca. 990. There is a translation
by William F. Skene in Chronicles of the Picts
and Scots, H.M. General Register House,
Edinburgh, 1867. VII.

Pliny the Elder (Gaius Plinius Secundus), Historiae
Naturalis, Book XVIII, ca. 77. There is an
edition with parallel French translation by
M. Ajasson de Grandsagne, C.L.F. Panckoucke,
Paris, 1832.

Plummer, C., Two of the Saxon Chronicles in Parallel,
v. 2, p. lxix, Clarendon Press, Oxford, 1899.

Plutarch, De Facie Quae in Orbe Lunae Apparet, ca.
90. There is a translation by Harold Cherniss
in Plutarch's Moralia, v. 12, H. Cherniss
and W.C. Helmbold, eds., Harvard Press, Cambridge, Mass., 1957.

___, Parallel Lives, ca. 100. An anonymous trans-
lation usually called "Dryden's" is reprinted
in part as v. 12 of Harvard Classics, Collier
and Son, New York, 1909.

Poole, Reginald L., Chronicles and Annals, Clarendon
Press, Oxford, 1926.

Praemonstratensis, Continuatio Chronica Sigeberti
Praemonstratensis, ca. 1155. There is an
edition by L.C. Bethmann in Monumenta Germaniae
Historica, Scriptores, v. VI, G.H. Pertz, ed.,
Hahn's, Hannover, 1844. X.

Pragenses, Annales, ca. 1220. There is an edition
by G.H. Pertz in Monumenta Germaniae Historica,
Scriptores, v. III, G.H. Pertz, ed., Hahn's,
Hannover, 1839. IX.

Pragensium, Pragensium Canonicorum Continuatio
Cosmae, ca. 1283. There is an edition by
Rudolfus Koepke in Monumenta Germaniae Historica,
Scriptores, v. IX, G.H. Pertz, ed., Hahn's,
Hannover, 1851. IX.

Procopius (Caesariensis), Historiae, ca. 554. There
are three histories called De Bello Persico,
De Bello Vandalico, and De Bello Gotthico.
They are edited together by W. Dindorf, with
a parallel translation into Latin by Claudius
Maltretus, in Corpus Scriptorum Historiae
Byzantinae, B.G. Niebuhr, ed., in 2 vols.,
Weber's, Bonn, 1833. XII.

Pruveningenses, Annales, ca. 1298. There is an
edition by G.H. Pertz in Monumenta Germaniae
Historica, Scriptores, v. XVII, G.H. Pertz,
ed., Hahn's, Hannover, 1861. XI.

Ptolemy, C., 'E Mathematike Syntaxis (The Almagest),
ca. 152. There is an edition with French
translation by M. Halma, Henri Grand Libraire,
Paris, 1813; there is a German translation by
K. Manitius, Leipzig, 1913 (no publisher
given).

Quedlinburgenses, Annales, ca. 1025. There is an
edition by G.H. Pertz in Monumenta Germaniae
Historica, Scriptores, v. III, G.H. Pertz,
ed., Hahn's, Hannover, 1839. XI.

Rainaldus, Chronica Domni Rainaldi Archidiaconi
S. Mauricii Andegavensis, ca. 1152. There
is an edition in Chroniques des Églises
d'Anjou, Paul Marchegay and Emile Mabille,
eds., La Société de l'Histoire de France,
Publ. No. 146, Chez Mme. Ve. Jules Renouard,
Paris, 1869. X.

Ralph (Radulphi) of Coggeshall, Chronicum Anglicanum,
ca. 1223. There is an edition with notes by
J. Stevenson in Rerum Britannicarum Medii Aevi
Scriptores, no. 66, H.M. Stationery Office,
London, 1875; reprinted by Kraus Reprint,
Ltd., 1965. VI.

Rampona, Cronaca, ca. 1425. There is an undated
edition by L.A. Muratori, revised and corrected
by Albano Sorbelli, in Rerum Italicarum Scriptores, v. XVIII, in 3 parts, Stamperia di S.
Lapi, Citta di Castello, 1906. XII.

Ratpertus, Casus S. Galli, ca. 883. There is an
edition by Ildephonsus von Arx in Monumenta
Germaniae Historica, Scriptores, v. II, G.H.
Pertz, ed., Hahn's, Hannover, 1829. IX.

Regii, Annales, ca. 1306. There is an edition by
Gustav Storm in Islandske Annaler indtil 1578,
Grøndahl & Sons, Christiana, 1888. There is
an edition which differs in some important
respects by Gudbrand Vigfusson in Sturlunga
Saga, v. 2, Clarendon Press, Oxford, 1878.
XIII.

Regino, Chronicon, ca. 906. There is an edition
by G.H. Pertz in Monumenta Germaniae Historica,
Scriptores, v. I, G.H. Pertz, ed., Hahn's,
Hannover, 1826. XI.

Reicherspergenses, Annales, ca. 1279. There is
an edition by W. Wattenbach in Monumenta
Germaniae Historica, Scriptores, v. XVII,
G.H. Pertz, ed., Hahn's, Hannover, 1861. IX.

Reseniani, Annales, ca. 1295. There is an edition
by Gustav Storm in Islandske Annaler indtil
1578, Grøndahl & Sons, Christiana, 1888. XIII.

Resnikoff, L.A., Jewish calendar calculations,
Scripta Mathematica, 9, in two parts; Part I,
pp. 191-195, Part II, pp. 274-277; 1943.

Richard of Devizes, Chronicle of Richard of Devizes
of the Time of Richard I, ca. 1192. There is
an edition with parallel translation by J.T.
Appleby in The Chronicle of Richard of Devizes,
Thomas Nelson and Sons, London, 1963. Richard
did not give his work a title, and various
editors have supplied various titles. VI.

Robertus Autissiodorensis, Chronicon Roberti Canonici
S. Mariani Autissiodorensis, ca. 1211. There
is an edition by O. Holder-Egger in Monumenta
Germaniae Historica, Scriptores, v. XXVI, G.
Waitz, ed., Hahn's, Hannover, 1882. X.

Robertus de Monte, Cronica, ca. 1186. There is an
edition by R. Howlett in Rerum Britannicarum
Medii Aevi Scriptores, no. 82, v. 4, Her Majesty's
Stationery Office, London, 1884. X.

Rodenses, Annales, ca. 1157. There is an edition
by G.H. Pertz in Monumenta Germaniae Historica,
Scriptores, v. XVI, G.H. Pertz, ed., Hahn's,
Hannover, 1859. XI.

Rodulfus Glaber, Historiarum Libris Quinque, ca.
1044. There is an anonymous edition in Recueil
des Historiens des Gaules et de la France, v.
X, Chez Martin, Guerin, Delatour, et Boudet,
Paris, 1760. X.

Roger of Hovenden (Rogeri de Houedene or Roger of
Howden), Chronica, ca. 1201. There is an
edition with notes by William Stubbs in Rerum
Britannicarum Medii Aevi Scriptores, no. 51,
in 4 vols., Longmans, Green, Reader, and Dyer,
London, 1868. VI.

Romanelli, Pietro, The Roman Forum, Istituto
Poligrafico Dello Stato, Roma, 1959.

Romani, Annales, ca. 1049. There is an edition by
G.H. Pertz in Monumenta Germaniae Historica,
Scriptores, v. V, G.H. Pertz, ed., Hahn's,
Hannover, 1844. XII.

Romualdus (Salernitanus), Chronicon, ca. 1178.
There is an edition by L.A. Muratori (1725),
revised and corrected by C.A. Garufi, in
Rerum Italicarum Scriptores, v. VII, part I,
Nicola Zanichelli, Bologna, 1935. XII.

Rosenveldenses, Annales, ca. 1131. There is an edi-
tion by G.H. Pertz in Monumenta Germaniae
Historica, Scriptores, v. XVI, G.H. Pertz, ed.,
Hahn's, Hannover, 1859. XI.

Rotomagensi, Chronicon, ca. 1174. There is an
anonymous edition in Recueil des Historiens
des Gaules et de la France, v. XII, Chez la
veuve Desaint, Paris, 1781. X.

Rupertus, Chronicon S. Laurentii Leodiensis, ca.
1095. There is an edition by W. Wattenbach
in Monumenta Germaniae Historica, Scriptores,
v. VIII, G.H. Pertz, ed., Hahn's, Hannover,
1846. VIII.

Ryenses, Annales, ca. 1288. There is an edition
by I.M. Lappenberg in Monumenta Germaniae
Historica, Scriptores, v. XVI, G.H. Pertz,
ed., Hahn's, Hannover, 1859. XIII.

S. Albini, Chronicae S. Albini Andegavensis, ca.
1357. There is an edition in Chroniques des
Églises d'Anjou, Paul Marchegay and Emile
Mabille, eds., La Société de l'Histoire de
France, Publ. No. 146, Chez Mme. Ve. Jules
Renouard, Paris, 1869. X.

S. Andreae, Chronicon S. Andreae Castri Cameracesii,
1133. There is an edition by L.C. Bethmann
in Monumenta Germaniae Historica, Scriptores,
v. VII, G.H. Pertz, ed., Hahn's, Hannover,
1846. X.

S. Benedicti, Chronicon S. Benedicti ad Ligerim,
ca. 1110. There is an anonymous edition for
the portion covering the years 1037-1072 in
Recueil des Historiens des Gaules et de la
France, v. XI, Chez L.F. Delatour et Cie.,
Paris, 1767. X.

S. Benigni, Annales S. Benigni Divionensis, ca.
1214. There is an edition by G. Waitz in
Monumenta Germaniae Historica, Scriptores,
v. V, G.H. Pertz, ed., Hahn's, Hannover,
1844. X.

S. Blasii, Annales, ca. 1147. There is an edition
by G.H. Pertz in Monumenta Germaniae Historica,
Scriptores, v. XVII, G.H. Pertz, ed., Hahn's,
Hannover, 1861. XI.

S. Denis, Chroniques Francoises de S. Denis, ca.
1150. The parts pertaining to the reigns of
Pepin and Charlemagne are edited by M. Bouquet
in Recueil des Historiens des Gaules et de la
France, v. V, Chez Martin, Coignard, Mariette,
Guerin, et Guerin, Paris, 1744. X.

S. Disibodi, Annales, ca. 1200. There is an edition
by G. Waitz in Monumenta Germaniae Historica,
Scriptores, v. XVII, G.H. Pertz, ed., Hahn's,
Hannover, 1861. XI.

S. Florentii, Breve Chronicon S. Florentii Salmur-
ensis, ca. 1236. There is an edition in
Chroniques des Églises d'Anjou, Paul Marchegay
and Emile Mabille, eds., La Société de l'Histoire de France, Publ. No. 146, Chez Mme. Ve.
Jules Renouard, Paris, 1869. X.

S. Iacobi, Annales S. Iacobi Leodiensis, ca. 1174.
There is an edition by G.H. Pertz in Monumenta
Germaniae Historica, Scriptores, v. XVI, G.H.
Pertz, ed., Hahn's, Hannover, 1859. VIII.

S. Maxentii, Chronicon S. Maxentii Pictavensis,
ca. 1140. There is an edition in Chroniques
des Églises d'Anjou, Paul Marchegay and Emile
Mabille, eds., La Société de l'Histoire de
France, Publ. No. 146, Chez Mme. Ve. Jules
Renouard, Paris, 1869. X.

S. Maximini, Annales S. Maximini Trevirensis, ca.
987. There is an edition by G.H. Pertz in
Monumenta Germaniae Historica, Scriptores,
v. IV, G.H. Pertz, ed., Hahn's, Hannover,
1841. XI.

S. Neots, Chronicle of the Church of (Chronicon
Fani S. Neoti), ca. 914. There is an edition
by W.H. Stevenson in Asser's Life of King
Alfred Together With the Annals of S. Neots
Erroneously Ascribed to Asser, Clarendon
Press, Oxford, 1904. VI.

S. Petri, Continuatio Clarii Chronici S. Petri
Senonensi, ca. 1180. There is an anonymous
edition in Recueil des Historiens des Gaules
et de la France, v. XII, Chez la veuve Desaint,
Paris, 1781. X.

S. Petri et Aquenses: See the discussion of
    Aquenses [ca. 1196] in Section XI.1.

S. Rudberti, Annales S. Rudberti Salisburgensis,
    ca. 1286. There is an edition by W. Watten-
    bach in Monumenta Germaniae Historica, Scrip-
    tores, v. IX, G.H. Pertz, ed., Hahn's, Hann-
    over, 1851. IX.

S. Rudberti, S. Rudberti Salisburgensis Annales Breves,
    ca. 1168. There is an edition by W. Wattenbach
    in Monumenta Germaniae Historica, Scriptores,
    v. IX, G.H. Pertz, ed., Hahn's, Hannover, 1851.
    IX.

S. Sergii, Chronicon S. Sergii Andegavensis, ca.
    1215. There is an edition in Chroniques des
    Églises d'Anjou, Paul Marchegay and Émile
    Mabille, eds., La Société de l'Histoire de
    France, Publ. No. 146, Chez Mme. Ve. Jules
    Renouard, Paris, 1869. X.

S. Stephani, Annales S. Stephani Frisingensis, ca.
    1448. There is an edition by G. Waitz in
    Monumenta Germaniae Historica, Scriptores,
    v. XIII, G. Waitz, ed., Hahn's, Hannover,
    1881. XI.

S. Trudperti, Annales, ca. 1246. There is an edi-
    tion by G.H. Pertz in Monumenta Germaniae
    Historica, Scriptores, v. XVII, G.H. Pertz,
    ed., Hahn's, Hannover, 1861. XI.

S. Vincentii, Annales S. Vincentii Mettensis, ca.
    1280. There is an edition by G.H. Pertz in
    Monumenta Germaniae Historica, Scriptores,
    v. III, G.H. Pertz, ed., Hahn's, Hannover,
    1839. X.

Sangallenses, Annales Sangallenses Maiores, ca.
    1056. There is an edition by Ildefonsus[†]
    von Arx in Monumenta Germaniae Historica,
    Scriptores, v. I, G.H. Pertz, ed., Hahn's,
    Hannover, 1826. IX.

[†] Comparison with the editor of Ratpertus [ca. 883]
shows that variety in spelling one's name was not
confined to Shakespeare.

Sangallensis, Annalium Alamannicorum Continuatio
Sangallensis, ca. 926. There is an edition
by G.H. Pertz in Monumenta Germaniae Historica,
Scriptores, v. I, G.H. Pertz, ed., Hahn's,
Hannover, 1826. IX.

Sasavensis, Monachi Sasavensis Continuatio Cosmae,
ca. 1170. There is an edition by Rudolfus
Koepke in Monumenta Germaniae Historica, Scriptores, v. IX, G.H. Pertz, ed., Hahn's, Hannover,
1851. IX.

Sayers, Dorothy, The Learned Adventure of the Dragon's
Head, which is Chapter IX in Lord Peter Views
the Body, Victor Gollancz, Ltd., London, 1928.

Scheftlarienses Maiores, Annales, ca. 1247. There
is an edition by Philip Jaffé in Monumenta
Germaniae Historica, Scriptores, v. XVII,
G.H. Pertz, ed., Hahn's, Hannover, 1861. XI.

Scheftlarienses Minores, Annales, ca. 1272. There
is an edition by Philip Jaffé in Monumenta
Germaniae Historica, Scriptores, v. XVII,
G.H. Pertz, ed., Hahn's, Hannover, 1861. XI.

Schove, D.J., The earliest British eclipse record
(A.D. 400-600), Jour. Brit. Astron. Assoc.,
65, pp. 37-43, 1954.

Schroeter, J. Fr., Spezieller Kanon der Zentralen
Sonnen- und Mondfinsternisse, Welche Innerhalb
des Zeitraums von 600 bis 1800 n. Chr. in
Europa Sichtbar Waren, Jacob Dybwad, Kristiana,
1923.

Scotorum, Chronicum Scotorum, ca. 1650. There is
an edition by W.M. Hennessy in Rerum Britannicarum Medii Aevi Scriptores, no. 46, Longmans, Green, Reader, and Dyer, London, 1866.
VII.

Sellar, W.C. and Yeatman, R.J., 1066 and All That,
E.P. Dutton & Co., New York, 1931.

Senonensis, Annales S. Columbae Senonensis, ca.
1203. There is an edition by G.H. Pertz in
Monumenta Germaniae Historica, Scriptores,
v. I, G.H. Pertz, ed., Hahn's, Hannover,
1826. X.

Shetelig, H., Norway, History of, Encyclopaedia
Britannica, v. 16, pp. 548-550, Encyclopaedia
Britannica, Inc., Chicago, Ill., 1958.

Sialandiae, Chronicon, ca. 1282. There is an edi-
tion by Jacob Langebek in Scriptores Rerum
Danicarum Medii Aevi, v. II, A.H. & F.C.
Godiche, Copenhagen, 1773. XIII.

Sigebertus (Sigebertus Gemblacensis), Chronica,
ca. 1111. There is an edition by L.C. Bethmann in Monumenta Germaniae Historica, Scriptores, v. VI, G.H. Pertz, ed., Hahn's, Hannover, 1844. VIII.

Simeon of Durham (Symeonis monachi Dunelmensis),
Historia Regum Anglorum et Dacorum, ca. 1129.
There is an edition with notes by Thomas
Arnold, ed., in Rerum Britannicarum Medii
Aevi Scriptores, no. 75, v. 2, Longmans and
Co., London, 1885. VI.

Simocatta, Theophylactus, Historiarum Libri Octo,
ca. 625. There is an edition by Immanuel
Bekker, with a parallel Latin translation by
J. Pontanus, in Corpus Scriptorum Historiae
Byzantinae, B.G. Niebuhr, ed., Weber's, Bonn,
1834. XV.

Sithienses, Annales, ca. 823. There is an edition
by G. Waitz in Monumenta Germaniae Historica,
Scriptores, v. XIII, G. Waitz, ed., Hahn's,
Hannover, 1881. X.

Skálholts, Annaler, ca. 1356. There is an edition
by Gustav Storm in Islandske Annaler indtil
1578, Grøndahl & Sons, Christiana, 1888. XIII.

Smith, Sidney, Chronology: Babylonian and Assyrian,
in Encyclopaedia Britannica, v. 5, p. 665,
Encyclopaedia Britannica, Inc., Chicago, 1958.

Smithsonian Astrophysical Observatory, Star Catalogue, Positions and Proper Motions of
258,997 Stars for the Epoch and Equinox of
1950.0, in 4 volumes, Smithsonian Institution,
Washington, 1966.

Snorri (Sturluson), Noregs Konunga Sogur (Heims-
kringla), ca. 1230. There is an edition by
Finnur Jonsson, G.E.C. Gads Forlag, Copenhagen,
1911. There is a translation into English
by L.M. Hollander, University of Texas Press,
Austin, Texas, 1964. XIII.

Spencer Jones, H., The rotation of the earth, and
the secular accelerations of the sun, moon,
and planets, Monthly Notices of the Royal
Astronomical Society, 99, pp. 541-558, 1939.

Stadenses: See Albertus [1256].

Storm, Gustav, Islandske Annaler indtil 1578,
Grøndahl & Sons, Christiana, 1888.

Stralius, Laurentius, Annales Danici, ca. 1314.
There is an edition by Jacob Langebek in
Scriptores Rerum Danicarum Medii Aevi, v.
III, A.H. & F.C. Godiche, Copenhagen, 1774.
XIII.

Suevicum, Chronicon Suevicum Universale, ca. 1043.
There is an edition by H. Bresslau in Monumenta
Germaniae Historica, Scriptores, v. XIII, G.
Waitz, ed., Hahn's, Hannover, 1881. XI.

Syncellus, Georgius, Chronographiae, ca. 810.
There is an edition by Guilielmus Dindorfius,
in 2 vols., in Corpus Scriptorum Historiae
Byzantinae, B.G. Niebuhr, ed., Weber's, Bonn,
1829. XV.

Tewkesbury (Theokesberia), Annales de, ca. 1263.
There is an edition in Annales Monastici, H.R.
Luard, ed.; in Rerum Britannicarum Medii
Aevi Scriptores, no. 36, v. 1, Longman, Green,
Longman, Roberts, and Green, London, 1864.
VI.

Theodorus (Palidensis), Annales Palidenses, ca.
    1182, with an anonymous continuation to 1421.
    There is an edition by G.H. Pertz in Monumenta
    Germaniae Historica, Scriptores, v. XVI, G.H.
    Pertz, ed., Hahn's, Hannover, 1859. XI.

Theophanes, Chronographia, ca. 813. There is an
    edition with parallel Latin translation by
    J. Classen in Corpus Scriptorum Historiae
    Byzantinae, B.G. Niebuhr, ed., Weber's, Bonn,
    1839. XV.

Thietmarus, Chronicon, ca. 1018. There is an edi-
    tion by J.M. Lappenberg in Monumenta Germaniae
    Historica, Scriptores, v. III, G.H. Pertz, ed.,
    Hahn's, Hannover, 1839. XI.

Thorndike, Lynn, A History of Magic and Experimental
    Science, in 4 vols., MacMillan Co., New York,
    1923.

Tilliani, Annales Tilliani, ca. 808. There is an
    edition by M. Bouquet in Recueil des Historiens
    des Gaules et de la France, v. V, Chez Martin,
    Coignard, Mariette, Guerin, et Guerin, Paris,
    1744. See Laurissenses [ca. 829].

Times Atlas of the World, Mid-Century Edition, in
    5 vols., The Times Office, London, 1955.

Trevirensis, Continuatio Reginonis Trevirensis,
    ca. 967. There is an edition by G.H. Pertz
    in Monumenta Germaniae Historica, Scriptores,
    v. I, G.H. Pertz, ed., Hahn's, Hannover,
    1826. XI.

Turoldus, La Chanson de Roland (Oxford manuscript),
    ca. 1100. There is a translation into modern
    French prose by Joseph Bédier published by
    H. Piazza, Paris; the 27th printing is dated
    1922. X.

Ulster, Annals of (frequently known as the Annals
    of Senat; A Chronicle of Irish Affairs),ca.
    1498. There is an edition by William M.
    Hennessy, H.M. Stationery Office, Dublin, in
    4 vols., dated from 1887 to 1901. (v. 1 has
    translation and notes by Hennessy, v. 2 and
    v. 3 have translation and notes by B. MacCarthy,
    and v. 4 has an introduction by MacCarthy.) VII.

Van Flandern, T.C., The secular acceleration of the moon, Astronomical Journal, 75, pp. 657-658, 1970.

Varignana, Cronaca, ca. 1425. There is an edition by L.A. Muratori, revised and corrected by Albano Sorbelli, in Rerum Italicarum Scriptores, v. XVIII, in 3 parts, Stamperia di S. Lapi, Citta di Castello, 1906. XII.

Vedastini, Annales, ca. 900. There is an edition by G.H. Pertz in Monumenta Germaniae Historica, Scriptores, v. I, G.H. Pertz, ed., Hahn's, Hannover, 1826. X.

Victoris Tonnennensis, Chronica, ca. 567. There is an edition by Theodor Mommsen in Chronica Minora, Saec. IV V VI VII, v. 2 (of 3 vols.). This volume itself is v. 11 of Monumenta Germaniae Historica, Auctorum Antiquissimorum, Weidmann's, Berlin, 1894.

Vigfusson, Gudbrand, Sturlunga Saga Including the Islendinga Saga of Lawman Sturla Thordsson and Other Works, in 2 vols., Clarendon Press, Oxford, 1878.

Villola, Pietro e Floriano da, Cronaca, ca. 1376. There is an undated edition by L.A. Muratori, revised and corrected by Albano Sorbelli, in Rerum Italicarum Scriptores, v. XVIII, in 3 parts, Stamperia di S. Lapi, Citta di Castello, 1906. XII.

Vindocinense, Chronicon Vindocinense seu de Aquaria, ca. 1251. There is an edition in Chroniques des Églises d'Anjou, Paul Marchegay and Émile Mabille, eds., La Société de l'Histoire de France, Publ. No. 146, Chez Mme. Ve. Jules Renouard, Paris, 1869. X.

Vita Hludowici Imperatoris, ca. 840. There is an edition by G.H. Pertz in Monumenta Germaniae Historica, Scriptores, v. II, G.H. Pertz, ed., Hahn's, Hannover, 1829. XI.

Waverley (Waverleia), Annales Monasterii de, ca.
    1291. There is an edition in Annales Monas-
    tici, H.R. Luard, ed., in Rerum Britannicarum
    Medii Aevi Scriptores, no. 36, v. 2, Longman,
    Green, Longman, Roberts, and Green, London,
    1864. VI.

Weingartenses, Annales, ca. 936. There is an edi-
    tion by G.H. Pertz in Monumenta Germaniae
    Historica, Scriptores, v. I, G.H. Pertz, ed.,
    Hahn's, Hannover, 1826. XI.

Weissemburgenses Maiores, Annales Weissemburgenses
    Maiores, ca. 1075. There is an edition by
    G.H. Pertz in Monumenta Germaniae Historica,
    Scriptores, v. III, G.H. Pertz, ed., Hahn's,
    Hannover, 1839. X.

Weissemburgenses Minores, Annales Weissemburgenses
    Minores, ca. 846. There is an edition by
    G.H. Pertz in Monumenta Germaniae Historica,
    Scriptores, v. I, G.H. Pertz, ed., Hahn's,
    Hannover, 1826. X.

Wendover, Roger of, Liber Qui Dicitur Flores
    Historiarum, ca. 1235. This is the form of
    the title that appears on the title page;
    the usual form of the title is Flores Histor-
    iarum. There is an edition with notes by
    H.G. Hewlett in Rerum Britannicarum Medii
    Aevi Scriptores, no. 84, in 3 vols., H.M.
    Stationery Office, London, 1886-1889. VI.

Widukindus, Res Gestae Saxonicae, ca. 973. There
    is an edition by G. Waitz in Monumenta Germaniae
    Historica, Scriptores, v. III, G.H. Pertz,
    ed., Hahn's, Hannover, 1839.

William of Malmesbury (Willelmi Malmesbiriensis),
    De Gestis Regum Anglorum, ca. 1125. There is
    an edition with preface and notes by William
    Stubbs in Rerum Britannicarum Medii Aevi Scrip-
    tores, no. 90, v. 1 and 2, H.M. Stationery
    Office, London, 1887.

___, *Historia Novella*, ca. 1142. There is an edition by William Stubbs in *Rerum Britannicarum Medii Aevi Scriptores*, no. 90, v. 2, H.M. Stationery Office, London, 1887. VI.

William of Newburgh, *Historia Rerum Anglicarum*, ca. 1198. There is an edition with preface and notes in *Chronicles of the Reigns of Stephen, Henry II, and Richard I*, Richard Howlett, ed., in *Rerum Britannicarum Medii Aevi Scriptores*, no. 82, v. 1, H.M. Stationery Office, London, 1884; reprinted by Kraus Reprint, Ltd., 1964.

Winchester (Wintonia), *Annales Monasterii de*, ca. 1277. There is an edition with notes in *Annales Monastici*, H.R. Luard, ed., in *Rerum Britannicarum Medii Aevi Scriptores*, no. 36, v. 2, Longman, Green, Longman, Roberts, and Green, London, 1864. VI.

Wirziburgenses, *Annales*, ca. 1101. There is an edition by G.H. Pertz in *Monumenta Germaniae Historica, Scriptores*, v. II, G.H. Pertz, ed., Hahn's, Hannover, 1829. XI.

Wisbyenses, *Annales Fratrum Minorum*, ca. 1340. There is an edition by Jacob Langebek in *Scriptores Rerum Danicarum Medii Aevi*, v. I, A.H. & F.C. Godiche, Copenhagen, 1772. XIII.

Wissegradensis, *Canonici Wissegradensis Continuatio Cosmae*, ca. 1142. There is an edition by Rudolfus Koepke in *Monumenta Germaniae Historica, Scriptores*, v. IX, G.H. Pertz, ed., Hahn's, Hannover, 1851. IX.

Worcester (Wigornia), *Annales Prioratus de*, ca. 1377. There is an edition with notes in *Annales Monastici*, H.R. Luard, ed., in *Rerum Britannicarum Medii Aevi Scriptores*, no. 36, v. 4, Longmans, Green, Reader, and Dyer, London, 1869. VI.

Wormald, Francis, *English Kalendars Before A.D. 1100*, v. 1, *Texts*, Harrison and Sons, London, 1934.

Wormatienses, Annales Breves, ca. 1295. There is
    an edition by G.H. Pertz in Monumenta Germaniae
    Historica, Scriptores, v. XVII, G.H. Pertz,
    ed., Hahn's, Hannover, 1861. XI.

Wykes, Thomas, Chronicon, ca. 1289. There is an
    edition with notes in Annales Monastici,
    H.R. Luard, ed., in Rerum Britannicarum Medii
    Aevi Scriptores, no. 36, v. 4, Longmans, Green,
    Reader, and Dyer, London, 1869. Wyke's Chron-
    icon is almost identical with the Annals of
    Osney (Annales Monasterii de Oseneia), and I
    have cited both sources as Wykes ⌊ca. 1289⌋.
    VI.

Xantenses, Annales, ca. 873. There is an edition
    by G.H. Pertz in Monumenta Germaniae Historica,
    Scriptores, v. II, G.H. Pertz, ed., Hahn's,
    Hannover, 1829. XI.

Zwetlense 255, Continuatio Zwetlensis Altera et
    Auctarium, ca. 1189. There is an edition
    by W. Wattenbach in Monumenta Germaniae His-
    torica, Scriptores, v. IX, G.H. Pertz, ed.,
    Hahn's, Hannover, 1851. IX.

Zwetlensis 102, Continuatio Zwetlensis Prima, ca.
    1170. There is an edition by W. Wattenbach
    in Monumenta Germaniae Historica, Scriptores,
    v. IX, G.H. Pertz, ed., Hahn's, Hannover,
    1851. IX.

Zwifaltenses, Annales, ca. 1221. There is an edi-
    tion by Otto Abel in Monumenta Germaniae His-
    torica, Scriptores, v. X, G.H. Pertz, ed.,
    Hahn's, Hannover, 1852. XI.

Added in proof:

Godellus, Chronicon, ca. 1173. There is an anony-
    mous edition in Recueil des Historiens des
    Gaules et de la France, Chez Martin, Guerin,
    Delatour, et Boudet, Paris, v. X, 1760, v. XI,
    1767.

INDEX

As it is elsewhere in this work, "S." is used to denote "Saint" and all its inflectional forms. For some index entries, "Saint" is an integral part of the name that is used in indexing. These entries are alphabetized in a separate section "S." between "R" and "S". A person who became a saint is indexed under his ordinary name, with "S." following. Thus, for example, the saint named Lambert (or Lambertus) appears under "Lambertus, S." while the cathedral dedicated to him appears under "S. Lambert's Cathedral".

The entries do not distinguish between a writer and the works that are cited under his name.

Angoulême, France, 294, 299, 327, 333ff, 567

Anjou, 329, 566, 727

Anna Comnena, 98, 521, 529, 552ff

Annales Maximi (alleged Roman annals), 46, 51

Annales Vetustissimi (Icelandic/Norse), 494, 495, 499, 504

annals, 57ff
    defined, 43
    described, Section III. 2
    errors in dates, Section III. 7
    relation to Easter tables, Section III. 2
    see also: Easter tables

Anselmus, dean of S. Lambert's Cathedral, 216, 224, 235ff

Anselmus (Gemblacensis), 216, 223, 224, 243, 657, 707, 714, 725, 745

anthelion (atmospheric phenomenon), 741

Antioch, 104, 115, 521, 537, 723, 727, 733, 734

Apulia, district of Italy, 731

Aquenses, Annales, 354, 358, 373, 379, 413, 419ff, 422, 730

Aquicense, Auctarium et Continuatio Chronica Sigeberti, 294, 295, 315, 347

Aquicinctina, Continuatio Chronica Sigeberti, 295, 315, 347, 348, 350, 660, 714, 749

Aquicinctus, island, France, 295

Aquisgranum = Aachen, Germany

Aquitania, 696

Arcadius, emperor of the East, 538

Argentinenses, Annales, 368

Ari the Historian, 492

Aristotle, 4, 103

Arnolf, duke of Bavaria, 375

Arnulfus, archbishop of Milan, 439

Arnulfus (of Milan), 439, 447, 468, 664

Arras, France, 297, 313, 329, 567

Arthur, King, 491

Ascalon, Holy Land, 559, 562-563

Asia Minor, 381, 383, 386, 418, 556, 734, 735

Assateague Island, Maryland, 80

Asser, monk of S. David's and bishop of Sherborne, 134, 141, 142, 145, 147, 150, 156-157, 204-205, 207, 210-211

Assyria, 51
    see also: Babylonia and Assyria

astrology, 85, 86, 165, 368

Athens, 418, 526, 541, 567

Attila, king of the Huns, 733

Augienses, Annales, 285, 287, 296, 354, 378, 379, 383, 386, 398, 400, 401, 677, 729, 750

Augsburg, Germany, 355, 409, 410, 411, 423, 430, 567, 729

Augusta = Augsburg

Augustani, Annales, 355, 379, 406, 408, 410, 411, 661, 678 697, 698, 714, 729, 751

Augustani Minores, Annales, 355, 362, 379, 423, 430, 436, 730

aurora, 97, 393, 705, Appendix VI

Austria, 87, 249, 257, 261, 440, 698

autumn, beginning of, 484

Autun, France, 301, 302, 344, 567, 631, 750,
    = Augustodunum

Auxerre, France, 308
    = Autesiodorum

B

Babylonia and Assyria, 88

Backhouse, J. M. , 156

Bad Kreuznach, Germany, 373

Baden, Germany, 376

Balderic, bishop of Noyon, 297

Balderic de Térouanne, 297

Baldericus, suggested author of
Cameracensium, 297

Baldric, bishop of Liège, 223

Baldwin, king of Jerusalem, 561

Baltic Sea, 501, 566

Bamberg, Germany, 359, 410,
414, 567

Bari, Italy, 443, 689

Bavaria, 400, 697, 698, 707,
729

Beauvais, France, 295, 346,
567

Beckman, N. , 68, 79, 492ff,
502, 594

Bede, 16, 21, 27, 30, 31, 32,
39, 40, 50, 54ff, 60, 96,
117ff, 126ff, 132, 134, 137,
138, 139, 141, 142, 143ff,
150ff, 154, 185, 190, 192,
193, 200, 201, 258, 261,
296, 304, 330, 363, 378,
380, 383, 388, 439, 440,
447, 449, 455ff, 459ff, 462
515, 654, 665, 671

Beirut, Lebanon, 733

Belgium, 87, 215, 228, 233, 235,
237, 243, 261, 492, 566

Belisarius, Byzantine general,
444, 455ff, 458

Bellovacense, Auctarium et
Continuatio Chronica
Sigeberti, 295, 315, 346,
660

Benedict II, Pope, 679

Benedict VIII, Pope, 239

Benedict IX, Pope, 469

Benedict of Peterborough, 142,
147

Beneventani, Annales, 109, 129,
439, 447, 466ff, 469, 680,
701, 731, 754

Benevento, Italy, 325, 439, 467,
468, 469, 567, 596, 731

Bergamo, Italy, 439, 465, 567

Bergen, Norway, 504, 567

Bermondsey, Annals of, 133,
158ff, 164, 175

Bermondsey, England, 132, 133,
143, 164, 175, 671

Bernoldus, monk of S. Blasien,
33, 158, 282-283, 286, 288-
289, 363, 631

Bertha, daughter of
Charlemagne, 306

Bertholdus, abbot of
Zwiefalten, 355-356, 379,
409, 714, 730

Bertholdus, of Reichenau (?),
356, 363

Besancon, France, 301

Besuense, monastery of, 295

Besuenses, Annales, 295, 296,
315, 329ff, 332, 334ff, 660,
675, 676

Bethmann, L. C. , 297, 299

Beze, France, 295, 304, 330,
335, 567
see also: Besuense

Bible, 59

bis-sextile year, 31

Blaise, France, 295; probably
equals Beze

Blandain, Belgium, 217, 238ff,
243, 245, 247, 248, 567

Blanden, Belgium, 217

Blandigny, Belgium (?), 216

Blandinienses, Annales, 72, 95,
216, 218, 224, 226, 227ff,
231ff, 238ff, 243, 245, 246ff,
325, 453-454, 657, 707

Blois, France, 328

Bodensee, 285
= Lake Constance

Bohemia, 278, 279, 707

Bologna, Italy, 446, 448, 449,
470, 472, 473, 567, 731

Bolognetti, Cronaca Detta Dei,
440, 446, 447, 470, 472

Boniface, S., 359

Bononensium, Corpus Chronicon,
446
see also: Varignana, Cronaca

Bordeaux, France, 727, 747

Borissus, Turkey, 527

Bouquet, Dom M., 291, 292,
296ff, 305, 308, 353, 366,
372, 396, 595

Bourbourg, France, 296, 343,
346, 567

Brauweiler, Germany, 356, 417,
420, 567
see also: Brunwilarenses,
Annales

Brechin (Brechne), Scotland,
202, 567

Brennus, legendary Gallic
leader, 51

Brescia, district of Italy, 753

Britain, British Isles, 87, 88,
89, 444, 493, 700, 724

Brittany, 723

Bruges, Belgium, 220

Brunsvicensium, Scriptores
Rerum, 353

Brunwilarenses, Annales, 356,
379-380, 417, 420, 678, 730

Brut y Tywysogion, 205, 210ff,
487, 654, 700, 706, 712, 723,
724

Bryennius, Nicephorus, 521

Buckler, Georgina, 552, 554-555

Burburgensis, Continuatio
Chronica Sigeberti, 296, 316,
343, 345, 660, 676

Burgundy, 230, 301, 308, 323,
566, 696, 697, 728, 747

Burton, Annales de, 133, 143,
157

Burton-upon-Trent, England,
133, 157, 567

Bury, J. B., 524, 537

Bury S. Edmunds, England, 134,
173, 174, 175, 178, 568

Byzantine Empire, iii, 59, 87,
112, 222, 292, 515, 644

C

Caesar, Augustus, 508

Caesar, Julius, 24, 55, 299,
458, 494

Calabria, district of Italy, 236,
237, 701

calendar
Byzantine, 515ff
Gregorian, 15, 20, 21, 24,
25, 38, 40, 119, 122
Hebrew (Jewish), 17ff, 31,
Section II.6, Section II.7
Italian, 437ff, 515
Julian, 11, 24ff, 41, 122
Roman, 12ff, 24, 515

Callixtus II, Pope, 158

Calor River, Italy, 754

Cambrai, France, 296, 297,
309, 340, 344, 568

Cambriae, Annales, Section
VII.3, 654, 656, 712, 723,
724

Cameracensium, Gesta
Episcoporum, 240, 241, 296,
297, 316, 340

Cameracensium et Atrebatensi,
Chronicon, 297

Campania, district of Italy, 443,
533, 753

cannibalism, 695, 697, 700

Canterbury, England, 134, 152,
153, 160, 163, 164, 168,
170ff, 174, 566, 568

Capua, Italy, 325

Caradog of Llancarvan, 205

Casinates, Annales, 440, 447,
465, 467, 664

Casinum = Monte Cassino, Italy

Cava = La Cava, Italy

Cavenses, Annales, 440, 447-
448, 462ff, 467ff, 471, 664,
680, 690, 701, 718, 731

Cedrenus, Georgius, 106, 116,
517, 522, 529, 535, 543ff,
548, 550, 551, 682, 683,
688, 700, 701, 706, 712,
713, 718, 733ff, 755

Celestine III, Pope, 271, 472

Celoria, G., 100, 101

Chabanais, France, 294

Charlemagne, 215, 283, 296,
306, 311, 324, 325, 358, 375,
377, 389, 394ff, 444, 595,
706

Charleroi, Belgium, 221

Charles, count of Flanders, 220,
242

Charles II (the Bald), king of
France, 307

Charles Martel, 300

China, 5, 88, 102, 103, 104, 395,
520

Christian era, Section III. 4, 508,
515ff

chronicles, 59ff
defined, 43

Cicero, M. Tullius, 51

Cilicia, 706, 734

circumzenithal arc, 741

Ciric: see Grig and Cyric, S.

Cistercian Order, 138

Clarius, monk of S. Peter's,
Sens, 298, 303, 311, 316, 342

Clemens, S. L., 552, 555

Clement III, Pope, 271

Clerke, A. M., 104

Clermont, France, 559

Clonmacnois, Annals of, 183

Cluny, France, 308, 309, 335,
340, 342, 568

coefficients of condition
defined, 603
tabulated, 604ff

Coggeshall, England, 137, 163,
166, 170, 172, 568

Cologne, Germany, 115, 216,
356, 368, 420, 422, 424, 426,
428, 429, 568, 730, 745

Colonienses Maximi, Annales,
111, 356, 380, 420, 422-423,
424, 426, 428, 429, 662,
678, 730, 752

Columba, S., 312

comet, 453, Appendix II
seen during eclipse, 453, 538

Conrad II, Holy Roman emperor,
340

Conrad of Austria, 265

Constance, Lake, 355
= Bodensee

Constans II, Byzantine emperor,
511, 543

Constantine I (the Great), Roman
emperor, 115, 523, 534

Constantine V Copronymus,
Byzantine emperor, 544, 706

Constantine VI, Byzantine
emperor, 545

Constantinople, 69, 75, 98, 106,
232, 233, 236, 237, 261, 389,
443, 519, 520, 522, 525,
526ff, 533, 535ff, 538,
540ff, 544ff, 548ff, 553,
557, 568, 592, 666, 700,
713, 733ff

Constantius I, Roman emperor,
34, 533

Constantius II, Roman emperor,
534-535

consular lists, 48ff, 56, 442, 443

Dorsetshire, 724

Dorylaeum: see Eskisehir

Douai, France, 294, 295, 347, 348, 350, 568

dragons, 140, 165, 279, 408, 705, 707, 708, 712

Dreyer, J. L. E., 103

Dubs, H. H., 520

Dungal, monk of S. Gall, 595-596

Dunstable (Dunstaplia), Annales Prioratus de, 140, 144, 708

Durham, England, 94, 138, 158, 568, 589

E

Eadberht, Saxon leader, 153

Eadmer, English chronicler, 671, 743

earthquakes, Appendix VIII

Easter
   compared with Passover, Section II. 7
   controversy, Section II. 2
   great Easter cycle, Section II. 8, 95, 125, 229, 299
   problem, 10, throughout Chapter II, 57, 61, 83, 261, 527
   relation of Easter tables to annals, 10, 61, Section III. 2
   tables, Section II. 9, 47, 53ff, 57ff, 121, 184, 229, 295, 299, 304, 307, 356, 370, 371, 374, 392, 440, 443, 494

Eastern Roman Empire: see Byzantine Empire

ecclesiastical equinox, 15, Section II. 3

ecclesiastical full moon, 15, Section II. 5

ecclesiastical moon, 15, Section II. 4, 39, 54, 83, 91, 118ff, 127ff, 240, 244, 247, 267, 268, 330, 337, 340, 369, 404, 410, 413, 421, 426, 435, 596

eclipse, lunar
   and simultaneous occultation, 91ff, 589ff, 626
   catalogue of records of, Appendix I
   reasons for not using most records of, 5, 649ff

eclipse, solar
   annular, 67ff
   basic properties of, Section IV. 2
   central, 66
   partial, 68, 78, 83
   reason for basing study on, 5
   total, 67ff, 78
   umbral, 64ff
   see also: comet, corona, dew, diamond ring, magnitude, Mercury, penumbra, penumbral eclipse, reliability, Venus

Edinburgh, 199

Egmond aan Zee, The Netherlands, 217, 244, 246, 568

Egmundani, Annales, 215, 217, 225, 239, 244, 246, 657, 673, 688, 689, 701, 725, 745

Egypt, 528

Eichstet, Germany, 360, 412, 568

Einhard, 357-358, 360, 365, 367, 394ff

Einhardi, Annales, 358, 365-366
   see also: Fuldenses, Laurissenses

Einsiedeln, Switzerland, 41

Ekkehardus Uraugiensis, 276, 359, 361, 362, 406

Elnon River, France, 298

Elnonenses Maiores, Annales, 298, 316, 350, 351, 660, 696, 708, 727, 749

Elnonenses Minores, Annales, 298, 316, 331, 337, 630

Elwangenses, Annales, 371

Engelberg, Switzerland, 283, 284, 289-290, 373, 420-421, 568
   monastery of, 373

Engelbergenses, Annales, 283,
284, 286, 289, 373, 374, 420,
658, 726

England, 87, 96, 151, 152, 155,
156, 162, 163, 165, 166, 168,
181, 204-205, 460, 481, 493,
500, 537, 546, 566, 597, 742

Engolismenses, Annales, 298,
307, 316, 327

epact, 119ff
Bede's epact, 120ff

eponym canon
eclipse, 52

equinox, 15, 22, 24ff, 27, 29
Roman, 22ff, 27
see also: ecclesiastical
equinox

Erchenfridus, abbot of Melk,
252, 253

Erfurt, Germany, 354, 358, 422,
568

Erphesfurdensis, Annales S.
Petri, 354, 358, 373, 380,
416, 421, 422

Eskinbek, Chronologia Rerum
Memorabilium ab Anno 1020
usque ad An. 1323, 476, 481,
484, 486ff, 490, 732

Eskisehir, Turkey, 556

Esrom, Lake, Denmark, 476,
484

Esromenses, Annales Rerum
Danicarum, 72, 227, 228,
230, 453, 476, 481, 483,
484, 486, 488ff, 665, 732

Essenbek, Denmark, 476, 479,
490, 568
monastery of, 476

Ethelwerd (Aethelwerd), relative
of King Alfred, 140, 141, 144

Eunomius, 527

Europe, 87, 88, 503, 629, 644,
695, 711, 721
central Europe, 87, 249

Eusebius (Pamphili or Caesarea),
18, 44, 49, 50, 52, 53, 56,
223, 300, 363, 376, 445,

505, 519, 523, 525, 527,
528, 529, 535-536

Evagrius (Scholasticus), 459,
733

Evière, abbey of, Vendôme, 314

Explanatory Supplement to the
Astronomical Ephemeris and
the American Ephemeris
and Nautical Almanac, 516-
517

F

famines, Appendix IV

Farfa, Italy, 441, 464, 568
monastery at, 441

Farfenses, Annales, 95, 441,
448, 453-454, 463, 466ff,
664, 731, 753, 754

Farnham, Surrey, England, 138
173, 179, 568

Fasti Consulares, 46
see also: Consularia Italica,
Fasti Vindobonenses

Fasti Vindobonenses, 48ff, 96,
132, 439, 441ff, 444, 448-
449, 451ff, 455ff, 679, 731,
753
see also: Consularia Italica

Fidis, S., 430

Fink, Harold, 559

Flanders, 215, 217, 219

Flatbøgens Annaler, 494

Flavigny, France, 299, 302,
324, 568

Flaviniacenses et Lausonenses,
Annales, 299, 316, 324, 659

Florence of Worcester, 134,
137, 138, 144-145, 146ff,
157, 158, 162, 165, 173,
174, 178, 211, 369, 655,
656, 671, 672, 721, 723,
724, 742, 743, 744

Floriacenses, Annales, 299,
316, 330, 332, 659, 675, 707,
727

Floriacum, France, 299
= S. Benoit-sur-Loire

Folcmar, abbot of Corvey, 414-415

Folcwinus of Thérouanne, S.,
300

Folcwinus, abbot of Lobbes,
300, 312, 316, 331, 748

Formoselenses, Annales, 218,
225, 241, 242, 657

Fosse, Belgium, 218, 219, 243,
247, 248, 569

Fossenses, Annales, 218, 220,
222, 225, 243, 247, 248

Fotheringham, J. K., 81

Four Masters, Annals of the
Kingdom of Ireland by the, 9,
Section VII.1, 742

France, 87, 160, 168, 215, 291,
301, 400, 697

Frankfurt, Germany, 503, 566

Frazer, J. G., 23, 24

"Fredegarius Scholasticus", 191,
230ff, 300, 305, 317, 323,
659, 675, 688, 696, 712, 747

Frederic I (Barbarossa), Holy
Roman emperor, 169

Frederic II, king of Denmark,
476

Freiburg, Germany, 374

Freising, Germany, 272, 374,
422, 426, 428, 430, 432, 435,
436, 569

Fribourg, Switzerland, 301

Frodoard, French chronicler,
303

Frowinus, abbot of Engelberg
and S. Blasien, 283, 284

Frutolf, prior of Michelsberg,
276, 359, 361, 362, 378,
380, 384, 406, 410, 469,
661

Fulcher of Chartres, 33, 559ff,
562, 666, 683, 701, 714, 735,
736

Fulda, Germany, 211, 359-360,
364, 367, 369, 395ff, 399,
400, 407, 569
monastery of, 359, 370

Fuldenses, Annales, 211, 233,
234, 283, 312-313, 325, 358,
359-360, 363, 367, 372, 381,
382, 384, 387, 390, 392,
394ff, 398, 400, 401, 661,
677, 678, 688, 697, 706,
729
see also: Necrologici
Fuldenses

G

Gaimar, Geffrei, 141, 145

Galatia, 735

Galbertus Brugensis, 220, 225,
242

Galicia, district of Spain, 505-506, 508ff, 566, 732

Gall, S., 284

Garstensis, Continuatio, 250,
253

Gascony, 707

Gaufredus, prior of Vigeois, 301,
317, 349, 696, 748, 749

Gaul, 696, 697, 725, 727, 729,
731

Gawain, 84, 115

Gemblacensis, Sigeberti
Continuatio, 99, 220, 223,
225, 244, 245, 600, 631

Gemblacum, 220, 223
= Gembloux, Belgium

Gembloux, Belgium, 99, 100,
216, 239, 243, 245ff, 569,
600
= Gemblacum

Geoffrey of Monmouth, 205, 491

Georgios Hamartolos, 106, 116,
523, 530, 533-534, 543, 544,
547, 548, 683, 700, 706

Georgius Monachus, 523, 524,
530, 547

Helena, S., 115

Heming, Saga of, 495, 501

Hennessy, W. M., 183, 197

Henry V, Holy Roman emperor, 484

Henry VI, Holy Roman emperor, 173, 247, 271

Henry, king of Austria, 265

Henry I, king of England, 160ff, 345, 722

Henry II, king of England, 135, 159

Henry III, king of England, 175

Henry I (the Fowler), king of Germany, 234

Henry of Huntingdon, 141, 145-146, 205, 688

Hepidannus, 107, 285
  see also: Sangallenses
  (Annales Sangallenses
  Maiores)

Heraclius, Byzantine emperor, 528, 543

Herbipolenses, Annales, 362, 381, 416, 418ff, 557, 718, 752

Heribertus, S., 222

Herigerus of Lobbes, 216

Herimannus Contractus
  (Hermannus Augiensis,
  Herman of Reichenau), 33,
  250, 251, 253, 254, 258,
  260, 264, 265, 283, 288,
  356, 362-363, 374, 375,
  378, 382, 386, 392, 402,
  406, 407, 729

Herodotus, 84, 552

Herolfesfeldenses, Annales, 364-365, 371, 372, 392, 397, 403, 463

Hersfeld, Germany, 364

Herzogenrath, Germany, 372, 413, 569

Hewlett, H. G., 135

Hieronymus = Jerome, S.

Hildesheim, Germany, 364, 403, 405, 408, 412, 569

Hildesheimenses, Annales, 242, 314, 363-364, 371, 372, 378, 382, 386, 392, 401, 403, 405, 407ff, 411, 412, 463, 662, 678, 697, 698, 707, 714, 729, 730, 751, 752

Hillel II, Jewish patriarch, 34, 36

Hipparchus, 104

histories, 59ff
  defined, 43

Hogarth, D. G., 491

Holland, 203, 743
  see also: Netherlands, The

Holland, A. W., 307

Holliday, C., v

Holy Land, iii, 87, 354, 515, 559, 655, 714
  see also: Palestine

Holy Roman Empire, 215, 265

Honorius II, Pope, 158

Honorius, emperor of the West, 452, 538, 539

Honorius Augustodunensis, 99, 267, 302, 317, 344, 631

Hour
  equal, 90
  of the day = unequal hour
  unequal, 89, 90

Høyers Annaler, 494, 495ff, 502, 504

Hradisch, Czechoslovakia, 275, 281, 569
  = Uherské Hradiste

Hsi, Tse-tsung, 102, 105ff

Hugh Capet, 332-333

Hugo, abbot of Farfa, 441

Hugo Flaviniacensis, 302-303, 317, 342

Hugo Floriacensis, 303, 310, 317, 326, 331, 675, 696, 713

Humboldt, A. von, 104, 107

Hungary, 274, 698, 729

Hydatius, bishop of Galicia, 21,
 56, 196, 209, 300, 376, 505ff,
 507, 508ff, 522, 529, 538,
 665, 681, 712, 732

I

Iceland, iii, 475ff, 490, 492ff,
 498, 502ff, 566, 594, 595

Idatius = Hydatius

Ieper (Ypres), Belgium, 218

Iliad, 59, 491

Illyrica, 525, 541

Imoto, S., 687

India, 5, 118, 492, 706

indiction, cycle of, 96, 456, 462,
 516ff, 539, 545

Ingelheim, Germany, 400

Ingolstadt, Germany, 400, 569

Inn River, 254

Innocent II, Pope, 484

Ireland, 87, 89, 117, Section
 VII.1, 368, 492, 566, 593,
 700

Irene, Byzantine empress, 116,
 545-546

Isaac I Comnenus, Byzantine
 emperor, 517, 522

Isar River, Germany, 375

Isidorus (Isidore of Seville), 60,
 295, 363, 461, 506, 507, 510,
 592-593

Islamic provenance, 88

Italy, 87, 96, 151, 236, 437,
 439, 441ff, 444, 452, 453,
 458, 461, 470, 521, 566,
 695, 697, 727, 729, 730,
 731

Izmit, Turkey, 733
 = Nicomedia

J

Jacobo Theopolo, doge of Venice,
 473

Jacobs, J., 34, 35, 37

Jaffé, Philip, 369-370

Japheth, son of Noah, 475

Jarrow, England, 138, 153, 154,
 460, 569

Jenkins, R. E., v, 80

Jerome, S., 44, 50, 223, 363,
 505, 523, 525, 535

Jerusalem, 172, 265, 281, 294,
 351, 488, 559, 561ff, 569,
 711, 735, 755

Jesus, 16, 49, 54, 532

Joannis Asserii, 141
 see also: S. Neots, Chronicle
 of the Church of

Jóhannesson, T., 494

John VIII, Pope, 400

John XIX, Pope, 239

John II Comnenus, Byzantine
 emperor, 522

John, king of England, 174, 206

John Bevere, 136
 = John of London

John de Cella, 135

John de Taxter, 134

John of Toledo, 368

John of Worcester, 134, 162

Johnson, S. J., 10, 81, 92, 591

Jones, Charles W., 16, 19, 21,
 23, 27, 29, 30, 34, 37, 40,
 45, 46, 48, 49ff, 53ff

Julian (the Apostate), Roman
 emperor, 526

Jupiter, 111, 691
 occultation of, 97, 393, 690

Justin I, Byzantine emperor, 106

Justinian I (the Great),
 Byzantine emperor, 525, 679

Justinian II, Byzantine emperor, 193, 543-544

Jutland, 476

Juvavenses, Annales, Maiores, 251, 257, 262ff

Juvavum, 251
= Salzburg, Austria

K

Karlsruhe, Germany, 377

Kelheim, Germany, 370

Kenneth, son of Malcolm, king of Scotland, 201

Kenneth Macalpin, king of Scotland, 201

Kent, England, 168, 175

King-Hele, D. G. , 82, 155

Klosterrad, Germany, 372, 413

Knoxville, Tennessee, 80

Knut IV, king of Denmark, 220

Köln: see Cologne

Konstanz, Germany, 282

Kozai, Y. , 2

Kratz, H. , v

Kremer, Gerhard = Mercator

Krems, Austria, 250

Krems River, Austria, 250

Kremsmünster, Austria, 250, 252, 271, 272, 569

Krimigis, S. M. , v, 113

L

La Cava, Italy, 440, 463, 467, 468, 472, 569

Ladizlaus, king of Hungary and Bohemia, 355

Laistner, M. L. W. , 460, 461

Lambacensis, Auctarium et Continuatio, 252, 254, 257, 269, 273, 726

Lambach, Austria, 252, 273, 274, 569

Lambertus, S. , 220, 248, 424

Lambertus, martyr, 220

Lambertus (Aschafnaburgensis), 314, 363, 364-365

Lambertus Parvus, 115, 220, 225, 244, 247, 363, 424

Landmark, J. D. , 502, 503

Langebek, Jacob, 476, 477, 479

Langius, Joannis, 488

Langton, Stephen, archbishop of Canterbury, 175

Laon, France, 303, 307, 343-344, 346, 569

La Rochelle, France, 311

Laubacenses, Annales, 221, 222, 225, 231, 232, 234, 284, 657, 745

Laubacum, 221, 233
= Lobbes, Belgium

Laubienses, Annales, 222

Laudunense, Auctarium Chronica Sigeberti, 303, 317, 343

Laureshamenses, Annales, 365, 382, 391

Laurissenses, Annales (Annals of Lorsch), 221, 232, 262, 312-313, 325, 326, 358, 360, 363, 365ff, 372, 381, 382, 383, 387, 390, 394ff, 448, 462, 464, 595, 661, 697, 729, 750

Leibniz, G. W. von, 353

Leipzig, Germany, 371

Lemovicenses, Annales, 303, 317, 338, 659

Leo I, Pope, 208

Leo VII, Pope, 234

Leo VIII, Pope, 239

Leo IX, Pope, 239, 445

Leo IV, Byzantine emperor, 116

Leo V (the Armenian), Byzantine
emperor, 532

Leo VI, Byzantine emperor,
525, 547

Leo Diaconus, 524, 530, 549-
550, 683, 707, 713, 735

Leodienses, Annales, 221-222,
225, 226, 238, 657, 673, 701,
725, 727

Leodiensis S. Laurentius,
monastery, 222

Leonis Imperatoris Imperium,
524, 530, 548-549, 683

Levison, W., 455

Licinius, emperor in the East,
534

Liège, Belgium, 216, 218, 220,
221ff, 236, 237, 238, 243,
246, 247, 380, 424, 569
= Luik

Liguria, Mediterranean district,
700

Lille, France, 298

limmu, 51ff

Limoges, France, 294, 301, 304,
333ff, 339, 569

Limousin, 301

Liudprandus, bishop of Cremona,
443, 525, 530, 550

Lobbes, Belgium, 221, 222, 300
= Laubacum

Loch Cè, Annals of, 183

locusts, 695, 696, 697, 700,
701, 753

Lodin (of Greenland?), 501

Lögmanns Annáll, 494, 495ff,
502, 504

Loire River, 328, 747, 748

Loiselianos, Annales, 366-367,
372, 383, 390, 392, 394, 396,
463, 661, 690, 712
see also: Laurissenses,
Annales

Lombardy, 723, 730, 753

London, 133, 159, 168, 170, 171,
175, 569, 742

Longou, Austria, 726

Lorsch, Germany, 365, 367,
390, 391, 394ff, 570, 595
monastery of, 365
Annals of, see: Laurissenses,
Annales

Lothar II (or III), Holy Roman
emperor, 344, 416

Louis I, king of France and
Germany, Holy Roman
emperor, emperor of the
West, 287, 307, 326, 327,
328, 358, 377, 398, 400,
464

Louis II, Holy Roman emperor,
679

Louis VII, king of France, 212,
526

Louvain, Belgium, 217

Luard, H. R., 133, 136

Lugdunenses, Annales, 295, 296,
304, 310, 317, 326, 329ff,
713

Lughaidh, 116

Luik, Belgium, 216
= Liège

Lund, Sweden, 477, 479, 485,
570

Lundense, Necrologium, 477,
478, 479, 481, 485, 500, 665

Lundenses, Annales, 478

Lüneburg, Germany, 729

"Lupus Barensis", 443, 449,
466ff, 680, 688

Lyon, France, 304, 327, 330,
570

M

MacCarthy, B., 86, 127, 182,
184, 187, 189, 192, 196, 357

MacMaghnusa Mag Uidhir, 181

Mercury
sunspot mistaken for a transit
of, 393
visibility during an eclipse, 81

Merseburg, Germany, 377, 403,
404, 405, 570

meteors and meteorites, 651,
669, Appendix III

Meton, 30

Metonic cycle, 30

Mettenses, Annales, 305, 317,
331, 366, 659, 696, 706, 727,
747

Metz, France, 223, 305, 306,
312, 351, 570

Michael I, Byzantine emperor,
546

Michael II, Byzantine emperor,
464

Michael VIII Palaeologus,
Byzantine emperor, 526

Michael Glycas, 520

Michelsberg Cloister, Bamberg,
Germany, 359

Middlesex, England, 133, 167-168

Milan, Italy, 115, 439, 468, 570

Millington, Maryland, 80

Minnaert, M., 739, 741

mock sun or moon, see: multiple
suns or moons

Modena, Italy, 708

Mommsen, Theodor, 188, 453,
455, 522, 534, 539

Monacensis, Annalium
Salisburgensium Additamentum,
369, 383, 423, 425, 730

Mont S. Michel, 4, 308, 347,
570

Monte Cassino, Italy, 440, 444,
466, 570
= Casinum

Monumenta Germaniae Historica,
292, 293, 299, 303, 308, 314,
353, 365, 370, 505

moon
lights reported on, 687, 690
"sign of the cross in", see:
cross in sun or moon

Moravia, 275

Mortemer, Normandy, 305

Mortui-Maris, Chronicon, 305,
309, 318, 345, 346, 707, 727,
748

Mosomagenses, Annales, 306,
318, 333, 339, 676, 748
see also: Pont-à-Mousson,
France

Mucius (P. Mucius Scaevola), 51

multiple suns or moons, 739ff,
throughout Appendix IX

Mundane era of Constantinople,
56, 516ff

Munich, Germany, 369, 374,
375, 423, 425, 570

Muratori, L. A., 437

Murbach (location unknown), 285

N

Naples, Italy, 439, 443, 455,
570, 754

Narbonne, France, 696

Naufragia Ratisbonensia, 369-
370, 383, 407, 432, 661, 690,
729, 751

Necrologici Fuldenses, Annales,
370, 383, 407

Needham, J., 102, 104, 670

Neresheim, Germany, 429, 430,
570

Neresheimenses, Annales, 370-
371, 383, 422, 429, 430, 435,
698, 730

Netherlands, The, 87, 215
see also: Holland

Neugebauer, O., 27, 491

Nevers, France, 307, 338, 341,
347, 570

Newton, F. P., 80
  L. M., 80
  R. C., 80
  R. R., 2, 3, 10, 64,
    84, 502

Nicaea, 526, 557, 733
  council of, 18, 19, 20, 24,
    25, 30, 34, 37, 523

Nicomedia, Turkey, 733, 734
  = Izmit

Nile River, 706

Nisan, 17ff, 33ff

Nithardus, grandson of
  Charlemagne, 115, 306-307,
  318, 328, 358, 659, 675,
  727

Nivernenses, Annales, 307, 318,
  327, 337, 341, 346-347, 659,
  727

Nogger, abbot of Zwiefalten, 409

Normandy, 345, 597

Norsemen, raids of, 312, 331,
  332, 493

Northumbria, England, 154

Norway, 475ff, 490, 492ff, 498,
  500, 502ff, 594

nova, Section V. 2, 669

Novaliciense, Chronicon, 679

O

O'Clerigh, Michael, 182

O'Donovan, John, 182ff, 192,
  198

O'Keefe, J. A., 181

Ó'Maílle, T., 181, 182

Oddveria Annall, 494, 496, 499,
  503ff

Odon de Deuil, 418, 526, 530,
  556-557

Olaf, S., king of Norway, 26, 84,
  97, 98, 198, 479, 493-494,
  500ff

Olympiads, 56, 456, 506, 527

Opatowitz, Czechoslovakia, 276

Oppolzer, T. R. von, 70ff, 83,
  93, 100, 168, 169, 172, 175,
  192, 193, 195ff, 214, 219,
  228, 229, 235, 238, 240,
  246, 264, 266, 270, 280,
  323, 335, 337, 338, 340,
  349, 350, 393, 400, 418,
  428, 458, 468, 469, 471,
  472, 490, 492, 499, 501ff,
  511ff, 534, 535, 537, 541,
  543ff, 553, 556, 562, 580,
  581, 589, 592, 595, 600,
  632, 653, 656, 664

Orléans, France, 299, 328, 566

Osborn, Wayne, 129

Oslo, Norway, 503, 566

Osney
  Annales Monasterii de, 139,
    141, 179
  monastery of, Oxfordshire,
    139, 141, 177, 179, 180, 570

Otto I (the Great), Holy Roman
  emperor, 141, 236, 288

Otto II, Holy Roman emperor,
  402, 525

Oxenedes, Joannis de, 136, 146,
  177, 178, 655, 656

Oxfordshire, England, 159, 566

Oxnead, England, 178, 570

P

Pachymeres, Georgios, 101,
  112ff, 526-527, 531, 557ff

Paderborn, Germany, 364, 411,
  412, 570, 752

Palestine, 270, 733, 735
  see also: Holy Land

Palidenses, Annales, see:
  Theodorus Palidensis

parhelic circle, 740, 741

Paris, 310, 431, 747

Parker, Matthew, archbishop of
  Canterbury, 142

Q

Quedlinburg, Germany, 97, 108,
129, 234, 235, 372, 392, 403,
571, 717, 718
nunnery of, 372

Quedlinburgenses, Annales, 97,
106, 108, 129, 235, 314, 363,
372, 382, 384, 386, 390,
392, 397, 403ff, 463, 677,
678, 688, 697, 698, 713,
717, 729, 750, 751

R

Rainaldus, archdeacon of S.
Maurice, 291, 307-308, 311,
312, 314, 318, 326, 328,
329, 675, 676, 696, 713,
727, 747

Rainard, abbot of S. Petrus
Senonensis (?), 345

Ralph of Coggeshall, 137, 146-
147, 148, 162, 166, 170, 172,
655, 671, 672, 714, 724,
743

Rampona, Cronaca, 445, 446,
449, 450, 451, 461, 466, 470,
472ff, 664, 680, 690, 707,
708, 712, 731, 753, 754

Ratisbon, see: Regensburg,
Germany

Ratisbonenses, Annales, 370

Ratpertus, monk of S. Gall,
284ff

Ravenna, Annals of, 46, see also:
Consularia Italica, Fasti
Vindobonenses

Ravenna, Italy, 438ff, 442, 460,
465, 571, 734, 754

Recueil des Historiens des
Gaules et de la France, see:
Bouquet, Don M.

Regensburg, Germany, 369-370,
371, 407, 571
= Ratisbon

Regii, Annales, 494, 495ff, 498,
499, 502ff, 681

Reginbold, bishop of Spirensus,
336

Regino, abbot of Prum, 366,
372, 374, 377, 384, 401

Reichenau
abbey of, 251, 283, 285, 355
island, 251, 283, 285, 354,
378, 398, 400, 403, 406,
571

Reichersberg, Austria, 254, 269,
273, 274, 571
monastery of, 254, 255

Reicherspergenses, Annales,
86, 254ff, 258-259, 268, 272,
274, 658, 701, 714, 746

Reims, France, 307

Reliability (of an eclipse record),
74, Section IV. 6, 583, 584
defined, Section IV. 6

Rerum Britannicarum Medii
Aevi Scriptores, 136

Rerum Italicarum Scriptores,
437

Reseniani, Annales, 494, 495,
497, 499, 502

Resnikoff, L. A., 36

Reykjavik, Iceland, 499

Rhine River, 365, 400

Rich, R. P., v

Richard I, king of England, 173,
247, 369, 425

Richard of Devizes, 137, 147,
150, 174

Roanoke, Virginia, 80

Robert Guiscard, 553-554

Robert of Normandy, 561

Robertus Autissiodorensis, 308,
318, 344, 348, 690

Robertus de Monte, 148, 163,
308, 319, 345, 347, 660, 728,
748

Rodenses, Annales, 354, 372-
373, 379, 384, 385, 412, 414,
419, 420

Rodulfus Glaber, 303, 308-309,
319, 334, 339, 341ff, 690

Roger of Hovenden, 142, 147,
211

Roland, 324, 390
Song of, 313, 320, 324

Romanelli, Pietro, 528

Romani, Annales, 445, 449, 469

Romanus II, Byzantine emperor,
525

Rome, 48, 51, 56, 325, 354,
356, 411, 441, 445, 459ff,
469, 521, 571, 700, 733,
752, 753

Romualdus, archbishop of
Salerno, 196, 326, 445ff, 450,
463ff, 468ff, 470ff, 664, 679,
680, 718, 731

Roncevalles (Roncevaux), Spain,
324

Rosan, L. J., 540

Rosenveldenses, Annales, 373,
384, 411

Roskilde, Denmark, 477

Rotomagensi, Chronicon, 309,
319, 345, 748

Rouen, France, 309, 345, 571

Rudbertus, S., 256

Ruhkloster = Rye, Germany

Rupertus, 222, 226, 237ff

Rus regium = Rye, Germany

Russia, 493

Rye, Germany, 373
monastery of, 373, 478, 479,
489, 490, 571

Ryenses, Annales, 373, 478, 480,
481ff, 485ff, 487ff, 701

S. (= Saint)[*]

S. Albans, England, 135, 136,
159, 166, 172, 174, 175ff,
571

S. Albini Andegavensis,
Chronicae, 309, 311, 312,
319, 343, 350-351, 660,
676, 688, 727, 728

S. Amand, Belgium, 298

S. Amand-les-Eaux, France,
298, 331, 337, 351, 571
abbey of, 298

S. Andreae Castri Cameracesii,
Chronicon, 309, 319, 344

S. Andreas, monastery of,
Cambrai, 309

S. Arnoul, monastery of, Metz,
305

S. Aubin, abbey of, Angers, 309

S. Benedicti ad Ligerim,
Chronicon, 309, 319, 338,
341, 659, 748
see also: S. Benoit-sur-
Loire

S. Benet Holme, Oxfordshire,
137

S. Benigni Divionensis, Annales,
296, 310, 319, 330, 336, 349
monastery of, 308

S. Benoit-sur-Loire, France,
299, 310, 330, 333, 338, 341,
571
= Floriacum

S. Bertini Sithiensis, abbey of,
300, 312

S. Blasien, Germany, 282, 289,
411, 417, 420-421, 571
monastery of, 283, 284

S. Blasii, Annales, 373, 379,
384, 410, 411, 417, 420, 661,
662, 678, 689, 730

S. Claude, France, 328

---

[*]See the note at the beginning of the index.

See the note at the beginning of the index.

S. Rudbertus, abbey of,
Salzburg, 270
= S. Peter, abbey of,
Salzburg

S. Salvatoris, monastery of,
Anchin, France, 295

S. Salvatoris, monastery of,
Schaffhausen, 282

S. Sergii Andegavensis
abbey of, Angers, 312
Chronicon, 312, 320, 348

S. Sophia, monastery of,
Benevento, 439

S. Stephan, monastery of,
Hradisch, 275

S. Stephani Frisingensis
Annales, 374, 385, 413-414,
421, 425ff, 427, 429, 432,
435, 662, 663, 678, 698,
714, 730
monastery of, Freising, 272,
374
= Weihenstephan

S. Swithun's Monastery,
Winchester, 137

S. Trudperti
Annales, 374, 385, 426, 434,
662, 752
monastery of (Freiburg), 374,
427, 572

S. Vaast, monastery of, Arras,
313

S. Vanne, monastery of, Verdun,
302

S. Vincentii Mettensis, Annales,
312, 320, 351, 696

S*

Saladin, 562

Salerno, Italy, 440, 445, 471,
572

Salinator, Roman consul, 451

Salisbury, England, 139, 742

Salzburg, Austria, 251, 252,
256, 263, 264, 266, 270, 271,
273ff, 423, 572, 574

Samothrace, island of, 532

Sanctae Crucis, Chronicon, 199

Sangallenses, Annales, Maiores,
106ff, 129, 284-285, 286,
287ff, 674, 688, 726, 746

Sangallensis, Annalium
Alamannicarum Continuatio,
221, 234, 284ff, 287, 380,
401, 658, 674

Sarton, George, 102

Sasavensis, Monachi,
Continuatio Cosmae, 276,
277ff, 280, 746

Saturn, 113, 114, 691, 701

Saumur, France, 311, 348, 349,
572

Saxony, 377, 404, 729, 730

Sayers, Dorothy, 274

Sazava, Czechoslovakia, 276,
280, 572

Sazava River, 278

Scandinavia, 87, 89, 271, 444,
475, 484, 501, 566

Scarpe River, France, 298

Schaffhausen, Switzerland, 282,
289, 572

Scheftlarienses Maiores, Annales,
375, 385, 424, 428, 433, 662,
698, 699, 730, 752

Scheftlarienses Minores, Annales,
375, 385, 434, 436

Scheftlarn, Germany, 375, 424,
428, 433, 434, 436, 572
monastery of, 375

Schleswig, 492

Schliemann, Heinrich, 491

Schove, D. J., 183, 190, 191,
209

Schroader, I. H., v

Schroeter, J. Fr., 70

Scotland, 87, 181, Section VII. 2

Scotorum, Chronicum, 9, 95,
  106, 116, 183, 185, 187ff,
  193, 197, 198, 199, 593,
  652, 654, 655, 707, 713,
  742

Scottish Text Society, 199

Scriptores Rerum Danicarum
  Medii Aevi, 478

Sellar, W. C., 84

Senonensis, Annales S.
  Columbae, 312, 320, 326,
  385, 395, 465, 675, 676,
  696, 713, 747, 748

Sens, France, 298, 312, 326,
  572, 696

Seville, Spain, 506, 511, 513,
  572

Sherborne, England, 204

Sherborne calendar, 26, 27, 90

Shetelig, H., 493

Sialandiae, Chronicon, 478,
  481ff, 488ff

Sicily, 701, 718, 731

Sigebertus Gemblacensis, 97, 109,
  123, 133, 138, 148, 157,
  215ff, 220, 223, 226ff, 230ff,
  233ff, 238, 239ff, 294ff,
  297, 303, 305ff, 308ff, 340ff,
  390, 441, 448, 462, 464,
  467, 492, 496, 499, 657,
  673, 688, 689, 713, 714,
  725

Sigvat the Scald, 97, 98, 198,
  501

Simeon of Durham, 91, 94, 132,
  137, 138, 145, 147, 157, 158,
  201, 654

Simocatta, Theophylactus, 528,
  531, 542, 706

Sithienses, Annales, 232, 300,
  312-313, 320, 325, 326, 360,
  366-367, 659, 675, 727

Sithiu Island, France, 312-313

Sjaelland, island of, 476, 477,
  478

Skálholt, Annalbrudstykke fra,
  494

Skálholts, Annaler, 494, 497,
  499, 503ff

Smith, Sydney, 51

Snorri (Sturluson), 97, 497, 499,
  501

Société de l'Histoire de France,
  291

Socrates, 49

solar eclipse, see: eclipse,
  solar

solstice, 23

Somerton, England, 152

Spain, 87, 311, 505, 511ff, 522,
  566, 592-593, 707

Spanish era, 508, 511, 512

Spencer Jones, H., 93, 576,
  642, 643

Stade, Germany, 432, 435, 572

Stadenses, Annales, see:
  Albertus, abbot of S. Maria
  Stadensis

Staffordshire, England, 133

Stephan (or Stephen), S., 374,
  415

Stephen, abbot of S. Petrus
  Senonensis (?), 345

Stevenson, Joseph, 92, 142

Stiklestad, Norway, 97, 479
  eclipse of, 26, 84, 97ff, 497,
  500

Storm, Gustav, 494, 498ff

Stralius, Laurentius, 479, 482,
  485ff, 487ff, 490

See the note at the beginning of the index.

Warren, J., 491

Washington, D. C., 80

Wattenbach, W., 255, 256

Waverley
  abbey of, 138, 170
  Annales Monasterii de, 138,
    139, 148, 157, 163, 170,
    172, 179, 208, 213, 214,
    723, 724, 743

Wazo, bishop of Liège, 236, 237

Wearmouth, England, monastery
  of, 460

Weihenstephan, monastery, see:
  S. Stephanus, monastery of

Weingarten, Germany, 377-378,
  398, 400, 572

Weingartenses, Annales, 355,
  377, 379, 383, 386, 398, 400

Weissemburgenses Maiores,
  Annales, 314, 321, 336, 364

Weissemburgenses Minores,
  Annales, 314, 321, 327, 713

Weltenburg, Germany,
  monastery of, 369-370, 433,
  572

Wendover, Roger of, 135, 136,
  138, 146ff, 159, 172, 174ff,
  655, 708, 714, 743, 744

Weser River, Germany, 356

Wido, archpraesul of Reims,
  341

Widukindus, 235, 678, 717, 718

Wilhelm, abbot of Hirsaugiensis,
  283

William Rufus, king of England,
  116

William II, king of Sicily, 445

William of Malmesbury, 7, 84,
  100, 115, 116, 133, 138, 139,
  148, 150, 161ff, 163ff, 166
  205, 207, 208, 211ff, 721,
  722, 723, 742, 743

William of Newburgh, 160

Williams, J., 670

Winchester, Annales Monasterii
  de, 137, 139, 148, 150, 162,
  164

Winchester, England, 116, 131,
  137, 174, 177, 573

Wirnher, abbot of S. Disibodi,
  374

Wirziburgense, Chronicon, 359

Wirziburgenses, Annales, 359,
  373, 378, 386, 406, 407ff,
  678

Wisbyenses, Annales Fratrum
  Minorum, 480, 483

Wissegrad, Bohemia, 110ff
  monastery of, 278

Wissegradensis, Canonici,
  Continuatio Cosmae, 110ff,
  276, 277ff, 279ff, 658, 701,
  707, 714, 739, 746

Wissembourg, France, 314, 327,
  336, 573

wolves, 739, 747, 752

Worcester, England, 134, 139,
  158, 165, 166, 169, 176, 177,
  573, 723

Worcester (Wigornia), Annales
  Prioratus de, 138, 139, 149,
  162, 166ff, 169, 172, 176,
  177, 655, 671, 691, 714,
  724, 742, 743, 744

Wormald, F., 26, 90, 204

Wormatienses, Annales Breves,
  378, 387, 427, 434

Worms, Germany, 365, 378,
  427, 434, 573, 729

Wratislaus, king of Bohemia,
  278

Württemberg, Germany, 376

Würzburg, Germany, 359, 362,
  416, 573
  monastery of, 359

Wykes, Thomas, 139, 141, 146,
  149, 158ff, 161, 172, 177,
  179, 656, 723, 724, 743,
  744

X

Xanten, Germany, 389, 399, 573, 630

Xantenses, Annales, 231, 233, 387, 388ff, 397, 399, 630, 661, 677, 697, 712, 713, 729, 750

Y

Yeatman, R. J., 84

Yoke, Ho Peng, 129

Ypres, Belgium, 218

Yugoslavia, 525

Z

zero, 117ff

Zonaras, Joannis, 521

Zozimus, Pope, 483

Zwetlense 255 (Continuatio Zwetlensis Altera et Auctarium), 254, 259, 260, 265, 267

Zwetlensis 102 (Continuatio Zwetlensis Prima), 254, 256, 259, 260, 267

Zwettl, Austria, 256, 260, 265, 267, 573

Zwiefalten, Germany, 355, 388, 409, 412, 421, 425, 427, 573
monastery of, 355, 388, 409

Zwifaltenses, Annales, 374, 387, 388, 409, 412, 421, 424, 427

THE JOHNS HOPKINS UNIVERSITY PRESS

Printed on 50-lb. Sebago MF Regular
by Universal Lithographers, Inc.

Bound in Joanna Arrestox
by L. H. Jenkins, Inc.